T0293541

Plant Endophytic Fungi

Plant Endophytic Fungi

Edited by Rhea Turner

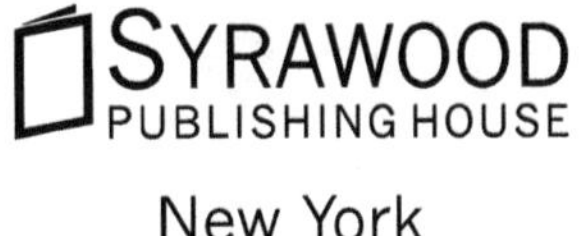

New York

Published by Syrawood Publishing House,
750 Third Avenue, 9th Floor,
New York, NY 10017, USA
www.syrawoodpublishinghouse.com

Plant Endophytic Fungi
Edited by Rhea Turner

International Standard Book Number: 978-1-64740-423-9 (Hardback)

Cataloging-in-publication Data

Plant endophytic fungi / edited by Rhea Turner.
p. cm.
Includes bibliographical references and index.
ISBN 978-1-64740-423-9
1. Endophytic fungi. 2. Plant physiology. I. Turner, Rhea.
QK604.2.E53 P53 2023
579.517 85--dc23

TABLE OF CONTENTS

Preface .... **VII**

Chapter 1 **Endophytic Fungi of Olive Tree** **1**
Rosario Nicoletti, Claudio Di Vaio and Chiara Cirillo

Chapter 2 **Cork Oak Endophytic Fungi as Potential Biocontrol agents against *Biscogniauxia mediterranea* and *Diplodia corticola*** **21**
Daniela Costa, Rui M. Tavares, Paula Baptista and Teresa Lino-Neto

Chapter 3 **Mangrove-Associated Fungi: A Novel Source of Potential Anticancer Compounds** **42**
Sunil K. Deshmukh, Manish K. Gupta, Ved Prakash and M. Sudhakara Reddy

Chapter 4 **Restoring Waning Production of Volatile Organic Compounds in the Endophytic Fungus *Hypoxylon* sp. (BS15)** **81**
Yuemin Wang and James K. Harper

Chapter 5 **From Concept to Commerce: Developing a Successful Fungal Endophyte Inoculant for Agricultural Crops** **93**
Brian R. Murphy, Fiona M. Doohan and Trevor R. Hodkinson

Chapter 6 **A Solvent-Free Approach for Converting Cellulose Waste into Volatile Organic Compounds with Endophytic Fungi** **104**
Tyler Maxwell, Richard G. Blair, Yuemin Wang, Andrew H. Kettring, Sean D. Moore, Matthew Rex and James K. Harper

Chapter 7 **Endophytic Fungi: A Source of Potential Antifungal Compounds** **119**
Sunil K. Deshmukh, Manish K. Gupta, Ved Prakash and Sanjai Saxena

Chapter 8 **The Emergence of Endophytic Microbes and their Biological Promise** **161**
Gary Strobel

Chapter 9 **Antiplasmodial Properties and Cytotoxicity of Endophytic Fungi from *Symphonia globulifera* (Clusiaceae)** **180**
Joël E. T. Ateba, Rufin M. K. Toghueo, Angelbert F. Awantu, Brice M. Mba'ning, Sebastian Gohlke, Dinkar Sahal, Edson Rodrigues-Filho, Etienne Tsamo, Fabrice F. Boyom, Norbert Sewald and Bruno N. Lenta

Chapter 10 **Endophytic Mycoflora and their Bioactive Compounds from *Azadirachta Indica*** **189**
Eyob Chukalo Chutulo and Raju Krishna Chalannavar

Chapter 11 **Antifungal Activities of Volatile Secondary Metabolites of Four *Diaporthe* Strains Isolated from *Catharanthus roseus*** **201**
Dong-Hui Yan, Xiaoyu Song, Hongchang Li, Tushou Luo, Guiming Dou and Gary Strobel

Chapter 12 **Fungi as Endophytes in *Artemisia thuscula*: Juxtaposed Elements of Diversity and Phylogeny**.......217
Andreea Cosoveanu, Samuel Rodriguez Sabina and Raimundo Cabrera

Permissions

List of Contributors

Index

PREFACE

This book was inspired by the evolution of our times; to answer the curiosity of inquisitive minds. Many developments have occurred across the globe in the recent past which has transformed the progress in the field.

An endophyte is an endosymbiont, which may be a bacterium or a fungus that lives within a plant for at least a part of its life cycle without causing any apparent disease. Fungal endophytes are quite common and highly diverse microorganisms. They may improve host plant's growth and nutrient acquisition along with enhancing the ability of the plants to tolerate abiotic stresses. These endophytes also help in reducing the biotic stress faced by plants by increasing their pathogen resistance. Endophytes can be transmitted either vertically or horizontally. There are two major classifications of endophytes, namely, systemic and non-systemic endophytes, and clavicipitaceous and non-clavicipitaceous endophytes. This book provides a detailed explanation of the significance of fungal endophytes in plants. A number of latest researches have also been included to keep the readers updated with the global concepts in this area of study. The book is appropriate for students seeking detailed information in this area as well as for experts.

This book was developed from a mere concept to drafts to chapters and finally compiled together as a complete text to benefit the readers across all nations. To ensure the quality of the content we instilled two significant steps in our procedure. The first was to appoint an editorial team that would verify the data and statistics provided in the book and also select the most appropriate and valuable contributions from the plentiful contributions we received from authors worldwide. The next step was to appoint an expert of the topic as the Editor-in-Chief, who would head the project and finally make the necessary amendments and modifications to make the text reader-friendly. I was then commissioned to examine all the material to present the topics in the most comprehensible and productive format.

I would like to take this opportunity to thank all the contributing authors who were supportive enough to contribute their time and knowledge to this project. I also wish to convey my regards to my family who have been extremely supportive during the entire project.

Editor

1

Endophytic Fungi of Olive Tree

Rosario Nicoletti [1,2], Claudio Di Vaio [2] and Chiara Cirillo [2,*]

1 Council for Agricultural Research and Economics, Research Centre for Olive, Fruit and Citrus Crops, 81100 Caserta, Italy; rosario.nicoletti@crea.gov.it
2 Department of Agricultural Sciences, University of Naples Federico II, 80055 Portici, Italy; divaio@unina.it
* Correspondence: chiara.cirillo@unina.it

Abstract: In addition to the general interest connected with investigations on biodiversity in natural contexts, more recently the scientific community has started considering occurrence of endophytic fungi in crops in the awareness of the fundamental role played by these microorganisms on plant growth and protection. Crops such as olive tree, whose management is more and more frequently based on the paradigm of sustainable agriculture, are particularly interested in the perspective of a possible applicative employment, considering that the multi-year crop cycle implies a likely higher impact of these symbiotic interactions. Aspects concerning occurrence and effects of endophytic fungi associated with olive tree (*Olea europaea*) are revised in the present paper.

Keywords: *Olea europaea*; endophytes; antagonism; defensive mutualism; plant growth promotion; bioactive compounds

---

1. Introduction

After evidence resulting from the manifold investigations carried out in the last decades, the awareness that endophytic fungi are constantly associated with plants and remarkably influence their ecological fitness has significantly increased. In fact, the original boost concerning natural ecosystems incited by the general theoretical intent to exploit all components of biodiversity, basically as a source of novel bioactive products, has more recently extended to crops. Within agricultural contexts, the role of the endophytic microbiota, or endosphere, is more consistent in orchards, where the time factor confers higher impact to the establishment of an equilibrium among the species which are part of the tree biocoenosis, as well as to its eventual disruption [1].

The extent at which the accumulating knowledge on the beneficial effects of endophytic microorganisms may have a practical impact in tree crop management, and further progresses can be achieved, is largely dependent on the opportunity by the scientific community and actors in the field to access it in an organized form. In this perspective, the state of the art of research concerning occurrence and effects of endophytic fungi associated with olive tree (*Olea europaea*) are revised in the present paper.

2. Relevance of Microorganisms for a Sustainable Management in Olive Growing

The Mediterranean Basin landscape and culture have been shaped by olive tree since ancient times, but the ecological importance of this tree has only recently been acknowledged [2,3]. In the semiarid Mediterranean agricultural lands, new approaches in fruit orchard management have been forced by environmental constrains, such as soil degradation and water shortage, and agronomical techniques that may be able to improve or preserve soil quality and fertility, other than plant health, have gained particular importance [4–7]. Modern intensification in olive cultivation practices is causing increased incidence and severity of olive pests and diseases; whereas sustainable management systems can positively affect soil biochemical characteristics and soil microbial biodiversity [8,9], and contribute to improve landscape stability, mainly in the rising condition of abandoned olive groves [3]. Thus, year by year a fast-growing percentage of the growers' incomes is invested in agrochemicals, to promote olive tree growth, to control plant pathogens, and to increase the olive yield and quality, simultaneously generating a great public concern on the negative effects of the agrochemicals use on the environment, on the ecosystem's biodiversity, and human and animal health [10]. Consequently, several efforts have been done on the development of eco-friendly cultivation practices suitable to sustainable disease control by ameliorating olive tree health and productivity through methods and strategies that promote soil biological processes, decrease agricultural inputs, and improve soil structure and fertility [9].

The diversity of microorganisms associated with plants may stimulate their growth and induce tolerance mechanisms helping plants to counteract adverse environmental conditions. In arid and semiarid environments, crops are facing environmental constraints due to climate-change-driven rising temperatures, changes in rainfall frequency, and occurrence of extreme events [11]. These habitat-elicited stresses may reduce crop productivity and lead to soil erosion and degradation. Plants dwelling in such environments have developed mechanisms helping them to mitigate and counteract abiotic stress. Microorganisms of the rhizosphere can play a pivotal role in health and growth of olive tree too, by establishing strong relationships with the root system that enable plants to grow in limiting conditions, such as water scarcity, salinity, low soil fertility, and so on. In addition to studies on the intrinsic ability of olive tree to adapt to adverse environmental conditions [12], a significant research activity has been performed on rhizosphere microbes providing increased tolerance to host plants under abiotic stress, mainly focusing on plant growth promoting rhizobacteria and arbuscular mycorrhizal fungi [13,14]. Moreover, fungi and actinomycetes have been recognized as able to use root exudates as a carbon source, supplying plants with promptly assimilable nitrates, and playing a crucial role in the maintenance of soil health, besides exerting antagonistic effects on root pathogens [5].

In the plant holobiont system, these beneficial effects are integrated by the microbial component of the endosphere. Endophyte colonization of plants has been recognized to involve a sequence of cross-talking signals that allow the onset of compatible interactions. Once the interrelation has established, endophytes increase stress tolerance through the stress-responsive gene induction/expression, reactive oxygen species and anti-stress metabolite synthesis [15]. Under abiotic stress conditions, endophytic fungi have been reported to produce plant hormones and compatible solutes that maintain integrity and promote growth of the host. Moreover, they are known to protect their host plants against biotic adversities through the production of bioactive compounds and the stimulation of the defense reaction [16]. As soon as interactions between endophytes and plants have been disclosed, it has been argued that they can be exploited for the development of innovative applications in sustainable but still highly productive cultivation systems ([17], and literature therein), similarly to the better known other groups of microbes. As a result of the rising demand for organic agricultural products, perspectives for the application of these microorganisms as potential biopesticides and biofertilizers have become more consistent in the olive sector too [18], along with an incremental interest for the search and identification of species-specific endophytes [19].

3. Occurrence and Ecological Implications of Endophytic Fungi of Olive Tree

Tables 1 and 2 list records concerning occurrence of endophytic fungi in olive tree as inferred from examination of the available literature and GenBank accessions. The first table, dedicated to Ascomycota, is much more numerous; in fact, it includes 245 entries, 116 of which (approx. 47%) are identified at the species level. Such a low proportion can be explained considering that rDNA-ITS sequences are not able to resolve species ascription within many fungal genera [20]. On the other hand, it could reflect the possible existence of novel species, which is quite a common outcome of investigations on endophytic fungi. The same inference could apply to the Basidiomycota series of records, where identification at the species level is occasional, and was achieved for just 7 out of

37 entries (about 19%). Finally, taxa belonging to the Mucoromycota appear to be quite infrequent (Table 2).

Table 1. Endophytic Ascomycota reported from *Olea europaea*.

Endophyte [1]	Plant Part	Country	Reference
Absconditella sp.	branch	Salento, Italy	[21]
Acaulium sp.	root	Córdoba, Spain	[22]
Acremonium sp.	leaf, twig	Sicily, Italy	[23]
	branch, leaf	Salento, Italy	[21]
Alternaria alternata	leaf, root	Bragança district, Portugal	[24]
	leaf	Trás-os-Montes, Portugal	[25]
	leaf	Alentejo, Portugal	[26]
	twig	Mirandela, Portugal	[27]
	fruit	Karaburun, Turkey	(GenBank)
Alternaria arborescens	leaf	Bragança district, Portugal	[24]
Alternaria brassicae	fruit	Mirandela, Portugal	[28]
Alternaria compacta	leaf	Alentejo, Portugal	[26]
Altenaria consortialis	root	Bragança district, Portugal	[24]
Alternaria infectoria	leaf	Alentejo, Portugal	[26]
	twig	Mirandela, Portugal	[27]
Alternaria murispora	leaf	Alentejo, Portugal	[26]
Alternaria preussii	leaf	Mirandela, Portugal	[29]
Alternaria solani	twig	Mirandela, Portugal	[27]
Alternaria sp.	stem	Majorca, Spain	[30]
	leaf, twig	Sicily, Italy	[23]
	leaf	Evora, Portugal	[31]
	fruit	Mirandela, Portugal	[28]
	leaf, twig	Mirandela, Portugal	[27,29]
	leaf	Alentejo, Portugal	[32]
	branch, leaf	Salento, Italy	[21]
Alternaria tenuissima	stem, xylem	Majorca, Spain	[30]
	twig	Mirandela, Portugal	[27]
Anthostomella leucospermi	leaf, twig	Mirandela, Portugal	[27,29]
Arcopilus aureus	leaf	Alentejo, Portugal	[26]
Arthrinium phaeospermum	stem, xylem	Majorca, Spain	[30]
Arthrinium sp.	leaf	Alentejo, Portugal	[32]
	leaf	Salento, Italy	[21]
Ascochyta sp.	leaf, twig	Sicily, Italy	[23]
Ascochytulina deflectens	stem, xylem	Majorca, Spain	[30]
Aspergillus sp.	leaf	Piracicaba, Brazil	[33]
	fruit	Bragança district, Portugal	[34]
	leaf, twig	Mirandela, Portugal	[27,29]
	leaf	Alentejo, Portugal	[26,32]
	root	Córdoba, Spain	[22]
	xylem	Apulia, Italy	[35]
	branch, leaf	Salento, Italy	[21]
Aspergillus stellatus	twig	Sicily, Italy	[23]
Aspergillus tubingensis	root	Siwa oasis, Egypt	[36]
Aureobasidium pullulans	stem, xylem	Majorca, Spain	[30]
	leaf, twig	Sicily, Italy	[23]
	leaf	Alentejo, Portugal	[26]
Aureobasidium sp.	leaf, twig	Mirandela, Portugal	[29]
	leaf	Alentejo, Portugal	[32]
	branch, leaf	Salento, Italy	[21]
Bartalinia sp.	fruit	Karaburun, Turkey	(GenBank)
Berkeleyomyces basicola	root	Bragança district, Portugal	[24]
Biatora sp.	branch	Salento, Italy	[21]

Table 1. *Cont.*

Endophyte [1]	Plant Part	Country	Reference
Biscogniauxia mediterranea	flower buds	Bragança district, Portugal	[34]
	fruit	Mirandela, Portugal	[28]
	leaf	Alentejo, Portugal	[26]
	leaf, twig	Mirandela, Portugal	[27,29]
Biscogniauxia nummularia	shoot	Ljubljana, Slovenia	[37]
Botryosphaeria sp.	leaf, twig	Sicily, Italy	[23]
Botrytis cinerea	leaf, twig	Sicily, Italy	[23]
	leaf	Alentejo, Portugal	[26]
	leaf, twig	Mirandela, Portugal	[27,29]
Botrytis sp.	leaf	Alentejo, Portugal	[32]
	branch	Salento, Italy	[21]
Cadophora luteo-olivacea	root	Bragança district, Portugal	[24]
Camarosporium sp.	leaf, twig	Sicily, Italy	[23]
	leaf, twig	Mirandela, Portugal	[27,29]
Canalisporium sp.	root	Córdoba, Spain	[22]
Candida sp.	twig	Sicily, Italy	[23]
	branch, leaf	Salento, Italy	[21]
Capnobotryella sp.	branch	Salento, Italy	[21]
Catenulostroma sp.	branch, leaf	Salento, Italy	[21]
Catillaria sp.	branch	Salento, Italy	[21]
Ceratocystis sp.	leaf, twig	Sicily, Italy	[23]
Cercospora sp.	branch	Salento, Italy	[21]
Ceuthospora sp.	xylem	Apulia, Italy	[35]
Chaetomium globosum	shoot	Ljubljana, Slovenia	[37]
Chaetomium sp.	leaf	Sicily, Italy	[23]
	leaf	Evora, Portugal	[31]
	shoot	Ljubljana, Slovenia	[37]
	leaf	Alentejo, Portugal	[32]
Chalara sp.	leaf, twig	Sicily, Italy	[23]
	leaf	Alentejo, Portugal	[32]
Chalastospora gossypii	leaf, twig	Mirandela, Portugal	[27,29]
Chromelosporium carneum	leaf, twig	Mirandela, Portugal	[27,29]
Ciboria sp.	leaf	Salento, Italy	[21]
Cladophialophora sp.	root	Córdoba, Spain	[22]
Cladosporium cladosporioides	leaf	Alentejo, Portugal	[26]
	fruit	Mirandela, Portugal	[28]
Cladosporium cucumerinum	fruit	Mirandela, Portugal	[28]
Cladosporium delicatulum	leaf	Alentejo, Portugal	[26]
Cladosporium herbarum	leaf	Alentejo, Portugal	[26]
Cladosporium pseudocladosporioides	leaf	Alentejo, Portugal	[26]
Cladosporium ramotenellum	fruit	Karaburun, Turkey	(GenBank)
Cladosporium sp.	leaf, twig	Sicily, Italy	[23]
	shoot	Ljubljana, Slovenia	[37]
	fruit	Mirandela, Portugal	[28]
	leaf, twig	Mirandela, Portugal	[27,29]
	xylem	Apulia, Italy	[35]
	leaf	Salento, Italy	[21]
Cladosporium sphaerospermum	shoot	Ljubljana, Slovenia	[37]
Cladosporium tenellum	leaf	Alentejo, Portugal	[26]
Cladosporium tenuissimum	stem, xylem	Majorca, Spain	[30]
Clonostachys rosea	root	Bragança district, Portugal	[24]
Colletotrichum acutatum	fruit	Gioia Tauro area, Italy	[38]
	twig	Portugal	(GenBank)
Colletotrichum nymphaeae	leaf	Alentejo, Portugal	[26]

Table 1. *Cont.*

Endophyte [1]	Plant Part	Country	Reference
Colletotrichum sp.	leaf	Piracicaba, Brazil	[33]
	leaf, twig	Mirandela, Portugal	[27,29]
	fruit	Mirandela, Portugal	[28]
Coniothyrium sp.	leaf, twig	Sicily, Italy	[23]
Coniozyma leucospermi	leaf, twig	Mirandela, Portugal	[27,29]
Coniozyma sp.	flower buds	Bragança district, Portugal	[34]
Cosmospora sp.	leaf, twig	Mirandela, Portugal	[27,29]
Cryptocoryneum sp.	leaf	Salento, Italy	[21]
Curvularia trifolii	root	Bragança district, Portugal	[24]
Cytospora pruinosa	leaf, twig	Bragança district, Portugal	[24]
Cytospora sp.	stem, xylem	Majorca, Spain	[30]
	xylem	Apulia, Italy	[35]
Dactylonectria pauciseptata	root	Bragança district, Portugal	[24]
Daldinia concentrica	leaf	Piracicaba, Brazil	[33]
	branch	Ha'Ela Valley, Israel	[39]
Dendrothyrium variisporum	leaf, twig	Mirandela, Portugal	[27,29]
Devriesia sp.	branch, leaf	Salento, Italy	[21]
Diaporthe ambigua	root	Bragança district, Portugal	[24]
Diaporthe columnaris	leaf, root, twig	Bragança district, Portugal	[24]
Diaporthe rudis	twig	Mirandela, Portugal	[27]
Diaporthe sp.	stem	Majorca, Spain	[30]
	leaf	Sicily, Italy	[23]
	leaf	Piracicaba, Brazil	[33]
	leaf	Evora, Portugal	[31]
	root	Bragança district, Portugal	[24]
	flower buds	Bragança district, Portugal	[34]
	leaf	Alentejo, Portugal	[32]
	leaf, twig	Mirandela, Portugal	[27,29]
Didymella macrostoma	leaf	Alentejo, Portugal	[26]
Didymella sp.	branch	Salento, Italy	[21]
Diplodia sp.	leaf, twig	Sicily, Italy	[23]
Discosia sp.	leaf, twig	Mirandela, Portugal	[27,29]
Dothiora oleae	fruit	Karaburun, Turkey	(GenBank)
Dothiorella iberica	twig	Mirandela, Portugal	[27]
Drechslera avenae	leaf	Alentejo, Portugal	[26]
Embellisia sp.	leaf, twig	Mirandela, Portugal	[29]
Endoconidioma populi	leaf, twig	Mirandela, Portugal	[27,29]
Epicoccum nigrum	stem	Majorca, Spain	[30]
	leaf	Evora, Portugal	[31]
	root, twig	Bragança district, Portugal	[24]
	leaf	Alentejo, Portugal	[26]
	leaf, twig	Mirandela, Portugal	[27,29]
Epicoccum sp.	leaf	Sicily, Italy	[23]
	leaf	Alentejo, Portugal	[32]
Eutypa tetragona	leaf, twig	Mirandela, Portugal	[27,29]
Eutypella sp.	fruit	Bragança district, Portugal	[34]
Exophiala sp.	root	Córdoba, Spain	[22]
	branch	Salento, Italy	[21]
Fimetariella rabenhorstii	leaf, twig	Bragança district, Portugal	[24,27,29]
Foliophoma sp.	leaf	Alentejo, Portugal	[32]
Fusarium lateritium	fruit	Mirandela, Portugal	[28]
	leaf	Alentejo, Portugal	[26]
	twig	Mirandela, Portugal	[27]

Table 1. *Cont.*

Endophyte [1]	Plant Part	Country	Reference
Fusarium musae	leaf	Alentejo, Portugal	[26]
Fusarium oxysporum	root	Bragança district, Portugal	[24,40]
	twig	Mirandela, Portugal	[27]
Fusarium sp.	leaf	Evora, Portugal	[31]
	leaf, twig	Mirandela, Portugal	[27,29]
	leaf	Alentejo, Portugal	[32]
	xylem	Apulia, Italy	[35]
Fusarium tricinctum	leaf	Alentejo, Portugal	[26]
Fusarium verticillioides	leaf	Alentejo, Portugal	[26]
Geopyxis sp.	leaf, twig	Mirandela, Portugal	[29]
Gibberella avenacea	fruit	Mirandela, Portugal	[28]
Gibberella sp.	twig	Mirandela, Portugal	[27]
	branch	Salento, Italy	[21]
Gloeosporium sp.	leaf	Sicily, Italy	[23]
Gloeotinia granigena	leaf	Alentejo, Portugal	[26]
Heydenia alpina	twig	Mirandela, Portugal	[27]
Heydenia sp.	leaf, twig	Mirandela, Portugal	[27,29]
Homortomyces sp.	leaf, twig	Mirandela, Portugal	[29]
Hormonema sp.	stem	Majorca, Spain	[30]
	leaf	Salento, Italy	[21]
Hortaea sp.	branch, leaf	Salento, Italy	[21]
Hyalodendriella betulae	leaf, twig	Mirandela, Portugal	[27,29]
Hyperphyscia sp.	branch	Salento, Italy	[21]
Hypoxylon sp.	stem	Majorca, Spain	[30]
Ilyonectria sp.	leaf, twig	Mirandela, Portugal	[29]
Kabatina sp.	xylem	Majorca, Spain	[30]
Lecania sp.	branch, leaf	Salento, Italy	[21]
Lecanora sp.	branch, leaf	Salento, Italy	[21]
Lecidella sp.	branch, leaf	Salento, Italy	[21]
Lecythophora sp.	leaf, twig	Mirandela, Portugal	[29]
Leimonis sp.	branch	Salento, Italy	[21]
Leptosphaerulina americana	leaf	Alentejo, Portugal	[26]
Leptosphaerulina australis	leaf	Alentejo, Portugal	[26]
Leptosphaerulina saccharicola	leaf	Alentejo, Portugal	[26]
Leptosphaerulina trifolii	leaf	Alentejo, Portugal	[26]
Libertasomyces platani	xylem	Apulia, Italy	[35]
Lophiostoma corticola	root	Bragança district, Portugal	[24]
Lophiostoma sp.	branch, leaf	Salento, Italy	[21]
Macrophomina phaseolina	root	Bragança district, Portugal	[24]
Macrophomina sp.	root	Córdoba, Spain	[22]
Microsphaeropsis arundinis	root	Bragança district, Portugal	[24]
Microsphaeropsis proteae	leaf	Mirandela, Portugal	[27,29]
Microsphaeropsis sp.	stem	Majorca, Spain	[30]
	leaf, twig	Mirandela, Portugal	[27,29]
Minimelanolocus sp.	root	Córdoba, Spain	[22]
Mycocalicium victoriae	xylem	Apulia, Italy	[35]
Mycosphaerella sp.	leaf, twig	Mirandela, Portugal	[29]
Naevala sp.	leaf	Salento, Italy	[21]
Nemania aenea	leaf	Mirandela, Portugal	[29]
Nemania sp.	leaf, twig	Mirandela, Portugal	[29]
Neocamarosporium sp.	leaf, twig	Mirandela, Portugal	[29]
Neocatenulostroma sp.	branch, leaf	Salento, Italy	[21]

Table 1. *Cont.*

Endophyte [1]	Plant Part	Country	Reference
Neocosmospora solani	root	Siwa oasis, Egypt	[36]
	twig	Mirandela, Portugal	[27]
Neodevriesia sp.	branch, leaf	Salento, Italy	[21]
Neofabraea kienholzii	leaf, twig	Mirandela, Portugal	[27,29]
Neofabraea sp.	twig	Sicily, Italy	[23]
	fruit	Mirandela, Portugal	[28]
	leaf, twig	Mirandela, Portugal	[27,29]
	leaf	Alentejo, Portugal	[26]
Neofabraea vagabunda	leaf	Alentejo, Portugal	[26]
	fruit	Mirandela, Portugal	[28,34]
	leaf, twig	Mirandela, Portugal	[27]
Neofusicoccum sp.	leaf	Alentejo, Portugal	[32]
Neophaeomoniella sp.	branch, leaf	Salento, Italy	[21]
Neosartorya sp.	leaf	Alentejo, Portugal	[26]
Neosetophoma sp.	leaf	Salento, Italy	[21]
Nigrospora oryzae	stem	Majorca, Spain	[30]
	leaf	Evora, Portugal	[31]
Nigrospora sp.	leaf	Piracicaba, Brazil	[33]
	leaf	Alentejo, Portugal	[32]
Ochrocladosporium sp.	leaf, twig	Mirandela, Portugal	[29]
Paecilomyces variotii	xylem	Bisignano, Italy	[41]
Paecilomyces verrucosus	root	Bragança district, Portugal	[24]
Paraconiothyrium sp.	leaf, twig	Mirandela, Portugal	[29]
	xylem	Apulia, Italy	[35]
Paraphaeosphaeria sporulosa	root	Bragança district, Portugal	[24]
Paraphoma chrysanthemicola	leaf, root	Bragança district, Portugal	[24]
Paraphoma sp.	root	Bragança district, Portugal	[24]
Parastagonospora avenae	twig	Mirandela, Portugal	[27]
Penicillium canescens	root, twig	Bragança district, Portugal	[24]
	leaf	Trás-os-Montes, Portugal	[25]
Penicillium chrysogenum	leaf	Siwa oasis, Egypt	[42]
	fruit	Karaburun, Turkey	(GenBank)
Penicillium citrinum	fruit	Siwa oasis, Egypt	[36]
Penicillium commune	leaf, twig	Bragança district, Portugal	[24,43]
	leaf	Trás-os-Montes, Portugal	[25]
Penicillium echinulatum	leaf	Alentejo, Portugal	[26]
Penicillium expansum	leaf	Alentejo, Portugal	[26]
Penicillium glabrum	twig	Mirandela, Portugal	[27]
Penicillium restrictum	leaf, root	Bragança district, Portugal	[24]
Penicillium roseopurpureum	root	Bragança district, Portugal	[24,40]
Penicillum sp.	stem, xylem	Majorca, Spain	[30]
	leaf, twig	Sicily, Italy	[23]
	leaf, twig	Mirandela, Portugal	[27,29]
	xylem	Apulia, Italy	[35]
	branch, leaf	Salento, Italy	[21]
Penicillium spinulosum	leaf	Alentejo, Portugal	[26]
Pestalotiopsis guepinii	stem	Majorca, Spain	[30]
Pestalotiopsis sp.	leaf, twig	Mirandela, Portugal	[27,29]
Phaeoacremonium sp.	root	Córdoba, Spain	[22]
Phaeococcomyces sp.	branch, leaf	Salento, Italy	[21]
Phaeohelotium sp.	branch	Salento, Italy	[21]
Phaeomoniella sp.	leaf, twig	Mirandela, Portugal	[27,29]
	branch, leaf	Salento, Italy	[21]
Phaeosphaeria sp.	leaf, twig	Mirandela, Portugal	[27,29]
Phaeothecoidea sp.	branch, leaf	Salento, Italy	[21]
Phoma herbarum	leaf	Alentejo, Portugal	[26]

Table 1. *Cont.*

Endophyte [1]	Plant Part	Country	Reference
Phoma sp.	stem, xylem	Majorca, Spain	[30]
	leaf, twig	Sicily, Italy	[23]
	leaf, twig	Mirandela, Portugal	[27,29]
Phyllosticta sp.	leaf	Piracicaba, Brazil	[33]
Pichia sp.	branch	Salento, Italy	[21]
Pithomyces chartarum	xylem	Apulia, Italy	[35]
Plectania rhytidia	twig	Mirandela, Portugal	[27]
Plectania sp.	leaf, twig	Mirandela, Portugal	[29]
Pleospora herbarum	stem	Majorca, Spain	[30]
Pleospora sp.	leaf, twig	Sicily, Italy	[23]
Pleurophoma sp.	stem	Majorca, Spain	[30]
Podospora sp.	root	Bragança district, Portugal	[24]
Preussia africana	leaf	Alentejo, Portugal	[26]
Preussia sp.	leaf	Sicily, Italy	[23]
	shoot	Ljubljana, Slovenia	[37]
	leaf, twig	Mirandela, Portugal	[29]
	branch, leaf	Salento, Italy	[21]
Prosthemium sp.	leaf, twig	Mirandela, Portugal	[27,29]
Pseudocamarosporium sp.	xylem	Apulia, Italy	[35]
Pseudocercospora sp.	leaf, twig	Mirandela, Portugal	[27,29]
	branch, leaf	Salento, Italy	[21]
Pseudocosmospora vilior	root	Bragança district, Portugal	[24]
Pseudophaeomoniella oleae	xylem	Apulia, Italy	[35]
Pseudophaeomoniella sp.	leaf, twig	Mirandela, Portugal	[29]
	xylem	Apulia, Italy	[35]
Purpureocillium lilacinum	root	Bragança district, Portugal	[24,40]
Purpureocillium sp.	root	Córdoba, Spain	[22]
Pycnidiophora sp.	leaf, twig	Mirandela, Portugal	[29]
Pyrenochaeta sp.	leaf, twig	Mirandela, Portugal	[27,29]
Pyronema domesticum	leaf, twig	Mirandela, Portugal	[27,29]
Pyrrhospora sp.	branch, leaf	Salento, Italy	[21]
Rachicladosporium sp.	branch	Salento, Italy	[21]
Ramularia sp.	stem	Majorca, Spain	[30]
	branch, leaf	Salento, Italy	[21]
Rhinocladiella similis	leaf, twig	Mirandela, Portugal	[27,29]
Rhinocladiella sp.	branch, leaf	Salento, Italy	[21]
Saccharata sp.	leaf	Alentejo, Portugal	[32]
Sarocladium sp.	branch, leaf	Salento, Italy	[21]
Scoliciosporum sp.	branch, leaf	Salento, Italy	[21]
Scutellinia sp.	root	Córdoba, Spain	[22]
Seimatosporium sp.	leaf, twig	Mirandela, Portugal	[29]
Seiridium sp.	twig	Sicily, Italy	[23]
Septoria sp.	leaf, twig	Sicily, Italy	[23]
	leaf, twig	Mirandela, Portugal	[27,29]
Sordaria macrospora	stem, xylem	Majorca, Spain	[30]
	twig	Mirandela, Portugal	[27]
Sordaria sp.	leaf, twig	Mirandela, Portugal	[27,29]
Sporormiella intermedia	stem, xylem	Majorca, Spain	[30]
Stagonosporopsis cucurbitacearum	fruit	Karaburun, Turkey	(GenBank)
Stemphylium solani	leaf	Alentejo, Portugal	[26]
Stemphylium sp.	leaf, twig	Sicily, Italy	[23]
	branch, leaf	Salento, Italy	[21]

Table 1. *Cont.*

Endophyte [1]	Plant Part	Country	Reference
Stemphylium vesicarium	leaf	Alentejo, Portugal	[26]
	fruit	Karaburun, Turkey	[GenBank]
Stigmatodiscus enigmaticus	xylem	Apulia, Italy	[35]
Talaromyces purpureogenus	root	Portugal	[GenBank]
Taphrina sp.	branch, leaf	Salento, Italy	[21]
Teratosphaeria sp.	branch, leaf	Salento, Italy	[21]
Tricharina sp.	leaf, twig	Mirandela, Portugal	[27,29]
Tricharina striispora	twig	Mirandela, Portugal	[27]
Trichoderma gamsii	root, twig	Bragança district, Portugal	[24]
Trichoderma koningii	fruit	Mirandela, Portugal	[28]
Trichoderma lixii	root	Bragança district, Portugal	[24,40]
Trichoderma polysporum	stem	Majorca, Spain	[30]
Trichoderma sp.	root, twig	Bragança district, Portugal	[24]
	fruit	Mirandela, Portugal	[28]
	leaf, twig	Mirandela, Portugal	[29]
Tumularia sp.	leaf, twig	Mirandela, Portugal	[27,29]
Valsa sp.	leaf, twig	Mirandela, Portugal	[29]
Valsaria sp.	leaf, twig	Mirandela, Portugal	[29]
Venturia sp.	flower buds	Bragança district, Portugal	[34]
Verticillium sp.	leaf, twig	Mirandela, Portugal	[29]
Wickerhamomyces sp.	branch, leaf	Salento, Italy	[21]
Xanthoparmelia sp.	leaf	Salento, Italy	[21]
Xanthoria sp.	branch, leaf	Salento, Italy	[21]
Xenosonderhenia sp.	branch, leaf	Salento, Italy	[21]
Xylaria sp.	leaf	Piracicaba, Brazil	[33]
	twig	Mirandela, Portugal	[27]
Zygoascus sp.	branch	Salento, Italy	[21]

[1] Species are reported according to the latest accepted name, which might not be the same as the one used in the corresponding reference.

Table 2. Other endophytic fungi reported from *Olea europaea*.

Endophyte [1]	Plant Part	Country	Reference
Basidiomycota			
Bullera sp.	leaf	Alentejo, Portugal	[32]
Chondrostereum purpureum	fruit	Mirandela, Portugal	[28]
Colacogloea sp.	branch	Salento, Italy	[21]
Conocybe sp.	root	Córdoba, Spain	[22]
Coprinellus sp.	leaf, twig	Mirandela, Portugal	[27,29]
Coriolopsis sp.	fruit	Bragança district, Portugal	[34]
Cryptococcus sp.	leaf, twig	Mirandela, Portugal	[29]
	leaf	Alentejo, Portugal	[32]
	branch, leaf	Salento, Italy	[21]
Cystofilobasidium sp.	branch	Salento, Italy	[21]
Dioszegia sp.	branch	Salento, Italy	[21]
Entoloma sp.	root	Córdoba, Spain	[22]
Erythrobasidium sp.	leaf	Alentejo, Portugal	[32]
	branch, leaf	Salento, Italy	[21]
Filobasidium sp.	branch	Salento, Italy	[21]
Kockovaella sp.	branch, leaf	Salento, Italy	[21]
Kondoa sp.	branch	Salento, Italy	[21]
Lepiota sp.	root	Córdoba, Spain	[22]
Malassezia sp.	root	Córdoba, Spain	[22]
	branch, leaf	Salento, Italy	[21]
Meira sp.	branch	Salento, Italy	[21]
Moniliophthora sp.	root	Córdoba, Spain	[22]

Table 2. *Cont.*

Endophyte [1]	Plant Part	Country	Reference
Peniophora cinerea	leaf	Alentejo, Portugal	[26]
Peniophora lycii	leaf	Alentejo, Portugal	[26]
Peniophora sp.	leaf, twig	Mirandela, Portugal	[29]
	branch	Salento, Italy	[21]
Phlebiopsis gigantea	leaf	Alentejo, Portugal	[26]
Porostereum sp.	leaf, twig	Mirandela, Portugal	[29]
Pseudomicrostroma sp.	branch	Salento, Italy	[21]
Quambalaria sp.	branch, leaf	Salento, Italy	[21]
Rhizoctonia sp.	stem	Majorca, Spain	[30]
	leaf, twig	Sicily, Italy	[23]
Rhodotorula mucilaginosa	leaf	Alentejo, Portugal	[26]
Sistotrema brinkmannii	shoot	Ljubljana, Slovenia	[37]
Sporobolomyces sp.	leaf	Alentejo, Portugal	[32]
	branch, leaf	Salento, Italy	[21]
Sporotrichum sp.	leaf	Sicily, Italy	[23]
Symmetrospora sp.	branch, leaf	Salento, Italy	[21]
Trametes sp.	leaf, twig	Mirandela, Portugal	[27,29]
	branch, leaf	Salento, Italy	[21]
Tremella sp.	branch	Salento, Italy	[21]
Tricholoma sp.	leaf, twig	Mirandela, Portugal	[29]
Vishniacozyma sp.	branch, leaf	Salento, Italy	[21]
Wallemia sp.	leaf	Salento, Italy	[21]
Xylobolus annosus	twig	Mirandela, Portugal	[27]
Mucoromycota			
Mucor plumbeus	stem, xylem	Majorca, Spain	[30]
Mucor racemosus	stem	Majorca, Spain	[30]
Rhizopus arrhizus	root	Bragança district, Portugal	[24]
Rhizopus sp.	leaf	Alentejo, Portugal	[26]
Rhizopus stolonifer	fruit	Karaburun, Turkey	(GenBank)
Umbelopsis vinacea	stem	Majorca, Spain	[30]

[1] Species are reported according to the latest accepted name, which might not be the same as the one used in the corresponding reference.

With reference to the geographic origin, not surprisingly the great majority of these records come from the Mediterranean region, where olive growing is absolutely dominant in statistical terms with approximately 10.2 million hectares in 2018, corresponding to more than 97% of the overall surface destined to this crop in the world [44]. In addition to geographic and climatic conditions, differences in the species assortment are related to several factors, such as plant organ (Table 1), phenological stage [34], cultivar [22,27,29,32,35], season and cardinal orientation of samplings [23,24,26,29], isolation procedure, and substrate employed [33].

In timeline terms, the great majority of records have been gathered in the last four years, with a significantly increasing trend. In fact, after just three previous investigations carried out in the island of Majorca (Spain) in 1992 [30], Sicily (Italy) in 2008 [23], and Brazil in 2013 [33], and a couple of extemporary findings from Italy [38,41], the available data have been integrated with 48 new records in 2016, 27 in 2017, 68 in 2018, 107 in 2019, and 73 in the first half of the current year. Several reasons can explain such an escalation. Particularly, the increasingly easier access to the molecular tools has remarkably enhanced the number of isolates which can be taxonomically identified, subverting the old classification procedures which basically relied on the investigators' mycological experience, or on the access to identification services. Previously infrequent or unknown taxa have started being reported thanks to this methodological improvement. However, in many of these records identification was limited at the genus level, particularly when rDNA-ITS sequences only were considered as the genetic marker [21,22,29]. Although acceptable, this reflects a lower significance of the reports, considering that different species within genera such as *Alternaria*, *Aspergillus*, *Cladosporium*, *Diaporthe*, *Fusarium*,

Penicillium, *Phoma*, etc., may play very different ecological roles. With reference to the identified species, just two common plant associates (*Alternaria alternata* and *Epicoccum nigrum*) have more than two records from different locations, indicating that for the time being no species seems to stand out for a regular endophytic association with olive tree.

Another explanation for this incremental trend is the previously introduced emerging awareness of a relationship between endophyte occurrence and pest and pathogen incidences in crops. In the case of olive tree, this concept has been particularly considered with reference to the outbreak of the quick decline syndrome incited by *Xylella fastidiosa* in southern Italy [45], although no significant associations of any identified endophytes with this bacterium were found in both a high and a low susceptible cultivar in a dedicated study [21]. In another study concerning relationships with another widespread bacterial pathogen, *Pseudomonas savastanoi* pv. *savastanoi* causing the olive knot disease, endophytic fungi were found to be more abundant in infected plants. However, this remark particularly involved potential pathogens, such as *Alternaria*, *Cladosporium*, *Pseudocercospora*, *Fusarium*, and other Nectriaceae [27].

3.1. Endophytic Fungi as Plant Disease Agents

The wilt agent *Verticillium dahliae* is probably the fungal pathogen of olive tree which could be considered to have more strict endophytic implications. Basically, it is considered a hemibiotrophic fungus which colonizes olive trees systemically by spreading through the xylem during a biotrophic phase where it causes no or minimal detrimental effects on plant physiology. Symptoms consisting in chlorotic leaves rolling inward, defoliation, necrosis, and branch desiccation become evident later on [46]. Considering that this more or less enduring latent stage is recognized as a crucial phase of the disease cycle, the recovery of *V. dahliae* from asymptomatic plants is conventionally not referred to a possible merely endophytic status. However, its finding in artificially inoculated plants which became asymptomatic after recovering from infection [47] raises questions on whether its occurrence within olive trees is necessarily related to pathogenicity.

Other known disease agents have been isolated from asymptomatic plants. An inventory of olive tree pathogens compiled in 2014 includes at least 12 species reported in Table 1, namely *A. alternata*, *Alternaria consortialis* (=*Ulocladium consortiale*), *Arthrinium phaeospermum*, *Berkeleyomyces basicola* (=*Thielaviopsis basicola*), *Botrytis cinerea*, *Colletotrichum acutatum*, *Dothiorella iberica*, *Epicoccum nigrum*, *Fusarium oxysporum*, *Macrophomina phaseolina*, *Neocosmospora solani* (=*Fusarium solani*) and *Neofabraea vagabunda* (=*N. alba*, *Phlyctema vagabunda*) [48]. In addition to the above-mentioned common associates of olive trees, *A. alternata* and *E. nigrum* are also known to develop epiphytically [41], for the other species, most of which are anyway not reported to cause relevant damage to this crop, it is not inferable if their endophytic presence documented in references of Table 1 was eventually preliminary to disease onset.

In addition to the above inventory, more taxa listed in Table 1 were recovered from diseased olive trees, albeit without verifying their involvement in etiology through the Koch's postulates. This is the case of the new species *Phaeomoniella oleae*, which was originally isolated from black-discolored xylem of wilting branch of a plant infected by *X. fastidiosa* [49], as well as *Conyozyma leucospermi*, *Nigrospora oryzae* and *Biscogniauxia mediterranea*, which were isolated from twig cankers in California [50]. However, the latter species has been very recently reported as the agent of a charcoal disease in Tunisia [51]. Other recent reports concern *Diaporthe ambigua* causing twig cankers in Italy [52] and *Neofabraea kienholzii* causing leaf and shoot lesions in California [53], while *Cytospora pruinosa* recovered from plants showing branch dieback symptoms in Spain was found not to be pathogenic [54]. Notwithstanding, reports concerning *Cytospora* [30,35] must be taken with caution, considering recent assessments of pathogenicity on olive tree by a few *Cytospora* spp., including the novel species *C. olivarum* [55,56].

The endophytic occurrence of more fungi only identified at the genus level should be better evaluated as well. In addition to *Venturia* and *Verticillium* found in Portugal [29,34], this applies to strains in the genera *Neofusicoccum* and *Phaeoacremonium*, with reference to records of *P. aleophilum* and several *Neofusicoccum* spp. as agents of branch dieback and decline of olive trees [50,54,57]. Similar considerations are valid for strains of *Diaporthe* (=*Phomopsis*), which are renowned canker agents of common endophytic occurrence on many woody plants [58], *Diplodia, Pestalotiopsis*, and *Phoma*, including species which can be disease agents of olive tree [48]. Above all, this concept concerns endophytic strains of *Colletotrichum* which are frequently reported in asymptomatic plants of both olive tree [27–29,33,59] and other crops, such as citrus [60]. Several *Colletotrichum* spp. are involved in etiology of olive anthracnose with various degrees of virulence and latency [61–63], including the new species *C. clavatus* to which the findings of *C. acutatum* as endophyte in olive drupes in Italy are likely to be referable [64]. It goes without saying that a more thorough assessment of their endophytic occurrence in the different cropping contexts is expected to have a remarkable impact on the management of anthracnose.

Unlike the above pathogens, the agent of the Dalmatian disease of olives, *Botryosphaeria dothidea* [65], is missing in the list of endophytic fungi, despite this species is best known as an endophyte of a high number of plants [66]. The fact that its documented occurrence is restricted to diseased drupes is possibly linked to the role as a vector by the cecidomyid *Lasioptera berlesiana*, a parasitoid of the olive fruit fly *Bactrocera oleae* [67], considering that *B. dothidea* is constantly associated with many cecidomyid midges [68].

3.2. Endophytic Fungi as Mutualists

Many species found as endophytes of olive tree are known to behave as mutualists in crops, based on their ability to contrast pests and pathogens, and/or to promote plant growth; a brief overview of such properties with reference to species included in Table 1 is proposed in this paragraph.

Strains of *Trichoderma*, mainly reported from Portugal and ascribed to at least four species, have already been experimentally evaluated on olive tree, with reference to both kinds of beneficial effects, particularly for the biocontrol of *V. dahliae* and *N. solani* [18,69–71]. The typical soil fungus *Penicillium restrictum* has been reported as an antagonist and mycoparasite of several plant pathogens [72], but other *Penicillium* spp. mentioned in Table 1 are also known as endophytic associates exhibiting antifungal effects [73]. Traditionally known as a biocontrol agent of the grey mold agent *B. cinerea* [74], *Clonostachys rosea* (=*Bionectria ochroleuca*) has also disclosed potential against insects and nematodes [75], likewise *Purpureocillium lilacinum* [76]. Already employed in the formulation of biopesticides, the latter species has been reported for antagonism against *V. dahliae* on eggplant, along with plant growth promoting effects [77]. Again, found in association with *V. dahliae* in southern Italy [41], *Paecilomyces variotii* is known for its antagonistic behavior against plant pathogenic fungi in vitro and in vivo [78], as well as nematodes [79]. Species of *Cosmospora* have been reported as mycoparasites [80], and pathogens of armored scales (Hemiptera, Diaspididae) [81]. Surprisingly, some species of common endophytic occurrence which are basically known as entomopathogens or display a dual biocontrol aptitude against both arthropods and fungi, such as *Beauveria bassiana, Metarhizium anisopliae*, and *Lecanicillium/Akanthomyces* spp. [82–84], have not been found as endophytes of olive tree so far.

Based on previous citations as fungal antagonists and producers of bioactive secondary metabolites, other species included in Table 1 may have a role in defensive mutualism [12,15,16,19], such as *Chaetomium globosum* [85], and species of *Paraconiothyrium* [86], *Alternaria* [87], and *Epicoccum* [88]. Strains of the two latter genera were found at a significantly higher rate in asymptomatic leaves in a survey carried out in Portugal considering the key leaf pathogens of olive tree *Venturia oleaginea* and *Pseudocercospora cladosporioides*, which may be indicative of an antagonistic role against the above disease agents [32]. Similar considerations have been advanced for *Chromelosporium carneum* and other taxa in the Pezizales (e.g., *Heydenia* and *Pyronema*); in fact, these fungi were more frequent in plants which did not show symptoms of the olive knot disease [27]. Indeed, their role in suppressing this bacterial disease deserves to be further investigated.

Finally, other endophytic associates of olive tree are more reputed for their plant growth promoting potential depending on improvement of nutrient availability and/or production of plant hormones, such as auxins by *Discosia* sp. [89], and gibberellins by *Phoma herbarum* [90]. Moreover, several yeast species are known for these effects, such as *Aureobasidium pullulans*, which stands out for its widespread occurrence on the olive phylloplane too [41,91].

3.3. Endophytic Fungi as Neutral Associates

The ecological role of the many identified endophytic fungi which were not mentioned in the previous sections requires further assessments. In the absence of any circumstantial evidence, such associations are usually defined as neutral. However, it is hard to accept that this approximate inference is valid for such a high number of taxa; rather, it is likely that at least some of them are going to disclose better defined ecological relationships in future. As an example, the ability by a plant to support endophytic development and reproduction of fungi which are pathogenic towards other plant species has been envisaged to possibly represent an ecological adaptation supporting the competitive attitude of the plant host [92]. In this respect the available data indicate that olive can be the host of species such as *Cadophora luteo-olivacea, Curvularia trifolii, Dactylonectria pauciseptata, Didymella macrostoma, Drechslera avenae, Eutypa tetragona, Paraphoma chrysanthemicola, Parastagonospora avenae, Pestalotiopsis guepinii, Pleospora herbarum, Stagonosporopsis cucurbitacearum*, and *Chondrostereum purpureum*, known as pathogens of various crops.

For other fungi already known for their endophytic attitude on several plant species, such as *Anthostomella leucospermi, Arcopilus aureus, Chalastospora gossypii, Daldinia concentrica, Endoconidioma populi, Fimetariella rabenhorstii, Nemania aenea, Paraphaeosphaeria sporulosa, Preussia africana*, and *Sporormiella intermedia*, the occurrence on olive tree may be rather interpreted as reflecting a general ecological adaptation to horizontal spread within the phytocoenoses [93].

Because of their recent taxonomic description, other species listed in Table 1 have no significant references in the literature considering their ecological role yet. This is the case of the Dothideomycetes *Leptosphaerulina saccharicola* [94], *Dendrothyrium variisporum* [95], *Libertasomyces platani* [96], and *Stigmatodiscus enigmaticus*, with the latter representing the founder of the new order Stigmatodiscales [97].

Although lichens are quite common epiphytes of olive trees, the endophytic occurrence of some lichenicolous species is to be remarked with reference to their exclusive finding in the mentioned study concerning the *X. fastidiosa* epidemic in Salento, Italy [21]. More in detail it is about the genera *Absconditella, Biatora, Lecania, Lecanora, Lecidella, Leimonis, Xanthoria, Catillaria, Hyperphyscia, Pyrrhospora, Scoliciosporum* and *Xanthoparmelia* (=*Karoowia*), with the last five only found in the cultivar Leccino which is known to be resistant to this bacterial disease.

Finally, a mention is deserved for some taxa which have been described as opportunistic human pathogens, such as *Cladophialophora, Exophiala, Fusarium musae, Hormonema, Lecythophora, Rhinocladiella similis*, and the yeasts *Candida, Pichia, Wickerhamomyces, Cryptococcus, Malassezia* and *Rhodotorula mucilaginosa* [98].

4. Biochemical Properties and Possible Biotechnological Applications

It is generally accepted that endophytes exert defensive mutualism through biochemical interactions with both the host plant and its pests and pathogens. The ability to release bioactive secondary metabolites and enzymes may ensure direct or indirect antagonistic effects and promote the host's defense reaction. Based on this paradigm, a huge amount of studies concerning microbial endophytes deal with the characterization of products and enzyme complexes and examine perspectives for their biotechnological exploitation [99,100].

So far, the available information concerning endophytic fungi of olive tree is quite limited, and basically concerns known compounds. The acidic terpenoid arundifungin was found as a product of an unidentified Coelomycetes strain recovered as an olive endophyte in Spain, possessing antifungal properties which derive from inhibition of glucan synthesis, likewise the better known echinocandin [101]. A strain of *Penicillium chrysogenum* was found to produce the bioactive indole alkaloids meleagrin, roquefortine C, and dehydrohistidyltryptophenyl-diketopiperazine [42]. Moreover, from the same location of the Egyptian oasis of Siwa, an isolate of *Penicillium citrinum* was reported to produce the polyketide mycotoxin citrinin, along with the pyrrolidine alkaloids 2-(hept-5-enyl)-3-methyl-4-oxo-6,7,8,8a-tetrahydro-4*H*-pyrrolo[2,1-*b*]-1,3-oxazine, scalusamide A, and perinadine A [36].

Endophytic fungi recovered from olive leaves and identified as *Alternaria* sp., *Chaetomium* sp., *Diaporthe* sp., *Fusarium* sp., *E. nigrum*, and *Nigrospora oryzae* were found to possess antagonistic properties against *C. acutatum* in vitro, with the latter species displaying the most consistent effects. Such effects were at least in part related to the production of volatile organic compounds (VOCs), such as phenylethyl alcohol, pyrazines, amine, and propanoic acid derivatives [31].

In other cases, investigations have been limited to a preliminary stage considering culture filtrates or extracts from the same. This is the case of culture filtrates obtained from endophytic strains of *E. nigrum* and *R. similis* displaying inhibitory effects against *P. savastanoi* pv. *savastanoi* [102,103]. Antibiotic activity against both Gram+ and Gram− bacteria has been reported in vitro for strains of *Penicillium canescens*, *Penicillium commune*, and *A. alternata*, with the latter also active against yeasts (*Candida* spp.) [25]. In the case of *A. alternata*, the antibiotic effect was also induced by the ethyl acetate extracts from both mycelium and the culture broth, while in the case of *P. commune*, bioactivity increased after the addition of an olive leaf extract to the medium.

Penicillium commune was again reported for in vitro antagonistic effects against the agent of anthracnose (*C. acutatum*); such effects were increased by the placement of an olive leaf in the test plates, indicating that interactions might be more dramatic in the contact with plant host [43]. The same research group pointed out inhibitory effects on mycelial growth of this pathogen by endophytic strains of *Chondrostereum purpureum*, with evident alterations in hyphal structure [28], and of both *C. acutatum* and *V. dahliae* by *Trichoderma lixii* and *P. lilacinum* [40]. Moreover, endophytic strains of *A. pullulans* applied at the blooming stage or just prior to harvest were respectively able to reduce incidence of the anthracnose agent during its latency and in post-harvest on the shelf [59]. The latter effect is confirmatory of reports referring to other fruit crops that VOCs released by endophytic strains of this yeast effectively inhibit spore germination of several post-harvest pathogens [104,105].

An olive endophytic strain identified as *Daldinia* cf. *concentrica* also showed inhibitory effects against a panel of plant pathogenic fungi and oomycota, which were basically dependent on the production of antimicrobial VOCs. Exposure of dried fruits and grains to these volatiles resulted in their full disinfection preventing the development of molds and suppressed infection by *Aspergillus niger* in peanuts [39]. In greenhouse experiments VOCs produced by this strain showed bionematicidal activity against the second-stage juveniles of *Meloidogyne javanica* (67% reduction in viability). Among these volatiles, 4-heptanone elicited the most consistent effect with 90% reduction in viability, while egg hatching decreased by 87%. Moreover, the application of a volatile mixture to soil inoculated with *M. javanica* eggs or juveniles significantly reduced galling index in susceptible tomato plants with no effect on root weight [106].

Finally, an interesting ground of investigation has been envisaged for endophytic fungi of olive tree in view of applicative perspectives in the management of *X. fastidiosa* epidemic spread, based on evidence of either repellent or attractant effects that some species possibly exert against its vector, the meadow spittlebug (*Philaenus spumarius*) [107].

5. Conclusions

Research on endophytic fungi is gradually evolving from a basically descriptive stage to the analysis of factors determining the structure of microbiomes, in the perspective that their manipulation may enable to increase plant protection and productivity. In this respect, it has been observed that a better comprehension of the genetic interactions with the host tree and other associated microbes is crucial for the success of practical applications of endophytic fungi in sustainable agriculture [108].

In the case of *O. europaea*, the increasing number of reports concerning endophytic fungi in the past few years confirm that their spatial–temporal distribution in olive trees has been poorly investigated. However, as for other crops, the accumulating data support the evidence of the substantial impact of this microbial component of biodiversity on fitness of olive tree. As a further example, the recalcitrance to sterilization protocols for in vitro propagation reported in a Slovenian study is indicative that at least some of these fungi are intimately associated to olive plants and adapted for a long-term survival and proliferation in their tissues [37].

Increasing knowledge about functions and dynamics of endophytic communities is fundamental in the aim to exploit use as biocontrol agents. The epidemiological relevance of these microorganisms is basically related to a modulatory role in the spread of cryptogamic diseases. Even when there is no apparent direct interaction with disease agents, the possible effect by endophytic fungi in stimulating plant defense reaction, or more in general to act as plant disease modifiers [109], should not be disregarded. In this respect, data concerning occasional isolations might as well disclose some relevance. Indeed, the role of microbial inoculants on protection, growth stimulation, and productivity is now recognized for olive tree too [14], and the expectancy is high that the applicative use of endophytic fungi may soon become an additional tool in the sustainable management of olive growing.

Author Contributions: Conceptualization, R.N.; resources, R.N. and C.C.; data curation, R.N. and C.C.; writing—original draft preparation, R.N., C.D.V., and C.C.; writing—review and editing, R.N., C.D.V., and C.C.; and funding acquisition, C.C. All authors have read and agreed to the published version of the manuscript.

Acknowledgments: The contribution of Veronica Vizzarri for some discussion hints is acknowledged.

References

1. Hirakue, A.; Sugiyama, S. Relationship between foliar endophytes and apple cultivar disease resistance in an organic orchard. *Biol. Control* **2018**, *127*, 139–144. [CrossRef]
2. Zornoza, R.; Mataix-Solera, J.; Guerrero, C.; Arcenegui, V.; Mataix-Beneyto, J. Comparison of soil physical, chemical, and biochemical properties among native forest, maintained and abandoned almond orchards in mountainous areas of Eastern Spain. *Arid Land Res. Manag.* **2009**, *23*, 267–282. [CrossRef]
3. Palese, A.M.; Magno, R.; Casacchia, T.; Curci, M.; Baronti, S.; Miglietta, F.; Crecchio, C.; Xiloyannis, C.; Sofo, A. Chemical, biochemical and microbiological properties of soils from abandoned and extensively cultivated olive orchards. *Sci. World J.* **2013**, 496278. [CrossRef] [PubMed]
4. Kushwaha, C.P.; Singh, K.P. Crop productivity and soil fertility in a tropical dryland agro-ecosystem: Impact of residue and tillage management. *Exp. Agric.* **2005**, *41*, 39–50. [CrossRef]
5. Govaerts, B.; Mezzalama, M.; Sayre, K.D.; Crossa, J.; Lichter, K.; Troch, V.; Vanherck, K.; De Corte, P.; Deckers, J. Long-term consequences of tillage, residue management, and crop rotation on selected soil micro-flora groups in the subtropical highlands. *Appl. Soil Ecol.* **2008**, *38*, 197–210. [CrossRef]
6. Di Vaio, C.; Marallo, N.; Marino, G.; Caruso, T. Effect of water stress on dry matter accumulation and partitioning in pot-grown olive trees (cv Leccino and Racioppella). *Sci. Hortic.* **2013**, *164*, 172–177. [CrossRef]
7. Cirillo, C.; Russo, R.; Famiani, F.; Di Vaio, C. Investigation on rooting ability of twenty olive cultivars from Southern Italy. *Adv. Hortic. Sci.* **2017**, *31*, 311–317.
8. Benitez, E.; Nogales, R.; Campos, M.; Ruano, F. Biochemical variability of olive-orchard soils under different management systems. *Appl. Soil Ecol.* **2006**, *32*, 221–231. [CrossRef]
9. Sofo, A.; Ciarfaglia, A.; Scopa, A.; Camele, I.; Curci, M.; Crecchio, C.; Xiloyannis, C.; Palese, A.M. Soil microbial diversity and activity in a Mediterranean olive orchard managed by a set of sustainable agricultural practices. *Soil Use Manag.* **2014**, *30*, 160–167. [CrossRef]

10. Ruano-Rosa, D.; Valverde-Corredor, A.; Gómez-Lama Cabanás, C.; Sesmero, R.; Mercado-Blanco, J. What lies beneath: Root-associated bacteria to improve the growth and health of olive trees. In *Soil Biological Communities and Ecosystem Resilience*; Lukac, M., Grenni, P., Gamboni, M., Eds.; Springer: Cham, Switzerland, 2017; pp. 107–122.
11. IPCC. *2014: Climate Change 2014: Synthesis Report. Contribution of Working Groups I. II and III to the Fifth Assessment Report of the Intergovernmental Panel on Climate Change*; IPCC: Geneva, Switzerland, 2014.
12. Sofo, A.; Manfreda, S.; Dichio, B.; Fiorentino, M.; Xiloyannis, C. The olive tree: A paradigm for drought tolerance in Mediterranean climates. *Hydrol. Earth Syst. Sci.* **2008**, *12*, 293–301. [CrossRef]
13. Choudhary, D.K. Microbial rescue to plant under habitat-imposed abiotic and biotic stresses. *Appl. Microbiol. Biotechnol.* **2012**, *96*, 1137–1155. [CrossRef] [PubMed]
14. Bizos, G.; Papatheodorou, E.M.; Chatzistathis, T.; Ntalli, N.; Aschonitis, V.G.; Monokrousos, N. The role of microbial inoculants on plant protection, growth stimulation, and crop productivity of the olive tree (*Olea europea* L.). *Plants* **2020**, *9*, 743. [CrossRef] [PubMed]
15. Lata, R.; Chowdhury, S.; Gond, S.; White, J.F. Induction of abiotic stress tolerance in plants by endophytic microbes. *Appl. Microbiol.* **2018**, *66*, 268–276. [CrossRef] [PubMed]
16. Yan, L.; Zhu, J.; Zhao, X.; Shi, J.; Jiang, C.; Shao, D. Beneficial effects of endophytic fungi colonization on plants. *Appl. Microbiol. Biotechnol.* **2019**, *103*, 3327–3340. [CrossRef]
17. Kasotia, A.; Choundhary, D.K. Role of endophytic microbes in mitigation of abiotic stress in plants. In *Emerging Technologies and Management of Crop Stress Tolerance*; Ahmad, P., Ed.; Elsevier: Amsterdam, The Netherlands, 2014; Volume 2, pp. 97–108.
18. Dini, I.; Graziani, G.; Gaspari, A.; Fedele, F.L.; Sicari, A.; Vinale, F.; Cavallo, P.; Lorito, M.; Ritieni, A. New strategies in the cultivation of olive trees and repercussions on the nutritional value of the extra virgin olive oil. *Molecules* **2020**, *25*, 2345. [CrossRef]
19. Fa, A.N. Endophytic fungi for sustainable agriculture. *Microb. Biosyst.* **2019**, *4*, 31–44.
20. Stielow, J.B.; Levesque, C.A.; Seifert, K.A.; Meyer, W.; Iriny, L.; Smits, D.; Renfurm, R.; Verkley, G.J.M.; Groenewald, M.; Chaduli, D.; et al. One fungus, which genes? Development and assessment of universal primers for potential secondary fungal DNA barcodes. *Persoonia* **2015**, *35*, 242–263. [CrossRef]
21. Vergine, M.; Meyer, J.B.; Cardinale, M.; Sabella, E.; Hartmann, M.; Cherubini, P.; De Bellis, L.; Luvisi, A. The *Xylella fastidiosa*-resistant olive cultivar "Leccino" has stable endophytic microbiota during the olive quick decline syndrome (OQDS). *Pathogens* **2020**, *9*, 35. [CrossRef]
22. Fernández-González, A.J.; Villadas, P.J.; Cabanás, C.G.L.; Valverde-Corredor, A.; Belaj, A.; Mercado-Blanco, J.; Fernández-López, M. Defining the root endosphere and rhizosphere microbiomes from the World Olive Germplasm Collection. *Sci. Rep.* **2019**, *9*, 20423. [CrossRef]
23. Ferraro, V.; Conigliaro, G.; Torta, L.; Burruano, S.; Moschetti, G. Preliminary investigation on the endophytic communities in *Olea europaea* in Sicily. In Proceedings of the 7th International Conference Integrated Fruit Production, Avignon, France, 27–30 October 2008; pp. 459–463.
24. Martins, F.; Pereira, J.A.; Bota, P.; Bento, A.; Baptista, P. Fungal endophyte communities in above-and belowground olive tree organs and the effect of season and geographic location on their structures. *Fungal Ecol.* **2016**, *20*, 193–201. [CrossRef]
25. Malhadas, C.; Malheiro, R.; Pereira, J.A.; de Pinho, P.G.; Baptista, P. Antimicrobial activity of endophytic fungi from olive tree leaves. *World J. Microbiol. Biotechnol.* **2017**, *33*, 46. [CrossRef] [PubMed]
26. Materatski, P.; Varanda, C.; Carvalho, T.; Bento Dias, A.; Campos, M.D.; Rei, F.; Félix, M.R. Spatial and temporal variation of fungal endophytic richness and diversity associated to the phyllosphere of olive cultivars. *Fungal Biol.* **2019**, *123*, 66–76. [CrossRef] [PubMed]
27. Gomes, T.; Pereira, J.A.; Lino-Neto, T.; Bennett, A.E.; Baptista, P. Bacterial disease induced changes in fungal communities of olive tree twigs depend on host genotype. *Sci. Rep.* **2019**, *9*, 5882. [CrossRef] [PubMed]
28. Preto, G.; Martins, F.; Pereira, J.A.; Baptista, P. Fungal community in olive fruits of cultivars with different susceptibilities to anthracnose and selection of isolates to be used as biocontrol agents. *Biol. Control* **2017**, *110*, 1–9. [CrossRef]
29. Gomes, T.; Pereira, J.A.; Benhadi, J.; Lino-Neto, T.; Baptista, P. Endophytic and epiphytic phyllosphere fungal communities are shaped by different environmental factors in a Mediterranean ecosystem. *Microb. Ecol.* **2018**, *36*, 668–679. [CrossRef]

30. Fisher, P.J.; Petrini, O.; Petrini, L.E.; Descals, E. A preliminary study of fungi inhabiting xylem and whole stems of *Olea europaea*. *Sydowia* **1992**, *44*, 117–121.
31. Landum, M.C.; Félix, M.R.; Alho, J.; Garcia, R.; Cabrita, M.J.; Rei, F.; Varanda, C.M. Antagonistic activity of fungi of *Olea europaea* L. against *Colletotrichum acutatum*. *Microbiol. Res.* **2016**, *183*, 100–108. [CrossRef]
32. Varanda, C.M.; Materatski, P.; Landum, M.; Campos, M.D.; Félix, M.D.R. Fungal communities associated with peacock and cercospora leaf spots in olive. *Plants* **2019**, *8*, 169. [CrossRef]
33. de Freitas Sia, E.; Marcon, J.; Mazzer Luvizotto, D.; Quecine, M.C.; Tsui, S.; Pereira, J.O.; Pizzirani-Kleiner, A.A.; Azevedo, J.L. Endophytic fungi from the Amazonian plant *Paullinia cupana* and from *Olea europaea* isolated using cassava as an alternative starch media source. *SpringerPlus* **2013**, *2*, 579.
34. Martins, F.; Pereira, J.A.; Baptista, P. Dynamics of fungal endophytes over different phenological stages of the olive tree host. In Proceedings of the 15th Congress of the Mediterranean Phytopathological Union, Cordoba, Spain, 20–23 June 2017.
35. Hanani, A.; Valentini, F.; Sanzani, S.M.; Gallo, M.; Davino, S.-W.; D'Onghia, A.M. Assessment of the endophytic fungal community in Apulian olive varieties with different potential susceptibilities to *Xylella fastidiosa*. *J. Plant Pathol.* **2019**, *101*, 811.
36. Mady, M.S.; Houssen, W.; Abdou, R.; Haggag, E.G.; El Sayed, K.A. Breast cancer migration and proliferation inhibitory and antibiotic secondary metabolites from the Egyptian olive tree endophytic fungus *Penicillium citrinum*. *J. Advan. Pharm. Res.* **2017**, *1*, 160–170. [CrossRef]
37. Oražem, P.; Celar, F.A.; Bohanec, B. Occurrence of endophytic fungi causing recalcitrance of olive cultivar 'Istrska belica' during shoot culture establishment. *Arch. Biol. Sci.* **2016**, *68*, 177–186. [CrossRef]
38. Agosteo, G.E.; Macrì, C.; Taccone, P. Susceptibility of olive cv Itrana to anthracnose. *J. Plant Pathol.* **2005**, *87*, 287.
39. Liarzi, O.; Bar, E.; Lewinsohn, E.; Ezra, D. Use of the endophytic fungus *Daldinia* cf. *concentrica* and its volatiles as bio-control agents. *PLoS ONE* **2016**, *11*, e0168242. [CrossRef] [PubMed]
40. Martins, F.; Pereira, J.A.; Bento, A.; Baptista, P. Potentialities of endophytic fungi of olive tree as biological control agents against *Colletotrichum acutatum* and *Verticillium dahliae*. In *Endophytes for Plant Protection: The State of the Art*; Schneider, C., Leifert, C., Feldmann, F., Eds.; Deutsche Phytomedizinische Gesellschaft: Braunschweig, Germany, 2013; p. 190.
41. Nicoletti, R.; Rinaldi, R. Indagine sulla micoflora del filloplano dell'olivo. *Riv. Patol. Veg.* **1993**, *3*, 41–47.
42. Mady, M.S.; Mohyeldin, M.M.; Ebrahim, H.Y.; Elsayed, H.E.; Houssen, W.E.; Haggag, E.G.; Soliman, R.F.; El Sayed, K.A. The indole alkaloid meleagrin, from the olive tree endophytic fungus *Penicillium chrysogenum*, as a novel lead for the control of c-Met-dependent breast cancer proliferation, migration and invasion. *Bioorg. Med. Chem.* **2016**, *24*, 113–122. [CrossRef]
43. Martins, F.; Pereira, J.A.; Bento, A.; Baptista, P. Plant-mediated effects on antagonistic activity of endophytic fungi towards olive fungal diseases. In *Endophytes for Plant Protection: The State of the Art*; Schneider, C., Leifert, C., Feldmann, F., Eds.; Deutsche Phytomedizinische Gesellschaft: Braunschweig, Germany, 2013; pp. 127–128.
44. FAOSTAT Online Database. Available online: http://faostat.fao.org/ (accessed on 1 July 2020).
45. Zicca, S.; De Bellis, P.; Masiello, M.; Saponari, M.; Saldarelli, P.; Boscia, D.; Sisto, A. Antagonistic activity of olive endophytic bacteria and of *Bacillus* spp. strains against *Xylella fastidiosa*. *Microbiol. Res.* **2020**, *236*, 126467. [CrossRef]
46. López-Escudero, F.J.; Mercado-Blanco, J. *Verticillium* wilt of olive: A case study to implement an integrated strategy to control a soil-borne pathogen. *Plant Soil* **2011**, *344*, 1–50. [CrossRef]
47. Mercado-Blanco, J.; Rodríguez-Jurado, D.; Pérez-Artés, E.; Jiménez-Díaz, R.M. Detection of the nondefoliating pathotype of *Verticillium dahliae* in infected olive plants by nested PCR. *Plant Pathol.* **2001**, *50*, 609–619. [CrossRef]
48. Chliyeh, M.; Touati, J.; Selmaoui, K.; Touhami, A.O.; Filali-Maltouf, A.; El Modafar, C.; Douira, A. Bibliographic inventory of the olive tree (*Olea europaea* L.) fungal diseases in the world. *Int. J. Pure Appl. Biosci.* **2014**, *2*, 46–79.
49. Crous, P.W.; Groenewald, J.Z.; Nigro, F.; Antelmi, I. Fungal planet description sheets 351—*Pseudophaeomoniella* Nigro, Antelmi & Crous, gen. nov. *Persoonia* **2015**, *34*, 224–227.
50. Úrbez-Torres, J.R.; Peduto, F.; Vossen, P.M.; Krueger, W.H.; Gubler, W.D. Olive twig and branch dieback: Etiology, incidence, and distribution in California. *Plant Dis.* **2013**, *97*, 231–244. [CrossRef] [PubMed]

51. Gharbi, Y.; Ennouri, K.; Bouazizi, E.; Cheffi, M.; Triki, M.A. First report of charcoal disease caused by *Biscogniauxia mediterranea* on *Olea europaea* in Tunisia. *J. Plant Pathol.* **2020**. [CrossRef]
52. Frisullo, S.; Elshafie, H.S.; Mang, S.M. First report of two *Phomopsis* species on olive trees in Italy. *J. Plant Pathol.* **2015**, *97*, 401.
53. Trouillas, F.P.; Nouri, M.T.; Lawrence, D.P.; Moral, J.; Travadon, R.; Aegerter, B.J.; Lightle, D. Identification and characterization of *Neofabraea kienholzii* and *Phlyctema vagabunda* causing leaf and shoot lesions of olive in California. *Plant Dis.* **2019**, *103*, 3018–3030. [CrossRef]
54. Moral, J.; Agustí-Brisach, C.; Pérez-Rodríguez, M.; Xaviér, C.; Raya, M.C.; Rhouma, A.; Trapero, A. Identification of fungal species associated with branch dieback of olive and resistance of table cultivars to *Neofusicoccum mediterraneum* and *Botryosphaeria dothidea*. *Plant Dis.* **2017**, *101*, 306–316. [CrossRef]
55. Lawrence, D.P.; Holland, L.A.; Nouri, M.T.; Travadon, R.; Abramians, A.; Michailides, T.J.; Trouillas, F.P. Molecular phylogeny of *Cytospora* species associated with canker diseases of fruit and nut crops in California, with the descriptions of ten new species and one new combination. *IMA Fungus* **2018**, *9*, 333–369. [CrossRef]
56. Úrbez-Torres, J.R.R.; Lawrence, D.P.; Peduto Hand, F.; Trouillas, F. Olive twig and branch dieback in California caused by *Cytospora oleicola* and the newly described species *Cytospora olivarum* sp. nov. *Plant Dis.* **2020**, *104*. [CrossRef]
57. Carlucci, A.; Raimondo, M.L.; Cibelli, F.; Phillips, A.J.; Lops, F. *Pleurostomophora richardsiae*, *Neofusicoccum parvum* and *Phaeoacremonium aleophilum* associated with a decline of olives in southern Italy. *Phytopathol. Medit.* **2013**, *52*, 517–527.
58. Udayanga, D.; Liu, X.; McKenzie, E.H.C.; Chukeatirote, E.; Bahkali, A.H.A.; Hyde, K.D. The genus *Phomopsis*: Biology, applications, species concepts and names of common phytopathogens. *Fungal Divers.* **2011**, *50*, 189–225. [CrossRef]
59. Nigro, F.; Antelmi, I.; Labarile, R.; Sion, V.; Pentimone, I. Biological control of olive anthracnose. *Acta Hortic.* **2018**, 439–444. [CrossRef]
60. Nicoletti, R. Endophytic fungi of citrus plants. *Agriculture* **2019**, *9*, 247. [CrossRef]
61. Cacciola, S.O.; Faedda, R.; Sinatra, F.; Agosteo, G.E.; Schena, L.; Frisullo, S.; di San Lio, G.M. Olive anthracnose. *J. Plant Pathol.* **2012**, *94*, 29–44.
62. Schena, L.; Mosca, S.; Cacciola, S.O.; Faedda, R.; Sanzani, S.M.; Agosteo, G.E.; Sergeeva, V.; di San Lio, G.M. Species of the *Colletotrichum gloeosporioides* and *C. boninense* complexes associated with olive anthracnose. *Plant Pathol.* **2014**, *63*, 437–446. [CrossRef]
63. Msairi, S.; Chliyeh, M.; Touhami, A.O.; El Alaoui, A.; Selmaoui, K.; Benkirane, R.; Filali-Maltouf, A.; El Modafar, C.; Douira, A. First report of *Colletotrichum lupini* causing anthracnose disease on the olive fruits in Morocco. *Plant Cell Biotechnol. Mol. Biol.* **2020**, *21*, 1–11.
64. Faedda, R.; Agosteo, G.E.; Schena, L.; Mosca, S.; Frisullo, S.; di San Lio, G.M.; Cacciola, S.O. *Colletotrichum clavatum* sp. nov. identified as the causal agent of olive anthracnose in Italy. *Phytopathol. Medit.* **2011**, *50*, 283–302.
65. Phillips, A.J.L.; Rumbos, I.C.; Alves, A.; Correia, A. Morphology and phylogeny of *Botryosphaeria dothidea* causing fruit rot of olives. *Mycopathologia* **2005**, *159*, 433–439. [CrossRef] [PubMed]
66. Slippers, B.; Wingfield, M.J. Botryosphaeriaceae as endophytes and latent pathogens of woody plants: Diversity, ecology and impact. *Fungal Biol. Rev.* **2007**, *21*, 90–106. [CrossRef]
67. Latinović, J.; Hrnčić, S.; Perović, T.; Latinović, N. *Botryosphaeria dothidea*–causal agent of olive fruit rot–pathogen of wounds or not? *IOBC-WPRS Bull.* **2014**, *108*, 35–38.
68. Zimowska, B.; Okoń, S.; Becchimanzi, A.; Krol, E.D.; Nicoletti, R. Phylogenetic characterization of *Botryosphaeria* strains associated with *Asphondylia* galls on species of Lamiaceae. *Diversity* **2020**, *12*, 41. [CrossRef]
69. Carrero-Carrón, I.; Trapero-Casas, J.L.; Olivares-García, C.; Monte, E.; Hermosa, R.; Jiménez-Díaz, R.M. *Trichoderma asperellum* is effective for biocontrol of *Verticillium* wilt in olive caused by the defoliating pathotype of *Verticillium dahliae*. *Crop Prot.* **2016**, *88*, 45–52. [CrossRef]
70. Ruano-Rosa, D.; Prieto, P.; Rincón, A.M.; Gómez-Rodríguez, M.V.; Valderrama, R.; Barroso, J.B.; Mercado-Blanco, J. Fate of *Trichoderma harzianum* in the olive rhizosphere: Time course of the root colonization process and interaction with the fungal pathogen *Verticillium dahliae*. *BioControl* **2016**, *61*, 269–282. [CrossRef]

71. Ben Amira, M.; Lopez, D.; Triki Mohamed, A.; Khouaja, A.; Chaar, H.; Fumanal, B.; Gousset-Dupont, A.; Bonhomme, L.; Label, P.; Goupil, P.; et al. Beneficial effect of *Trichoderma harzianum* strain Ths97 in biocontrolling *Fusarium solani* causal agent of root rot disease in olive trees. *Biol. Control* **2017**, *110*, 70–78. [CrossRef]
72. Nicoletti, R.; De Stefano, M. *Penicillium restrictum* as an antagonist of plant pathogenic fungi. *Dyn. Biochem. Process Biotechnol. Mol. Biol.* **2012**, *6*, 61–69.
73. Nicoletti, R.; Fiorentino, A.; Scognamiglio, M. Endophytism of *Penicillium* species in woody plants. *Open Mycol. J.* **2014**, *8*, 1–26. [CrossRef]
74. Sutton, J.C.; Li, D.W.; Peng, G.; Yu, H.; Zhang, P.G.; Valdebenito-Sanhueza, R.M. *Gliocladium roseum*—A versatile adversary of *Botrytis cinerea* in crops. *Plant Dis.* **1997**, *81*, 316–328. [CrossRef]
75. Sun, Z.B.; Li, S.D.; Ren, Q.; Xu, J.L.; Lu, X.; Sun, M.H. Biology and applications of *Clonostachys rosea*. *J. Appl. Microbiol.* **2020**. [CrossRef]
76. Fiedler, Ż.; Sosnowska, D. Nematophagous fungus *Paecilomyces lilacinus* (Thom) Samson is also a biological agent for control of greenhouse insects and mite pests. *BioControl* **2007**, *52*, 547–558. [CrossRef]
77. Lan, X.; Zhang, J.; Zong, Z.; Ma, Q.; Wang, Y. Evaluation of the biocontrol potential of *Purpureocillium lilacinum* QLP12 against *Verticillium dahliae* in eggplant. *BioMed Res. Int.* **2017**, *2017*, 4101357. [CrossRef]
78. Rodrigo, S.; Santamaria, O.; Halecker, S.; Lledó, S.; Stadler, M. Antagonism between *Byssochlamys spectabilis* (anamorph *Paecilomyces variotii*) and plant pathogens: Involvement of the bioactive compounds produced by the endophyte. *Ann. Appl. Biol.* **2017**, *171*, 464–476. [CrossRef]
79. Al-Qasim, M.; Abu-Gharbieh, W.; Assas, K. Nematophagal ability of Jordanian isolates of *Paecilomyces variotii* on the root-knot nematode *Meloidogyne javanica*. *Nematol. Medit.* **2009**, *37*, 53–57.
80. Herrera, C.S.; Hirooka, Y.; Chaverri, P. Pseudocospeciation of the mycoparasite *Cosmospora* with their fungal hosts. *Ecol. Evol.* **2016**, *6*, 1504–1514. [CrossRef] [PubMed]
81. Mauchline, N.; Hallett, I.; Hill, G.; Casonato, S. Process of infection of armored scale insects (Diaspididae) by an entomopathogenic *Cosmospora* sp. *J. Invert. Pathol.* **2011**, *108*, 46–51. [CrossRef] [PubMed]
82. McKinnon, A.C.; Saari, S.; Moran-Diez, M.E.; Meyling, N.V.; Raad, M.; Glare, T.R. *Beauveria bassiana* as an endophyte: A critical review on associated methodology and biocontrol potential. *BioControl* **2017**, *62*, 1–17. [CrossRef]
83. Behie, S.W.; Jones, S.J.; Bidochka, M.J. Plant tissue localization of the endophytic insect pathogenic fungi *Metarhizium* and *Beauveria*. *Fungal Ecol.* **2015**, *13*, 112–119. [CrossRef]
84. Nicoletti, R.; Becchimanzi, A. Endophytism of *Lecanicillium* and *Akanthomyces*. *Agriculture* **2020**, *10*, 205. [CrossRef]
85. Aswini, C. A review on *Chaetomium globosum* is versatile weapons for various plant pathogens. *J. Pharmocognosy Phytochem.* **2019**, *8*, 946–949.
86. Verkley, G.J.; da Silva, M.; Wicklow, D.T.; Crous, P.W. *Paraconiothyrium*, a new genus to accommodate the mycoparasite *Coniothyrium minitans*, anamorphs of *Paraphaeosphaeria*, and four new species. *Stud. Mycol.* **2004**, *50*, 323–335.
87. Lou, J.; Fu, L.; Peng, Y.; Zhou, L. Metabolites from *Alternaria* fungi and their bioactivities. *Molecules* **2013**, *18*, 5891–5935. [CrossRef]
88. Braga, R.M.; Padilla, G.; Araújo, W.L. The biotechnological potential of *Epicoccum* spp.: Diversity of secondary metabolites. *Crit. Rev. Microbiol.* **2018**, *44*, 759–778. [CrossRef]
89. Rahi, P.; Vyas, P.; Sharma, S.; Gulati, A.; Gulati, A. Plant growth promoting potential of the fungus *Discosia* sp. FIHB 571 from tea rhizosphere tested on chickpea, maize and pea. *Indian J. Microbiol.* **2009**, *49*, 128–133. [CrossRef] [PubMed]
90. Hamayun, M.; Khan, S.A.; Khan, A.L.; Rehman, G.; Sohn, E.Y.; Shah, A.A.; Kim, S.K.; Joo, G.J.; Lee, I.J. *Phoma herbarum* as a new gibberellin-producing and plant growth-promoting fungus. *J. Microbiol. Biotechnol.* **2009**, *19*, 1244–1249. [PubMed]
91. Abdelfattah, A.; Li Destri Nicosia, M.G.; Cacciola, S.O.; Droby, S.; Schena, L. Metabarcoding analysis of fungal diversity in the phyllosphere and carposphere of olive (*Olea europaea*). *PLoS ONE* **2015**, *10*, e0131069. [CrossRef]
92. Aschehoug, E.T.; Metlen, K.L.; Callaway, R.M.; Newcombe, G. Fungal endophytes directly increase the competitive effects of an invasive forb. *Ecology* **2012**, *93*, 3–8. [CrossRef]

93. Rodriguez, R.J.; White, J.F., Jr.; Arnold, A.E.; Redman, A.R.A. Fungal endophytes: Diversity and functional roles. *New Phytol.* **2009**, *182*, 314–330. [CrossRef]
94. Phookamsak, R.; Liu, J.K.; Chukeatirote, E.; McKenzie, E.H.; Hyde, K.D. Phylogeny and morphology of *Leptosphaerulina saccharicola* sp. nov. and *Pleosphaerulina oryzae* and relationships with *Pithomyces. Cryptogam. Mycol.* **2013**, *34*, 303–319. [CrossRef]
95. Verkley, G.J.M.; Dukik, K.; Renfurm, R.; Göker, M.; Stielow, J.B. Novel genera and species of coniothyrium-like fungi in Montagnulaceae (Ascomycota). *Persoonia* **2014**, *32*, 25–51. [CrossRef] [PubMed]
96. Crous, P.W.; Wingfield, M.J.; Burgess, T.I.; Hardy, G.E.S.J.; Crane, C.; Barrett, S.; Cano-Lira, J.F.; Le Roux, J.J.; Thangavel, R.; Guarro, J.; et al. Fungal planet description sheets: 469–557. *Persoonia* **2016**, *37*, 218–403. [CrossRef]
97. Voglmayr, H.; Gardiennet, A.; Jaklitsch, W.M. *Asterodiscus* and *Stigmatodiscus*, two new apothecial dothideomycete genera and the new order Stigmatodiscales. *Fungal Divers.* **2016**, *80*, 271–284. [CrossRef]
98. Sherrington, S.L.; Kumwenda, P.; Kousser, C.; Hall, R.A. Host sensing by pathogenic fungi. *Advan. Appl. Microbiol.* **2018**, *102*, 159–221.
99. Nicoletti, R.; Fiorentino, A. Plant bioactive metabolites and drugs produced by endophytic fungi of Spermatophyta. *Agriculture* **2015**, *5*, 918–970. [CrossRef]
100. Yan, L.; Zhao, H.; Zhao, X.; Xu, X.; Di, Y.; Jiang, C.; Shi, J.; Shao, D.; Huang, Q.; Yang, H.; et al. Production of bioproducts by endophytic fungi: Chemical ecology, biotechnological applications, bottlenecks, and solutions. *Appl. Microbiol. Biotechnol.* **2018**, *102*, 6279–6298. [CrossRef] [PubMed]
101. Cabello, A.M.; Platas, G.; Collado, J.; Díez, T.M.; Martín, I.; Vicente, F.; Meinz, M.; Onishi, J.C.; Thompson, C.D.J.; Kurtz, M.B.; et al. Arundifungin, a novel antifungal compound produced by fungi: Biological activity and taxonomy of the producing organisms. *Int. Microbiol.* **2001**, *4*, 93–102. [CrossRef] [PubMed]
102. Gomes, T.; Pereira, J.A.; Lino-Neto, T.; Baptista, P. Endophytic and epiphytic fungal community associated to olive tree differ in antagonistic activity against *Pseudomonas savastanoi* pv. *savastanoi*. In Proceedings of the 15th Congress Mediterranean Phytopathological Union, Cordoba, Spain, 20–23 June 2017; pp. 202–203.
103. Berardo, C.; Bulai, I.M.; Venturino, E.; Baptista, P.; Gomes, T. Modeling the endophytic fungus *Epicoccum nigrum* action to fight the "olive knot" disease caused by *Pseudomonas savastanoi* pv. *savastanoi* (Psv) bacteria in *Olea europaea* L. Trees. In *Trends in Biomathematics: Modeling, Optimization and Computational Problems*; Springer: Cham, Switzerland, 2018; pp. 189–207.
104. Di Francesco, A.; Ugolini, L.; Lazzeri, L.; Mari, M. Production of volatile organic compounds by *Aureobasidium pullulans* as a potential mechanism of action against postharvest fruit pathogens. *Biol. Control* **2015**, *81*, 8–14. [CrossRef]
105. Don, S.Y.; Schmidtke, L.M.; Gambetta, J.M.; Steel, C.C. *Aureobasidium pullulans* volatilome identified by a novel, quantitative approach employing SPME-GC-MS, suppressed *Botrytis cinerea* and *Alternaria alternata in vitro*. *Sci. Rep.* **2020**, *10*, 4498.
106. Liarzi, O.; Bucki, P.; Miyara, S.B.; Ezra, D. Bioactive volatiles from an endophytic *Daldinia* cf. *concentrica* isolate affect the viability of the plant parasitic nematode *Meloidogyne javanica*. *PLoS ONE* **2016**, *11*, e0168437.
107. Cameirão, C.; Fernandes, G.; Martins, F.; Pereira, J.A.; Baptista, P. The effect of *Philaenus spumarius* feeding on the endophytic fungi community of *Coleostephus myconis* and its possible applicability in the insect biocontrol. In Proceedings of the 8th Meeting of the IOBC-WPRS Working Group 'Integrated Protection of Olive Crops', Florence, Italy, 4–7 June 2018.
108. Schlaeppi, K.; Bulgarelli, D. The plant microbiome at work. *Mol. Plant Microbe Interact.* **2015**, *28*, 212–217. [CrossRef]
109. Busby, P.E.; Ridout, M.; Newcombe, G. Fungal endophytes: Modifiers of plant disease. *Plant Mol. Biol.* **2016**, *90*, 645–655. [CrossRef]

2

Cork Oak Endophytic Fungi as Potential Biocontrol agents against *Biscogniauxia mediterranea* and *Diplodia corticola*

Daniela Costa [1], Rui M. Tavares [1], Paula Baptista [2] and Teresa Lino-Neto [1,*]

1 BioSystems & Integrative Sciences Institute (BioISI), Plant Functional Biology Centre, University of Minho, Campus de Gualtar, 4710-057 Braga, Portugal; danielacosta@bio.uminho.pt (D.C.); tavares@bio.uminho.pt (R.M.T.)

2 Centro de Investigação de Montanha (CIMO), Instituto Politécnico de Bragança, Campus de Santa Apolónia, 5300-253 Bragança, Portugal; pbaptista@ipb.pt

* Correspondence: tlneto@bio.uminho.pt

Abstract: An increase in cork oak diseases caused by *Biscogniauxia mediterranea* and *Diplodia corticola* has been reported in the last decade. Due to the high socio-economic and ecologic importance of this plant species in the Mediterranean Basin, the search for preventive or treatment measures to control these diseases is an urgent need. Fungal endophytes were recovered from cork oak trees with different disease severity levels, using culture-dependent methods. The results showed a higher number of potential pathogens than beneficial fungi such as cork oak endophytes, even in healthy plants. The antagonist potential of a selection of eight cork oak fungal endophytes was tested against *B. mediterranea* and *D. corticola* by dual-plate assays. The tested endophytes were more efficient in inhibiting *D. corticola* than *B. mediterranea* growth, but *Simplicillium aogashimaense, Fimetariella rabenhorstii, Chaetomium* sp. and *Alternaria alternata* revealed a high potential to inhibit the growth of both. *Simplicillium aogashimaense* caused macroscopic and microscopic mycelial/hyphal deformations and presented promising results in controlling both phytopathogens' growth in vitro. The evaluation of the antagonistic potential of non-volatile and volatile compounds also revealed that *A. alternata* compounds could be further explored for inhibiting both pathogens. These findings provide valuable knowledge that can be further explored in in vivo assays to find a suitable biocontrol agent for these cork oak diseases.

Keywords: fungi; endophytes; *Biscogniauxia mediterranea*; *Diplodia corticola*; biocontrol; cork oak

1. Introduction

Cork oak (*Quercus suber* L.) is an evergreen tree species that covers between 1.7 and 2.7 million ha in the western Mediterranean, distributed among Portugal, Spain, France, Italy, Morocco, Tunisia and Algeria [1]. Cork oak forests display a high socio-economic and ecological importance, being mainly explored for cork production [2]. The synthesis of cork, a homogeneous tissue of phellem cells, depends on the activity of the cork cambium, which covers the trunk and branches of cork oak. Due to the interesting and unique set of physical, biological and chemical properties, such as sealing and insulating features, cork is mainly used for the production of bottle stoppers [3]. Every year, 80% of worldwide produced cork comes from the Iberian Peninsula, in which Portugal is responsible for almost half of the total global production [2,4].

In recent years, cork oak forests have been declining in the Mediterranean region, which will be further enhanced by the combined effect of global warming and drought [5]. Indeed, the climate change predictions reveal the Mediterranean region as one of the most affected regions, and Mediterranean forests as one of the most vulnerable ecosystems to the combined effect of temperature increase and precipitation decrease [6,7]. Not only the geographic distribution of plant pathogens is predicted to be reshuffled [8], but also the severity of plant diseases and the rate at which endophytes switch from mutualistic to pathogenic behavior are expected to increase [9,10]. Indeed, during the last decades, an increase in cork oak diseases has been reported [11], including charcoal disease (caused by *Biscogniauxia mediterranea* (De Not.) Kuntze; Xylariales) and bot canker (caused by *Diplodia corticola* A.J.L. Phillips, A. Alves and J. Luque; Botryosphaeriales). Both pathogenic fungi have an endophytic lifestyle and opportunistic behavior, which account for the increase in diseases incidence in cork oak trees under environmental stress [12]. These diseases cause high economic losses due to the development of symptoms in cork oak trunk that affect both cork production and quality, and eventually lead to tree death [13].

Few preventive measures for cork oak charcoal disease and bot canker are currently known. Good phytosanitary practices in cork oak forests are essential to prevent the spreading of diseases through spore release and colonization from tree wounds derived from pruning or cork extraction [11]. Thiophanate-methyl and carbedazim fungicides have been found effective as a preventive measure against *D. corticola* [14,15]. However, the application of fungicides has been increasingly restricted due to the potential negative impact on human health and the environment. Therefore, more environmentally friendly strategies, such as the use of biological control agents, are now being explored for controlling many forest tree diseases [16,17]. Indeed, promising results have been obtained for restricting cork oak diseases. For example, the use of *Fusarium tricinctum* resulted in a reduced mortality of *Q. cerris* and *Q. pubescens* seedlings inoculated with *D. corticola* [18]. Further, *Trichoderma* spp. presented a high in vitro antagonistic potential against *B. mediterranea* and *D. corticola* [19]. In particular, a cork oak endophytic strain of *Trichoderma citrinoviride* produces peptaibols capable of inhibiting cork oak pathogens' in vitro growth [20]. Spray suspensions of *T. atroviride* and *T. viride* combined with sanitation and scarification practices were recently suggested for decreasing charcoal disease incidence in *Q. castaneifolia* trees [21]. In the present study, we hypothesize that cork oak fungal endophytes are able to control cork oak pathogens' (*B. mediterranea* and *D. corticola*) growth and we intend to highlight the most promising cork oak endophytes to be used as a biocontrol strategy. Specifically, we aimed to identify endophytic fungal species of cork oak that can be used to control *B. mediterranea* and *D. corticola*, by studying their antifungal activities against both pathogens and by characterizing the interactions between endophytes and pathogens.

2. Materials and Methods

2.1. Cork Oak Forests Sampling and Endophyte Recovery

Cork oak samples were collected from eight different forests in Portugal (Table S1). Two forests were sampled from the National Park of Peneda-Gerês (PG-ER and PG-RC) and Herdade da Contenda (HC-CT and HC-MA). A single forest was sampled from Limãos (LI), Alcobaça (AL), Gavião (GV) and Grândola (GR). Between April and October of 2017, five to six trees were sampled from each cork oak forest, considering trees at different disease severity levels (Table 1; Figure 1). Disease severity levels were grouped into three categories and determined by considering defoliation (5 levels: 0–10%—no damage; 11–25%—light damage; 26–50%—moderate damage; 51–90%—severe damage; > 91%—extreme damage), as well as canopy and trunk damages (3 levels: 0—no damage; 1—moderate damage; 2—severe damage), for different factors (dried, wilting and decolorated leaves, presence of cankers, decolorated trunk, presence of exudates and visible sporulation). Trees were considered as healthy, presenting mild symptoms or declining. Healthy trees presented no or light defoliation, displaying low canopy and trunk damage (maximum of two factors with 1 damage level, but not affecting overall vigor of the tree). When tree vigor was moderately affected by a combination of factors (moderate defoliation and/or some canopy and trunk damage), trees were considered as displaying mild symptoms. Declining trees presented a clear decline in their vigor (accentuated or very accentuated defoliation that could be coupled with more than three factors classified with 1 and/or 2 damage levels).

Table 1. Collection of cork oak branches in sampled forests. The number of trees sampled from each disease severity level is presented. For more details of sampled forests, see Table S1.

Location	Cork Oak Stand	Collection Date	Disease Severity Level		
			Healthy	Mild Symptoms	Declining
Peneda-Gerês	PG-ER	May, 2017	2	3	0
	PG-RC	July, 2017	4	1	0
Limãos	LI	April, 2017	0	5	1
Gavião	GV	July, 2017	0	4	2
Alcobaça	AL	May, 2017	2	2	2
Grândola	GR	May, 2017	0	3	3
Herdade	HC-CT	October, 2017	2	2	2
Contenda	HC-MA	October, 2017	2	2	2

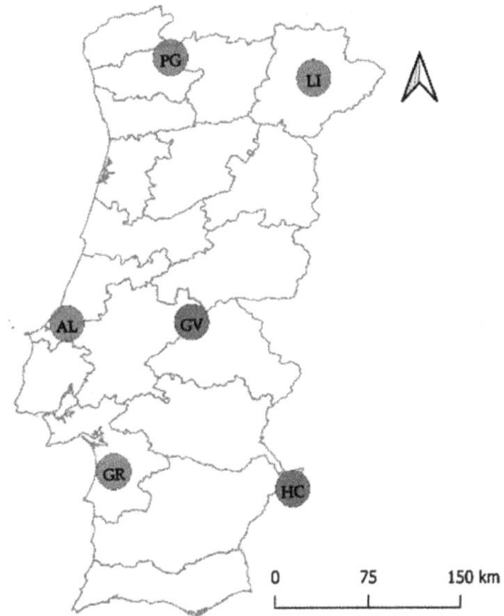

Figure 1. Distribution of cork oak stands sampled in Portugal. Letters represent sampled location (PG: Peneda-Gerês; LI: Limãos; GV: Gavião; AL: Alcobaça; GR: Grândola; HC: Herdade Contenda).

Five to seven branches were collected from each cork oak tree. Twigs were collected from each branch and thoroughly washed in tap water. To obtain only endophytes, a surface sterilization was performed based on the method described by Martins et al. [22]. Twigs were sequentially immersed in ethanol 70% (*v*/*v*) for 2 min, bleach (3–5% chlorine) for 6 min and ethanol 70% (*v*/*v*) for 1 min, followed by three washes in sterile deionized water (1 min each) and drying. Sterile twigs were cut into segments (4–5 cm) and transferred to Potato Dextrose Agar (PDA) medium (5 segments/plate). Three replicates were used for each branch. Sterilization controls were performed by spreading the last washing water (10 µL) onto PDA medium. Incubation was performed in the dark, at room temperature (21–23 °C). The outgrowing fungi were recognized as endophytic fungi and were successively subcultured in fresh PDA medium until pure cultures were obtained.

2.2. *Endophytic Fungi Identification and Selection of Potential Antagonistic Fungi*

Endophytic fungi were grouped into morphotypes, according to their cultural features (color, shape, elevation and margins) [23]. From each morphotype, at least three isolates were used for DNA extraction, using the Quick-DNA Fungal/Bacterial Miniprep Kit (Zymo Research, Irvine, CA, USA). The fungal rDNA-ITS region was amplified using universal primer pairs *ITS1F* (5′-CTTGGTCATTTAGAGGAAGTAA-3′) and *ITS2* (5′-GCTGCGTTCTTCATCGATGC-3′), or *ITS1F* and *ITS4* (5′-TCCTCCGCTTATTGATATGC-3′) [24]. PCR mixtures (25 µL) contained 1× Complete NH_4 Reaction Buffer (BIORON GmbH, Germany), 200 µM of each dNTP (NZYTech, Portugal), 1 µM of each primer, 1 µL of DNA template and 1.25 U of DFS-Taq DNA Polymerase (BIORON GmbH, Germany). Amplifications were performed using the following protocol: initial denaturation 5 min at 94 °C; 35 cycles of 30 s at 94 °C, 30 s at 52 °C (or 54 °C with *ITS1F-ITS4*) and 60 s at 72 °C; final elongation at 72 °C for 10 min. PCR products were run on a 1% (*w/v*) agarose gel, stained with Green Safe Premium (NZYTech, Portugal). PCR products were purified using isopropanol 75% (*v/v*) and sequenced by Macrogen, Inc services (Madrid, Spain). DNA sequences were trimmed by sequencing quality and alignments were performed in Geneious 2010.4.8.5 (https://www.geneious.com). All sequences were identified using the UNITE [25] and NCBI GenBank databases and taxonomic classification was assigned for those with a similarity higher than 97%. Alignments with NCBI deposited sequences were used to confirm taxonomic classification and to attempt taxonomic classification of those sequences with similarity between 80 and 97%. Fungal identifications were supported by cultural and morphological features of fungal cultures. Fungal sequences were deposited in GenBank (www.ncbi.nlm.nih.gov/genbank/) under accession numbers MT819608–MT819946.

For understanding the ecological roles of fungal communities, the identified fungi were categorized into functional groups (pathogenic and/or beneficial), based on the bibliographic research [26,27]. Identified operational taxonomic units (OTUs) were assigned as beneficial when described as promoting plant growth and/or protecting the host against biotic or abiotic stress, while the pathogenic group contained fungi causing disease to its host and latent pathogens. Fungi belonging to other functional groups (mutualism, commensalism, etc.) were assigned to other groups and those with unknown function to unknown. Fungi belonging to more than one group (for example, pathogenic–beneficial) were added to both groups. For proceeding to the antagonism assays, endophytes were selected based on their potential beneficial role and availability. All groups were considered with the exception of those fungi exclusively described as pathogenic and all cork oak pathogens. Eight fungal isolates were selected to be tested against the pathogens in antagonistic assays due to their cultural readiness, namely *Simplicillium aogashimaense* isolate Gr67, *Coniothyrium carteri* isolate Gv5, *Diaporthe passiflorae* isolate Erm6, *Fimetariella rabenhorstii* isolate Br33, *Fusarium oxysporum* isolate Cab77, *Chaetomium* sp. isolate Erm52, *Alternaria alternata* isolate Cab37 and *Penicillium olsonii* isolate Gv63. *B. mediterranea* and *D. corticola* isolates were inoculated on cork oak stems to confirm symptoms development, completing Koch's postulates.

2.3. Antagonistic Assay In Vitro by Dual-Plate and Categorization of Fungal Interactions

The eight selected endophytic fungi were tested in vitro against phytopathogens *Biscogniauxia mediterranea* isolate Gr13 and *Diplodia corticola* isolate Gr23, both isolated from cork oak trees showing mild symptoms. Antagonistic assays were performed by the dual-culture method in PDA medium. Fungal plugs (5 mm) of an actively growing endophyte and pathogen were placed 3 cm apart from each other (using 9 cm diameter Petri dishes) and incubated in the dark, at 28 °C (±2 °C). Those endophytes displaying a slow growth rate (<0.1 cm^2/h; Table S2) were inoculated 72 h before the pathogen. In parallel, control plates were similarly prepared but using a single pathogen (or endophyte) plug. All plates were photographed 72 h after being inoculated with the endophyte and/or pathogen, and growth area was measured using *ImageJ 1.50i* software [28]. Growth areas were used to determine the percentage of mycelial growth inhibition, according to the formula: (Ac−Ai)/Ac × 100, where Ac is the area of fungal growth in the control plate and Ai is the area of fungal growth with the interacting fungus. Three independent experiments were performed (with at least 3 replicates each) for all endophyte/pathogen combinations. Statistical analysis was performed using ANOVA in GraphPad Prism 7.00 software (La Jolla, CA, USA) to determine the impact of endophytes on pathogen growth.

Fungal interactions between endophytes and pathogens were categorized based on Tuininga [29]. Considering that (-) corresponds to mycelial growth decrease, (+) to growth increase and (0) to similar mycelial growth, interactions among endophyte/pathogen could be defined as co-antagonism (-/-), antagonism (-/0), agonism (-/+), co-habitation (0/0), commensalism (0/+) and mutualism (+/+). For describing the fungal interactions, dual-culture plates were observed daily for 15 days. Mycelial interactions were also recorded based on Badalyan et al. [30], who defined the following classes: A—deadlock with mycelial contact; B—deadlock at distance; C—replacement, overgrowth without deadlock; CA1 and CA2—partial and complete replacement after initial deadlock with mycelial contact;

and CB1 and CB2—partial and complete replacement after initial deadlock at a distance. Finally, hyphal interactions were observed using mycelia from the interacting region of dual-culture plates, after 15 days of interaction, and compared with mycelia taken from controls. Photographs were taken using a Leica MC170 HD digital camera attached to a Leica S9 D stereomicroscope (Leica Microsystems, Germany) or using an automated Leica DM5000B microscope (Leica Microsystems, Germany).

2.4. *Antifungal Non-Volatile Compounds Assay*

The antagonistic effect of the metabolites produced by the eight selected fungal endophytes was tested using a method adapted from Campanile et al. [18]. Three mycelial fungal plugs (5 mm) of actively growing endophytes were inoculated in Erlenmeyer flasks (250 mL) containing 50 mL of sterile Potato Dextrose Broth (BD Difco™, Switzerland). Culture flasks were placed in an orbital incubator at 28 °C (±2 °C), with 150 rpm agitation, for 12 days. Liquid cultures were transferred to falcon tubes and centrifuged for 15 min at 12,000 rpm. The supernatant was collected and filtered through a 0.22 μM membrane filter to remove hyphal residues and conidia. Fungal filtrates were added to sterile warm PDA medium [20% (*v*/*v*)] and poured into 9 cm Petri dishes. A mycelial plug of an actively growing pathogen (*B. mediterranea* or *D. corticola*) was placed in the center of the plate. Control plates were similarly prepared but containing PDA without a fungal filtrate. Mycelial growth areas were determined as previously referred and mycelial growth inhibition was determined according to the formula: $(Ac-Af)/Ac \times 100$, where Ac is the area of fungal growth in the control plate and Af is the area of fungal growth in the presence of the filtrate. Three independent experiments were performed (with at least 3 replicates each) for all endophyte/pathogen combinations. Statistical analysis was performed using ANOVA in GraphPad Prism 7.00 software (La Jolla, CA, USA) to understand the impact of endophytic compounds on pathogen growth.

2.5. *Antifungal Volatile Compounds Assay*

The effect of volatile compounds produced by the eight selected fungal endophytes in inhibiting pathogen growth was evaluated by an inverted plate method [31]. For each endophyte/pathogen combination, two PDA plates were inoculated with a single mycelial plug (5 mm) of an actively growing endophyte (or pathogen). Then, the PDA plate with the pathogen was inverted on the top of the endophyte plate and both were sealed with parafilm. Incubation occurred at 28 °C (±2 °C), for 72 h. Control plates were performed without inoculation of the endophyte in the bottom plate. As previously described, those endophytes displaying a slow growth rate (<0.1 cm^2/h; Table S2) were inoculated 72 h before the pathogen. The percentage of inhibition was calculated as mentioned before. Three independent experiments were performed (with at least 3 replicates each) for all endophyte/pathogen combinations. Statistical analysis was performed using ANOVA in GraphPad Prism 7.00 software (La Jolla, CA, USA) to understand the impact of endophytic volatiles on pathogen growth.

3. Results and Discussion

3.1. *Endophytic Fungal Community of Cork Oak*

Endophytic fungi of cork oak twigs were recovered from eight different forests in Portugal. From a total of 1117 fungal isolates, 440 were molecularly identified and grouped into 128 OTUs comprising 18 orders, 38 families, 45 genera and 39 species (Table 2). Only OTUs classified up to genus and species were considered in this work (70 OTUs). From these, and to the best of our knowledge, 54 fungal OTUs have never been reported as cork oak endophytes before (e.g., *Fimetariella rabenhorstii*, *Discosia* sp.), including 18 that have never been described as plant endophytes (e.g., *Caliciopsis beckhausii*, *Diaporthe passiflorae*, *Proliferodiscus* sp.) (Table 2). Among previously undescribed cork oak endophytes, 28 were only recovered from a single sampled cork oak forest, but others (26) were recovered from different locations, which strengthens their role as cork oak endophytes. For example, *Discosia* sp., *Cryphonectria naterciae* and *Neocucurbitaria* sp. were recovered from different Portuguese forests, or from forests displaying the highest (e.g., *F. rabenhorstii*) or the lowest (e.g., *Plectania rhytidia*) precipitation levels.

Some OTUs were more widespread throughout Portuguese cork oak forests than others. *B. mediterranea* was the only fungus identified in all cork oak stands and *Fusarium* sp. was present in all forests, with the exception of LI and GR. Further, *Penicillium* sp., *Sarocladium kiliense* and *Neocucurbitaria* sp. were recovered from almost all cork oak stands. In contrast, there were fungi (such as *Diplodia corticola*, *Nonappendiculata quercina*, *Ciboria* sp. and *Pezicula cinnamomea*) that were only recovered from one sampling location. Since many fungi are difficult to be cultured, culture-dependent methods are known to underestimate fungal communities [32]. The used approach only targeted those endophyte fungi that can be easily cultured on artificial media and have a rapid growth rate. Furthermore, as many endophytes do not sporulate in culture, their morphotype discrimination through cultural features is challenging and might have been underestimated. For these reasons, we are aware that the performed endophyte survey through cultural methods (and based on morphotypes discrimination) misrepresented the endophyte diversity in cork oak forests, as already reported elsewhere for other plant hosts [32]. Accordingly, our study reported a strong dominance of fungi belonging to Ascomycota (only one OTU belonged to the Basidiomycota phylum), like previously reported in other studies using culture-dependent methods, e.g., [33,34]. Culture-independent methods could have provided a different picture of fungal communities. For example, in grapevine, besides Ascomycota (described using cultural methods), Basidiomycota and Zygomycota fungi were additionally detected recurring to a metabarcoding approach [33]. Despite the recognized limitations, the used culture-dependent approach provided the availability of endophyte isolates to proceed in searching for biocontrol strategies.

The ecological role of each identified fungal species was determined based on the literature. As the functional role of certain endophytes could change according to their plant host genotype, during different stages of the plant life cycle or in extreme conditions [26], many endophytes were included in mixed groups (such as the pathogenic–beneficial group). Furthermore, within certain genera, there are species known to be pathogenic and others beneficial, being impossible to consider a single functional role. Despite these constrains, the number of identified OTUs displaying a phytopathogenic role was higher, when compared to other functional groups (Table 2). A total of 39 OTUs were considered as displaying a phytopathogenic role (including exclusive phytopatogenic (26), phytopathogenic–beneficial (10) and phytopathogenic–other (3) functional groups), while only 21 displayed a beneficial role (including beneficial (10), phytopathogenic–beneficial (10) and beneficial–other (1) functional groups). The richness of fungi displaying a potential phytopathogenic role was also higher than any other functional group in all cork oak stands and whatever the cork oak tree disease severity level (Figure 2). Among them, few cork oak pathogens were found: *B. mediterranea* [12], *Coryneum* sp. [35], *Cryphonectria naterciae* [36], *D. corticola* [37], *D. quercivora* [38], *Discula quercina* [39] and *Neofusicoccum parvum* [35]. Interestingly, *B. mediterranea*, *Coryneum* sp. and *C. naterciae* were isolated from trees in all disease severity levels, while *D. corticola*, *D. quercivora* and *Discula quercina* were only isolated from trees with declining symptoms (Figure 3). In agreement, cork oak pathogens (*B.mediterranea* and *Coryneum* sp.) were isolated from healthy cork oak trees in Italy [12,40]. The presence of *D. corticola*, *D. quercivora* and *Discula quercina* in declining cork oak trees agrees with their role as emerging pathogens to *Quercus* spp. in different regions of the world [11,41], which is emphasized by their risk to cork oak health as previously reported [39,42]. Concerning the fungi displaying a beneficial role, AL forest presented the highest number of OTUs with a potential beneficial role (all described as pathogenic–beneficial), but GV forest was the richest with exclusive beneficial fungi (Figure 2A). GV forest corresponded to one of the forests displaying the highest number of declining trees (results not presented), which agrees with the finding that trees with declining symptoms displayed a higher richness of exclusive beneficial fungi (a non-significant 1.5-fold increase in relation to healthy trees; Figure 2B). The role of stress-affected plants in recruiting beneficial microorganisms is still under debate and a "cry-for-help" hypothesis was recently proposed, in which plants are able to recruit plant-protective microbes when they are under attack by pathogens [43,44]. For example, tomato plants under stress produce root exudates to signal the beneficial *Trichoderma harzianum* T22 strain to direct growth toward the plant host [45]. However, further studies on endophyte distribution among healthy and diseased cork oak trees are still needed for providing clear evidence that support a "cry-for-help" strategy for the sustainability of threatened cork oak forests.

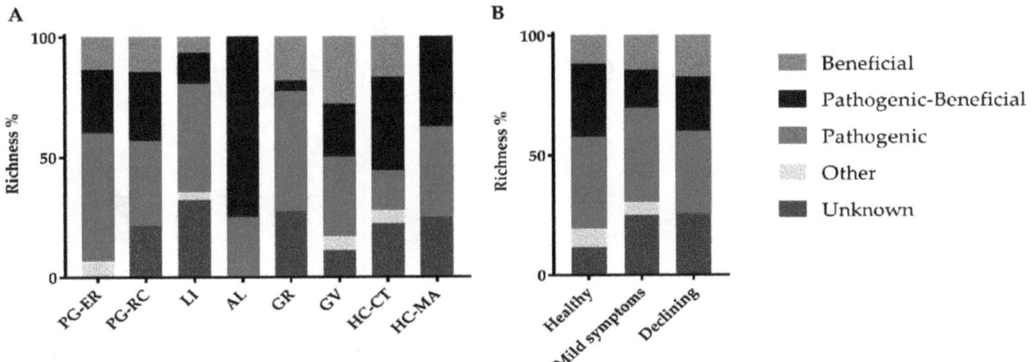

Figure 2. Cork oak endophytic fungal functional groups presented in cork oak forests (**A**) and disease severity level (**B**). Operational taxonomic units (OTUs) belonging to more than one functional group were added to both, except for the pathogenic–beneficial group that is represented as such.

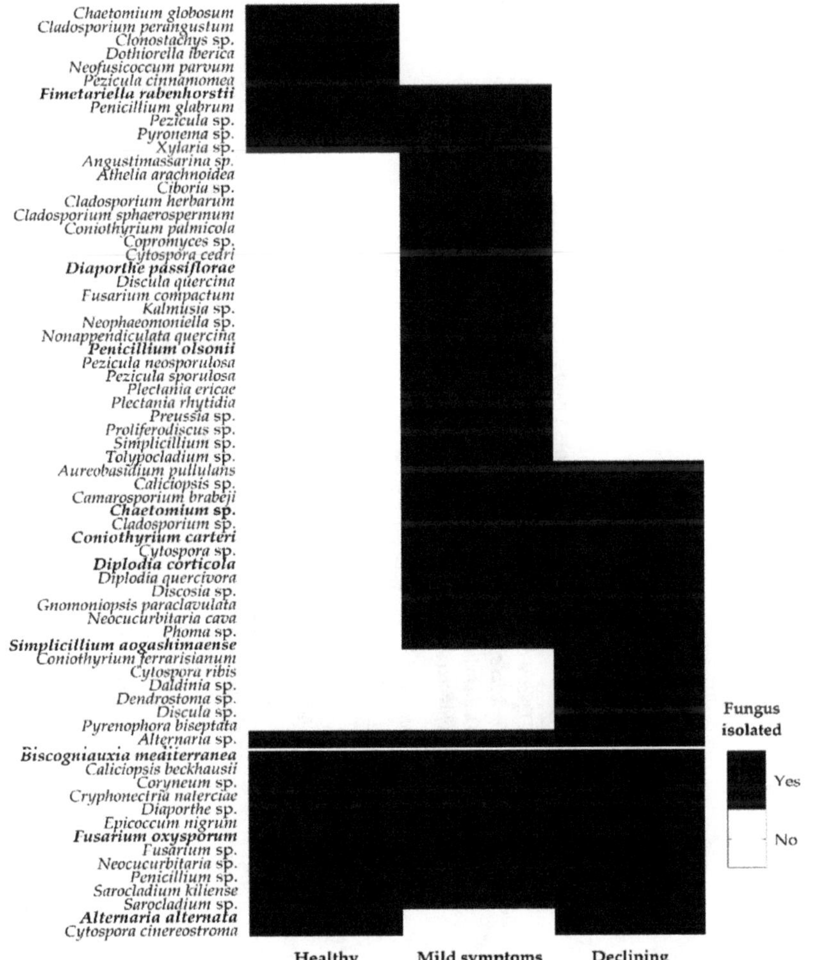

Figure 3. Endophytic fungal OTUs recovered from cork oak trees at different disease severity levels. Black/white color indicates fungal isolation/no isolation from trees with different disease severity levels. Isolates used in this work are depicted in bold.

Table 2. Endophytic fungal OTUs recovered from twigs of cork oak trees located in different forests. Taxonomic classification (with closest match identity in brackets) and their respective functional group are revealed. Reports about their endophytic behavior are referred to. When existing, references related to cork oak are presented (bold). Isolates used in this work are depicted in red. Functional groups are represented as: P—pathogenic; B—beneficial; O—other; U—unknown.

Taxonomic Classification		Closest Match GenBank	Cork Oak Forests								Functional Group	Identified Endophyte
			PG-ER	PG-RC	LI	AL	GV	GR	HC-CT	HC-MA		
Ascomycota												
Amphisphaeriales	*Discosia* sp.	KU325138.1 (100%)			x				x		P [46]/B [47]	[48]
	Nonappendiculata quercina	MH554025.1 (98.78%)			x						U	-
Botryosphaeriales	*Diplodia corticola*	MT015621.1 (100%)						x			P [37]	**[40]**
	Diplodia quercivora	JX894205.1 (97.72%)						x			**P [38]**	**[11]**
	Dothiorella iberica	MT261024.1 (100%)							x		**P [49]**	**[12]**
	Neofusicoccum parvum	MT645697.1 (99.3%)	x								**P [35]**	[50]
Capnodiales	*Cladosporium herbarum*	LT854669.1 (99.22%)			x						P [51]	[52]
	Cladosporium perangustum	MK111614.1 (99.10%)								x	P [53]	[54]
	Cladosporium sp.	MN879328.1 (100%)			x			x			P [55]/B [56]	**[40]**
	Cladosporium sphaerospermum	MT645920.1 (99.51%)					x				B [57]	[57]
Coryneliales	*Caliciopsis beckhausii*	NR_132090.1 (99.57%)			x					x	U	-
	Caliciopsis sp.	NR_132090.1 (91.91%)			x			x			P [58]	-
Diaporthales	*Coryneum* sp.	MH674330.1 (95.54%)		x	x		x	x			P **[35]**	**[40]**
	Cryphonectria naterciae	MT645942.1 (100%)		x	x		x	x			P **[36]**	-
	Cytospora cedri	MN871816.1 (100%)			x			x	x		P [59]	-
	Cytospora cinereostroma	KY051964.1 (100%)						x		x	U	-
	Cytospora ribis	KP641138.1 (100%)			x						U	[60]
	Cytospora sp.	MK656248.1 (100%)			x			x			P [61]	**[40]**
	Dendrostoma sp.	MN447228.1 (99.66%)			x						P [62]	-
	Diaporthe passiflorae	NR_120155.1 (99.82%)	x								O [63]/P [64]	-
	Diaporthe sp.	MT561408.1 (99.48%)	x			x				x	P [65]	[65]
	Discula quercina	MH758705.1 (99.18%)			x						P **[39]**	**[40]**
	Discula sp.	KY367498.2 (94.23%)			x						P [66]	[67]
	Gnomoniopsis paraclavulata	MH863162.1 (100%)			x			x			U	-
Dothideales	*Aureobasidium pullulans*	MT645930.1 (99.57%)			x						O [68]/B [69]	**[40]**

Table 2. *Cont.*

Taxonomic Classification		Closest Match GenBank	Cork Oak Forests								Functional Group	Identified Endophyte
			PG-ER	PG-RC	LI	AL	GV	GR	HC-CT	HC-MA		
Eurotiales	*Penicillium glabrum*	MT582777.1 (100%)		x			x	x			P [70]	[71]
	Penicillium olsonii	MT582783.1 (100%)					x				B [72]	[73]
	Penicillium sp.	LN901128.1 (99.54%)			x	x	x		x	x	O [74]/P [70]/B [72]	**[40]**
Helotiales	*Ciboria* sp.	KF545322.1 (94.59%)			x						P [75]	-
	Pezicula cinnamomea	MK907714.1 (100%)	x								P [76]	[77]
	Pezicula neosporulosa	KR859231.1 (100%)	x								P [78]	[79]
	Pezicula sporulosa	MH862573.1 (98.66%)					x				O[80]	[81]
	Pezicula sp.	MG098317.1 (100%)	x		x						O [82]/P [76]	[67]
	Proliferodiscus sp.	MN901941.1 (95.50%)			x						U	-
Hypocreales	*Clonostachys* sp.	MK789204.1 (91.84%)		x							B [83]	[84]
	Fusarium compactum	KJ562364.1 (98.53%)					x				P [85]	[86]
	Fusarium oxysporum	MT530243.1 (100%)	x	x		x			x		P [87]/B [88]	[89]
	Fusarium sp.	MT645120.1 (100%)	x	x		x	x		x	x	P [87]/B [18]	[18]
	Sarocladium kiliense	MK789203.1 (100%)	x			x	x		x	x	P [90]/B [91]	[92]
	Sarocladium sp.	MT645143.1 (99.36%)	x			x			x	x	P [93]/B [91]	[94]
	Simplicillium aogashimaense	MK685280.1 (99.82%)					x	x			U	-
	Simplicillium sp.	MH859771.1 (99.12%)	x				x		x		B [95]	[95]
	Tolypocladium sp.	KX034386.1 (100%)	x								O [96]	[97]
Pezizales	*Plectania rhytidia*	MH003435.1 (98.99%)						x	x		U	[27]
	Pseudoplectania ericae	MT498082.1 (99.65%)					x				U	-
	Pyronema sp.	MT556695.1 (100%)			x				x		O [98]	[99]
Phaeomoniellales	*Neophaeomoniella* sp.	MK646052.1 (96.14%)			x						P [100]	[101]
Pleosporales	*Alternaria alternata*	MT635274.1 (100%)		x					x		P [102]/B [103]	**[40]**
	Alternaria sp.	MT557456.1 (100%)		x	x		x			x	P [104]/B [103]	**[40]**
	Angustimassarina sp.	MN963689.1 (100%)			x						U	-
	Camarosporium brabeji	LN714529.1 (97.76%)							x	x	U	[105]
	Coniothyrium carteri	KX359604.1 (99.82%)					x	x	x		B [106]	[106]
	Coniothyrium ferrarisianum	MH860854.1 (100%)			x						U	[107]
	Coniothyrium palmicola	JX681086.1 (99.53%)						x			U	-
	Epicoccum nigrum	MT548679.1 (100%)			x			x	x		B **[27]**	**[40]**
	Kalmusia sp.	MK796143.1 (100%)					x	x			P [108]	[109]
	Neocucurbitaria cava	MK796144.1 (100%)		x	x				x		U	[110]

Table 2. *Cont.*

Taxonomic Classification		Closest Match GenBank	Cork Oak Forests								Functional Group	Identified Endophyte
			PG-ER	PG-RC	LI	AL	GV	GR	HC-CT	HC-MA		
	Neocucurbitaria sp.	MH858303.1 (93.78%)		x	x			x	x	x	U	[110]
	Phoma sp.	KX815489.1 (100%)			x					x	P [111]	**[40]**
	Preussia sp.	MN696547.1 (100%)						x			B [112]	[113]
	Pyrenophora biseptata	MH864748.1 (100%)								x	P [114]	-
Sordariales	*Chaetomium globosum*	MT588864.1 (100%)		x							B [115]	[116]
	Chaetomium sp.	MN153902.1	x				x	x			B [117]	[118]
	Copromyces sp.	(100%)		x							U	-
	Fimetariella rabenhorstii	MN555335.1 (100%)	x	x	x						O [119]/P [120]	[121]
Xylariales	*Biscogniauxia mediterranea*	MT862330.1 (100%)	x	x	x	x	x	x	x	x	P **[12]**	**[40]**
	Daldinia sp.	MN341734.1 (97.83%)			x						B [122]	[123]
	Xylaria sp.	JQ761730.1 (99.78%)					x			x	P [124]/B [125]	[126]
Basidiomycota Atheliales	*Athelia arachnoidea*	MH860510.1 (100%)								x	P [127]	-

3.2. Interactions of Endophytes against B. mediterranea and D. corticola

Eight endophyte isolates were selected for testing their ability to inhibit *B. mediterranea* and *D. corticola* growth, based on their potential beneficial ecological role, availability and culture readiness. *D. corticola* mycelial growth was persistently inhibited by the presence of these cork oak endophytes, while *B. mediterranea* was differentially inhibited by endophytes (Figure 4). Dual-plate experiments revealed that *B. mediterranea* growth was inhibited by *F. rabenhorstii* (33.4%, $p \leq 0.001$), *A. alternata* (19.3%), *Chaetomium* sp. (13.6%) and *S. aogashimaense* (12.6%), whereas the other fungal endophytes promoted or did not affect pathogen growth (Figure 4). Regarding endophytes inhibiting *B. mediterranea* and considering the effect of pathogens on endophyte growth (Figure S1), *S. aogashimaense* was the only endophyte not inhibited by this pathogen, displaying a typical antagonist interaction (0/-) (Table 3). Although *F. rabenhorstii* strongly inhibited *B. mediterranea,* there was a negative impact of the pathogen on this endophyte growth (co-antagonism; -/-). Both endophytes displayed a similar interaction with the *D. corticola* pathogen, resulting in antagonism/co-antagonism with pathogen inhibition (40.8%, $p \leq 0.001$ by *S. aogashimaense*; 42.2%, $p \leq 0.001$ by *F. rabenhorstii*). In contrast, *A. alternata* and *Chaetomium* sp. increased their growth while inhibiting *B. mediterranea,* thus displaying agonist interactions (+/-) with this pathogen. Both endophytes displayed a distinct interaction with *D. corticola*. *A. alternata* was not affected (antagonism; 0/-) and *Chaetomium* sp. was inhibited by this pathogen (co-antagonism; /). Except for *F. oxysporum* (antagonism; 0/-), all the other tested endophytes also revealed a co-antagonism interaction with *D. corticola.*

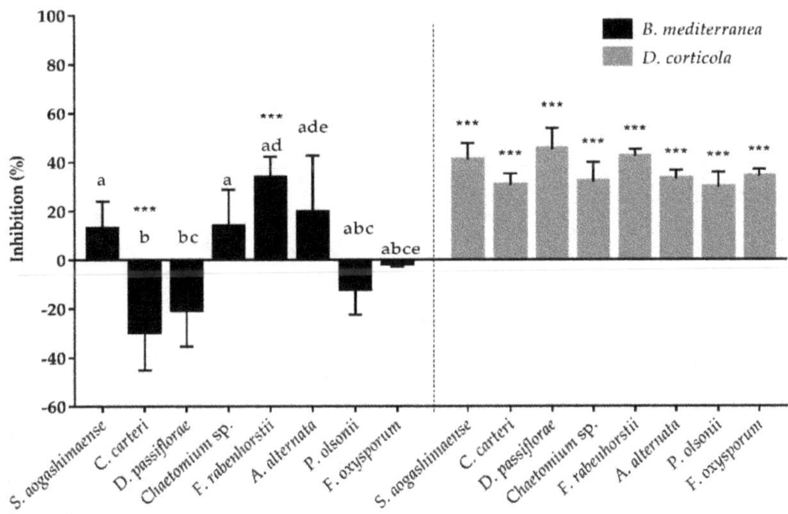

Figure 4. Inhibition of *B. mediterranea* and *D. corticola* growth caused by the tested endophytic fungi in dual-culture assay. Negative values indicate that the area of the plate covered by the interacting pathogen was higher than that of the control. Different letters represent statistical significance ($p \leq 0.005$) between endophytes in each antagonistic assay and *** ($p \leq 0.001$) represents statistical significance between the control and co-culture of a pathogen.

Fungal interactions were further evaluated by following the macro- and microscopic modifications of mycelia in the interaction region. When interacting with both pathogens, *S. aogashimaense* revealed a deadlock at distance (B) interaction type (Figure S2A,B) and caused visible modifications on hyphae of both pathogens (Figure 5A–D). In the presence of this endophyte, *B. mediterranea* showed typical hyphal deformations caused by the interacting partner, such as the presence of coiled hyphae and production of vesicle-like structures (Figure 5A,B). Further, *D. corticola* suffered mycelial modifications caused by interaction with *S. aogashimaense,* which included hyphal coiling and vacuolization, as well as the production of vesicle-like structures (Figure 5C,D). Such alterations have been frequently reported in different incompatibility systems [128,129] and have been related to programmed cell death (PCD) events occurring during interaction [130]. The *A. alternata–B. mediterranea* interaction also revealed hyphal deformations, such as hyphal vacuolization, production of vesicle-like structures and hyphal penetration that resemble a mycoparasitism interaction (Figure 5E–G). Accordingly, instead of a deadlock at distance interaction type, in this interaction, there was a partial replacement of mycelia after an initial deadlock with mycelial contact (CA1; Table 3 and Figure S2C). Interestingly, although the inhibitory activity of an *A. alternata* isolate from *Q. cerris* against *D. corticola* has been reported, both in dual-culture and in planta [18], we have not detected hyphal distortions in the *A. alternata–D. corticola* interaction. All the other studied interactions did not reveal hyphal distortions, even though similar interaction types have been detected after 15 days of interaction (partial replacement after initial deadlock with mycelial contact; CA1), as well as a deadlock with mycelial contact (A) (Table 3).

Table 3. Classification of fungal interactions occurring between tested endophytes and pathogens on dual-culture assay. The interaction type was classified based on endophyte/pathogen growth by "+" for higher growth, "-" for less growth and "0" for equal growth in relation to control. Underlined mycelial interactions denote endophyte replacement by the pathogen.

Endophyte	*B. Mediterranea*		*D. Corticola*	
	Type of Interaction	Mycelial Interaction	Type of Interaction	Mycelial Interaction
S. aogashimaense	antagonism (0/-)	B	antagonism (0/-)	B
C. carteri	agonism (-/+)	CA2	co-antagonism (-/-)	CA2
D. passiflorae	agonism (-/+)	CB1	co-antagonism (-/-)	CB1
F. rabenhorstii	co-antagonism (-/-)	A	co-antagonism (-/-)	CA1
F. oxysporum	commensalism (+/0)	CA1	antagonism (0/-)	CA1
Chaetomium sp.	agonism (+/-)	CA1	co-antagonism (-/-)	CA1
A. alternata	agonism (+/-)	CA1	antagonism (0/-)	CA1
P. olsonii	agonism (-/+)	CB1	co-antagonism (-/-)	CA1

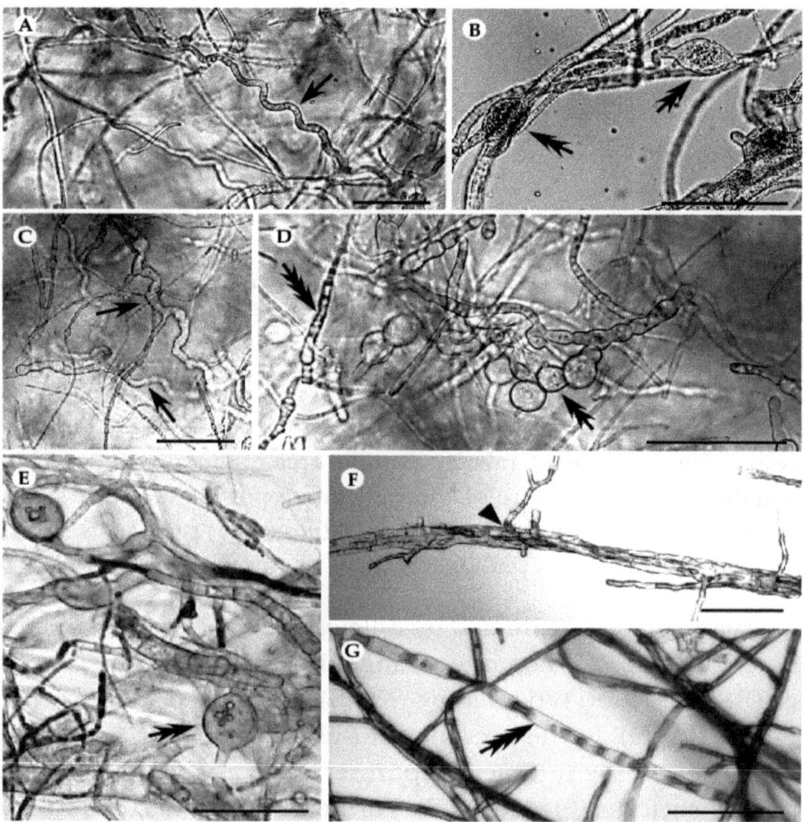

Figure 5. Hyphal modifications produced in the interaction between *S. aogashimaense* and *B. mediterranea* (**A,B**), *S. aogashimaense* and *D. corticola* (**C,D**) and *A. alternata* and *B. mediterranea* (**E–G**). Single arrows designate coiled hyphae, double arrows vesicle-like structures and triple arrows hyphal vacuolization. Black line represents 50 µM scale.

3.3. Fungal Inhibitors Production by Cork Oak Endophytes

For understanding the production of inhibitors by the tested endophytes, their non-volatile and soluble compounds (produced in liquid culture) and volatile emissions were tested against *B. mediterranea* and *D. corticola* pathogens. Endophytic soluble compounds seem to have more impact on inhibiting mycelial growth than volatile compounds, mainly against *B. mediterranea* (Figure 6). However, this primary conclusion should be taken with some precaution as these inhibitory compounds were tested by using different strategies for their collection and assay. The endophytes that revealed the greatest inhibitory activity against *B. mediterranea* in the dual-culture method (*F. rabenhorstii, A. alternata, Chaetomium* sp. and *S. aogashimaense*) produced a soluble extract with high anti-fungal activity against both pathogens. These endophytes inhibited from 28% to 51% of *B. mediterranea* growth (*F. rabenhorstii* $p \leq 0.05$ and *A. alternata* $p \leq 0.01$) and from 26% to 72% of *D. corticola* growth (*S. aogashimaense* and *A. alternata* $p \leq 0.001$; *F. rabenhorstii* $p \leq 0.01$). The volatile emissions from the same endophytes also revealed inhibitory activities against *B. mediterranea* and *D. corticola* (ranging from 1% to 8% for *B. mediterranea*, and 18% to 39% for *D. corticola*). Multiple studies have already revealed the antifungal activity of genera from tested endophytes, in particular of *Chaetomium* spp., e.g., [131], *Coniothyrium* spp., e.g., [132], *Diaporthe* spp., e.g., [133], *Penicillium* spp., e.g., [134], and *Simplicillium* spp., e.g., [135], as well as their ability to produce a battery of antifungal compounds. However, less information is available on the inhibitory activity of *Fimetariella* spp. or the antifungal activities of *A. alternaria* and *F. oxysporum* species, which have been mainly recognized as important plant pathogens. This work provides new information related to the production of antimicrobial compounds, mainly from *F. rabenhorstii* and *A. alternata*, against cork oak pathogens.

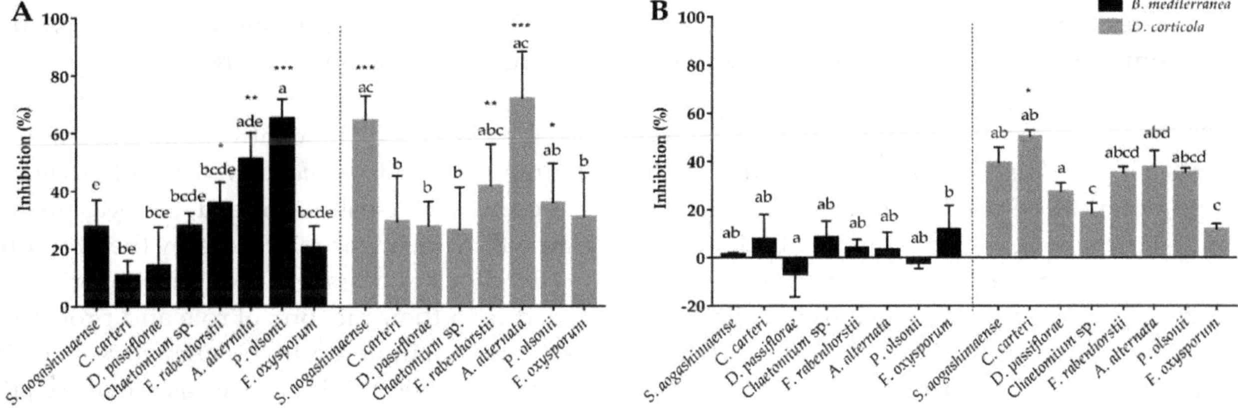

Figure 6. Inhibition of *B. mediterranea* and *D. corticola* growth caused by non-volatile (**A**) and volatile compounds (**B**) produced by the tested endophytic fungi. Negative values indicate that the area of the plate covered by the interacting pathogen was higher than that of the control. Different letters represent statistical significance ($p < 0.005$) between endophytes in each antagonistic assay and * ($p \leq 0.05$), ** ($p \leq 0.01$) and *** ($p \leq 0.001$) represent statistical significance between the control and co-culture of a pathogen.

Interestingly, endophytes with no inhibitory activity against *B. mediterranea* in dual-plate methods also revealed the production of inhibitors for this phytopathogen. For example, the soluble compounds of *P. olsonii* revealed the highest inhibition (65%, $p \leq 0.001$) of all against *B. mediterranea*, and volatiles from *C. carteri* and *F. oxysporum* also exhibited high inhibitory activity against the same pathogen (8% and 12%, respectively). These results suggest that the inhibitory effect of a specific fungal isolate against a phytopathogen may be due to the production of multiple compounds that could act in a synergistic or antagonistic way [136]. Accordingly, the production of inhibitory compounds such as volatiles, antibiotics and other secondary metabolites is gaining biotechnological interest for the control of phytopathogens. For example, the potential of *S. coffeanum* volatile compounds against *Aspergillus* species was reported by Gomes et al. [95] and *S. lamellicola* was used to produce a fungicide against *Botrytis cinerea* [137]. The production of inhibitory compounds is gaining interest when produced by potential pathogens. For example, the potential of *Alternaria* sp. was already reported for the control of fungal and bacterial growth [103], but some species are widely known as phytopathogens [104], representing a disadvantage for field application. The same is described for the *F. oxysporum* phytopathogen [88,138]. Therefore, the recognition of non-volatile or volatile inhibitory compounds could represent a biotechnological advantage for using those isolates as biocontrol agents.

4. Conclusions

The use of naturally adapted endophytes (for a specific plant host/environment) in a biocontrol strategy has gained increasing interest for restricting plant diseases. In this cork oak endophyte survey, we detected a high number of fungal OTUs from trees displaying different disease severity levels. The number of OTUs belonging to the pathogenic functional group (including described pathogens to other plant species) was high, even when considering healthy trees. Isolates of potential pathogens (namely, *Alternaria alternata* and *Fusarium oxysporum*) have revealed a strong in vitro inhibitory effect against cork oak pathogens (*B. mediterranea* and *D. corticola*). In particular, an *A. alternata* isolate revealed a high inhibitory activity against both pathogens, promoting hyphal deformations on *B. mediterranea*. Although this was not the case when interacting with *D. corticola*, this pathogen was similarly inhibited by *A. alternata* and *F. oxysporum*, displaying a mycelial interaction type based on the partial replacement of mycelia after an initial deadlock with mycelial contact. Non-volatiles and volatiles obtained from these isolates (particularly from *A. alternata*) revealed inhibitory activity and their potential to be used in a biocontrol strategy for restraining cork oak diseases should be further explored.

The colonization of plants by beneficial endophytes has been a useful biocontrol strategy. This work suggests *Simplicillium aogashimaense* as an antagonistic fungus towards *B. mediterranea* and *D. corticola* with potential to be used as a biocontrol agent against cork oak diseases. Indeed, *S. aogashimaense* presented promising results inhibiting both pathogens' growth, which was reinforced by the promotion of pathogens' hyphae deformations during interaction (deadlock at distance). A high ecological and economical value has been given to *Simplicillium* species due to their biocontrol role and production of bioactive compounds. For example, soybean plants when inoculated with *S. lanosoniveum* before infection with soybean rust pathogen (*Phakopsora pachyrhizi*) revealed reduced disease severity [139]. Although we believe *S. aogashimaense* could similarly be explored to control cork oak diseases, better understanding of its potential role as a biocontrol agent is still required. We thus conclude that cork oak endophytes could be further explored as biocontrol agents against cork oak diseases.

Table S1: Characterization of sampled cork oak stands; Table S2: Growth rates of cork oak endophytic fungi used for the antagonistic assays; Figure S1: Inhibition of fungal endophytes growth caused by the *B. mediterranea* and *D. corticola* in dual culture assay; Figure S2: Interaction in dual culture of *S. aogashimaense-B. mediterranea* (A), *S. aogashimaense-D. corticola* (B) and *A. alternata- B. mediterranea* (C) 72 h and 15 days after inoculation.

Author Contributions: Conceived and designed the experiments: T.L.-N., P.B., R.M.T. and D.C. Sampling: T.L.-N. and D.C. Conducted the main experiments: D.C. Analyzed the data: T.L.-N. and D.C. Wrote the original draft: D.C. Reviewed, edited and prepared the MS for submission: T.L.-N., P.B. and R.M.T. All authors have read and agreed to the published version of the manuscript.

Acknowledgments: The authors thank Maria Carolina Varela for the help in identification of disease severity levels of trees and Pinto Moreira (HC forests), Eng. Carlos Barroso Dias (PNPG), Eng. Luís Dias (Herdade das Barradas da Serra—GR) and José Lino Neto (GV forest) for providing access to cork oak forests.

References

1. Gauquelin, T.; Michon, G.; Joffre, R.; Duponnois, R.; Génin, D.; Fady, B.; Dagher-Kharrat, M.B.; Derridj, A.; Slimani, S.; Badri, W.; et al. Mediterranean forests, land use and climate change: A social-ecological perspective. *Reg. Environ. Chang.* **2018**, *18*, 623–636. [CrossRef]
2. FAO and Plan Bleu. *State of Mediterranean Forests 2018*; Food and Agriculture Organization of the United Nations, Food and Agriculture Organization of the United Nations, Rome and Plan Bleu Marseille, France, 2018.
3. Costa, R.; Lourenço, A.; Oliveira, V.; Pereira, H. Chemical characterization of cork, phloem and wood from different *Quercus suber* provenances and trees. *Heliyon* **2019**, *5*, 02910. [CrossRef] [PubMed]
4. APCOR. *APCOR's Cork Yearbook 2018/2019*; Portuguese Cork Association: Santa Maria de Lamas, Portugal, 2019.
5. Touhami, I.; Chirino, E.; Aouinti, H.; El Khorchani, A.; Elaieb, M.T.; Khaldi, A.; Nasr, Z. Decline and dieback of cork oak (*Quercus suber* L.) forests in the Mediterranean basin: A case study of Kroumirie, Northwest Tunisia. *J. For. Res.* **2020**, *31*, 1461–1477. [CrossRef]
6. Giorgi, F. Climate change hot-spots. *Geophys. Res. Lett.* **2006**, *33*. [CrossRef]
7. Rego, F.C.; Rocha, M.S. Climatic patterns in the Mediterranean region. *Ecol. Mediterr.* **2014**, *40*, 49–59. [CrossRef]
8. Shaw, M.W.; Osborne, T.M. Geographic distribution of plant pathogens in response to climate change. *Plant Pathol.* **2011**, *60*, 31–43. [CrossRef]
9. La Porta, N.; Capretti, P.; Thomsen, I.M.; Kasanen, R.; Hietala, A.M.; Von Weissenberg, K. Forest pathogens with higher damage potential due to climate change in Europe. *Can. J. Plant Pathol.* **2008**, *30*, 177–195. [CrossRef]
10. Elad, Y.; Pertot, I. Climate Change Impacts on Plant Pathogens and Plant Diseases. *J. Crop. Improv.* **2014**, *28*, 99–139. [CrossRef]
11. Moricca, S.; Linaldeddu, B.T.; Ginetti, B.; Scanu, B.; Franceschini, A.; Ragazzi, A. Endemic and Emerging Pathogens Threatening Cork Oak Trees: Management Options for Conserving a Unique Forest Ecosystem. *Plant Dis.* **2016**, *100*, 2184–2193. [CrossRef]
12. Linaldeddu, B.T.; Sirca, C.; Spano, D.; Franceschini, A. Variation of endophytic cork oak-associated fungal communities in relation to plant health and water stress. *For. Pathol.* **2011**, *41*, 193–201. [CrossRef]
13. Linaldeddu, B.T.; Sirca, C.; Spano, D.; Franceschini, A. Physiological responses of cork oak and holm oak to infection by fungal pathogens involved in oak decline. *For. Pathol.* **2009**, *39*, 232–238. [CrossRef]
14. Luque, J.; Pera, J.; Parladé, J. Evaluation of fungicides for the control of *Botryosphaeria corticola* on cork oak in Catalonia (NE Spain). *For. Pathol.* **2008**, *38*, 147–155. [CrossRef]
15. Serrano, M.S.; Romero, M.A.; Jiménez, J.J.; De Vita, P.; Ávila, A.; Trapero, A.; Sánchez, M.E. Preventive control of *Botryosphaeria* canker affecting *Quercus suber* in southern Spain. *Forestry* **2015**, *88*, 500–507. [CrossRef]
16. Terhonen, E.; Kovalchuk, A.; Zarsav, A.; Asiegbu, F.O. Biocontrol potential of forest tree endophytes. In *Endophytes of Forest Trees*; Springer: Cham, Switzerland, 2018; pp. 283–318. [CrossRef]
17. Martín-García, J.; Zas, R.; Solla, A.; Woodward, S.; Hantula, J.; Vainio, E.J.; Mullett, M.; Morales-Rodríguez, C.; Vannini, A.; Martínez-Álvarez, P.; et al. Environmentally friendly methods for controlling pine pitch canker. *Plant Pathol.* **2019**, *68*, 843–860. [CrossRef]
18. Campanile, G.; Ruscelli, A.; Luisi, N. Antagonistic activity of endophytic fungi towards *Diplodia corticola* assessed by in vitro and in planta tests. *Eur. J. Plant Pathol.* **2007**, *117*, 237–246. [CrossRef]
19. Linaldeddu, B.T.; Maddau, L.; Franceschini, A. Preliminary in vitro investigation on the interactions among endophytic fungi isolated from *Quercus* spp. *IOBC WPRS Bull.* **2005**, *28*, 101.
20. Maddau, L.; Cabras, A.; Franceschini, A.; Linaldeddu, B.T.; Crobu, S.; Roggio, T.; Pagnozzi, D. Occurrence and characterization of peptaibols from *Trichoderma citrinoviride*, an endophytic fungus of cork oak, using electrospray ionization quadrupole time-of-flight mass spectrometry. *Microbiology* **2009**, *155*, 3371–3381. [CrossRef]
21. Karami, J.; Kavosi, M.R.; Babanezhad, M.; Kiapasha, K. Integrated management of the charcoal disease by silviculture, chemical and biological methods in forest parks. *J. Sustain. For.* **2018**, *37*, 429–444. [CrossRef]
22. Martins, F.; Pereira, J.A.; Bota, P.; Bento, A.; Baptista, P. Fungal endophyte communities in above- and belowground olive tree organs and the effect of season and geographic location on their structures. *Fungal Ecol.* **2016**, *20*, 193–201. [CrossRef]

23. Costa, D.; Tavares, R.; Baptista, P.; Lino-Neto, T. Diversity of fungal endophytic community in *Quercus suber* L. under different climate scenarios. *Rev. Ciências Agrárias* **2018**, *41*, 41–50. [CrossRef]
24. White, T.J.; Bruns, T.; Lee SJ, W.T.; Taylor, J.L. Amplification and direct sequencing of fungal ribosomal rna genes for phylogenetics. *PCR Protoc. Guid. Methods Appl.* **1990**, *18*, 315–322. [CrossRef]
25. Nilsson, R.H.; Larsson, K.-H.; Taylor, A.F.S.; Bengtsson-Palme, J.; Jeppesen, T.S.; Schigel, D.; Kennedy, P.; Picard, K.; Glöckner, F.O.; Tedersoo, L.; et al. The UNITE database for molecular identification of fungi: Handling dark taxa and parallel taxonomic classifications. *Nucleic Acids Res.* **2019**, *47*, D259–D264. [CrossRef] [PubMed]
26. Hardoim, P.R.; Van Overbeek, L.S.; Berg, G.; Pirttilä, A.M.; Compant, S.; Campisano, A.; Döring, M.; Sessitsch, A. The Hidden World within Plants: Ecological and Evolutionary Considerations for Defining Functioning of Microbial Endophytes. *Microbiol. Mol. Biol. Rev.* **2015**, *79*, 293–320. [CrossRef] [PubMed]
27. Gomes, T.; Pereira, J.A.; Lino-Neto, T.; Bennett, A.E.; Baptista, P. Bacterial disease induced changes in fungal communities of olive tree twigs depend on host genotype. *Sci. Rep.* **2019**, *9*, 1–10. [CrossRef]
28. Schneider, C.A.; Rasband, W.S.; Eliceiri, K.W. NIH Image to ImageJ: 25 years of image analysis. *Nat. Methods* **2012**, *9*, 671–675. [CrossRef]
29. Tuininga, A. Interspecific interaction terminology: From mycology to general ecology. In *The Fungal Community: Its Organization and Role in the Ecosystem*; Dighton, J., White, J.P.O., Eds.; CRC Press: Boca Raton, FL, USA, 2005; pp. 274–280.
30. Badalyan, S.M.; Innocenti, G.; Garibyan, N.G. Antagonistic Activity of Xylotrophic Mushrooms against Pathogenic Fungi of Cereals in Dual Culture. *Phytopathol. Mediterr.* **2002**, *41*, 220–225. [CrossRef]
31. Dennis, C.; Webster, J. Antagonistic properties of species-groups of Trichoderma. *Trans. Br. Mycol. Soc.* **1971**, *57*, 41–48. [CrossRef]
32. Wu, B.; Hussain, M.; Zhang, W.; Stadler, M.; Liu, X.; Xiang, M. Current insights into fungal species diversity and perspective on naming the environmental DNA sequences of fungi. *Mycology* **2019**, *10*, 127–140. [CrossRef]
33. Dissanayake, A.J.; Purahong, W.; Wubet, T.; Hyde, K.D.; Zhang, W.; Xu, H.; Zhang, G.; Fu, C.; Liu, M.; Xing, Q.; et al. Direct comparison of culture-dependent and culture-independent molecular approaches reveal the diversity of fungal endophytic communities in stems of grapevine (*Vitis vinifera*). *Fungal Divers.* **2018**, *90*, 85–107. [CrossRef]
34. Gomes, T.; Pereira, J.A.; Benhadi, J.; Lino-Neto, T.; Baptista, P. Endophytic and Epiphytic Phyllosphere Fungal Communities Are Shaped by Different Environmental Factors in a Mediterranean Ecosystem. *Microb. Ecol.* **2018**, *76*, 668–679. [CrossRef]
35. Bragança, H.; Machado, H.; Inácio, L.; Henriques, J.; Diogo, E.; Moreira, C. Detecção de agentes potencialmente patogénicos em sobreiro e azinheira. In *Abstracts of the Congresso Florestal Nacional*; Vila Real/Bragança: Bragança, Portugal, 2013.
36. Smahi, H.; Belhoucine-Guezouli, L.; Bouhraoua, R.T.; Franceschini, A.; Linaldeddu, B.T. First Report of Branch Canker and Dieback Caused by *Cryphonectria naterciae* on *Quercus suber* in Algeria. *Plant Dis.* **2018**, *102*, 251. [CrossRef]
37. Luque, J.; Parladé, J.; Pera, J. Pathogenicity of fungi isolated from *Quercus suber* in Catalonia (NE Spain). *For. Pathol.* **2000**, *30*, 247–263. [CrossRef]
38. Bragança, H.; Neno, J.; Henriques, J.; Diogo, E.; Alves, A. First Report of *Diplodia quercivora* Causing Dieback on *Quercus suber* and in Europe. *Plant Dis.* **2016**, *100*, 2166. [CrossRef]
39. Ragazzi, A.; Turco, E.; Marianelli, L.; Dellavalle, I.; Moricca, S. Disease gradient of the anthracnose agent *Apiognomonia quercina* in a natural oak stand. *Phytopathol. Mediterr.* **2007**, *46*, 295–303. [CrossRef]
40. Franceschini, A.; Linaldeddu, B.T.; Marras, F. Occurrence and distribution of fungal endophytes in declining cork oak forests in Sardinia (Italy). *IOBC WPRS Bull.* **2005**, *28*, 67–74.
41. Ferreira, S.L.; Stauder, C.M.; Martin, D.; Kasson, M.T. Morphological and Phylogenetic Resolution of *Diplodia corticola* and *D. quercivora*, Emerging Canker Pathogens of Oak (*Quercus* spp.), in the United States. *Plant Dis.* **2020**. [CrossRef]
42. Linaldeddu, B.T.; Franceschini, A.; Alves, A.; Phillips, A.J.L. *Diplodia quercivora* sp. nov.: A new species of *Diplodia* found on declining *Quercus canariensis* trees in Tunisia. *Mycologia* **2013**, *105*, 1266–1274. [CrossRef]
43. Bakker, P.A.; Pieterse, C.M.; De Jonge, R.; Berendsen, R.L. The Soil-Borne Legacy. *Cell* **2018**, *172*, 1178–1180. [CrossRef]

44. Liu, H.; Macdonald, C.A.; Cook, J.; Anderson, I.C.; Singh, B.K. An Ecological Loop: Host Microbiomes across Multitrophic Interactions. *Trends Ecol. Evol.* **2019**, *34*, 1118–1130. [CrossRef]
45. Lombardi, N.; Vitale, S.; Turrà, D.; Reverberi, M.; Fanelli, C.; Vinale, F.; Marra, R.; Ruocco, M.; Pascale, A.; D'Errico, G.; et al. Root Exudates of Stressed Plants Stimulate and Attract Trichoderma Soil Fungi. *Mol. Plant-Microbe Interact.* **2018**, *31*, 982–994. [CrossRef]
46. Crous, P.; Wingfield, M.J.; Guarro, J.; Cheewangkoon, R.; Van Der Bank, M.; Swart, W.J.; Stchigel, A.M.; Cano-Lira, J.F.; Roux, J.; Madrid, H.; et al. Fungal Planet description sheets: 154–213. *Persoonia Mol. Phylogeny Evol. Fungi* **2013**, *31*, 188–296. [CrossRef] [PubMed]
47. Rahi, P.; Vyas, P.; Sharma, S.; Gulati, A.; Gulati, A. Plant growth promoting potential of the fungus *Discosia* sp. FIHB 571 from tea rhizosphere tested on chickpea, maize and pea. *Indian J. Microbiol.* **2009**, *49*, 128–133. [CrossRef] [PubMed]
48. Szink, I.; Davis, E.L.; Ricks, K.D.; Koide, R.T. New evidence for broad trophic status of leaf endophytic fungi of *Quercus gambelii*. *Fungal Ecol.* **2016**, *22*, 2–9. [CrossRef]
49. Smahi, H.; Belhoucine-Guezouli, L.; Berraf-Tebbal, A.; Chouih, S.; Arkam, M.; Franceschini, A.; Linaldeddu, B.T.; Phillips, A.J.L. Molecular characterization and pathogenicity of *Diplodia corticola* and other Botryosphaeriaceae species associated with canker and dieback of *Quercus suber* in Algeria. *Mycosphere* **2017**, *8*, 1261–1272. [CrossRef]
50. Li, H.; Li, Z.; Ruan, G.; Yu, Y.; Liu, X. Asymmetric reduction of acetophenone into R-(+)-1-phenylethanol by endophytic fungus *Neofusicoccum parvum* BYEF07 isolated from *Illicium verum*. *Biochem. Biophys. Res. Commun.* **2016**, *473*, 874–878. [CrossRef]
51. Barbosa, M.A.G.; Rehn, K.G.; Menezes, M.; Mariano, R.D.L.R. Antagonism of *Trichoderma* species on *Cladosporium herbarum* and their enzimatic characterization. *Braz. J. Microbiol.* **2001**, *32*, 98–104. [CrossRef]
52. Larran, S.; Perelló, A.; Simón, M.R.; Moreno, V. The endophytic fungi from wheat (*Triticum aestivum* L.). *World J. Microbiol. Biotechnol.* **2007**, *23*, 565–572. [CrossRef]
53. Oliveira, R.R.; Aguiar, R.L.; Tessmann, D.J.; Nunes, W.M.C.; Santos, A.F.; Vida, J.B. First Report of Leaf Spot Caused by *Cladosporium perangustum* on *Syagrus oleracea* in Brazil. *Plant Dis.* **2014**, *98*, 280. [CrossRef]
54. Ashkezari, S.J.; Fotouhifar, K.-B. Diversity of endophytic fungi of common yew (*Taxus baccata* L.) in Iran. *Mycol. Prog.* **2017**, *16*, 247–256. [CrossRef]
55. Jones, D.A.; Thomas, C.M.; Hammond-Kosack, K.E.; Balint-Kurti, P.J.; Jones, J.D.G. Isolation of the tomato Cf-9 gene for resistance to *Cladosporium fulvum* by transposon tagging. *Science* **1994**, *266*, 789–793. [CrossRef]
56. Wang, X.; Gul, W.; Taráwneh, A.H.; Gao, J.; Wedge, D.E.; Rosa, L.H.; Cutler, H.G.; Cutler, S.J. Antifungal Activity against Plant Pathogens of Metabolites from the Endophytic Fungus *Cladosporium cladosporioides*. *J. Agric. Food Chem.* **2013**, *61*, 4551–4555. [CrossRef] [PubMed]
57. Hamayun, M.; Khan, S.A.; Ahmad, N.; Tang, D.-S.; Kang, S.-M.; Na, C.-I.; Sohn, E.-Y.; Hwang, Y.-H.; Shin, D.-H.; Lee, B.-H.; et al. *Cladosporium sphaerospermum* as a new plant growth-promoting endophyte from the roots of *Glycine max* (L.) Merr. *World J. Microbiol. Biotechnol.* **2009**, *25*, 627–632. [CrossRef]
58. Pascoe, I.; (Maher), P.M.; Smith, I.; Dinh, S.-Q.; Edwards, J. *Caliciopsis pleomorpha* sp. nov. (Ascomycota: Coryneliales) causing a severe canker disease of *Eucalyptus cladocalyx* and other eucalypt species in Australia. *Fungal Syst. Evol.* **2018**, *2*, 45. [CrossRef] [PubMed]
59. Panteleev, S.V.; Baranov, O.Y.; Rubel, I.E.; Yarmolovich, V.A.; Dishuk, N.G.; Seredich, M.O. Diseases of Container-Grown Conifers in the Nurseries of Mogilev Area According to Molecular Phytopathological Survey. *Proceedings of BSTU* **2016**, *1*, 95–97.
60. Alidadi, A.; Kowsari, M.; Javan-Nikkhah, M.; Jouzani, G.R.S.; Rastaghi, M.E. New pathogenic and endophytic fungal species associated with Persian oak in Iran. *Eur. J. Plant Pathol.* **2019**, *155*, 1017–1032. [CrossRef]
61. Lawrence, D.P.; Travadon, R.; Pouzoulet, J.; Rolshausen, P.E.; Wilcox, W.F.; Baumgartner, K. Characterization of *Cytospora* isolates from wood cankers of declining grapevine in North America, with the descriptions of two newCytosporaspecies. *Plant Pathol.* **2016**, *66*, 713–725. [CrossRef]
62. Jiang, N.; Fan, X.-L.; Crous, P.W.; Tian, C. Species of *Dendrostoma* (Erythrogloeaceae, Diaporthales) associated with chestnut and oak canker diseases in China. *MycoKeys* **2019**, *48*, 67–96. [CrossRef]
63. Li, H.; Yu, S.; Tang, W.; Miao, M.; Liu, Y. First Report of *Diaporthe passiflorae* and *Diaporthe nobilis* Causing a Postharvest Kiwifruit Rot in Sichuan Province, China. *Plant Dis.* **2019**, *103*, 771. [CrossRef]
64. Elfar, K.; Torres, R.; Díaz, G.A.; Latorre, B.A. Characterization of *Diaporthe australafricana* and Diaporthe spp. Associated with Stem Canker of Blueberry in Chile. *Plant Dis.* **2013**, *97*, 1042–1050. [CrossRef]

65. Gomes, R.R.; Glienke, C.; Videira, S.I.R.; Lombard, L.; Groenewald, J.Z.; Crous, P. *Diaporthe*: A genus of endophytic, saprobic and plant pathogenic fungi. *Pers. Mol. Phylogeny Evol. Fungi* **2013**, *31*, 1–41. [CrossRef]
66. Venkatasubbaiah, P.; Chilton, W.S. Toxins Produced by the Dogwood Anthracnose Fungus *Discula* sp. *J. Nat. Prod.* **1991**, *54*, 1293–1297. [CrossRef]
67. Ganley, R.J.; Brunsfeld, S.J.; Newcombe, G. A community of unknown, endophytic fungi in western white pine. *Proc. Natl. Acad. Sci. USA* **2004**, *101*, 10107–10112. [CrossRef] [PubMed]
68. Castoria, R.; De Curtis, F.; Lima, G.; Caputo, L.; Pacifico, S.; De Cicco, V. *Aureobasidium pullulans* (LS-30) an antagonist of postharvest pathogens of fruits: Study on its modes of action. *Postharvest Biol. Technol.* **2001**, *22*, 7–17. [CrossRef]
69. Wachowska, U.; Głowacka, K. Antagonistic interactions between *Aureobasidium pullulans* and *Fusarium culmorum*, a fungal pathogen of winter wheat. *BioControl* **2014**, *59*, 635–645. [CrossRef]
70. Bardas, G.A.; Tzelepis, G.D.; Lotos, L.; Karaoglanidis, G.S. First Report of *Penicillium glabrum* Causing Fruit Rot of Pomegranate (*Punica granatum*) in Greece. *Plant Dis.* **2009**, *93*, 1347. [CrossRef] [PubMed]
71. Hammerschmidt, L.; Wray, V.; Lin, W.; Kamilova, E.; Proksch, P.; Aly, A.H. New styrylpyrones from the fungal endophyte *Penicillium glabrum* isolated from *Punica granatum*. *Phytochem. Lett.* **2012**, *5*, 600–603. [CrossRef]
72. Demirci, E.; Dane, E.; Eken, C. In vitro antagonistic activity of fungi isolated from sclerotia on potato tubers against *Rhizoctonia solani*. *Turk. J. Biol.* **2011**, *35*, 457–462. [CrossRef]
73. Vega, F.E.; Posada, F.; Peterson, S.W.; Gianfagna, T.J.; Chaves, F. *Penicillium* species endophytic in coffee plants and ochratoxin A production. *Mycologia* **2006**, *98*, 31–42. [CrossRef]
74. Stierle, A.A.; Stierle, D.B. Bioactive Compounds from four Endophytic *Penicillium* sp. of a Northwest Pacific Yew Tree. In *Studies in Natural Products Chemistry*; Elsevier: Amsterdam, The Netherlands, 2000; Volume 24, pp. 933–977.
75. Whetzel, H.H.; Wolf, F.A. The Cup Fungus, *Ciboria carunculoides*, Pathogenic on Mulberry Fruits. *Mycologia* **1945**, *37*, 476. [CrossRef]
76. Kehr, R.D. *Pezicula* canker of *Quercus rubra* L., caused by *Pezicula cinnamomea* (DC.) Sacc. I. Symptoms and pathogenesis. *Eur. J. For. Pathol.* **1991**, *21*, 218–233. [CrossRef]
77. Bissegger, M.; Sieber, T.N. Assemblages of Endophytic Fungi in Coppice Shoots of *Castanea sativa*. *Mycologia* **1994**, *86*, 648. [CrossRef]
78. Chen, C.; Verkley, G.J.M.; Sun, G.; Groenewald, J.Z.; Crous, P.W. Redefining common endophytes and plant pathogens in *Neofabraea*, *Pezicula*, and related genera. *Fungal Biol.* **2016**, *120*, 1291–1322. [CrossRef] [PubMed]
79. Yuan, Z.; Verkley, G.J.M. *Pezicula neosporulosa* sp. nov. (Helotiales, Ascomycota), an endophytic fungus associated with *Abies* spp. in China and Europe. *Mycoscience* **2015**, *56*, 205–213. [CrossRef]
80. McMullin, D.R.; Green, B.D.; Prince, N.C.; Tanney, J.B.; Miller, J.D. Natural Products of *Picea* Endophytes from the Acadian Forest. *J. Nat. Prod.* **2017**, *80*, 1475–1483. [CrossRef]
81. Liu, K.H.; Ding, X.; Deng, B.W.; Chen, W. Isolation and characterization of endophytic taxol-producing fungi from *Taxus chinensis*. *J. Ind. Microbiol. Biotechnol.* **2009**, *36*, 1171–1177. [CrossRef]
82. Schulz, B.; Sucker, J.; Aust, H.J.; Krohn, K.; Ludewig, K.; Jones, P.G.; Döring, D. Biologically active secondary metabolites of endophytic *Pezicula* species. *Mycol. Res.* **1995**, *99*, 1007–1015. [CrossRef]
83. Xue, A.G. Biological Control of Pathogens Causing Root Rot Complex in Field Pea Using *Clonostachys rosea* Strain ACM941. *Phytopathology* **2003**, *93*, 329–335. [CrossRef]
84. Cannon, P.F.; Simmons, C.M. Diversity and host preference of leaf endophytic fungi in the Iwokrama Forest Reserve, Guyana. *Mycologia* **2002**, *94*, 210–220. [CrossRef]
85. Madar, Z.; Kimchi, M.; Solel, Z. *Fusarium* canker of Italian cypress. *Eur. J. For. Pathol.* **1996**, *26*, 107–112. [CrossRef]
86. Manici, L.M.; Kelderer, M.; Franke-Whittle, I.H.; Rühmer, T.; Baab, G.; Nicoletti, F.; Caputo, F.; Topp, A.; Insam, H.; Naef, A. Relationship between root-endophytic microbial communities and replant disease in specialized apple growing areas in Europe. *Appl. Soil Ecol.* **2013**, *72*, 207–214. [CrossRef]
87. Di Pietro, A.; García-Maceira, F.I.; Méglecz, E.; Roncero, M.I.G. A MAP kinase of the vascular wilt fungus *Fusarium oxysporum* is essential for root penetration and pathogenesis. *Mol. Microbiol.* **2001**, *39*, 1140–1152. [CrossRef] [PubMed]

88. Bolwerk, A.; Lagopodi, A.L.; Lugtenberg, B.J.J.; Bloemberg, G.V. Visualization of Interactions Between a Pathogenic and a Beneficial *Fusarium* Strain During Biocontrol of Tomato Foot and Root Rot. *Mol. Plant-Microbe Interact.* **2005**, *18*, 710–721. [CrossRef] [PubMed]
89. Kour, A.; Shawl, A.S.; Rehman, S.; Sultan, P.; Qazi, P.H.; Suden, P.; Khajuria, R.K.; Verma, V. Isolation and identification of an endophytic strain of *Fusarium oxysporum* producing podophyllotoxin from *Juniperus recurva*. *World J. Microbiol. Biotechnol.* **2008**, *24*, 1115–1121. [CrossRef]
90. Fernández-Silva, F.; Capilla, J.; Mayayo, E.; Sutton, D.; Guarro, J. In VitroEvaluation of Antifungal Drug Combinations against *Sarocladium* (*Acremonium*) *kiliense*, an Opportunistic Emergent Fungus Resistant to Antifungal Therapies. *Antimicrob. Agents Chemother.* **2014**, *58*, 1259–1260. [CrossRef]
91. Campos, L.A. *Caracterização de Leveduras Promotoras do Crescimento de Plantas*; RIUFSC: Florianópolis, Brazil, 2017.
92. Yuan, W.H.; Jiang, N.; Dong, C.H.; Wei, Z.W.; Wu, H.K.; Chen, C.F.; Zhao, Y.X.; Zhou, S.L.; Zhang, M.M.; Zheng, W.F. Lasiodiplodin analogues from the endophytic fungus Sarocladium kiliense. *Chem. Pharm. Bull.* **2013**, *61*, 363–365. [CrossRef]
93. Tschen, J.S.M.; Chen, L.L.; Hsieh, S.T.; Wu, T.S. Isolation and phytotoxic effects of helvolic acid from plant pathogenic fungus Sarocladium oryzae. *Bot. Bull. Acad. Sin.* **1997**, *38*, 251–256.
94. Potshangbam, M.; Indira, S.; Sahoo, D.; Strobel, G.A. Functional Characterization of Endophytic Fungal Community Associated with *Oryza sativa* L. and *Zea mays* L. *Front. Microbiol.* **2017**, *8*, 325. [CrossRef]
95. Gomes, A.A.M.; Pinho, D.B.; Cardeal, Z.D.L.; Menezes, H.C.; De Queiroz, M.V.; Pereira, O.L. *Simplicillium coffeanum*, a new endophytic species from Brazilian coffee plants, emitting antimicrobial volatiles. *Phytotaxa* **2018**, *333*, 188–198. [CrossRef]
96. Herrero, N.; Zabalgogeazcoa, I. Mycoviruses infecting the endophytic and entomopathogenic fungus *Tolypocladium cylindrosporum*. *Virus Res.* **2011**, *160*, 409–413. [CrossRef]
97. Hanada, R.E.; Pomella, A.W.V.; Costa, H.S.; Bezerra, J.L.; Loguercio, L.L.; Pereira, J.O. Endophytic fungal diversity in *Theobroma cacao* (cacao) and *T. grandiflorum* (cupuaçu) trees and their potential for growth promotion and biocontrol of black-pod disease. *Fungal Biol.* **2010**, *114*, 901–910. [CrossRef]
98. Deng, Z.; Li, C.; Luo, D.; Teng, P.; Guo, Z.; Tu, X.; Zou, K.; Gong, D. A new cinnamic acid derivative from plant-derived endophytic fungus *Pyronema* sp. *Nat. Prod. Res.* **2017**, *31*, 2413–2419. [CrossRef] [PubMed]
99. Botella, L.; Diez, J.J. Phylogenic diversity of fungal endophytes in Spanish stands of *Pinus halepensis*. *Fungal Divers.* **2011**, *47*, 9–18. [CrossRef]
100. Ferreira, A.B.M.; Leite, L.G.; Hernandes, J.L.; Harakava, R.; Padovani, C.R.; Bueno, C.J. Colonization of vines by Petri disease fungi, susceptibility of rootstocks to *Phaeomoniella chlamydospora* and their disinfection. *Arquivos Instituto Biológico* **2018**, *85*. [CrossRef]
101. Lacerda, L.T.; Gusmão, L.F.P.; Rodrigues, A. Diversity of endophytic fungi in *Eucalyptus microcorys* assessed by complementary isolation methods. *Mycol. Prog.* **2018**, *17*, 719–727. [CrossRef]
102. Peever, T.L.; Ibañez, A.; Akimitsu, K.; Timmer, L.W. Worldwide Phylogeography of the Citrus Brown Spot Pathogen, *Alternaria alternata*. *Phytopathology* **2002**, *92*, 794–802. [CrossRef]
103. Soltani, J.; Moghaddam, M.S.H. Antiproliferative, Antifungal, and Antibacterial Activities of Endophytic *Alternaria* Species from Cupressaceae. *Curr. Microbiol.* **2014**, *69*, 349–356. [CrossRef] [PubMed]
104. Thomma, B.P.H.J. *Alternaria* spp.: From general saprophyte to specific parasite. *Mol. Plant Pathol.* **2003**, *4*, 225–236. [CrossRef]
105. Jinu, M.; Jayabaskaran, C. Diversity and anticancer activity of endophytic fungi associated with the medicinal plant Saraca asoca. *Curr. Res. Environ. Appl. Mycol.* **2015**, *5*, 169–179. [CrossRef]
106. Qadri, M.; Rajput, R.; Abdin, M.Z.; Vishwakarma, R.A.; Riyaz-Ul-Hassan, S. Diversity, Molecular Phylogeny, and Bioactive Potential of Fungal Endophytes Associated with the Himalayan Blue Pine (*Pinus wallichiana*). *Microb. Ecol.* **2014**, *67*, 877–887. [CrossRef]
107. Ibrahim, A.; Sørensen, D.; Jenkins, H.A.; Ejim, L.; Capretta, A.; Sumarah, M.W. Epoxynemanione A, nemanifuranones A–F, and nemanilactones A–C, from *Nemania serpens*, an endophytic fungus isolated from Riesling grapevines. *Phytochemistry* **2017**, *140*, 16–26. [CrossRef]
108. Abed-Ashtiani, F.; Narmani, A.; Arzanlou, M. Analysis of *Kalmusia variispora* associated with grapevine decline in Iran. *Eur. J. Plant Pathol.* **2019**, *154*, 787–799. [CrossRef]
109. Ghobad-Nejhad, M.; Asgari, B.; Dokhaharani, S.C. Notes on some endophytic fungi isolated from *Quercus brantii* in Dena Region of Kohgiluyeh and Boyer-Ahmad Province. *Iran. Mycol. Iran.* **2017**, *4*, 1–12. [CrossRef]

110. Kwaśna, H.; Szewczyk, W.; Behnke-Borowczyk, J. Fungal root endophytes of *Quercus robur* subjected to flooding. *For. Pathol.* **2015**, *46*, 35–46. [CrossRef]

111. Strobel, G.; Singh, S.K.; Riyaz-Ul-Hassan, S.; Mitchell, A.; Geary, B.; Sears, J. An endophytic/pathogenic Phoma sp. from creosote bush producing biologically active volatile compounds having fuel potential. *FEMS Microbiol. Lett.* **2011**, *320*, 87–94. [CrossRef] [PubMed]

112. Weber, H.A.; Gloer, J.B. The preussomerins: Novel antifungal metabolites from the coprophilous fungus *Preussia isomera* Cain. *J. Org. Chem.* **1991**, *56*, 4355–4360. [CrossRef]

113. Pulina, M.A.; Linaldeddu, B.T.; Franceschini, A. Topoclimats et communautés des champignons endophytiques dans des bois de chênes-lièges dépéris et non dépéris en Sardaigne (Italie). Proceddings of XIXe COLLOQUE INTERNATIONAL DE CLIMATOLOGIE, Epernay, France, 6–9 September 2006; pp. 474–479.

114. Summerell, B.A. Diseases of Proteaceae. In *Handbook of Plant Disease Management*; McGovern, R.E.W., Ed.; Springer: Cham, Switzerland, 2017; pp. 1–20.

115. Zhao, S.-S.; Zhang, Y.-Y.; Yan, W.; Cao, L.-L.; Xiao, Y.; Ye, Y.-H. *Chaetomium globosum* CDW7, a potential biological control strain and its antifungal metabolites. *FEMS Microbiol. Lett.* **2016**. [CrossRef]

116. Qin, C.; Tao, J.; Liu, T.; Liu, Y.; Xiao, N.; Li, T.; Gu, Y.; Yin, H.; Meng, D. Responses of phyllosphere microbiota and plant health to application of two different biocontrol agents. *AMB Express* **2019**, *9*, 1–13. [CrossRef]

117. Tveit, M.; Wood, R.K.S. The control of *Fusarium* blight in oat seedlings with antagonistic species of chaetomium. *Ann. Appl. Biol.* **1955**, *43*, 538–552. [CrossRef]

118. Fisher, P.; Petrini, O.; Petrini, L. Endophytic ascomycetes and deuteromycetes in roots of Pinus sylvestris. *Nov. Hedwig.* **1991**, *52*, 11–15.

119. Deng, L.; Niu, S.; Liu, X.; Che, Y.; Li, E. Coniochaetones E–I, new 4H-chromen-4-one derivatives from the Cordyceps-colonizing fungus *Fimetariella* sp. *Fitoterapia* **2013**, *89*, 8–14. [CrossRef]

120. Bashiri, S.; Abdollahzadeh, J.; Di Lecce, R.; Alioto, D.; Górecki, M.; Pescitelli, G.; Masi, M.; Evidente, A. Rabenchromenone and Rabenzophenone, Phytotoxic Tetrasubstituted Chromenone and Hexasubstituted Benzophenone Constituents Produced by the Oak-Decline-Associated Fungus *Fimetariella rabenhorstii*. *J. Nat. Prod.* **2020**, *83*, 447–452. [CrossRef] [PubMed]

121. Tao, M.H.; Li, D.-L.; Zhang, W.-M.; Tan, J.-W.; Wei, X.-Y. [Study on the chemical constituents of endophytic fungus *Fimetariella rabenhorstii* isolated from *Aquilaria sinensis*]. *Zhong Yao Cai* **2011**, *34*, 221–223. [PubMed]

122. Liarzi, O.; Bar, E.; Lewinsohn, E.; Ezra, D. Use of the Endophytic Fungus *Daldinia* cf. *concentrica and Its Volatiles as Bio-Control Agents. PLoS ONE* **2016**, *11*, e0168242. [CrossRef]

123. Higginbotham, S.J.; Arnold, A.E.; Ibañez, A.; Spadafora, C.; Coley, P.D.; Kursar, T.A. Bioactivity of Fungal Endophytes as a Function of Endophyte Taxonomy and the Taxonomy and Distribution of Their Host Plants. *PLoS ONE* **2013**, *8*, e73192. [CrossRef] [PubMed]

124. Ko, W.H.; Kunimoto, R.K. Quick decline of macadamia trees: Association with *Xylaria arbuscula*. *Plant Pathol.* **1991**, *40*, 643–644. [CrossRef]

125. Park, J.H.; Choi, G.J.; Lee, H.B.; Kim, K.M.; Jung, H.S.; Lee, S.W.; Jang, K.S.; Cho, K.Y.; Kim, J.C. Griseofulvin from *Xylaria* sp. Strain F0010, an endophytic fungus of *Abies holophylla* and its antifungal activity against plant pathogenic fungi. *J. Microbiol. Biotechnol.* **2005**, *15*, 112–117.

126. Liu, X.; Dong, M.; Chen, X.; Jiang, M.; Lv, X.; Yan, G. Antioxidant activity and phenolics of an endophytic *Xylaria* sp. from *Ginkgo biloba*. *Food Chem.* **2007**, *105*, 548–554. [CrossRef]

127. Adams, G.C.; Kropp, B.R. *Athelia arachnoidea*, the sexual state of *Rhizoctonia carotae*, a pathogen of carrot in cold storage. *Mycologia* **1996**, *88*, 459–472. [CrossRef]

128. Kotasthane, A.S.; Agrawal, T.; Kushwah, R.; Rahatkar, O.V. In-vitro antagonism of *Trichoderma* spp. against *Sclerotium rolfsii* and *Rhizoctonia solani* and their response towards growth of cucumber, bottle gourd and bitter gourd. *Eur. J. Plant Pathol.* **2014**, *141*, 523–543. [CrossRef]

129. Ujor, V.C.; Adukwu, E.C.; Okonkwo, C.C. Fungal wars: The underlying molecular repertoires of combating mycelia. *Fungal Biol.* **2018**, *122*, 191–202. [CrossRef]

130. Escano-Calderón, C.; Rotem, N.; Harris, R.; Vela-Corcía, D.; Levy, M. *Pseudozyma aphidis* activates reactive oxygen species production, programmed cell death and morphological alterations in the necrotrophic fungusBotrytis cinerea. *Mol. Plant Pathol.* **2019**, *20*, 562–574. [CrossRef] [PubMed]

131. Fatima, N.; Muhammad, S.A.; Khan, I.; Qazi, M.A.; Shahzadi, I.; Mumtaz, A.; Hashmi, M.A.; Khan, A.K.; Ismail, T. *Chaetomium* endophytes: A repository of pharmacologically active metabolites. *Acta Physiol. Plant.* **2016**, *38*, 1–18. [CrossRef]
132. Yang, N. Secondary Metabolites Isolated from *Coniothyrium* Species. *Nat. Prod. J.* **2017**, *7*, 248–254. [CrossRef]
133. Tanapichatsakul, C.; Monggoot, S.; Gentekaki, E.; Pripdeevech, P. Antibacterial and Antioxidant Metabolites of *Diaporthe* spp. Isolated from Flowers of *Melodorum fruticosum*. *Curr. Microbiol.* **2018**, *75*, 476–483. [CrossRef] [PubMed]
134. Gandía, M.; Monge, A.; Garrigues, S.; Orozco, H.; Giner-Llorca, M.; Marcos, J.F.; Manzanares, P. Novel insights in the production, activity and protective effect of *Penicillium expansum* antifungal proteins. *Int. J. Biol. Macromol.* **2020**, *164*, 3922–3931. [CrossRef]
135. Chen, R.-S.; Huang, C.-C.; Li, J.-C.; Tsay, J.-G. Evaluation of characteristics of *Simplicillium lanosoniveum* on pathogenicity to aphids and in vitro antifungal potency against plant pathogenic fungi. *Int. J. Environ. Agric. Res.* **2017**, *3*, 55–61.
136. Caesar, L.K.; Cech, N.B. Synergy and antagonism in natural product extracts: When 1 + 1 does not equal 2. *Nat. Prod. Rep.* **2019**, *36*, 869–888. [CrossRef]
137. Shin, T.S.; Yu, N.H.; Lee, J.; Choi, G.J.; Kim, J.-C.; Shin, C.S. Development of a Biofungicide Using a Mycoparasitic Fungus *Simplicillium lamellicola* BCP and Its Control Efficacy against Gray Mold Diseases of Tomato and Ginseng. *Plant Pathol. J.* **2017**, *33*, 337–344. [CrossRef]
138. Fravel, D.; Olivain, C.; Alabouvette, C. *Fusarium oxysporum* and its biocontrol. *New Phytol.* **2003**, *157*, 493–502. [CrossRef]
139. Gauthier, N.A.W.; Robertson, C.L.; Chanda, A.K.; Schneider, R.W. Effects of *Simplicillium lanosoniveum* on *Phakopsora pachyrhizi*, the Soybean Rust Pathogen, and Its Use as a Biological Control Agent. *Phytopathology* **2012**, *102*, 749–760. [CrossRef]

3

Mangrove-Associated Fungi: A Novel Source of Potential Anticancer Compounds

Sunil K. Deshmukh [1,*], Manish K. Gupta [1], Ved Prakash [2] and M. Sudhakara Reddy [3]

1 TERI-Deakin Nano Biotechnology Centre, The Energy and Resources Institute (TERI), Darbari Seth Block, IHC Complex, Lodhi Road, New Delhi 110003, India; manish.gupta@teri.res.in

2 Department of Biotechnology, Motilal Nehru National Institute of Technology, Allahabad 211004, India; ved.mits@gmail.com

3 Department of Biotechnology, Thapar Institute of Engineering & Technology, Patiala, Punjab 147004, India; msreddy@thapar.edu

* Correspondence: sunil.deshmukh@teri.res.in or sunil.deshmukh1958@gmail.com

Abstract: Cancer is the second leading cause of death worldwide, and the number of cases is increasing alarmingly every year. Current research focuses on the development of novel chemotherapeutic drugs derived from natural as well as synthetic sources. The abundance and diversity in natural resources offer tremendous potential for the discovery of novel molecules with unique mechanisms for cancer therapy. Mangrove-derived fungi are rich source of novel metabolites, comprising novel structure classes with diverse biological activities. Across the globe, coastal areas are primarily dominated by mangrove forests, which offer an intensely complex environment and species that mostly remain unexplored. In recent years, many structurally diverse compounds with unique skeletons have been identified from mangrove fungi and evaluated for their antiproliferative properties. These compounds may serve as lead molecules for the development of new anticancer drugs. Mangrove endophytes can be modulated using epigenetic means or culture optimization methods to improve the yield or to produce various similar analogs. The present review provides an insight into the bioactive metabolites from mangrove endophytes reported during the period from 2012 to 2018 (up to April, 2018) along with their cytotoxic properties, focusing on their chemical structures and mode of action, as indicated in the literature.

Keywords: endophytic fungi; anticancer compounds; mangroves; co-culture; epigenetic modification

---

1. Introduction

Mangroves are salt-tolerant forest ecosystems, representing a lively ecosystem with an amalgam of land-dwelling and marine habitats with high biodiversity and socio-economic importance [1]. Marine fungi are among the most prominent species existing in mangrove forests, and support nutrient replenishment [2]. As per reports on marine fungi, in this ecological niche mangrove fungi make up the second largest group [3]. These fungi may occur as saprophytes, symbiotically or as a parasites in the mangrove ecosystem. In addition, these fungi belong to both the lower class, such as oomycetes and thraustochytrids, and the upper class, such as ascomycetes and basidiomycetes. Fungal secondary metabolites are structurally quite diverse, and their functions mostly depend on self-defense against other microorganisms [4]. Most often mangrove fungi flourish in challenging habitats, making them a rich source of bioactive metabolites. Endophytes are one of the various groups of mangrove fungi that have resulted in the identification of a large number of new bioactive metabolites of nutraceutical and pharmaceutical importance. These include antibiotic, anticancer, antidiabetic, antioxidant, antiviral, anti-inflammatory and immunosuppressive drugs, along with other pharmaceutical agents [5].

Cancer affects different organs, and is identified by the unchecked proliferation of abnormal cells that invade other healthy tissue. The treatment is primarily confined to chemotherapy. Besides being expensive, chemotherapy is known to cause severe side effects, making treatment problematic. The non-effectiveness of many existing drugs along with multi-drug resistance further aggravates the problem, making cancer treatment difficult. For medicinal chemists, the primary goal remains the discovery and identification of chemotherapeutic agents derived from natural products. Secondary metabolites have opened new avenues for the development of novel therapeutic agents [6,7]. Endophytic fungi, which are a less-explored area of the microbial community, have a tremendous potential to produce new metabolites that can be used for pharmaceutical applications. Since the initial report of the identification of paclitaxel, derived from an endophyte associated with Northwest Pacific yew by Stierle et al. [8], scientists have identified many other crucial anticancer molecules from fungal endophytes [6]. Many researchers were attracted to marine mangrove fungi because of their diversity, which may lead to the discovery of several novel natural products. With the remarkable advancements in spectroscopic techniques, separation methods and microplate-based sensitive in vitro assays, the natural product exploration of mangrove fungi has attracted special attention regarding novel and unexplored chemical scaffolds [9]. Of the various existing groups of mangrove fungi, endophytes have been identified as producers of new bioactive metabolites with pharmaceutical and nutraceutical importance.

Most of the endophytes have the potential to produce novel bioactive metabolites, which will undoubtedly boost novel drug discovery. However, higher similarity among microbes leads to the frequent identification of the same compound in the endophytes. During axenic cultivation, a specific portion of the biosynthetic genes are expressed while growing in vitro, and various genes stay masked or silent and do not express in laboratory conditions. For this reason, the routine method of fermentation yields metabolites without chemical diversity. Co-cultivation could help to overcome this problem and is preferred, with two or more microbes allowed to grow together. This approach offers a better competitive environment, allowing the increased production of constitutive as well as cryptic compounds that are not traced out in axenic cultures [10]. Several co-cultivation strategies such as different combinations of fungi, the co-cultivation of fungi with bacteria and the co-cultivation of different bacteria have been reported for the enhancement of the chemical diversity of marine-derived microorganisms [10].

The development of methodologies to induce the expression of biosynthetic transcription as well as the suppression of these genes plays a vital role in the search for new secondary metabolites. The regulation of the enzymes that control metabolite production can be achieved by changing epigenetic mechanisms such as DNA methylation and histone modifications (acetylation and phosphorylation) by using epigenetic modifiers [11]. As an example of the importance of epigenetic modulation in producing unknown natural products, when *Aspergillus niger* is cultivated over a two-week period in vermiculite-based semi-solid medium treated with suberoylanilide hydoxamic acid (SAHA), it leads to the isolation of a new fungal metabolite nygerone [12]. Hence the epigenetic approach can be a game changer in the production/enhancement of secondary metabolites.

The present review provides a comprehensive overview of the bioactive metabolites identified from mangrove endophytes during the period from 2012 to 2018 (up to April, 2018) including eighty novel compounds of the total 181 reported. The total number of compounds as well as novel compounds isolated from mangrove fungi during this period is presented in Figure 1 . The origin, chemical structure of the biological targets and efficacies of these compounds are also discussed where available. The anticancer properties of many of these compounds are presented in Table 1. They are arranged based on the broader category of the taxonomic class of the cytotoxic compounds producing fungi. An attempt has also been made to review recent developments such as co-cultivation and epigenetic modifications in endophytic fungi to enhance the secondary metabolite production.

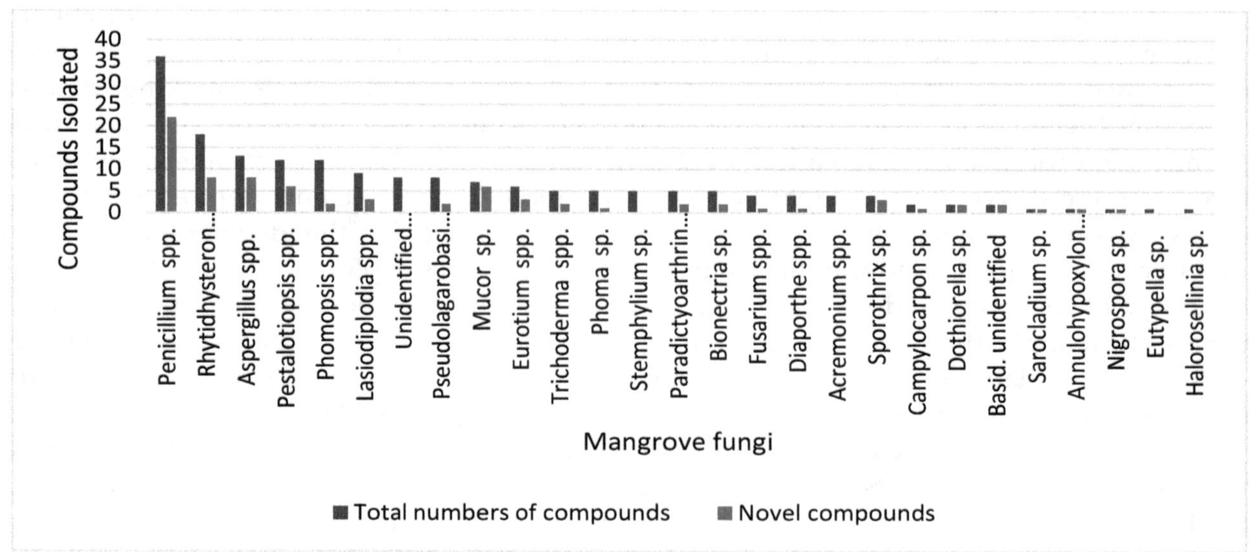

Figure 1. Novel anticancer bioactive compounds reported from mangrove fungi.

2. Bioactive Compounds in Mangrove Plants

2.1. Compounds Produced by Coelomycetes

Pestalotiopsis is the most noteworthy coelomycetous fungi and the species of *Pestalotiopsis* are known to produce the diverse array of novel compounds. Strobel and Long [13] described *Pestalotiopsis* as the "*E. coli* of the temperate and tropical rainforest systems". The species *Pestalotiopsis* is widely recognized to be a prolific producer of a diverse array of metabolites that include alkaloids, chromones, coumarins, isocoumarin derivatives, lactones, peptides, phenols, phenolic acids, quinones, semiquinones, xanthones, terpenoids and xanthone derivatives along with an array of antimicrobial, antifungal, antitumor, antiviral, antineoplastic, and antioxidant compounds [14,15]. Some of the cytotoxic compounds reported from this genus such as demethylincisterol A3 (**1**), ergosta-5,7,22-trien-3-ol (**2**), stigmastan-3-one (**3**), stigmast-4-en-3-one (**4**), stigmast4-en-6-ol-3-one (**5**), and flufuran (**6**) (Figure 2), were discovered from *Pestalotiopsis* spp., associated with Chinese mangrove *Rhizophora mucronata*. Compounds **2–6** showed cytotoxicity against human cancer cell lines HeLa, A549, and HepG, with IC_{50} values in the range of 11.44–102.11 μM. Compound **1** had the most potential, with IC_{50} values reaching the nM activity level from 0.17 to 14.16 nM. Flow cytometric investigation demonstrated that compound 1 inhibited the cell cycle at the G0/G1 phase in a dose-dependent manner with a significant induction of apoptosis on the three tested cell lines. The involvement of the mitochondria in compound-**1**-induced apoptosis was demonstrated using MMP [16].

The compounds 7-*O*-methylnigrosporolide (**7**), pestalotioprolides D–F (**8**, **9**, **10**) (Figure 2), were extracted from the *Pestalotiopsis microspora*, endophytic fungus obtained from the fruits of *Drepanocarpus lunatus* collected from Douala, Cameroon. An approximately ten-fold increase in the yield of compounds **9** and **10** compared to axenic fungal control was observed when *P. microspora* was co-cultured with *Streptomyces lividans*. Compounds **7–10** exhibited cytotoxicity against the L5178Y cell line with IC_{50} values of 0.7, 5.6, 3.4, and 3.9 μM, respectively, and compound **9** also showed potent activity against the A2780 cell line displaying an IC_{50} value of 1.2 μM [17].

Another study by Hemphill et al. reported a new compound pestalpolyol I (**11**) (Figure 2) of the polyketide group from *Pestalotiopsis clavispora*, the endophytic fungus obtained from petioles of the *Rhizophora harrisonii*, growing in Port Harcourt (Nigeria). Compound **11** showed cytotoxicity against the L5178Y cell line with an IC_{50} value of 4.10 μM [18].

A new aromatic amine, pestalamine A (**12**) (Figure 2), was isolated from *P. vaccinia* from a branch of *Kandelia candel*, a viviparous mangrove species widely distributed in coastal and estuarine areas of southern China. The structure of pestalamine A **12** was determined by spectroscopic methods, especially 2D NMR analyses. Compound **12** showed moderate cytotoxicity against MCF-7, HeLa, and HepG2 human cancer cell lines with IC_{50} values of 40.3, 22.0, and 32.8 μM, respectively [19].

Phomazines B (**13**), epicorazine A (**14**), epicorazine B (**15**), epicorazine C (**16**), exserohilone A (**17**) (Figure 2), were isolated from an endophytic fungus, *Phoma* sp. OUCMDZ-1847 associated with the fruit of *Kandelia candel* collected in Wenchang, Hainan Province, China. Compounds **13–17** showed cytotoxicity against the HL-60, HCT-116, K562, MGC-803, and A549 cell lines with IC_{50} values in the range of 0.05 to 8.5 μM [20].

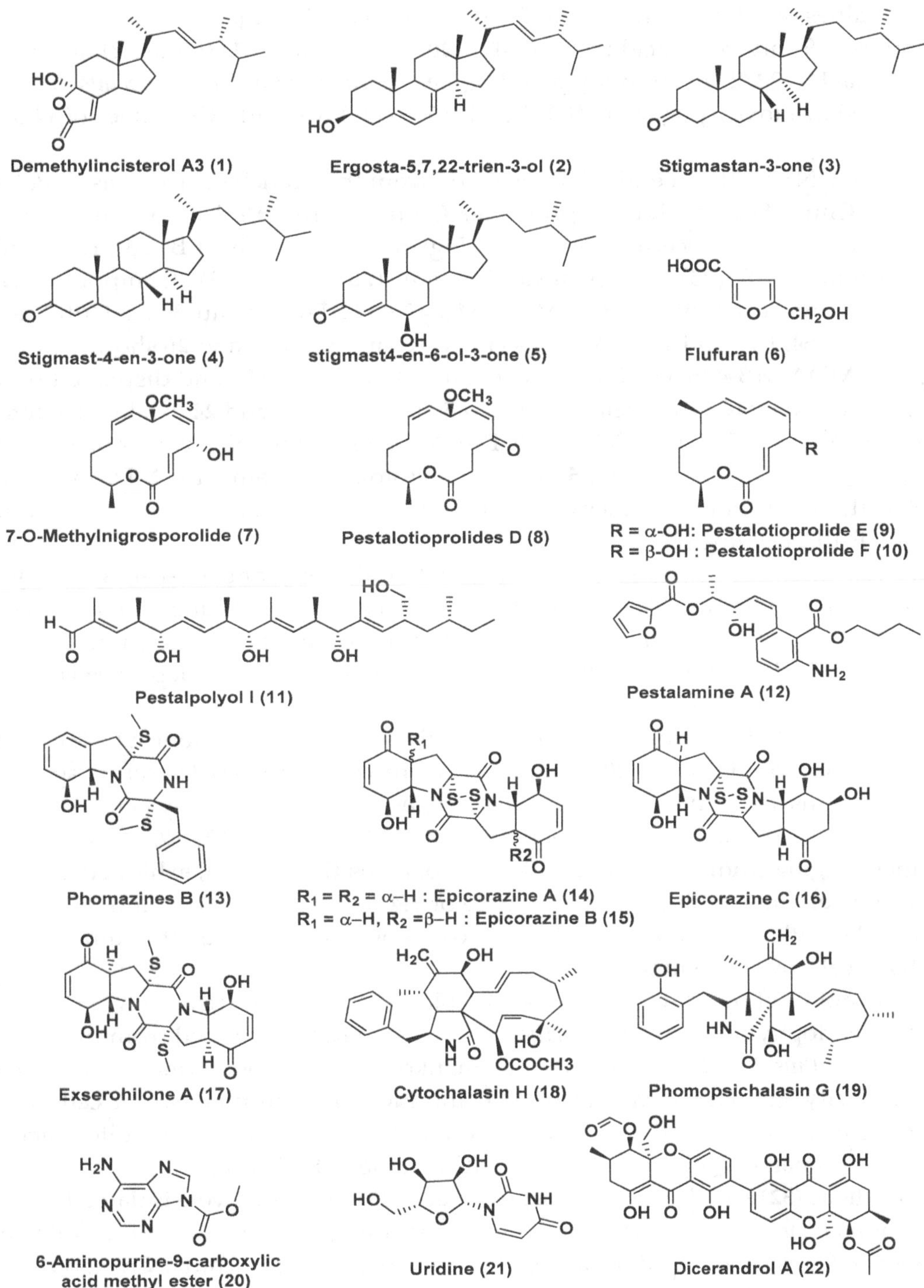

Figure 2. Structures of metabolites isolated from Coelomycetes (**1–22**).

Cytochalasin H (**18**) (Figure 2), was identified form *Phomopsis* sp., an endophytic fungus of mangrove origin in Zhanjiang, China. Compound **18** was shown to inhibit the cell cycle of A549 cells at the G2/M phase. Additionally, DNA fragmentation along with a decrease in the transmembrane potential of mitochondria was observed in A549 cells. It was also shown to regulate the expression level of Bax, P53, Bcl-xL, and Bcl-2. On treatment with Cytochalasin H, the migration capability was impaired in a dose-dependent manner [21].

A new cytochalasin, phomopsichalasin G (**19**) (Figure 2), was isolated from *Phomopsis* sp. xy21, and xy22, associated with *Xylocarpus granatum*, collected in Trang Province, Thailand. Compound **19** exhibited inhibitory activities against HCT-8, HCT-8/T, A549, MDA-MB-231, and A2780 cancer cell lines with IC_{50} values of 7.5, 8.6, 6.4, 3.4, and 7.1 μM, respectively [22].

6-Aminopurine-9-carboxylic acid methyl ester (**20**) and uridine (**21**) (Figure 2), were isolated from *Phomopsis longicolla* HL-2232, an endophyte of *Bruguiera sexangula* var. *rhynchopetala*. The compounds **20** and **21** exhibited cytotoxicity against MCF-7 and A549 cell lines with IC_{50} values of 14.9 and 8.6 μM, respectively [23].

Phomopsis sp. HNY29-2B, an endophyte isolated from the branch of *Acanthus llicifolius* collected from the South China Sea in Hainan province, China, is reported as a source of the known phomoxanthones, including dicerandrol A (**22**) (Figure 2), dicerandrol B (**23**), dicerandrol C (**24**), diacetylphomoxanthone B (**25**) and penexanthone A (**26**) (Figure 3). Compound **22** exhibited broad-spectrum cytotoxic activity against MDA-MB-435, HCT-116, Calu-3 and Huh7 cell lines with IC_{50} values of 3.03, 2.64, 1.76 and 4.19 μM, respectively. Compound **23** and **26** showed potent cytotoxic activities against MDA-MB-435, HCT-116, and Calu-3 (IC_{50} < 10 μM), and displayed poor cytotoxic activity effects against the MCF-10A cell line (IC_{50} > 50 μM). Compound **24** exhibited cytotoxic activity against the MDA-MB-435, HCT-116, Calu-3, MCF-10A cell lines with IC_{50} values of 44.10, 42.63, 36.52, and 33.05 μM, respectively. Compound **25** exhibited cytotoxicity against the MDA-MB-435, HCT-116, Calu-3, Huh7 cell lines with IC_{50} values of 14.40, 7.12, 4.14 and 29.20 μM and showed no cytotoxic effect on the MCF-10A cell line [24].

A new xanthone *O*-glycoside, 3-*O*-(6-*O*-α-L-arabinopyranosyl)-β-D–glucopyranosyl-1,4-dimethoxyxanthone (**27**) (Figure 3), was isolated from an endophytic fungus, *Phomopsis* sp. (ZH76) obtained from the stem of *Excoecaria agallocha* of a mangrove from Dong Sai on the South China Sea coast. Compound **27** had an inhibitory effect on the growth of HEp-2 and HepG2 cells displaying IC_{50} values of 9 and 16 μM, respectively [25].

A highly oxygenated chloroazaphilone derivative, isochromophilone D (**28**), and a known analogue epi-Isochromophilone II (**29**) (Figure 3), were isolated from *Diaporthe* sp. SCSIO 41011, the endophytic fungus associated with *Rhizophora stylosa*, collected in Sanya city, Hainan Province, China. Compound **29** exhibited cytotoxic activity against the ACHN, OS-RC-2, and 786-O cell lines with IC_{50} values ranging from 3.0 to 4.4 μM. Sorafenib, a positive control, exhibited cytotoxic activity against the tested cell lines with IC_{50} in the range of 3.4 to 7.0 μM. Compound **28** showed activity against 786-O cells with an IC_{50} of 8.9 μM. In a dose- and time-dependent manner, apoptosis was induced in 786-O cells by compound **28** [26].

New chromeno[3,2-c] pyridine, 5-deoxybostrycoidin (**30**), and fusaristatin A (**31**) (Figure 3), were isolated from an endophytic fungus *Diaporthe phaseolorum* SKS019 isolated from the branches of the mangrove plant *Acanthus ilicifolius*, collected from Shankou in Guangxi province, China. Compound **30** exhibited cytotoxicity against the MDA-MB-435, and NCI-H460 human cancer cell lines with IC_{50} values of 5.32 and 6.57 μM, respectively, and compound **31** showed growth-inhibitory activity against the MDA-MB-435 human cancer cell line with an IC_{50} value of 8.15 μM [27].

Mycoepoxydiene (**32**) and deacetylmycoepoxydiene (**33**) (Figure 3) were isolated from a mangrove endophytic fungus *Phomosis* sp. A818 isolated from the foliage of *Kandelia candel*, collected from the Mangrove Nature Conservation Area of Fugong, Fujian Province, China. Compounds **32** and **33** had IC_{50} values of 7.85 and 14.61 μM, respectively, against MDA-MB-435 [28]. Mycoepoxydiene (**32**) (Figure 3), significantly suppressed antigen-stimulated degranulation and cytokine production in mast cells and IgE-mediated passive cutaneous anaphylaxis in mice. Compound **32** suppressed antigen-induced activation of Syk, and subsequently inhibited the phosphorylation of PLCγ1, Akt, and MAPKs such as extracellular signal-regulated kinase, c-jun N-terminal kinase, and p38 in mast cells. Mycoepoxydiene can inhibit the activation of mast cells and protect mice from a mast-cell-mediated allergic response through inhibiting the activation of Syk [29].

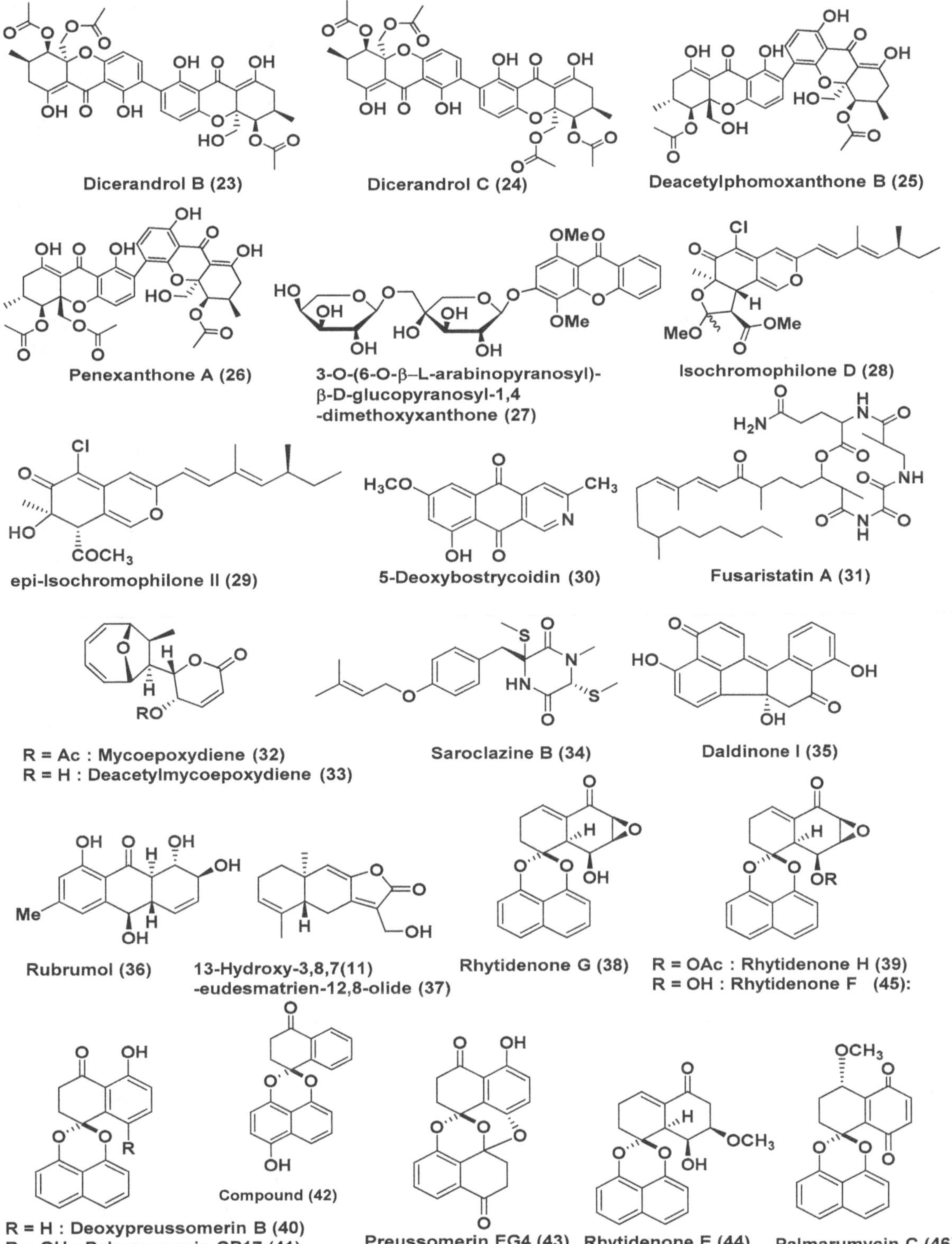

Figure 3. Structures of metabolites isolated from Coelomycetes (**23**–**33**) and Ascomycetes (**34**–**46**).

2.2. Compounds Produced by Ascomycetes

A new diketopiperazine derivative, saroclazine B (**34**) (Figure 3), was isolated from the mangrove-derived fungus *Sarocladium kiliense* HDN11-84 isolated from the rhizosphere soil of the mangrove plant *Thespesia populnea*, collected in Guangxi Province, China. Compound **34** showed cytotoxicity against HeLa cell lines with an IC_{50} value of 4.2 μM[30].

Benzofluoranthene metabolites and daldinone I (**35**) (Figure 3) were extracted from *Annulohypoxylon* sp., an endophytic fungus associated with *Rhizophora racemose*, collected in Cameroon. Compound **35** exhibited average to potent cytotoxicity with IC_{50} values of 14.1 and 6.6 μM, against Jurkat J16 and Ramos cell lines, respectively. It was reported that compound **35** induces apoptotic cell death caused by the induction of intrinsic apoptosis [31].

A new anthraquinone rubrumol (**36**) (Figure 3) with poly-hydroxyl groups was isolated from a halo-tolerant endophytic fungus *Eurotium rubrum*, isolated from the salt-tolerant wild plant *Suaeda salsa* L. collected from the "BoHai" seaside, China. The biological effect of compound **36** on Topo I to relax supercoiled pBR322 DNA was investigated in the cleavable complex assay. The results indicated that compound **36** displayed biological activity compared to the positive control camptothecin. The relaxation activity of rubrumol (**36**) was stronger than that of camptothecin at the concentration of 100 μM. The band backward shifting and trailing of rubrumol (**36**) was observed at 100, 50, 10, 5 and 1 μM. Compound **36** also exhibited cytotoxic activities against A549, MDA-MB-231, PANC-1 and HepG2 human cancer cell lines, by MTT method. The inhibition rate for compound **36** against these four cancer cell line was less than 60% at 100 μg/mL, which implied that it displayed no significant cytotoxic activity [32].

The 13-Hydroxy-3,8,7(11)-eudesmatrien-12, 8-olide (**37**) (Figure 3) was isolated from *Eutypella* sp. 1–15 isolated from the soil of the mangrove rhizosphere in Jimei, Fujian Province, China. Compound **37** exhibited potent anticancer activity against JEKO-1 and HepG2 with IC_{50} values of 8.4 and 28.5 Mm, respectively [33].

Rhytidenones G (**38**), H (**39**), deoxypreussomerin B (**40**), palmarumycin CP17 (**41**), 1-oxo-1,4-dihydronapthalene-4-spiro-20-naptho[400-hydroxy-100,800-de][10,30]-dioxine (**42**), preussomerin EG4 (**43**), rhytidenone E (**44**), rhytidenone F (**45**), palmarumycin C5 (**46**), (Figure 3), and 4,8-dihydroxy- 3,4-dihydronaphthalen-1(2*H*)-one (**47**) (Figure 4), were isolated from *Rhytidhysteron rufulum* AS21B, an endophytic fungus associated with the leaves of *Azima sarmentosa*, collected from the mangrove forest in Samutsakhon province, Thailand. The culture in acidic medium enhanced the production of compounds **41** and **45**, with a four-fold and eight-fold increase, respectively, which are present in minor quantities under normal culture condition. Compounds **38–47** exhibited cytotoxicity against Ramos cells with IC_{50} values of 17.98, 0.018, 18.00, 33.1, 15, 82.9, 0.461, 0.048, 31.7 and 23.1 μM, respectively, while the control Ibrutinib exhibited cytotocity of 28.7 μM against the same cell line. Compounds **38–39**, **44–45**, and **47** exhibited cytotoxicity against the H1975 cell lines with an IC_{50} value of 7.3, 0.252, 10.24, 1.17 and 50 μM, respectively, while control afatinib exhibited cytotoxicity of 1.97 μM against the same cell line [34].

Lasiodiplodia sp. 318#, an endophytic fungus associated with *Excoecaria agallocha*, collected from Guangdong Province, China, was the source of compound 2,4-Dihydroxy-6-nonylbenzoate (**48**)(Figure 4). Compound **48** exhibited cytotoxicity against the MMQ and GH3 cell lines with IC_{50} values of 5.2 and 13.0 μM, respectively [35]. Previously, a new lasiodiplodin—ethyl-2,4-dihydroxy-6-(8′-hydroxynonyl)-benzoate (**49**)—was obtained from the same fungus. Compound **49** exhibited average cytotoxicity against the MDA-MB-435, HepG2, HCT-116, A549, and leukaemia THP1 cell lines with IC_{50} values of 10.13, 12.50, 11.92, 13.31 and 39.74 μM, respectively [36].

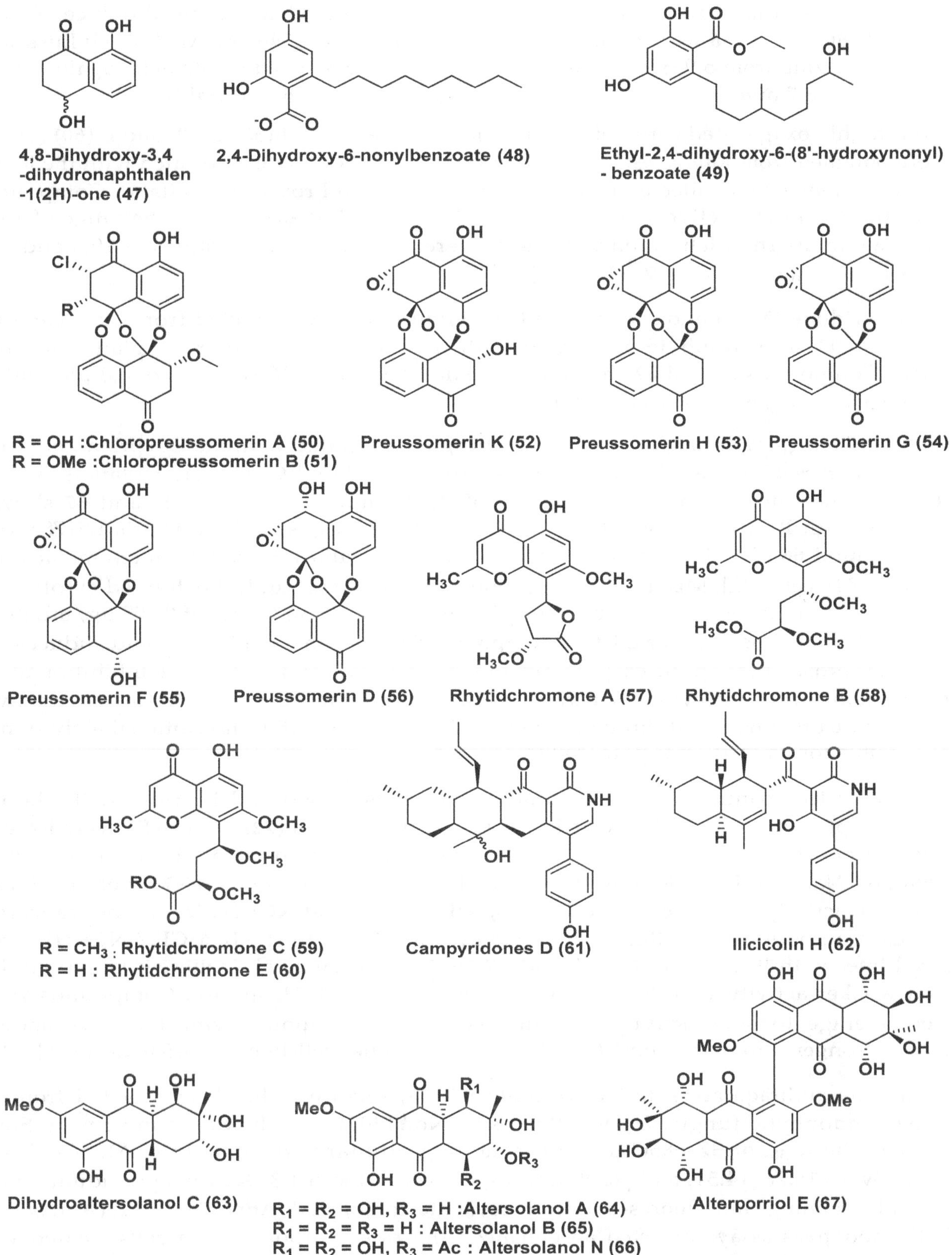

Figure 4. Structures of metabolites isolated from Ascomycetes (**47**–**67**).

Two new chlorinated preussomerins, chloropreussomerins A (**50**), and B (**51**), and a known preussomerin analog preussomerin K (**52**), preussomerin H (**53**), preussomerin G (**54**), preussomerin F (**55**), preussomerin D (**56**) (Figure 4), were obtained from the endophytic fungus *Lasiodiplodia theobromae*

ZJ-HQ1 isolated from *Excoecaria agallocha* collected from Guangdong Province, China. Compounds **50–51** and **56** were found to be active against the A549 and MCF-7 cell lines with IC_{50} values ranging from 5.9–8.9 μM, and compounds **52–55** showed cytotoxicity against A549, HepG2, and MCF-7 human cancer cell lines with IC_{50} values of 2.5–9.4 μM[37].

Four highly oxygenated chromones—rhytidchromone A (**57**),B(**58**),C(**59**) and E (**60**) (Figure 4)—were obtained from *Rhytidhysteron rufulum* BG2-Y, an endophyte associated with the leaves of *Bruguiera gymnorrhiza* collected from Prachuab Kiri Khan Province, Thailand. Compounds **57–60** were found to be active against Kato-3 cell lines, with IC_{50} values in the range of 16.0–23.3 μM, while the rhytidchromones A and C were active, with IC_{50} values of 19.3 and 17.7 μM respectively, against MCF-7 cell lines [38].

Campyridone D (**61**) and ilicicolin H (**62**) (Figure 4) were extracted from *Campylocarpon* sp. HDN13-307, an endophyte associated with the root of the mangrove plant *Sonneratia caseolaris*. Compounds **61** and **62** showed cytotoxicity, with the IC_{50} values of 8.8 μM and 4.7 μM, respectively, against HeLa cells [39].

Stemphylium globuliferum, an endophytic fungus associated with the Egyptian mangrove plant *Avicennia marina*, was the source of dihydroaltersolanol C (**63**), altersolanols A, B, N (**64**, **65**, **66**), and alterporriol E (**67**) (Figure 4)[40]. Compounds **63**, **64**, **65**, and **67** showed cytotoxicity with IC_{50} values of 3.4, 2.5, 3.7 and 6.9 μM, respectively, towards L5178Y cells [41]. Compound **66** also showed good activity, with IC_{50} values in the low micro-molar range towards L5178Y cells [42]. Mishra et al. [43] reported that compound **64** exhibited cytotoxicity against 34 human cancer cell lines in vitro, with mean IC_{50} (IC_{70}) values of 0.005 μg/mL (0.024 μg/mL). It has also been reported that compound **64** is a kinase inhibitor and induces cell death by apoptosis through the caspase-dependent pathway, and that kinase inhibition might be the mechanism for the cytotoxic activity [44]. The pro-apoptotic and anti-invasive activity of compound **64** that occurred through the inhibition of the NF-κB transcriptional activity may be responsible for its antitumor potential [45].

Two new hydroanthraquinones, paradictyoarthrins A (**68**) and B (**69**) and the known compounds, preussomerin C (**70**), ymf 1029C (**71**) and altenusin (**72**) (Figure 5), were isolated from *Paradictyoarthrinium diffractum* BCC 8704, an endophyte associated with mangrove wood in Laem Son National Park, Ranong Province, Thailand. Compounds **68–72** were evaluated for cytotoxic activity against cancer cell-lines, KB, MCF-7, and NCI-H187, and noncancerous Vero cells. Compound **69** exhibited moderate cytotoxicity against KB, MCF-7 NCI-H187 and Vero cell lines with IC_{50} of 3.1, 3.8, 9.5 and 5.6 μg/mL, respectively, whereas Compound **68** showed weaker activity with IC_{50} values in the range of 23–31 μg/mL. Compounds **70–72** showed average to poor activity in tested cell lines. Compound ymf 1029C **71** showed relatively stronger cytotoxicity on NCI-H187 cells than other cell-lines (IC_{50} 5.0 μg/mL) [46].

A marine anthraquinone derivative SZ-685C (**73**) (Figure 5) has been isolated from the mangrove endophytic fungus *Halorosellinia* sp. (No. 1403), which was found in the South China Sea. The IC_{50s} of SZ-685C in nonfunctioning pituitary adenoma (NFPA), MMQ, and RPC cells were 18.76, 14.51, and 56.09 μM, respectively. Hoechst 33342 dye/propidium iodide (PI) double staining and fluorescein isothiocyanate-conjugated Annexin V/PI (Annexin V-FITC/PI) apoptosis assays detected an enhanced the rate of apoptosis in cells treated with SZ-685C. Enhanced expression levels of caspase 3 and phosphate and tensin homologs were determined by Western blotting. The protein expression levels of Akt were decreased when the primary human NFPA cells were treated with SZ-685C. It has been observed that SZ-685C (**73**) induces the apoptosis of human NFPA cells through the inhibition of the Akt pathway in vitro. These findings suggest that SZ-685C may be a potentially promising Akt inhibitor and anti-cancer agent for the treatment of NFPA [47].

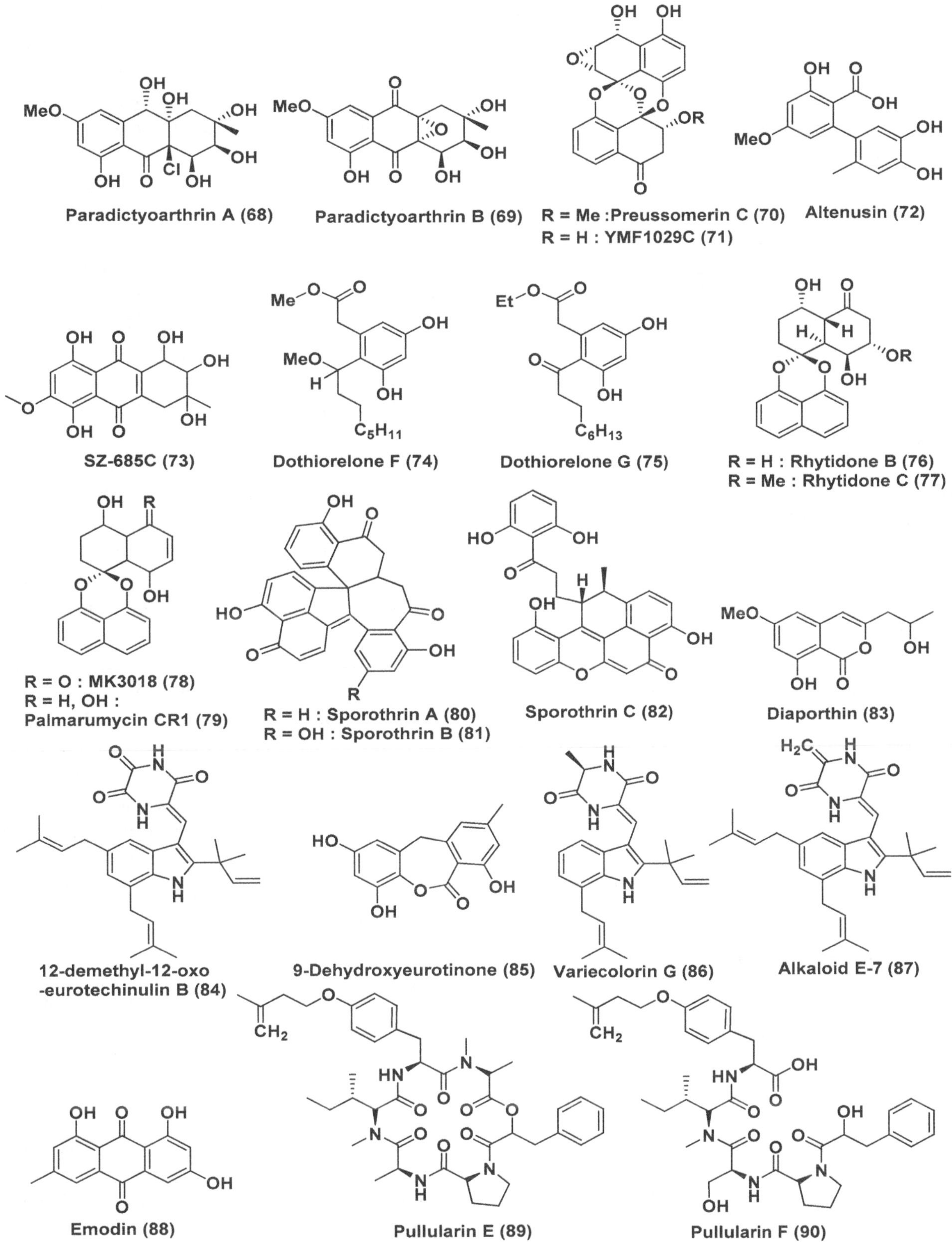

Figure 5. Structures of metabolites isolated from Ascomycetes (**68–90**).

SZ-685C (**73**), was previously reported to inhibit the proliferation of certain tumor cells. SZ-685C inhibited MMQ cell growth in a dose-dependent manner but showed little toxicity toward rat pituitary cells. The IC_{50} of SZ-685C in MMQ cells and RPCs were 13.2 and 49.1 μM, respectively. Increasing numbers of apoptotic cells were observed in response to escalating concentrations of SZ-685C, and the expression level of prolactin was inhibited. Nevertheless, the level of prolactin mRNA was unchanged. Additionally, miR-200c was upregulated in MMQ cells compared with RPCs, and downregulation of miR-200c was observed in SZ-685C-treated MMQ cells. Furthermore, the overexpression of miR-200c weakened the effect of the SZ-685C-induced apoptosis of MMQ cells. It has been suggested that SZ-685C induces MMQ cell apoptosis in a miR-200c-dependent manner [48].

Two new polyketides, named dothiorelons F (**74**) and G (**75**) (Figure 5), were isolated from *Dothiorella* sp., an endophytic fungus associated with the bark of the mangrove tree *Aegiceras corniculatum* at the estuary of Jiulong River, Fujian Province, China. Compounds **74** and **75** showed significant cytotoxicity against the Raji cancer cell line, with an IC_{50} value of 2 μg/mL [49].

New spironaphthalenes, rhytidones B–C (**76, 77**) and known MK3018 (**78**), palmarumycin CR1 (**79**) (Figure 5), were extracted from *Rhytidhysteron* sp. an endophytic fungus associated with the leaves of *Azima sarmentosa*, collected from the mangrove forest in Samutsakhon province, Thailand. Compound **76** was found to be poorly active against CaSki cells, with an IC_{50} value of 22.81 μM, while compounds **77–79** showed average activity against the MCF-7 and CaSkicell lines with IC_{50} values in the range of 14.47 and 25.59 μM [50].

Sporothrix sp. isolated from the bark of an inshore mangrove sample *Kandelia candel* in the South China Sea was the source of Sporothrin A (**80**), sporothrin B (**81**), sporothrin C (**82**), diaporthin (**83**) (Figure 5). Compounds **81–83** were found to show weak cytotoxic activity with IC_{50} values of 20, 23, and 23 g/mL, respectively. Compound **80** exhibited strong inhibition of AChE in vitro (IC_{50} was 1.05 μM) [51].

A new dioxopiperazine alkaloid, 12-demethyl-12-oxo-eurotechinulin B (**84**), and one new anthraquinone derivative 9-dehydroxyeurotinone (**85**), and known compounds, variecolorin G (**86**), alkaloid E-7 (**87**), and emodin (**88**) (Figure 5), were isolated from *Eurotium rubrum*, an endophytic fungus associated with the inner tissue of *Hibiscus tiliaceus*, collected from Hainan Island, China. Compounds **84–88** displayed cytotoxic activity against one or two of the MCF-7, SW1990, HepG2, NCI-H460, SMMC7721, Hela, and Du145 cell lines in the range of 15–30 mg/mL [52].

Two new peptides—pullularins E (**89**) and F (**90**) (Figure 5)—and three known compounds—pullularins A (**91**) and C (**92**) and verticillin D (**93**) (Figure 6)—were extracted from *Bionectria ochroleuca*, an endophytic fungus associated with the inner leaf tissues of *Sonneratia caseolaris* from Hainan island, China. Compound **93** showed potent to moderate activity against L5178Y cell lines with an EC_{50} value of <0.1 μg/mL. Compounds **89–92** also exhibited potent to average activity against the L5178Y cell lines, with EC_{50} values ranging between 0.1 and 6.7 μg/mL [53].

2.3. Compounds Produced by Hyphomycetes

New isocoumarin derivatives, aspergisocoumrins A (**94**) and B (**95**) (Figure 6) were obtained from the culture of the endophytic fungus *Aspergillus* sp. HN15-5D derived from the fresh leaves of the mangrove plant *Acanthus ilicifolius* collected from Dongzhaigang Mangrove National Nature Reserve on Hainan Island, China. Compounds **94** and **95** exhibited cytotoxicity against MDA-MB-435, with IC_{50} values of 5.08 and 4.98 μM, respectively [54].

Two new 6,8(14),22-hexadehydro-5α,9α-epidioxy-3,15-dihydroxy sterols, nigerasterols A (**96**) and B (**97**) (Figure 6), were isolated from *Aspergillus niger* MA-132, an endophytic fungus associated with *Avicennia marina*. Compounds **96–97** displayed cytotoxicity against the HL60 cell line with an IC_{50} value of 0.30 and 1.50 μM, respectively. Compounds **96–97** also exhibited potent activity against the A549 cell line with IC_{50} values of 1.82 and 5.41 μM, respectively [55].

Four new quinazolinone alkaloids, namely, aniquinazolines A–D (**98–101**) (Figure 6), were identified from *Aspergillus nidulans* MA-143, an endophytic fungus associated with the leaves of the marine mangrove plant *Rhizophora stylosa*. Compounds **98–101** showed potent lethality against brine shrimp with LD_{50} values of 1.27, 2.11, 4.95 and 3.42 μM, respectively, which were stronger than that of the positive control colchicine (with the LD_{50} value of 88.4 μM) [56].

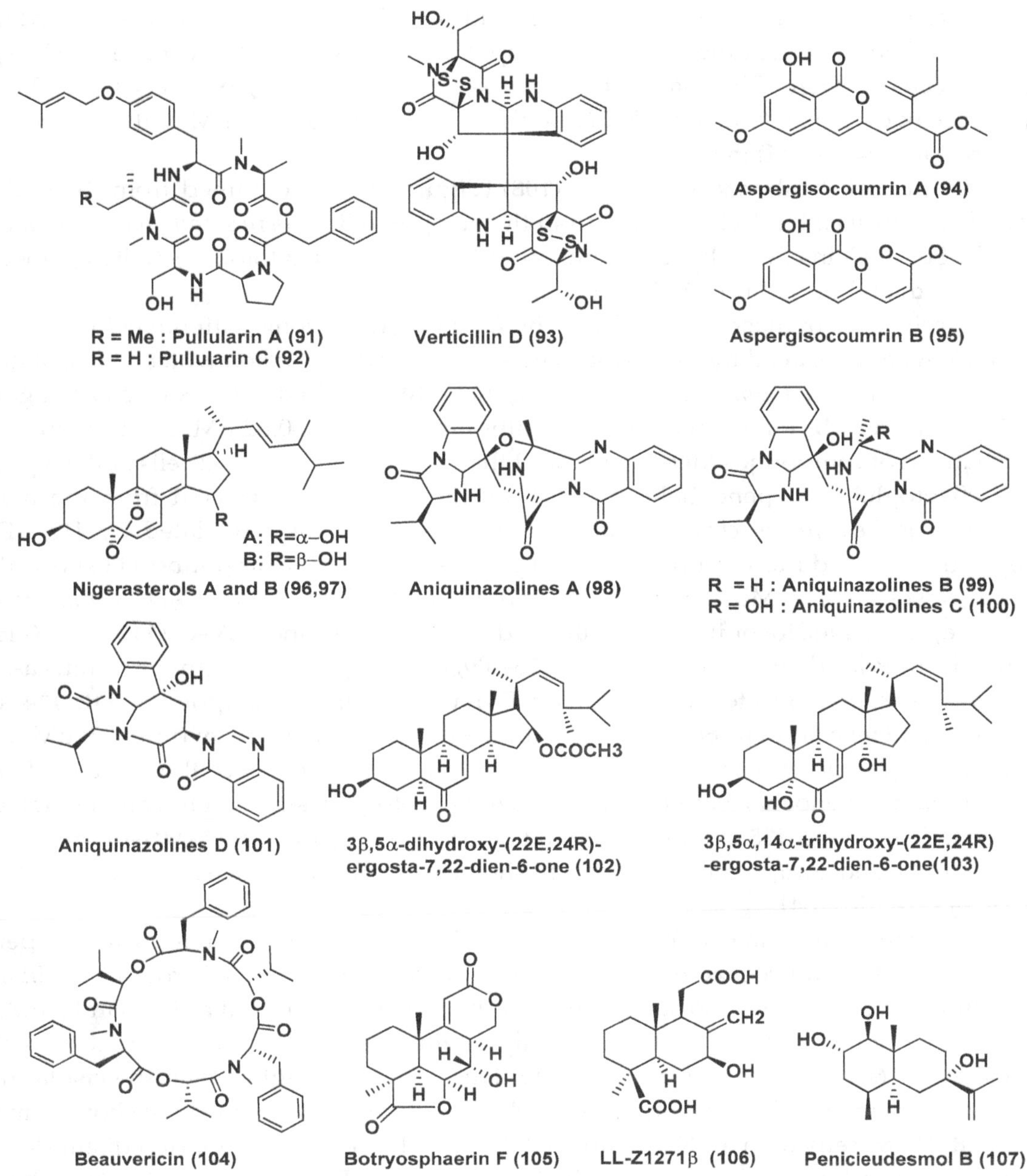

Figure 6. Structures of metabolites isolated from Ascomycetes (**91–93**) and Hyphomycetes (**94–107**).

The compounds 3β,5α-Dihydroxy-(22*E*,24*R*)-ergosta-7,22-dien-6-one (**102**), 3β,5α,14α-trihydroxy-(22*E*,24*R*)-ergosta-7,22-dien-6-one (**103**), and beauvericin (**104**) (Figure 6), were extracted from *Aspergillus terreus* (No. GX7-3B), a mangrove endophytic fungus isolated from the leaves of *Rhizophora stylosa*. Compounds **102** and **104** exhibited good or moderate cytotoxic activity against MCF-7, A549, HeLa and KB cell lines with IC_{50} values of 4.98, 1.95, 0.68, 1.50 and 2.02, 0.82, 1.14, 1.10 μM, respectively; compound **103** exhibited poor activity against the cell lines tested, namely MCF-7 (25.4 μM), A549 (27.1 μM), HeLa (24.4 μM) and KB (19.4 μM) [57].

A new compound, botryosphaerin F (**105**), and a known compound, LL-Z1271β (**106**) (Figure 6), were obtained from the mangrove fungus *Aspergillus terreus* (No. GX7-3B) isolated from the branch of *Brugnieria gymnoihiza*, growing in a coastal salt marsh of the South China Sea in Guangxi Province, China. Compound **105** showed potent inhibiting activity towards the MCF-7 and HL-60 cancer cell lines with IC_{50} values of 4.49 and 3.43 μM, respectively, and compound **106** exhibited promising activity against the HL-60 cell line with an IC_{50} value of 0.6 μM [58].

A new eudesmane-type sesquiterpenoid, penicieudesmol B (**107**) (Figure 6), was isolated from the mangrove-derived endophytic fungus *Penicillium* sp. J-54 associated with the leaves of *Ceriops tagal*, which were collected in Dong Zhai Gang Mangrove Reserve in Hainan province, China. Compound **107** exhibited weak cytotoxicity against K-562 with an IC_{50} value of 90.1 μM, with paclitaxel as the positive control (IC_{50} = 9.5 μM) [59].

A new compound Penibenzophenone B (**108**) (Figure 7), was obtained from the endophytic fungus *Penicillium citrinum* HL-5126 isolated from the mangrove *Bruguiera sexangula* var. *rhynchopetala* collected in the South China Sea. The new compound **108** displayed cytotoxic activity against human A549 cell lines with an IC_{50} value of 15.7 μg/mL [60].

Five new derivatives of macrolide antibiotic Brefeldin A (**109**), along with Brefeldin A 7-*O*-acetate (**110**) (Figure 7), were produced by an endophytic fungus, *Penicillium* sp., which was isolated from the healthy root of *Panax notoginseng*. Compounds **109–110** exhibited cytotoxic activity against the 293, HepG2, Huh7 and KB cell line with an ID_{50} values from 0.024 to 0.62 μM. Further, studies of the cellular mechanism of compounds **109–110** showed that they arrested HepG2 cells at the S phase [61].

A new chaetoglobosin, penochalasin K (**111**) (Figure 7), was extracted from the mangrove endophytic fungus *Penicillium chrysogenum* V11. Its structure was elucidated by 1D, 2D NMR spectroscopic analysis and high resolution mass spectroscopic data. Compound **111** showed strong cytotoxicity against the MDA-MB-435, SGC-7901 and A549 cell lines with an IC_{50} less than 10 μM [62].

The new epipolythiodioxopiperazine alkaloids, penicisulfuranols A–C (**112–114**) (Figure 7), were obtained from *Penicillium janthinellum* HDN13-309, the mangrove endophytic fungus associated with *Sonneratia caseolaris* collected from Hainan Province, China. Compounds **112–114** showed cytotoxicity against HeLa and HL-60 cell lines, with IC_{50} values ranging from 0.1 to 3.9 μM [63].

Penicillium chrysogenum V11, a mangrove endophytic fungus was the source of a novel chaetoglobosin named penochalasin I (**115**), along with chaetoglobosins A (**116**), and cytoglobosin C (**117**) (Figure 7). Compound **115** exhibited marked cytotoxicity against MDA-MB-435 and SGC-7901 cells (IC_{50} < 10 μM), and compounds **116** and **117** showed potent cytotoxicity against SGC-7901 and A549 cells (IC_{50} < 10 μM) [64].

Using the one strain many compounds (OSMAC) approach, new diketopiperazines, spirobrocazine C (**118**) and brocazine G (**119**) (Figure 7) were characterized from *Penicillium brocae* MA-231, an endophytic fungus associated with *Avicennia marina* collected at Hainan Island, China. Compound **119** exhibited potent cytotoxic activity against the A2780 and A2780 CisR cell lines, with IC_{50} values of 664 and 661 nM, respectively. This activity is higher than the cisplatin where the IC_{50} values were reported as 1.67 and 12.63 μM, respectively. Compound **118** showed moderate activity against A2780 cells with an IC_{50} value of 59 μM [65]. Previously reported disulfide-bridged diketopiperazine derivatives, brocazines A (**120**), B (**121**), E (**122**), F (**123**) (Figure 7) were isolated from the same fungus. Compounds **120–123** displayed cytotoxic activity against the Du145, HeLa, HepG2, MCF-7, NCI-H460, SGC-7901, SW1990, SW480, and U251 cell lines with IC_{50} values ranging from 0.89 to 9.0 μM. Compounds **121** and **122** exhibited strong activity against the SW480 cell line with IC_{50} values of 2.0 and 1.2 μM, respectively, while compound **123** showed potent activity against the DU145 and NCI-H460 cell lines, with IC_{50} values of 1.7 and 0.89 μM, respectively [66].

The fungal metabolites TMC-264 (**124**), and PR-toxin (**125**) (Figure 7) were obtained from *Penicillium chermesinum* strain HLit-ROR2, an endophytic fungus associated with the root of *Heritiera littoralis*, collected from Samut Sakhon province, Thailand. Compound **124** showed cytotoxicity against T47D and MDA-MB231 cell lines with the IC_{50} value of 1.08 and 2.81 μM, respectively, while a positive control, doxorubicin, showed activity with IC_{50} of 1.55 and 2.24 μM respectively. Compound **124** also showed cytotoxic activity against the HepG2 cell line, with an IC_{50} value of 3.27 μM, which was 11 times more potent than that of etoposide, an anticancer drug (IC_{50} 35.66 μM). Besides, compound **124** selectively exhibited cytotoxic activity toward MOLT-3 and T47D cancer cells with the IC_{50} values of 1.36 and 1.08 μM and with selectivity index (SI) values of 9 and 11, respectively (IC_{50} 12.64 μM for the normal cell line, MRC-5). Compound **125** exhibited cytotoxicity against the HuCCA-1, HeLa, T47D, and MDA-MB231 cell lines with IC_{50} values in the range of 0.81–2.19 μM, which was comparable to that of doxorubicin (IC_{50} 0.26–2.24 μM). Interestingly, the compound **125** exhibited cytotoxicity against the HL-60 cell line with an IC_{50} value of 0.06 μM, which is superior to doxorubicin (IC_{50} 1.21 μM). Compound **125** selectively showed cytotoxic activity against MOLT-3 and HL-60 cancer cell lines with IC_{50} of 0.09 and 0.06 μM, respectively, with respective SI values of 40 and 61 (IC_{50} 3.66 μM for the MRC-5 cell line) [67].

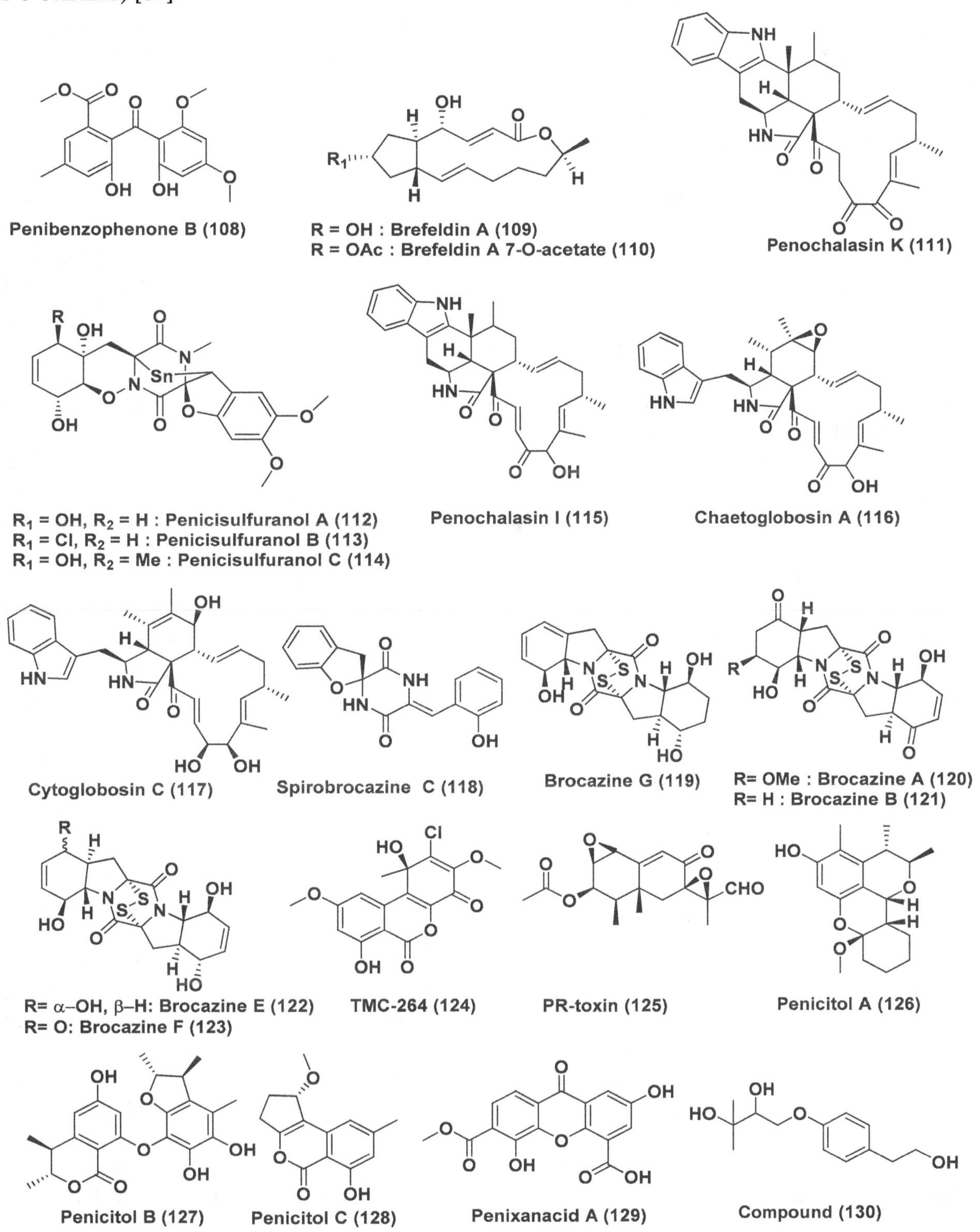

Figure 7. Structures of metabolites isolated from Hyphomycetes (**108–130**).

Three new citrinin analogs, penicitols A–C (**126–128**), and one new xanthone derivative, penixanacid A (**129**) (Figure 7), were isolated from *Penicillium chrysogenum* HND11–24 isolated from the rhizosphere soil of *Acanthus ilicifolius*. Compounds **126–127** showed good activity against the HeLa, BEL-7402, HEK-293, HCT-116, and A549 cell lines with IC_{50} values of 4.6−10.5 and 3.4−9.6 μM, respectively. Compounds **128–129** exhibited cytotoxicity against the HeLa, BEL-7402, HEK-293, HCT-116, and A549 cell lines with IC_{50} values of 10–40.5 μM [68].

Two new metabolites, compounds **130** (Figure 7) and **131** (Figure 8), were obtained from *Penicillium* sp. FJ-1, associated with *Avicennia marina* collected in Fujian, China. Compound **131** exhibited anti-proliferative activity against Tca8113, MG-63 and the normal liver cell line WRL-68 with an IC_{50} value of 10 μM, 55 nM and 58 μM, respectively. Compound **131** has also shown a significant inhibition of tumor growth of human osteosarcoma when tested against nude mice. Compound **130** exhibited weak activity against Tca8113 and MG-63 cells with the IC_{50} values of 26 μM and 35 μM, respectively, in the anti-proliferative assay. Taxol, the positive control, showed activity against Tca8113 and MG-63 cell lines with IC_{50} values of 46 nM and 10 nM, respectively [69].

Meleagrin (**132**) (Figure 8) was isolated from the endophytic fungus *Penicillium* sp. GD6, associated with the Chinese mangrove *Bruguiera gymnorrhiza* collected off the coasts of Zhanjiang, China. Compound **132** showed potent cytotoxic activity against two tumor cell lines, HL60 and A549, with IC_{50} values of 9.7 and 8.3 μM, respectively [70].

Mangrove endophytic fungus, *Penicillium* 303# obtained from the sea water in Zhanjiang Mangrove National Nature Reserve in Guangdong Province, China, was the source of compounds **133**, **134**, **135** and **136** (Figure 8). Compounds **133–135** exhibited cytotoxicity against MDA-MB-435, HepG2, HCT-116, and A549 with IC_{50} value of 11.9–37.82 μg/mL. Compound **136** showed strong cytotoxic activity against the MDA-MB-435 cell lines with IC_{50} values of 7.13 μM and average to low activity against the HepG2 and HCT-116 cell lines with IC_{50} values of 39.64 and 27.80 μM, respectively [71].

A new isobenzofuranone, 4-(methoxymethyl)-7-methoxy-6-methyl-1(3*H*)-isobenzofuranone (**137**) (Figure 8), was isolated from the mangrove endophytic fungus, *Penicillium* sp. ZH58, which was associated with the leaves of mangrove tree *Avicennia* from Dong Sai, Hainan, on the South China Sea coast. Compound **137** showed cytotoxic activity against the KB and KB_V200 cell lines with IC_{50} values of 6 and 10 μg/mL, respectively [72].

The new and rare sulfur-containing curvularin derivatives sumalarins A, B, (**138**, **139**) C (**140**) (Figure 8), and dehydrocurvularin (**141**) (Figure 8) were obtained from *Penicillium sumatrense* MA-92, isolated from the rhizosphere of the mangrove *Lumnitzera racemose* collected at Wen Chang on Hainan Island, China. Compounds **138–141** displayed cytotoxic activities against the Du145, HeLa, Huh 7, MCF-7, NCI-H460, SGC-7901, and SW1990 cell lines, with IC_{50} values in the range of 3.8 to 10 μM [73].

Penicillium sp. ZH16, a mangrove endophytic fungus, obtained from the South China Sea, was the source of the furanocoumarin 5-methyl-8-(3-methylbut-2-enyl) furanocoumarin (**142**) (Figure 8). Compound **142** exhibited cytotoxicity against the KB and KB_V200 cell lines with IC_{50} values of 5 and 10 μg/mL, respectively [74].

Zhang et al. [75] reported the production of the compounds (3*S*)-6-oxo-de-*O*-methyllasiodiplodin (**143**), (3*R*)-de-*O*-methyllasiodiplodin (**144**), (3*R*)-nordinone (**145**) (Figure 8), by the co-culturing of mangrove endophytic fungus *Trichoderma* sp. 307 isolated from the stem bark of *Clerodendrum inerme*, and the aquatic pathogenic bacterium *Acinetobacter johnsonii* B2 isolated from the stem bark of *Clerodendrum inerme*, collected in Zhanjiang Mangrove National Nature Reserve in Guangdong Province, China. Compound **144** exhibited good cytotoxic activity against the GH3 and MMQ cell lines with IC_{50} values of 6.44 and 6.58 μM, respectively, and against rat normal pituitary cells (RPC) with an IC_{50} value of 6.94 μM. Compound **145** displayed average cytotoxic activity against the GH3 and MMQ cell lines with IC_{50} values of 12.33 and 10.13 μM, respectively, and against RPC cell lines with an IC_{50} value of 100.03 μM. Compound **143** was less active against the GH3 and MMQ cell lines with IC_{50} values of 21.42 and 13.59 μM, respectively, and against the RPC cell lines with an IC_{50} value of 142.8 μM as a positive control [75].

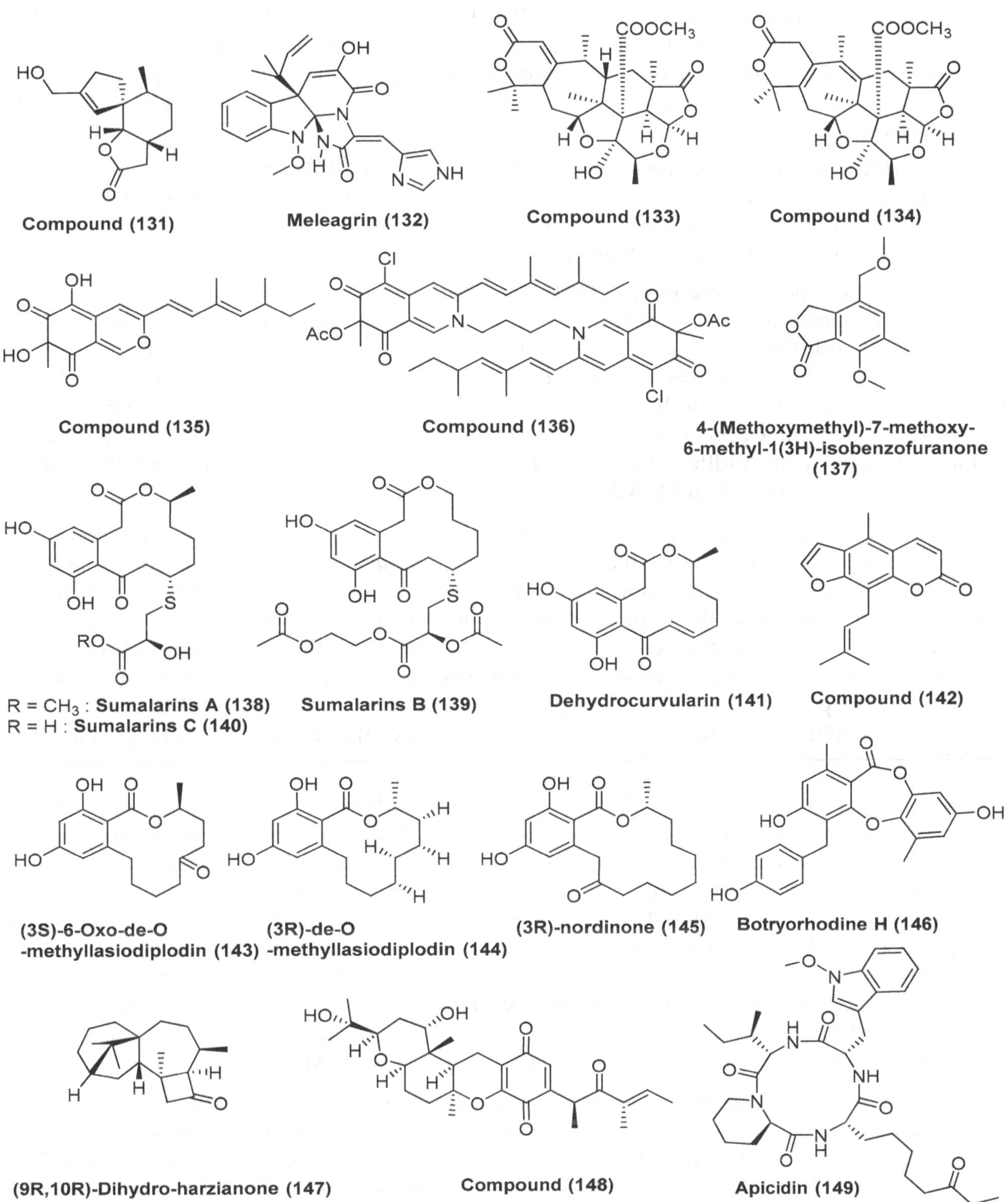

Figure 8. Structures of metabolites isolated from Hyphomycetes (**131–149**).

Zhang et al. [76] reported the production of a new depsidone, botryorhodine H (**146**) (Figure 8), by co-culturing mangrove endophytic fungus *Trichoderma* sp. 307 and *Acinetobacter johnsonii* B2. Compound **146** exhibited good cytotoxic activity against the MMQ and GH3 cell lines with IC_{50} values of 3.09 and 3.64 μM, respectively.

A new harziane diterpenoid, named (9*R*,10*R*)-dihydro-harzianone (**147**) (Figure 8), was isolated from *Trichoderma* sp. Xy24, an endophytic fungus residing in the leaves, stems, and peels of the mangrove plant *Xylocarpus granatum*, collected in the Sanya district of Hainan province, China. Compound **147** was active against the HeLa and MCF-7 cell lines with IC_{50} values of 30.1 μM and 30.7 μM, respectively [77].

The endophytic fungus *Nigrospora* sp. MA75 associated with *Pongamia pinnata*, which led to the isolation of a new compound 2,3-didehydro-19a-hydroxy-14-epicochlioquinone B (**148**) (Figure 8) when grown in medium containing 3.5% NaI. Compound **148** showed potent cytotoxicity against the MCF-7, SW1990, and SMMC7721 cell lines, with IC_{50} values of 4, 5, and 7 μg/mL, respectively [78].

A known cyclic peptide, beauvericin (**104**) (Figure 6), was obtained from *Fusarium* sp. (No. DZ27) an endophytic fungus residing inside the bark of *Kandelia candel* from Dongzhai mangrove, Hainan, China, in the South China Sea. Compound (**104**) showed cytotoxic activity against the KB and KBv200 cell lines with IC_{50} values of 5.76 and 5.34 μM, respectively. It induces apoptosis through the mitochondrial pathway, including the decrease of relative oxygen species generation, the loss of mitochondrial membrane potential, the release of cytochrome c, the activation of Caspase-9 and -3, and the cleavage of PARP. Additionally, the regulation of Bcl-2 or Bax was not involved in the apoptosis induced by beauvericin in KB and KBv200 cells [79].

An inhibitor of histone deacetylase, Apicidin (**149**) (Figure 8), was isolated from *Fusarium* sp., an endophytic fungus associated with the leaf of mangrove *Kandelia candel* planted at Dongzhai Harbor on Hainan Island, China. Apicidin showed good cytotoxic activity against GLC-82 cells with the IC_{50} value of 6.94 ± 0.27 μM. Apicidin suppressed proliferation and invasion, and induced apoptosis via the mitochondrial pathway in GLC-82 cells, including the loss of $\Delta\Psi m$, the release of cytochrome c from mitochondria, the activation of caspase-9 and -3, and the cleavage of poly-ADP-ribose polymerase [80]. Apicidin **149**, was previously isolated from the mangrove endophytic fungus ZZF42 from the South China Sea and exhibited selective in vitro cytotoxicity towards KB and KBv200 with IC_{50} values of less than 0.78 μg/mL [81].

An unusual alkaloid 2-acetyl-1,2,3,4-tetrahydro-β-carboline (**150**), fusamine (**151**), and 3-(1-aminoethylidene)-6-methyl-2*H*-pyran-2,4(3*H*)-dione (**152**) (Figure 9), were isolated from the culture broth of *Fusarium incarnatum* (HKI0504), an endophytic fungus of the mangrove plant *Aegiceras corniculatum*. Compound **150** exhibite weak antiproliferative activity against HUVEC and the K-562 cell line with GI_{50} values of 41.1 and 33.3, respectively. Compounds **151** and **152** exhibit anti-proliferative activity against HUVEC and the GI50 K-562 cell line with GI_{50} values of 37.3, 37.6, and 41.1, 33.3, respectively. The compounds **150**, **151** and **152** displayed cytotoxic activity against HeLa cells with a CC_{50} value of 23.8, 23.3, and 23.8 μM, respectively. The standard imatinib exhibited anti-proliferative activity against HUVEC and the GI50 K-562 cell line with a GI_{50} value of 18.5 and 0.17, respectively, and cytotoxic activity against HeLa cells with a CC_{50} value of 65.8 μM [82].

Acremonium sp., an endophytic fungus residing inside the leaves of *Sonneratia caseolaris* collected from Hainan, China, was the source of a dimeric anthracene derivative torrubiellin B (**153**) (Figure 9). Compound **153** exhibited potent cytotoxic activity against the cisplatin-sensitive cell lines Cal27, Kyse510, HCC38, A2780, MDA-MB-231with IC_{50} values ranging from of 0.3 to 1.5 μM, and against the cisplatin-resistant cell lines Cal27, Kyse510, HCC38, A2780, MDA-MB-231 with IC_{50} values ranging from 0.2 to 2.6 μM. The positive control cisplatin exhibited cytotoxic activity against the tested cell lines with IC_{50} values in the range of 1.5–38.1 μM [83].

Waol A (**154**), pestalotiopene A (**155**), and cytosporone E (**156**) (Figure 9) were obtained from the endophytic fungus *Acremonium strictum*, isolated from the mangrove tree *Rhizophora apiculate*, collected on the island of CatBa, Vietnam. Compounds **154**–**156** showed moderate cytotoxic activity against human cisplatin-sensitive (IC_{50} values 27.1, 76.2, and 8.3 μM, respectively) and resistant A2780 cell lines (IC_{50} values 12.6, 30.1, and 19.0 μM, respectively) [84].

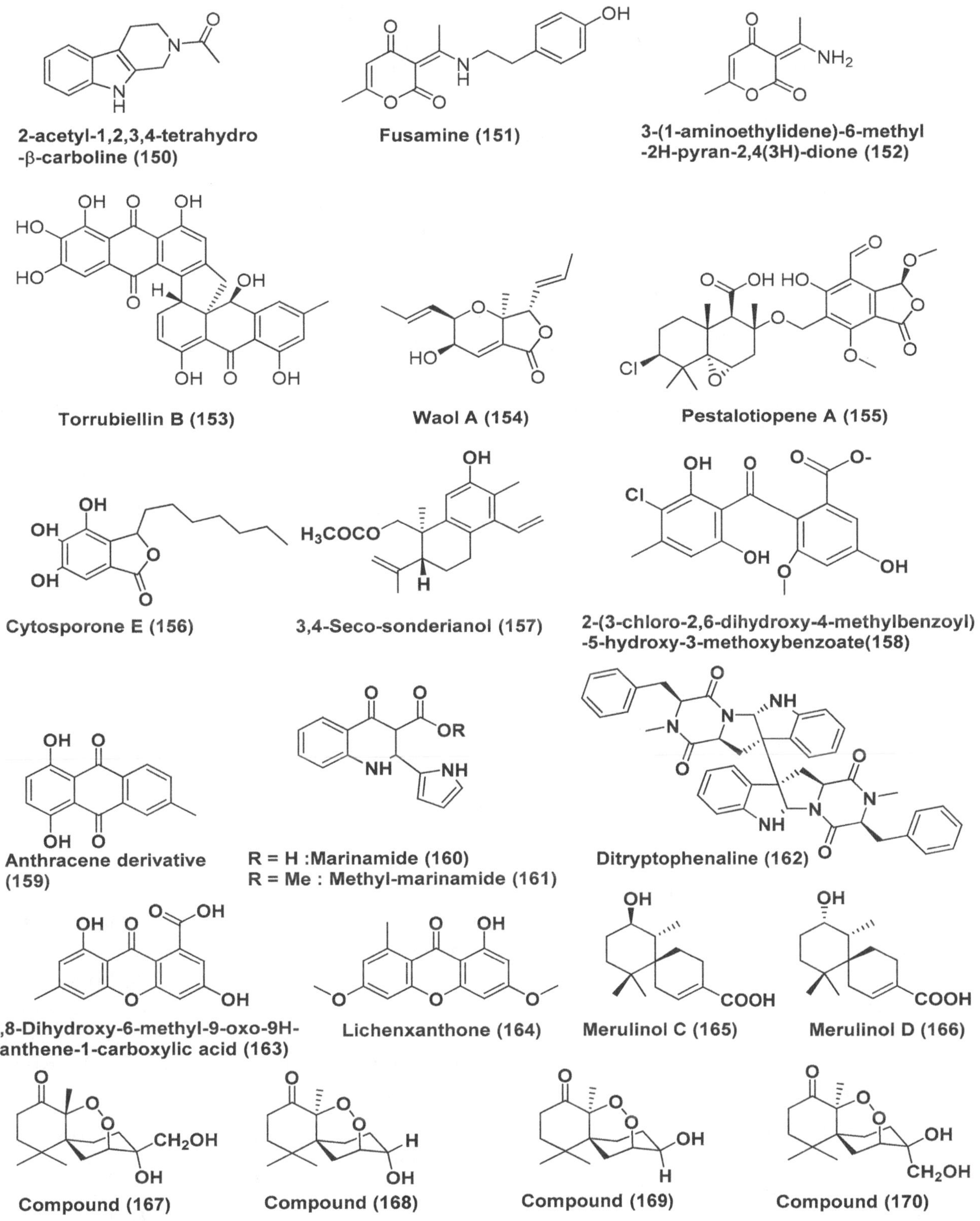

Figure 9. Structures of metabolites isolated from Hyphomycetes (**150–170**).

The compound 3,4-seco-sonderianol (**157**) (Figure 9), a known diterpenoid, was obtained from the endophytic fungus J3 of *Ceriops tagal*, collected in Hainan province, China. Compound **157** showed cytotoxic activity against the K562, SGC-7901, and BEL-7402 cell lines with IC_{50} values of 9.2, 15.7, and 25.4 μg/mL, respectively. The positive control paclitaxel displayed activity against the K562, SGC-7901 and BEL-7402 cell lines with IC_{50} values of 5.1, 1.6 and 6.3 μg/mL, respectively [85].

The compound 2-(3-chloro-2,6-dihydroxy-4-methylbenzoyl)-5-hydroxy-3-methoxybenzoate (**158**) (Figure 9), was extracted from the endophytic fungus No. ZH-3 from the South China Sea. Compound **158** showed cytotoxic activity against the hepG2 cell line (IC_{50} = 25 μg/mL) [86].

Mangrove endophytic fungus No.5094, which was collected in the South China Sea was the source of anthracene derivative (**159**) (Figure 9). Compound **159** exhibited potent activity towards the KB and KBv200 cell lines with LD_{50} values of 5.5 and 10.2 μM, respectively [87].

Marinamide (**160**) and methyl marinamide (**161**) (Figure 9) were obtained by co-cultures of two marine-derived mangrove endophytic fungi (strains Nos. 1924 and 3893) from the South China Sea coast. Their structures were elucidated using comprehensive spectra methods. Compound **160** was found to be cytotoxic with IC_{50} values of 7.0, 0.4, 91 nM and 0.529 μM, respectively against HepG2, 95-D, MGC832 and HeLa cells. For the similar cell lines, compound **161** exhibited cytotoxicity with IC_{50} values of 2.52, 1.54, 13.0 and 0.110 μM, respectively [88].

Mangrove endophytic fungus No. Gx-3a in the South China Sea was the source of ditryptophenaline (**162**) (Figure 9). Compound **162** exhibited potent cytotoxic activity against KB and KBv200 cells, with LD_{50} values of 8.0 μM and 12.0 μM [89].

The 3, 8-Dihydroxy-6-methyl-9-oxo-9*H*-xanthene-1-carboxylate (**163**) and lichenxanthone (**164**) (Figure 9), were isolated from the mangrove endophytic fungus No·SK7RN3G_1 from the South China Sea. Compounds **163–164** exhibited cytotoxicity against the HepG2 cell line with IC_{50} values of 20 and 25 μg/mL, respectively [90].

2.4. *Compounds Produced by Basidiomycetes*

Two new chamigrane sesquiterpenes, merulinols C and D (**165**, **166**) (Figure 9), were isolated from the culture of the basidiomycetous fungus XG8D, isolated from the healthy leaves of *Xylocarpus granatum*, collected in Samutsakorn province, Thailand. Compounds **165** and **166** selectively displayed cytotoxicity against KATO-3 cells with IC_{50} values of 35.0 and 25.3 μM, respectively [91].

Compounds **167**, **168**, **169** and **170** (Figure 9) were extracted from *Pseudolagarobasidium acaciicola*, associated with *Bruguiera gymnorrhiza* from Samut Sakhon province, Thailand. These compounds were evaluated for their cytotoxicity against the HuCCA-1, A549, MOLT-3, HepG2, HL-60, MDA-MB231, T47D, and HeLa cancer cell lines and normal human embryonic lung cell lines (MRC-5). Compound **167** displayed cytotoxicity against the HuCCA-1, A549, MOLT-3, HepG2, HL-60, MDA-MB231, and T47D cancer cell lines with IC_{50} values in the range of 0.28–37.46 μM and against the MRC-5 normal cell line with an IC_{50} value of 17.92 μM. Compound **167** selectively showed activity against the HL-60 cell line with an IC_{50} value of 0.28 μM and a selectivity index (SI) value of 64.0. Compound **168** showed activity against the A549, MOLT-3, HepG2, HL-60, MDA-MB231, T47D, HeLa and MRC-5 cells IC_{50} values in the range of 12.09–170.08 μM. Compounds **169** and **170** were active against the cell lines tested, IC_{50} 15.20–76.97 μM for **169** and IC_{50} 18.31–154.51 μM for **170**. Compound **170** also selectively exhibited cytotoxic activity toward the HL-60 cell line (IC_{50} 18.31 μM) with an SI value of 4.4 [92].

A new nor-chamigrane endoperoxide, 3-*epi*-Steperoxide A (**171**), along with the known sesquiterpenes steperoxide A (**172**), merulin B (**173**), and merulin C (**174**) (Figure 10) were isolated from *Pseudolagarobasidium acaciicola*, an endophyte residing inside the mangrove plant *Bruguiera gymnorrhiza*. Compounds **171** and **172** showed strong cytotoxic activity with IC_{50} ranges of 0.68–3.71 and 0.67–5.25 μg/mL, respectively, against MOLT-3, HuCCA-1, A549, HepG2, HL-60, MDA-MB-231, T47D, and HeLa cells. Compound **173** exhibited weak activity against MOLT-3, A549, HepG2, HL-60, MDA-MB-231 and T47D cells with an IC_{50} ranging from 11.94–49.08 μg/mL, but was inactive toward HuCCA-1 and HeLa cells at 50 μg/mL. Compound **174** showed the most potent cytotoxic activity against HL60 cancer cells, with an IC_{50} value of 0.08 μg/mL, whereas it displayed activity toward the MOLT-3, HuCCA-1, A549, HepG2, MDA-MB-231, T47D, and HeLa cell lines with an IC_{50} range of 0.19–3.75 μg/mL [93].

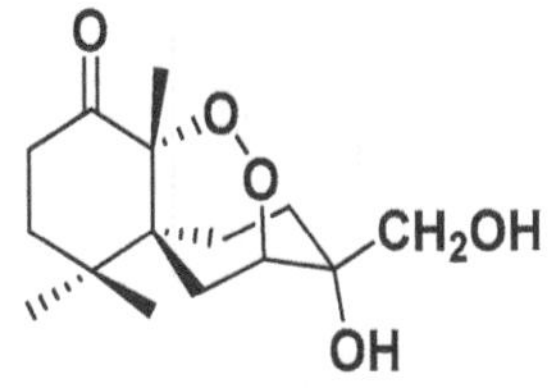

Compound (171) 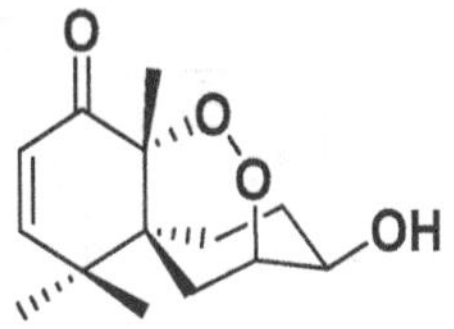 Steperoxide A (172) 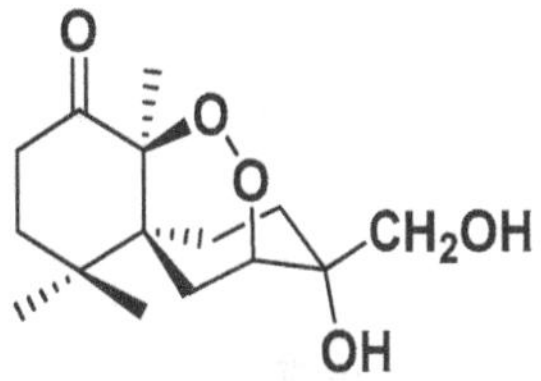 Merulin B (173) 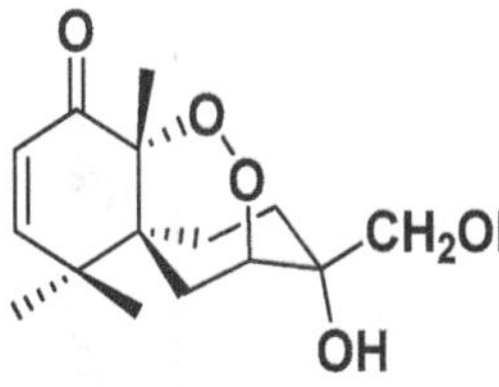 Merulin C (174)

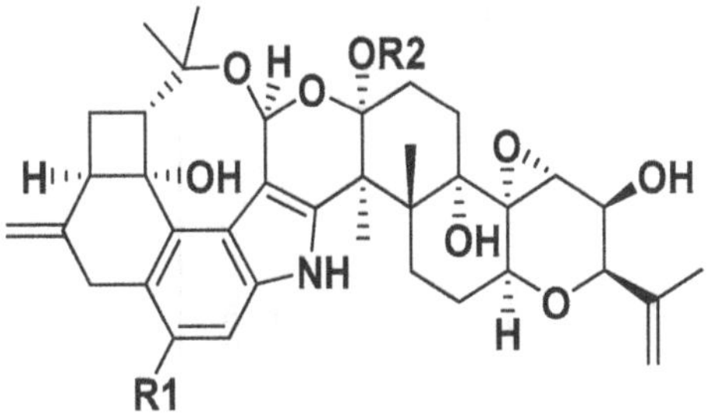

A: R_1=Cl, R_2=H
B: R_1=Cl, R_2=CH_3
Rhizovarins A, B (175,176)

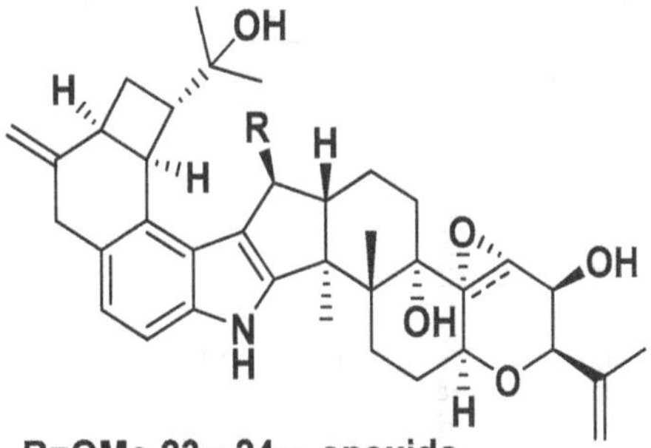

R=OMe,23α,24α–epoxide
Rhizovarin E (177)

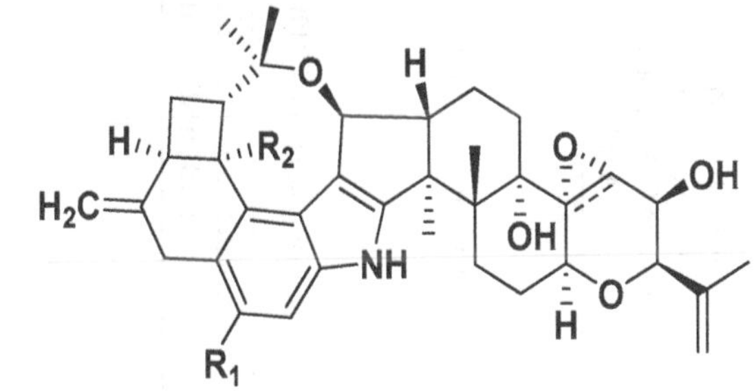

R1=Cl, R2=OH,23α,24α–epoxide : Penitrem A,(178)
R1=Cl, R2=H, 23α,24α–epoxide : Penitrem B (179)
R1=Cl, R2=H, 23α,24α–epoxide :Penitrem F (180)

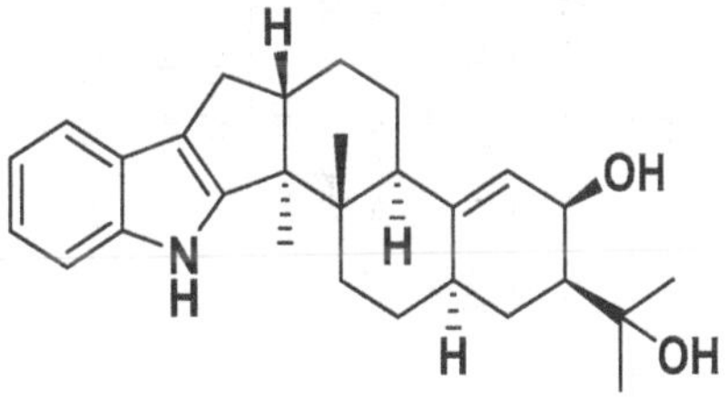

3β-hydroxy- 4β-desoxypaxilline (181)

Figure 10. Structures of metabolites isolated from Basidiomycetes (**171–174**) and from Zygomycetes (**175–181**).

2.5. *Compounds Produced by Zygomycetes*

The rhizovarins A, B, E (**175–177**), penitrems A, C, F (**178–180**) and 3β-hydroxy-4β-desoxypaxilline (**181**) (Figure 10) were obtained from *Mucor irregularis* QEN-189, an endophytic fungus residing inside the inner tissue of *Rhizophora stylosa*, collected in Hainan Island, China. Compounds **175–181** showed cytotoxicity against the human A-549 cell lines with IC_{50} values of 11.5, 6.3, 9.2, 8.4, 8.0, 8.2, and 4.6 μM, while compounds **175, 176, 178–181** were cytotoxic against the human HL-60 cell lines with IC_{50} values of 9.6, 5.0, 7.0, 4.7, 3.3 and 2.6 μM, respectively. Adriamycin, a positive control, exhibited activity against A-549 and HL-60 cell lines with IC_{50} values of 0.30 and 0.06 μM, respectively [94].

Table 1. Novel anticancer bioactive compounds reported from mangrove fungi.

Sr. No.	Fungus	Host Plant(s)	Plant Part or Tissue Locality of Host Plants	Compounds Isolated	Cell Line	IC_{50}/EC_{50}/Inhibition	Refs.
				Compounds Produced by *Coelomycetes*			
1	*Pestalotiopsis* sp.	*Rhizophora mucronata*	Not reported	Demethylincisterol A3 (**1**)	HeLa, A549 and HepG	In the range of 0.17 to 14.16 nM	[16]
				Ergosta-5,7,22-trien-3-ol (**2**), stigmastan-3-one (**3**), stigmast-4-en-3-one (**4**), stigmast4-en-6-ol-3-one (**5**), flufuran (**6**)	HeLa, A549 and HepG	In the range of 11.44–102.11 μM	
2	*Pestalotiopsis microspore*	*Drepanocarpus lunatus*	Cameroon	Compound (**7**) Pestalotioprolide D–F (**8–10**)	L5178Y	0.7, 5.6, 3.4, and 3.9 μM	[17]
				Pestalotioprolide E (**9**)	A2780	1.2 μM	
3	*Pestalotiopsis clavispora*	*Rhizophora harrisonii*	Port Harcourt, Nigeria	Pestalpolyol I (**11**)	L5178Y	4.10 μM	[18]
4	*Pestalotiopsis vaccinia*	*Kandelia candel*	China	Pestalamine A (**12**)	MCF-7, HeLa, and HepG2	40.3, 22.0, and 32.8 μM	[19]
5	*Phoma* sp. OUCMDZ-1847	Fruit sample of *Kandelia candel*	Wenchang, Hainan Province, China	Phomazines B (**13**), epicorazine A (**14**), epicorazine B (**15**), epicorazine C (**16**), exserohilone A (**17**)	HL-60, HCT-116, K562, MGC-803, and A549 Cells	In the range 0.05 to 8.5 μM	[20]
6	*Phomopsis* sp.		Zhanjiang, China	Cytochalasin H (**18**)	A549 cells	Arrested A549 cells at the G2/M phase, inhibited the migration ability of A549 cells in a dose-dependent manner	[21]
7	*Phomopsis* spp. xy21 and xy22	Leaves *Xylocarpus granatum*	Trang Province, Thailand	Phomopsichalasin G (**19**)	HCT-8, HCT-8/T, A549, MDA-MB-231, and A2780 cancer Cells	7.5, 8.6, 6.4, 3.4, and 7.1 μM	[22]
8	*Phomopsis longicolla* HL-2232	*Bruguiera sexangula* var. *rhynchopetala*		6-aminopurine-9-carboxylic acid Me ester (**20**), uridine (**21**)	B16F10, A549, HL-60 and MCF-7 Cells	14.9 and 8.6 μM	[23]

Table 1. *Cont.*

Sr. No.	Fungus	Host Plant(s)	Plant Part or Tissue Locality of Host Plants	Compounds Isolated	Cell Line	IC_{50}/EC_{50}/Inhibition	Refs.
				Compounds Produced by *Coelomycetes*			
9	*Phomopsis* sp. HNY29-2B	Branch of *Acanthus llicifolius*	South China Sea, Hainan province, China	Dicerandrol A (**22**)	MDA-MB-435, HCT-116, Calu-3 and Huh7 Cells	3.03, 2.64, 1.76 and 4.19 µM	[24]
				Dicerandrol B (**23**), Penexanthone A (**26**)	MDA-MB-435, HCT-116 and Calu-3	<10 µM	
				Dicerandrol C (**24**)	MDA-MB-435, HCT-116, Calu-3, MCF-10A Cells	44.10, 42.63, 36.52, and 33.05 µM	
				diacetyl phomoxanthone B (**25**)	MDA-MB-435, HCT-116, Calu-3, Huh7 Cells	14.40, 7.12, 4.14 and 29.20 µM	
10	*Phomopsis* sp. (ZH76)	*Excoecaria agallocha*	Dong Sai, South China Sea coast	3-*O*-(6-*O*-α-L-arabinopyranosyl)-β-D-glucopyranosyl-1,4-dimethoxyxanthone (**27**)	HEp-2 and HepG2	9 and 16 µM	[25]
11	*Diaporthe* sp. SCSIO 41011	*Rhizophora stylosa*	Sanya city, Hainan Province, China	Isochromophilone D (**28**)	786-O cells	8.9 µM	[26]
				epi-Isochromophilone II (**29**)	ACHN, OS-RC-2, and 786-O cells,	In the range of 3.0 to 4.4 µM, Sorafenib (3.4 to 7.0 µM)	
12	*Diaporthe phaseolorum* SKS019	Branches of *Acanthus ilicifolius*	Shankou, Guangxi province, China	5-deoxybostrycoidin (**30**)	MDA-MB-435 and NCI-H460 cancer cells	5.32 and 6.57 µM	[27]
				Fusaristatin A (**31**)	MDA-MB-435 cancer cells	8.15 µM	
13	*Phomosis* sp. A818	Foliage of *Kandelia candel*	Fujian Province, China	Mycoepoxydiene (**32**), deacetylmycoepoxydiene (**33**)	MDA-MB-435	7.85 and 14.61 µM,	[28]
14	*Phomosis* sp. A818	Foliage of *Kandelia candel*	Fujian Province, China	Mycoepoxydiene (**32**)	Suppress antigen-stimulated degranulation and cytokine production in mast cells and IgE-mediated passive cutaneous anaphylaxis in mice		[29]
15	*Sarocladium kiliense* HDN11-84	rhizosphere soil of *Thespesia populnea,*	Guangxi Province, China	Saroclazine B (**34**)	HeLa Cells	4.2 µM	[30]

Table 1. *Cont.*

Sr. No.	Fungus	Host Plant(s)	Plant Part or Tissue Locality of Host Plants	Compounds Isolated	Cell Line	IC_{50}/EC_{50}/Inhibition	Refs.
				Compounds Produced by *Ascomycetes*			
16	*Annulohypoxylon* sp.	*Rhizophora racemosa*	Cameroon	Daldinone I (**35**)	Ramos and Jurkat J16	6.6 and 14.1 μM, Potently blocks autophagy, a potential pro-survival pathway for cancer cells	[31]
17	*Eurotium rubrum*	*Suaeda salsa*	"BoHai" seaside, China	Rubrumol (**36**)	A549, MDA-MB-231, PANC-1 and HepG2	Cytotoxic	[32]
					Topo I	Relaxation activity The band backward shifting and trailing of rubrumol (36) was observed at 100, 50, 10, 5 and 1 μM	
18	*Eutypella* sp. 1–15	Soil of mangrove rhizosphere in Jimei, Fujian Province, China	Not reported	13-Hydroxy-3,8,7(11)-eudesmatrien-12,8-olide (**37**)	JEKO-1 and HepG2	8.4 and 28.5 μM	[33]
19	*Rhytidhysteron rufulum* AS21B	Leaves of *Azima sarmentosa*	Samutsakhon province, Thailand	Rhytidenones G (**38**), H (**39**)), deoxypreussomerin B (**40**), palmarumycin CP17 (**41**), 1-oxo-1,4-dihydronapthalene-4-spiro-20-naptho[400-hydroxy-100,800-de][10,30]-dioxine (**42**), preussomerin EG4 (**43**), rhytidenone E (**44**), rhytidenone F (**45**), palmarumycin C5 (**46**), and 4,8-dihydroxy-3,4-dihydronaphthalen- 1(2*H*)-one (**47**)	Ramos lymphoma	17.98, 0.018, 18.00, 33.1, 15, 82.9, 0.461, 0.048, 31.7 and 23.1 μM, (Ibrutinib 28.7 μM)	[34]
				Compounds (**38**), (**39**), (**44**), (**45**), and (**47**)	H1975 Cell	7.3, 0.252, 10.24, 1.17 and 50 μM (afatinib 1.97 μM)	

Table 1. *Cont.*

Sr. No.	Fungus	Host Plant(s)	Plant Part or Tissue Locality of Host Plants	Compounds Isolated	Cell Line	IC_{50}/EC_{50}/Inhibition	Refs.
				Compounds Produced by *Ascomycetes*			
20	*Lasiodiplodia* sp. 318#	*Excoecaria agallocha*	Guangdong Province, China	2,4-Dihydroxy-6-nonylbenzoate (**48**)	MMQ and GH3 Cells	5.2 and 13.0 µM	[35]
21	*Lasiodiplodia* sp. 318#	*Excoecaria agallocha*	Guangdong Province, China	Ethyl-2,4-dihydroxy-6-(80-hydroxynonyl)-benzoate (**49**)	MDA-MB-435, HepG2, HCT-116, A549 and leukaemia THP1 Cells	0.13, 12.50, 11.92, 13.31 and 39.74 µM	[36]
22	*Lasiodiplodia theobromae* ZJ-HQ1	*Acanthus ilicifolius*	Guangdong Province, China	Chloropreussomerins A and B (**50**, **51**) Preussomerin D (**56**)	A549 and MCF-7	In the range of 5.9–8.9 µM	[37]
				Preussomerin K (**52**), Preussomerin H (**53**), Preussomerin G (**54**), Preussomerin F (**55**),	A549, HepG2, MCF-7	In the range of 2.5–9.4 µM	
23	*Rhytidhysteron rufulum* BG2-Y	Leaves of *Bruguiera gymnorrhiza*	Pak Nam Pran, Prachuab Kiri Khan Province, Thailand	Rhytidchromone A (**57**), B (**58**), C (**59**), and E (**60**)	Kato-3 Cells	In the range of 16.0–23.3 µM	[38]
					MCF-7 cells	19.3–17.7 µM	
24	*Campylocarpon* sp. HDN13-307	Root of *Sonneratia caseolaris*	China	Campyridone D (**61**), and ilicicolin H (**62**)	HeLa	8.8 and 4.7 µM	[39]
25	*Stemphylium globuliferum*	*Avicennia marina*	Hurghada, Egypt	Dihydroaltersolanol C (**63**), Altersolanol A (**64**), Altersolanol B (**65**), Alterporriol E (**67**)	L5178Y	3.4, 2.53, 3.78 and 6.9 µM	[40–42]
				Altersolanol N (**66**)	L5178Y	Low micromolar range (% growth-1.4)	
26	*Paradictyoarthrinium diffractum* BCC 8704	Associated with mangrove wood	Laem SonNational Park, Ranong Province, Thailand	Paradictyoarthrins A (**68**)	KB, MCF-7, NCI-H187, Vero Cells, KB, MCF-7, NCI-H187, Vero Cells	In the range of 23–31 µg/mL	[46]
				Paradictyoarthrin B (**69**)	KB, MCF-7, NCI-H187, Vero Cells	3.1, 3.8, 9.5, and 5.6 µg/mL	
				Preussomerin C (**70**), ymf 1029C (**71**) and altenusin (**72**)	KB, MCF-7, NCI-H187, Vero Cells	Moderate to poor activity	
				ymf 1029C (**71**)	NCI-H187 cells	5.0 µg/mL	

Table 1. *Cont.*

Sr. No.	Fungus	Host Plant(s)	Plant Part or Tissue Locality of Host Plants	Compounds Isolated	Cell Line	IC_{50}/EC_{50}/Inhibition	Refs.
			Compounds Produced by *Ascomycetes*				
27	*Halorosellinia* sp. (No. 1403)	–	South China Sea	SZ-685C (**73**)	NFPA, MMQ and RPC cells	18.76, 14.51, and 56.09 μM	[47]
28	*Dothiorella* sp.	*Aegiceras corniculatum*	Fujian Province, China	Dothiorelone F (**74**), Dothiorelone G (**75**)	Raji cancer	2 μg/mL	[49]
29	*Rhytidhysteron* sp.	Leaves of *Azima sarmentosa*	Samutsakhon province, Thailand	Rhytidones B–C (**76, 77**), MK3018 (**78**), palmarumycin CR1 (**79**)	MCF-7 and CaSki Cells	In the range of 14.47 and 25.59 μM	[50]
				Rhytidones B (**76**)	CaSki	22.81 μM	
30	*Sporothrix* sp.	Bark, *Kandelia candel*	South China Sea	Sporothrin A (**80**)	Inhibition of AChE in vitro	1.05 μM	[51]
				sporothrin B (**81**), sporothrin C (**82**), diaporthin (**83**)	HepG2	20, 23, and 23 μg/mL	
31	*Eurotium rubrum*	Semi-mangrove plant *Hibiscus tiliaceus*	Hainan Island, China	12-demethyl-12-oxo-eurotechinulin B (**84**), 9-dehydroxyeurotinone (**85**), variecolorin G (**86**), alkaloid E-7 (**87**), and emodin (**88**)	HepG2, MCF-7, SW1990, HepG2, NCI-H460, SMMC7721, HeLa, and Du145	In the range of 15–30 μg/mL	[52]
32	*Bionectria ochroleuca*	Inner leaf tissues of the plant *Sonneratia caseolaris*	Hainan island, China	Pullularins E (**89**), F (**90**), pullularins A (**91**), and C (**92**)	L5178Y	EC_{50} values In the range of 0.1 and 6.7 μg/mL	[53]
				verticillin D (**93**)	L5178Y	<0.1 μg/mL	
33	*Aspergillus* sp. HN15-5D	Leaves, *Acanthus ilicifolius*	Hainan Island, China	Aspergisocoumrins A–B (**94–95**)	MDA-MB-435	5.08 and 4.98 μM	[54]
34	*Aspergillus niger* MA-132	*Avicennia marina*	Hainan, China	Nigerasterol A (**96**) Nigerasterol B (**97**)	HL60	0.30 μM, 1.50 μM	[55]
				Nigerasterol A (**96**) Nigerasterol B (**97**)	A549	1.82 and 5.41 μM	
35	*Aspergillus nidulans* MA-143	Leaves, *Rhizophora stylosa*		Aniquinazolines A–D (**98–101**)	Brine shrimp	LD_{50} 1.27, 2.11, 4.95 and 3.42 μM, (Colchicine LD_{50} 88.4 μM	[56]

Table 1. *Cont.*

Sr. No.	Fungus	Host Plant(s)	Plant Part or Tissue Locality of Host Plants	Compounds Isolated	Cell Line	IC_{50}/EC_{50}/Inhibition	Refs.
				Compounds Produced by *Ascomycetes*			
36	*Aspergillus terreus* (No. GX7-3B)	Branch of *Bruguiera gymnoihiza* (Linn.)	South China Sea	3β,5α-dihydroxy-(22*E*,24*R*)-ergosta-7,22-dien-6-one (**102**), Beauvericin (**104**)	MCF-7, A549, HeLa and KB	4.98 and 2.02, 1.95 and 0.82, 0.68 and 1.14, 1.50 and 1.10 μM	[57]
				3β,5α,14α-trihydroxy-(22*E*,24*R*)-ergosta-7, 22-dien-6-one (**103**)	MCF-7, A549, HeLa and KB	25.4, 27.1, 24.4, 19.4 μM	
37	*Aspergillus terreus* (No. GX7-3B)	Branch of *Bruguiera gymnoihiza* (Linn.)	South China Sea	Botryosphaerin F (**105**) and LL-Z1271β (**106**)	MCF-7 and HL-60	4.49 and 3.43 μM	[58]
					HL-60	0.6 μM	
38	*Penicillium* sp. J-54	Leaves, *Ceriops tagal*	Hainan province, China	Penicieudesmol B (**107**),	K-562	90.1 μM, (paclitaxel, 9.5 μM)	[59]
39	*Penicillium citrinum* HL-5126	*Bruguiera sexangula* var. *rhynchopetala*	South China Sea	Penibenzophenone B (**108**)	A549 Cells	15.7 μg/mL	[60]
40	*Penicillium* sp.	*Panax notoginseng*	Wenshan, Yunnan province, China	Brefeldin A (**109**), Brefeldin A 7-*O*-acetate (**110**)	293, HepG2, Huh7 and KB cell line	LD_{50} value from 0.024 to 0.62 μM. Both the compounds arrested HepG2 cells at the S phase	[61]
41	*Penicillium chrysogenum* V11	Vein of *Myoporum bontioides*	Leizhou Peninsula, China	Penochalasin K (**111**)	MDA-MB-435, SGC-7901 and A549 cells	<10 μM	[62]
42	*Penicillium janthinellum* HDN13-309	*Sonneratia caseolaris*	Hainan Province, China	Penicisulfuranols A–C (**112–114**)	HeLa and HL-60 Cells	In the range of 0.1 to 3.9 μM	[63]
43	*Penicillium chrysogenum* V11	Not reported	Not reported	Penochalasin I (**115**), chaetoglobosins A (**116**), and cytoglobosin C (**117**)	MDA-MB-435 and SGC-7901 cells	<10 μM	[64]
				Compounds (**116**), and (**117**)	SGC-7901 and A549 cells	<10 μM	
44	*Penicillium brocae* MA-231	Mangrove plant *Avicennia marina*	Hainan Island, China	Spirobrocazine C (**118**)	A2780	59 μM	[65]
				Brocazine G (**119**)	A2780 and A2780 CisR	664 nM, 661 nM (cisplatin 1.67 and 12.63 μM)	

Table 1. *Cont.*

Sr. No.	Fungus	Host Plant(s)	Plant Part or Tissue Locality of Host Plants	Compounds Isolated	Cell Line	IC_{50}/EC_{50}/Inhibition	Refs.
				Compounds Produced by *Ascomycetes*			
45	*Penicillium brocae* MA-231	Mangrove plant *Avicennia marina*	Hainan Island, China	Brocazines A (**120**), B (**121**), E (**122**), F (**123**)	Du145, Hela, HepG2, MCF-7, NCI-H460, SGC-7901, SW1990, SW480, and U251	from 0.89 to 9.0 μM	[66]
				Compounds (**120**) and (**121**)	SW480 tumor cell line	2.0 and 1.2 μM	
				Compound (**123**)	DU145 and NCI-H460 Cells,	1.7 and 0.89 μM	
46	*Penicillium chermesinum* strain HLit-ROR2	*Heritiera littoralis,*	Samut Sakhon province, Thailand	TMC-264 (**124**)	T47D and MDA-MB231	1.08 and 2.81 μM (doxorubicin 1.55 and 2.24 μM)	[67]
					HepG2	3.27 μM (Etoposide, 35.66 μM)	
					MOLT-3	1.36 μM	
					T47D	1.08 μM	
				PR-toxin (**125**)	HuCCA-1, HeLa, T47D, and MDA-MB231	0.81–2.19 μM (doxorubicin, 0.26–2.24 μM)	
					HL-60 cell line	0.06 μM (doxorubicin, 1.21 μM)	
					MOLT-3 and HL-60	0.09 μM, 0.06 μM	
47	*Penicillium chrysogenum* HND11-24	The rhizosphere soil of the mangrove plant *Acanthus ilicifolius*	China	Penicitols A (**126**)	HeLa, BEL-7402, HEK-293, HCT-116, and A549 Cells	4.6−10.5 μM	[68]
				Penicitols B (**127**)		3.4−9.6 μM	
				Penicitols C (**128**) and Penixanacid A (**129**)		In the range of 10–40.5 μM	
48	*Penicillium* sp. FJ-1	*Avicennia marina*	Fujian, China	Compound (**130**)	Tca8113 and MG-63 cells	26 and 35 μM (Taxol, 46 and 10 nM)	[69]
				Compounds (**131**)	Tca8113 and WRL-68	10 and 58 μM	
				Compounds (**131**)	MG-63 cells	55 nM	

Table 1. *Cont.*

Sr. No.	Fungus	Host Plant(s)	Plant Part or Tissue Locality of Host Plants	Compounds Isolated	Cell Line	IC_{50}/EC_{50}/Inhibition	Refs.
				Compounds Produced by *Ascomycetes*			
49	*Penicillium* sp. GD6,	*Bruguiera gymnorrhiza*	Zhanjiang, China	Meleagrin (**132**)	HL60 and A549	9.7 and 8.3 µM	[70]
50	*Penicillium* 303#	Sea water	Guangdong Province, China	5S, 7R, 9S, 10S, 11R, 12S, 13R, 22R, and 23R. (**133**), 7R, 9S, 10S, 11R, 12S, 13R, 22R, and 23R (**134**)	MDA-MB-435, HepG2, HCT-116, and A549	In the range of 11.9–37.82 µg/mL	[71]
				Compounds (**135**)	MDA-MB-435	7.13	
				Compound (**136**)	HepG2 and HCT-116	39.64 and 27.80 µM	
51	*Penicillium* sp. ZH58	Leaves, *Avicennia* sp.	Dong Sai, Hainan of the South China Sea coast	4-(methoxymethyl)-7-methoxy-6-methyl-1(3*H*)-isobenzofuranone (**137**)	KB and KBv200 cells	6 and 10 µg/mL	[72]
52	*Penicillium sumatrense* MA-92	Rhizosphere, *Lumnitzera racemose*	WenChang in Hainan Island, China	Sumalarins A–C (**138**, **139**, **140**), and dehydrocurvularin (**141**)	Du145, HeLa, Huh 7, MCF-7, NCI-H460, SGC-7901, and SW1990 Cells	In the range of 3.8 to 10 µM	[73]
53	*Penicillium* sp. ZH16	*Avicennia* sp.	South China Sea	5-methyl-8-(3-methylbut-2-enyl) furanocoumarin (**142**)	KB and KBv200	5 and 10 µg/mL	[74]
54	*Trichoderma* sp. 307	Stem bark, *Clerodendrum inerme*	Guangdong Province, China	(3*S*)-6-oxo-de-*O*-methyllasiodiplodin (**143**)	GH3 and MMQ Cells RPC	21.42 and 13.59 µM, 142.8 µM	[75]
	Co cultured with *Acinetobacter johnsonii* B2			(3*R*)-de-*O*-methyllasiodiplodin (**144**)		6.44 and 6.58 µM, 6.94 µM	
				(3*R*)-nordinone (**145**)		12.33 and 10.13 µM, 100.03 µM.	
55	*Trichoderma* sp. 307 co-culturing with *Acinetobacter johnsonii* B2	Stem bark of *Clerodendrum inerme*	Guangdong Province, China	Botryorhodine H (**146**)	MMQ GH3 Cells	3.09 and 3.64 µM	[76]
56	*Trichoderma* sp. Xy24	Leaves, stems and peels of *Xylocarpus granatum*	Hainan province, China	(9*R*,10*R*)-dihydro-harzianone (**147**)	HeLa and MCF-7 Cells	30.1 µM and 30.7 µM	[77]

Table 1. *Cont.*

Sr. No.	Fungus	Host Plant(s)	Plant Part or Tissue Locality of Host Plants	Compounds Isolated	Cell Line	IC_{50}/EC_{50}/Inhibition	Refs.
				Compounds Produced by *Ascomycetes*			
57	*Nigrospora* sp. MA75	*Pongamia pinnata*	Guangxi Zhuang Autonomous Region of China	2,3-didehydro-19α-hydroxy-14-epicochlioquinone B (**148**)	MCF-7, SW1990, and SMMC7721	4, 5, and 7 μg/mL	[78]
58	*Fusarium* sp. (No. DZ27)	Bark of *Kandelia candel*	Dongzhai mangrove, Hainan, China	Beauvericin (**104**)	KB and KBv200 cells	5.76 and 5.34 μM	[79]
59	*Fusarium* sp.	Leaf of mangrove *Kandelia candel*	Dongzhai Harbor of Hainan Island, China	Apicidin (**149**)	GLC-82 cells	6.94 μM	[80]
60	Unidentified fungus ZZF42	South China Sea	Not reported	Apicidin (**149**)	KB and KBv200	0.78 μg/mL	[81]
61	*Fusarium incarnatum* (HKI0504)	*Aegiceras corniculatum*	Not reported	2-acetyl-1,2,3,4-tetrahydro-β-carboline (**150**)	HUVEC and K-562	GI_{50} 41.1 and 33.3	[82]
					HeLa cell	CC_{50} 23.8 μM	
	Fusarium incarnatum (HKI0504)	*Aegiceras corniculatum*	Not reported	Fusamine (**151**)	HUVEC and K-562	GI_{50} 37.3 and 37.6	[82]
					HeLa cell	CC_{50} 23.3 μM	
	Fusarium incarnatum (HKI0504)	*Aegiceras corniculatum*	Not reported	3-(1-aminoethylidene)-6-methyl-2*H*-pyran-2,4(3*H*)- dione (**152**)	HUVEC and K-562	GI_{50} 41.1 and 33.3	[82]
					HeLa cell	CC_{50} 23.8 μM	
62	*Acremonium* sp.	leaves of *Sonneratia caseolaris* c	Dong Zhai Gang Mangrove Garden, Hainan, China	Torrubiellin B (**153**)	Cisplatin sensitive Cal27, Kyse510, HCC38, A2780, MDA-MB-231	In the range of 0.3 to 1.5 μM	[83]
					Cisplatin resistant, Cal27, Kyse510, HCC38, A2780, MDA-MB-231	In the range of 0.2 to 2.6 μM	
63	*Acremonium strictum*	*Rhizophora apiculata*	Island of Cat Ba, Vietnam	Waol A (**154**), Pestalotiopene A (**155**) Cytosporone E (**156**)	Cisplatin-sensitive, A2780	27.1, 76.2, and 8.3 μM	[84]
					Cisplatin-Resistant A2780	12.6, 30.1, and 19.0 μM	
64	Endophytic fungus J3	*Ceriops tagal*	Hainan province, China	3,4-seco-sonderianol (**157**)	K562, SGC-7901, and BEL-7402 Cells	9.2, 15.7, and 25.4 μg/mL	[85]
65	Endophytic fungus No. ZH-3	Not reported	South China Sea	2-(3-chloro-2, 6-dihydroxy-4-methylbenzoyl)-5-hydroxy-3-methoxybenzoate (**158**)	HepG2 cell line	25 μg/mL	[86]

Table 1. *Cont.*

Sr. No.	Fungus	Host Plant(s)	Plant Part or Tissue Locality of Host Plants	Compounds Isolated	Cell Line	IC_{50}/EC_{50}/Inhibition	Refs.
				Compounds Produced by *Ascomycetes*			
66	Endophytic fungus No. 5094	Not reported	South China Sea	Anthracene derivative (**159**)	KB and KBv200	LD_{50} values of 5.5 and 10.2 μM	[87]
67	Co-cultures of two mangrove endophytic fungi (strains Nos. 1924 and 3893)	Not reported		Marinamide (**160**)	HepG2, 95-D, MGC832 and HeLa Cells	7.0, 0.4, 91 nM and 0.529 μM	[88]
				Methyl marinamide (**161**)	HepG2, 95-D, MGC832 and HeLa Cells	2.52, 1.54 13, 0.110 μM	
68	Endophytic fungus No·Gx-3a	Not reported	South China sea	Ditryptophenaline (**162**)	KB, KBv200	8.0 and 12.0 μM	[89]
69	Mangrove endophytic fungus No·SK7RN3G1	Not reported	South China Sea	3,8-dihydroxy-6-methyl-9-oxo-9*H*-xanthene-1-carboxylate (**163**), Lichenxanthone (**164**),	HepG2 cell line	20 and 25 μg/mL	[90]
				Compounds Produced by *Basidiomycetes*			
70	Basidiomycetous fungus XG8D	leaves of *Xylocarpus granatum*	Samutsakorn province, Thailand	Merulinols C and D (**165**, **166**)	KATO-3 cells	35.0 and 25.3 μM	[91]
71	*Pseudolagarobasidium acaciicola*,	*Bruguiera gymnorrhiza*	Samut Sakhon province, Thailand	Compound (**167**)	HuCCA-1, A549, MOLT-3, HepG2, MDA-MB231, T47D	0.28–37.46 μM	[92]
					MRC-5	IC_{50} 17.92 μM	
					HL-60 cell line	IC_{50} 0.28 μM	
				Compound (**168**)	A549, MOLT-3, HepG2, HL-60, MDA-MB231, T47D, HeLa cancer cell, MRC-5	12.09–170.08 μM	
				Compound (**169**)	HuCCA-1, A549, MOLT-3, HepG2, HL-60, MDA-MB231, T47D, HeLa cancer cell	15.20–76.97 μM	
				Compound (**170**)	HuCCA-1, A549, MOLT-3, HepG2, MDA-MB231, T47D, HeLa cancer cell	18.31–154.51 μM	
					HL-60	18.31 μM	

Table 1. *Cont.*

Sr. No.	Fungus	Host Plant(s)	Plant Part or Tissue Locality of Host Plants	Compounds Isolated	Cell Line	IC_{50}/EC_{50}/Inhibition	Refs.
			Compounds Produced by *Basidiomycetes*				
72	*Pseudolagarobasidium acaciicola*	*Bruguiera gymnorrhiza*	Not reported	Endoperoxide (**171**), Steperoxide A (**172**)	MOLT-3, HuCCA-1, A549, HepG2, HL-60, MDA-MB-231, T47D, and HeLa cancer Cells	In the range of 0.68–3.71 and 0.67–5.25 μg/mL	[93]
				Merulin B (**173**)	MOLT-3, A549, HepG2, HL-60, MDA-MB-231 and T47D Cells	In the range of 11.94–49.08 μg/mL	
				Merulin C (**174**)	HL60 cancer cells	0.08 μg/mL	
					MOLT-3, HuCCA-1, A549, HepG2, MDA-MB-231, T47D, and HeLa Cells	In the range of 0.19–3.75 μg/mL	
73	*Mucor irregularis* QEN-189	*Rhizophora stylosa*	Hainan Island, China	Rhizovarins A, B, E (**175**, **176**, **177**) Penitrems A, C, F (**178**, **179**, **180**) and 3β-hydroxy-4β- desoxypaxilline (**181**)	A-549	11.5, 6.3, 9.2, 8.4, 8.0, 8.2 and 4.6 μM	[94]
				Rhizovarins A, B, (**175**, **176**), Penitrems A, C, F (**178**, **179**, **180**) and 3β-hydroxy-4β-desoxypaxilline (**181**)	HL-60	9.6, 5.0, 7.0, 4.7, 3.3 and 2.6 μM	

3. Methods Used for the Activation of Silent Biosynthetic Genes

Recent studies in the marine-based microorganisms have shown that these microorganisms are a rich source for novel bioactive compounds. Salinosporamide A (marizomib), a microbial compound isolated from marine *Salinispora* bacteria with proteasome inhibitory activity is expected to be a future anti-cancer drug, and is presently under clinical trials [95]. However, the reoccurrence of the same compound as discovered in terrestrial sources, in marine microorganisms often leads to serious issues. Advances in molecular biology have enhanced our understanding regarding how to exploit the genetic potential of bacteria and fungi to produce newer chemical entities apart from those that are currently known, which have yet to be explored [96,97]. It has been reported that under laboratory conditions, biosynthetic genes are not expressed as such, as only limited bioactive compounds are produced by these microbes. To overcome these limitations, different strategies have been proposed, including culturing promising strains in varying culture media and under a variety of culture conditions [98], mixing cultures of two or more microbe variants and epigenetic modifications that treat microbes with epigenetic modifiers such as histone deacetylase inhibitors or DNA methyl transferase to initiate the transcription of silent genes [99,100] to enhance the variation and diversity of the produced metabolites.

3.1. *The Co-Culture Strategy*

Microbes in natural ecosystem conditions always harbor and flourish in co-existence with a variety of microbes. Antagonism and competition for limited resources often lead to high competition among species, and microbes adopt various defense strategies, which favor the production of important bioactive secondary metabolites [101]. The co-culturing of two or more different microbes at the laboratory scale might mimic the ecological setting and induce the cascade of genes responsible for biosynthesis that are normally are masked under optimum culture parameters. Co-cultivation of two *Aspergillus* species derived from mangroves produced the new alkaloid aspergicin and the previously recognized compounds neoaspergillic acid and ergosterol, with antibacterial activity [102]. Li et al. [103] co-cultured two mangrove epiphytes and identified a novel xanthone derivative compound that showed antifungal activity. Two new alkaloids, marinamide, and marinamide methylether, were reported from mangrove-derived endophytic fungi with a cytotoxic effect when grown in mixed fermentation [88]. Pestalone, a chlorinated prenylsecoanthraquinone, was produced by the marine-derived fungus *Pestalotia* sp. when grown in the presence of the marine-derived bacterium *Thalassopia* sp., which belongs to the Gram-negative group. [104]. When *Libertella* sp., a marine-based fungi, were cultured in the presence of the bacteria *Thalassopia* sp., it resulted in the production of diterpenoid libertellenones of fungal origin [105]. In another set of studies, when the bacterium *Sphingomonas* sp. was grown in the presence of *Aspergillus fumigatus*, a novel compound glionitrin A, a diketopiperazine disulfide, was identified and appeared to show strong cytotoxicity against HCT-116, A549, AGS and DU145 cells [106]. These studies suggest that co-cultivation has tremendous potential to generate novel chemical entities from microbes when cultured under laboratory conditions.

3.2. *Epigenetic Modification*

The addition of epigenetic modifiers to fungi would allow us to induce cryptic fungal gene clusters. This technique can be applied to any fungal strain and does not require strain-dependent genetic manipulation. Williams et al. [107] reported that epigenetic modifiers could be rationally employed to access silent natural product pathways. Histone deacetylase (HDAC) or DNA methyltransferase (DMAT) are often used as epigenetic agents to change the transcription rate of some genes [108]. Henrikson et al. [12] reported the identification of nygerone A from *A. niger* when grown with suberoylanilide hydoxamic acid (SAHA). Wang et al. [109] reported induced metabolite generation in *Penicillium citreonigrum* when grown in the presence of methyl transferase inhibitor, 5-azacytidine (5-AZA). When *Hypoxylon* sp., an endophytic fungi, was treated with the epigenetic modifiers SAHA and AZA it enhanced the production of volatile organic compounds (VOCs) [110].

The marine endophytic fungus *Leucostoma persoonii* from *Rhizophora mangle* enhanced the production of cytosporones B, C, E and R in HDAC inhibited fermentation [111]. These studies provide evidence that the use of epigenetic modifiers modulate secondary metabolite production, resulting in different gene expressions.

4. Conclusions

Mangrove fungi are a ubiquitous source of novel bioactive metabolites with the potential to display anticancer properties. It is interesting to observe the chemical diversity in these metabolites, which include simple glycoside (**27**) and peptide molecules (pullularins E, **89**; F, **90** and apicidin, **149**) as well as complex stereospecific structures such as cytochalasin H (**18**), phomopsichalasin G (**19**), aniquinazolines A–D (**98–101**) and penitrem A, B and F (**178–180**). Chemical diversity plays an important role in the drug discovery pipeline, as this provides structurally diverse scaffolds that display similar activity via different modes and/or mechanisms of action. This phenomenon is also observed in mangrove fungal metabolites, as they show potent anticancer activity via different mechanisms of action such as apoptotic cell death (SZ-685C, **73**; beauvericin, **104**), the inhibition of kinase proteins involved in signal transduction pathways (Mycoepoxydiene, **32**; Altersolanol A, **64**; and the inhibition of topoisomerase I (**36**). Although many metabolites demonstrated moderate cytotoxic activities against cancer cell lines, only a few displayed superior activity than the standard anticancer drugs (**98–101**, **119**, **124**). It can be suggested that the rational derivatization of metabolites may provide molecules with better activity against a wide range of cancer cell lines. In addition, the identified metabolites with broad-spectrum anticancer activity need to be investigated to establish their mechanisms of action and to develop as novel anticancer therapeutics.

Author Contributions: The manuscript was critically evaluated by S.K.D., V.P., M.S.R. and M.K.G. The chemical structures were drawn by V.P. and M.K.G. and they assisted in the preparation of Table 1. The manuscript has been read and approved by all named authors.

Acknowledgments: The authors are thankful to Alok Adholeya, Sustainable Agriculture Division, The Energy and Resources Institute (India) for continuous support.

References

1. Gopal, B.; Chauhan, M. Biodiversity and its conservation in the *Sundarban mangrove* ecosystem. *Aquat. Sci.* **2006**, *68*, 338–354. [CrossRef]
2. Shearer, C.A.; Descals, E.; Kohlmeyer, B.; Kohlmeyer, J.; Marvanová, L.; Padgett, D.; Porter, D.; Raja, H.A.; Schmit, J.P.; Thorton, H.A.; et al. Fuangal diversity in aquatic habitats. *Biodivers. Conserv.* **2007**, *16*, 49–67. [CrossRef]
3. Sridhar, K.R. Mangrove fungi in India. *Curr. Sci.* **2004**, *86*, 1586–1587.
4. Sun, H.H.; Mao, W.J.; Chen, Y.; Guo, S.D.; Li, H.Y.; Qi, X.H.; Chen, Y.L.; Xu, J. Isolation, chemical characteristics and antioxidant properties of the polysaccharides from marine fungus *Penicillium* sp. F23-2. *Carbohydr. Polym.* **2009**, *78*, 117–124. [CrossRef]
5. Balagurunathan, R.; Radhakrishnan, M. Exploiting the less explored-microbial endophytes. *Adv. Biotechnol.* **2007**, *6*, 20–23.
6. Kharwar, R.N.; Mishra, A.; Gond, S.K.; Stierle, A.; Stierle, D. Anticancer compounds derived from fungal endophytes: Their importance and future challenges. *Nat. Prod. Rep.* **2011**, *28*, 1208–1228. [CrossRef] [PubMed]
7. Xu, J. Bioactive natural products derived from mangrove associated microbes. *RSC Adv.* **2015**, *5*, 841–892. [CrossRef]
8. Stierle, A.; Strobel, G.; Stierle, D. Taxol and taxane production by *Taxomyces andreanae*, an endophytic fungus of Pacific yew. *Science* **1993**, *260*, 214–216. [CrossRef] [PubMed]

9. Thatoi, H.; Behera, B.C.; Mishra, R.R. Ecological role and biotechnological potential of mangrove fungi: A review. *Mycology* **2013**, *4*, 54–71.
10. Marmann, A.; Aly, A.H.; Lin, W.; Wang, B.; Proksch, P. Co-cultivation—A powerful emerging tool for enhancing the chemical diversity of microorganisms. *Mar. Drugs* **2014**, *12*, 1043–1065. [CrossRef] [PubMed]
11. Fisch, K.M.; Gillaspy, A.F.; Gipson, M.; Henrikson, J.C.; Hoover, A.R.; Jackson, L.; Najar, F.Z.; Wagele, H.; Cichewicz, R.H. Chemical induction of silent pathway transcription in *Aspergillus niger*. *J. Ind. Microbiol. Biotechnol.* **2009**, *36*, 1199–1213. [CrossRef] [PubMed]
12. Henrikson, J.C.; Hoover, A.R.; Joyner, P.M.; Cichewicz, R.H. A chemical epigenetics approach for engineering the in situ biosynthesis of a cryptic natural product from *Aspergillus niger*. *Org. Biomol. Chem.* **2009**, *7*, 435–438. [CrossRef] [PubMed]
13. Strobel, G.A.; Long, D.M. Endophytic microbes embody pharmaceutical potential. *Am. Soc. Microbiol. News* **1998**, *64*, 263–268.
14. Xu, J.; Ebada, S.S.; Proksch, P. *Pestalotiopsis* a highly creative genus: Chemistry and bioactivity of secondary metabolites. *Fungal Divers.* **2010**, *44*, 15–31. [CrossRef]
15. Deshmukh, S.K.; Prakash, V.; Ranjan, N. Recent advances in the discovery of bioactive metabolites from *Pestalotiopsis*. *Phytochem. Rev.* **2017**, *16*, 883–920. [CrossRef]
16. Zhou, J.; Li, G.; Deng, Q.; Zheng, D.; Yang, X.; Xu, J. Cytotoxic constituents from the mangrove endophytic *Pestalotiopsis* sp. induce G0/G1 cell cycle arrest and apoptosis in human cancer cells. *Nat. Prod. Res.* **2017**, 1–5. [CrossRef] [PubMed]
17. Liu, S.; Dai, H.; Makhloufi, G.; Heering, C.; Janiak, C.; Hartmann, R.; Mándi, A.; Kurtán, T.; Müller, W.E.; Kassack, M.U.; et al. Cytotoxic 14-membered macrolides from a mangrove-derived endophytic fungus *Pestalotiopsis microspora*. *J. Nat. Prod.* **2016**, *79*, 2332–2340. [CrossRef] [PubMed]
18. Hemphill, C.F.P.; Daletos, G.; Liu, Z.; Lin, W.; Proksch, P. Polyketides from the mangrove-derived fungal endophyte *Pestalotiopsis clavispora*. *Tetrahedron Lett.* **2016**, *57*, 2078–2083. [CrossRef]
19. Zhou, X.; Lin, X.; Ma, W.; Fang, W.; Chen, Z.; Yang, B.; Liu, Y. A new aromatic amine from fungus *Pestalotiopsis vaccinia*. *Phytochem. Lett.* **2014**, *7*, 35–37. [CrossRef]
20. Kong, F.; Wang, Y.; Liu, P.; Dong, T.; Zhu, W. Thiodiketopiperazines from the marine-derived fungus *Phoma* sp. OUCMDZ-1847. *J. Nat. Prod.* **2014**, *77*, 132–137. [CrossRef] [PubMed]
21. Ma, Y.; Xiu, Z.; Liu, X.; Huang, B.; Hu, L.; Liu, J.; Zhou, Z.; Tang, X.; Wu, X. Cytochalasin H isolated from mangrove-derived endophytic fungus induces apoptosis and inhibits migration in lung cancer cells. *Oncol. Rep.* **2018**, *39*, 2899–2905. [CrossRef] [PubMed]
22. Luo, Y.F.; Zhang, M.; Dai, J.G.; Pedpradab, P.; Wang, W.J.; Wu, J. Cytochalasins from mangrove endophytic fungi *Phomopsis* sp. xy21 and xy22. *Phytochem. Lett.* **2016**, *17*, 162–166. [CrossRef]
23. Song, X.; Zhou, X.; Li, X.; Zheng, C.; Huang, G.; Yu, Z.; Song, X.; Chen, G. Secondary metabolites of a *Bruguiera sexangula* var. *Rhynchopetala*-derived fungus *Phomopsis longicolla* HL-2232. *Youji Huaxue* **2015**, *35*, 2102–2107.
24. Ding, B.; Yuan, J.; Huang, X.; Wen, W.; Zhu, X.; Liu, Y.; Li, H.; Lu, Y.; He, L.; Tan, H.; et al. New dimeric members of the phomoxanthone family: Phomolactonexanthones A, B and deacetylphomoxanthone C isolated from the fungus *Phomopsis* sp. *Mar. Drugs* **2013**, *11*, 4961–4972. [CrossRef] [PubMed]
25. Huang, Z.; Yang, J.; Lei, F.; She, Z.; Lin, Y. A new xanthone O-glycoside from the mangrove endophytic fungus *Phomopsis* sp. *Chem. Nat. Compd.* **2013**, *49*, 27–30. [CrossRef]
26. Luo, X.; Lin, X.; Tao, H.; Wang, J.; Li, J.; Yang, B.; Zhou, X.; Liu, Y. Isochromophilones A-F, cytotoxic chloroazaphilones from the marine mangrove endophytic fungus *Diaporthe* sp. SCSIO 41011. *J. Nat. Prod.* **2018**, *81*, 934–941. [CrossRef] [PubMed]
27. Cui, H.; Yu, J.; Chen, S.; Ding, M.; Huang, X.; Yuan, J.; She, Z. Alkaloids from the mangrove endophytic fungus *Diaporthe phaseolorum* SKS019. *Bioorg. Med. Chem. Lett.* **2017**, *27*, 803–807. [CrossRef] [PubMed]
28. Zhang, W.; Zhao, B.; Du, L.; Shen, Y. Cytotoxic polyketides with an oxygen-bridged cyclooctadiene core skeleton from the mangrove endophytic fungus *Phomosis* sp. A818. *Molecules* **2017**, *22*, 1547. [CrossRef] [PubMed]
29. Xia, X.C.; Chen, Q.; Liu, K.; Mo, P.L.; Zhu, J.W.; Zhuang, M.Q.; Shen, Y.M.; Yu, C.D. Mycoepoxydiene inhibits antigen-stimulated activation of mast cells and suppresses IgE-mediated anaphylaxis in mice. *Int. Immunopharmacol.* **2013**, *17*, 336–341. [CrossRef] [PubMed]

30. Li, F.; Guo, W.; Wu, L.; Zhu, T.; Gu, Q.; Li, D.; Che, Q. Saroclazines A-C, thio-diketopiperazines from mangrove-derived fungi *Sarocladium kiliense* HDN11-84. *Arch. Pharm. Res.* **2018**, *41*, 30–34. [CrossRef] [PubMed]
31. Liu, Y.; Stuhldreier, F.; Kurtan, T.; Mandi, A.; Arumugam, S.; Lin, W.; Stork, B.; Wesselborg, S.; Weber, H.; Henrich, B.; et al. Daldinone derivatives from the mangrove-derived endophytic fungus *Annulohypoxylon* sp. *RSC Adv.* **2017**, *7*, 5381–5393. [CrossRef]
32. Zhang, Y.; Jia, A.; Chen, H.; Wang, M.; Ding, G.; Sun, L.; Li, L.; Dai, M. Anthraquinones from the saline-alkali plant endophytic fungus *Eurotium rubrum*. *J. Antibiot.* **2017**, *70*, 1138–1141. [CrossRef] [PubMed]
33. Wang, Y.; Wang, Y.; Wu, A.A.; Zhang, L.; Hu, Z.; Huang, H.; Xu, Q.; Deng, X. New 12,8-Eudesmanolides from *Eutypella* sp. 1–15. *J. Antibiot.* **2017**, *70*, 1029–1032. [CrossRef] [PubMed]
34. Siridechakorn, I.; Yue, Z.; Mittraphab, Y.; Lei, X.; Pudhom, K. Identification of spirobisnaphthalene derivatives with anti-tumor activities from the endophytic fungus *Rhytidhysteron rufulum* AS21B. *Bioorg. Med. Chem.* **2017**, *25*, 2878–2882. [CrossRef] [PubMed]
35. Huang, J.; Xu, J.; Wang, Z.; Khan, D.; Niaz, S.I.; Zhu, Y.; Lin, Y.; Li, J.; Liu, L. New lasiodiplodins from mangrove endophytic fungus *Lasiodiplodia* sp. 318. *Nat. Prod. Res.* **2017**, *31*, 326–332. [CrossRef] [PubMed]
36. Li, J.; Xue, Y.; Yuan, J.; Lu, Y.; Zhu, X.; Lin, Y.; Liu, L. Lasiodiplodins from mangrove endophytic fungus *Lasiodiplodia* sp. 318#. *Nat. Prod. Res.* **2016**, *30*, 755–760. [PubMed]
37. Chen, S.; Chen, D.; Cai, R.; Cui, H.; Long, Y.; Lu, Y.; Li, C.; She, Z. Cytotoxic and antibacterial preussomerins from the mangrove endophytic fungus *Lasiodiplodia theobromae* ZJ-HQ1. *J. Nat. Prod.* **2016**, *79*, 2397–2402. [CrossRef] [PubMed]
38. Chokpaiboon, S.; Choodej, S.; Boonyuen, N.; Teerawatananond, T.; Pudhom, K. Highly oxygenated chromones from mangrove-derived endophytic fungus *Rhytidhysteron rufulum*. *Phytochemstry* **2016**, *122*, 172–177. [CrossRef] [PubMed]
39. Zhu, M.; Zhang, X.; Feng, H.; Che, Q.; Zhu, T.; Gu, Q.; Li, D. Campyridones A-D, pyridone alkaloids from a mangrove endophytic fungus *Campylocarpon* sp. HDN13-307. *Tetrahedron* **2016**, *72*, 5679–5683. [CrossRef]
40. Moussa, M.; Ebrahim, W.; El-Neketi, M.; Mandi, A.; Kurtan, T.; Hartmann, R.; Lin, W.; Liu, Z.; Proksch, P. Tetrahydroanthraquinone derivatives from the mangrove-derived endophytic fungus *Stemphylium globuliferum*. *Tetrahedron Lett.* **2016**, *57*, 4074–4078. [CrossRef]
41. Liu, Y.; Marmann, A.; Abdel-Aziz, M.S.; Wang, C.Y.; Müller, W.E.G.; Lin, W.H.; Mandi, A.; Kurtan, T.; Daletos, G.; Proksch, P. Tetrahydroanthraquinone derivatives from the endophytic fungus *Stemphylium globuliferum*. *Eur. J. Org. Chem.* **2015**, *2015*, 2646–2653. [CrossRef]
42. Debbab, A.; Aly, A.H.; Edrada-Ebel, R.; Wray, V.; Pretsch, A.; Pescitelli, G.; Kurtan, T.; Proksch, P. New anthracene derivatives–structure elucidation and antimicrobial activity. *Eur. J. Org. Chem.* **2012**, 1351–1359. [CrossRef]
43. Mishra, P.D.; Verekar, S.A.; Deshmukh, S.K.; Joshi, K.S.; Fiebig, H.H.; Kelter, G. Altersolanol A: A selective cytotoxic anthraquinone froma *Phomopsis* sp. *Lett. Appl. Microbiol.* **2015**, *60*, 387–391. [CrossRef] [PubMed]
44. Debbab, A.; Aly, A.H.; Edrada-Ebel, R.; Wray, V.; Muller, W.E.G.; Totzke, F.; Zirrgiebel, U.; Schachtele, C.; Kubbutat, M.H.G.; Lin, W.; et al. Bioactive metabolites from endophytic fungus *Stemphylium globuliferum* isolated from *Mentha pulegium*. *J. Nat. Prod.* **2009**, *72*, 626–631. [CrossRef] [PubMed]
45. Teiten, M.H.; Mack, F.; Debbab, A.; Aly, A.H.; Dicato, M.; Proksch, P.; Diederich, M. Anticancer effect of altersolanol A, a metabolite produced by the endophytic fungus *Stemphylium globuliferum*, mediated by its proapoptotic and anti-invasive potential via the inhibition of NF-kB activity. *Bioorg. Med. Chem.* **2013**, *21*, 3850–3858. [CrossRef] [PubMed]
46. Isaka, M.; Chinthanom, P.; Rachtawee, P.; Srichomthong, K.; Srikitikulchai, P.; Kongsaeree, P.; Prabpai, S. Cytotoxic hydroanthraquinones from the mangrove-derived fungus *Paradictyoarthrinium diffractum* BCC 8704. *J. Antibiot.* **2015**, *68*, 334–338. [CrossRef] [PubMed]
47. Wang, X.; Tan, T.; Mao, Z.G.; Lei, N.; Wang, Z.M.; Hu, B.; Chen, Z.Y.; She, Z.G.; Zhu, Y.H.; Wang, H.J. The marine metabolite SZ-685C induces apoptosis in primary human nonfunctioning pituitary adenoma cells by inhibition of the Akt pathway in vitro. *Mar. Drugs* **2015**, *13*, 1569–1580. [CrossRef] [PubMed]
48. Chen, C.H.; Xiao, W.W.; Jiang, X.B.; Wang, J.W.; Mao, Z.G.; Lei, N.; Fan, X.; Song, B.B.; Liao, C.X.; Wang, H.J.; et al. A novel marine drug, SZ-685C, induces apoptosis of MMQ pituitary tumor cells by downregulating miR-200c. *Curr. Med. Chem.* **2013**, *20*, 2145–2154. [CrossRef] [PubMed]

49. Du, X.P.; Su, W.J. Two new polyketides from mangrove endophytic fungus *Dothiorella* sp. *Chem. Nat. Compd.* **2014**, *50*, 214–216. [CrossRef]
50. Pudhom, K.; Teerawatananond, T.; Chookpaiboon, S. Spirobisnaphthalenes from the mangrove-derived fungus *Rhytidhysteron* sp. AS21B. *Mar. Drugs* **2014**, *12*, 1271–1280. [CrossRef] [PubMed]
51. Wen, L.; Wei, Q.; Chen, G.; Cai, J.; She, Z. Chemical constituents from the mangrove endophytic fungus *Sporothrix* sp. *Chem. Nat. Compd.* **2013**, *49*, 137–140. [CrossRef]
52. Yan, H.J.; Li, X.M.; Li, C.S.; Wang, B.G. Alkaloid and anthraquinone derivatives produced by the marine-derived endophytic fungus *Eurotium rubrum*. *Helv. Chim. Acta* **2012**, *95*, 163–168. [CrossRef]
53. Ebrahim, W.; Kjer, J.; El Amrani, M.; Wray, V.; Lin, W.; Ebel, R.; Lai, D.; Proksch, P. Pullularins E and F, two new peptides from the endophytic fungus *Bionectria ochroleuca* isolated from the mangrove plant *Sonneratia caseolaris*. *Mar. Drugs* **2012**, *10*, 1081–1091. [CrossRef] [PubMed]
54. Wu, Y.; Chen, S.; Liu, H.; Huang, X.; Liu, Y.; Tao, Y.; She, Z. Cytotoxic isocoumarin derivatives from the mangrove endophytic fungus *Aspergillus* sp. HN15-5D. *Arch. Pharm. Res.* **2018**. [CrossRef] [PubMed]
55. Liu, D.; Li, X.M.; Li, C.S.; Wang, B.G. Nigerasterols A and B, antiproliferative sterols from the mangrove-derived endophytic fungus *Aspergillus niger* MA-132. *Helv. Chim. Acta* **2013**, *96*, 1055–1061. [CrossRef]
56. An, C.Y.; Li, X.M.; Luo, H.; Li, C.S.; Wang, M.H.; Xu, G.M.; Wang, B.G. 4-Phenyl-3,4-dihydroquinolone derivatives from *Aspergillus nidulans* MA-143, an endophytic fungus isolated from the mangrove plant *Rhizophorastylosa*. *J. Nat. Prod.* **2013**, *76*, 1896–1901. [CrossRef] [PubMed]
57. Deng, C.M.; Liu, S.X.; Huang, C.H.; Pang, J.Y.; Lin, Y.C. Secondary metabolites of a mangrove endophytic fungus *Aspergillus terreus* (No. GX7-3B) from the South China Sea. *Mar. Drugs* **2013**, *11*, 2616–2624. [CrossRef] [PubMed]
58. Deng, C.; Huang, C.; Wu, Q.; Pang, J.; Lin, Y. A new sesquiterpene from the mangrove endophytic fungus *Aspergillus terreus* (No. GX7-3B). *Nat. Prod. Res.* **2013**, *27*, 1882–1887. [CrossRef] [PubMed]
59. Qiu, L.; Wang, P.; Liao, G.; Zeng, Y.; Cai, C.; Kong, F.; Guo, Z.; Dai, H.; Mei, W.; Qiu, L.; et al. New eudesmane-type sesquiterpenoids from the mangrove-derived endophytic fungus *Penicillium* sp. J-54. *Mar. Drugs* **2018**, *16*, 108. [PubMed]
60. Zheng, C.J.; Liao, H.X.; Mei, R.Q.; Huang, G.L.; Yang, L.J.; Zhou, X.M.; Shao, T.M.; Chen, G.Y.; Wang, C.Y. Two new benzophenones and one new natural amide alkaloid isolated from a mangrove-derived Fungus *Penicillium citrinum*. *Nat. Prod. Res.* **2018**. [CrossRef] [PubMed]
61. Xie, J.; Wu, Y.Y.; Zhang, T.Y.; Zhang, M.Y.; Zhu, W.W.; Gullen, E.A.; Wang, Z.J.; Cheng, Y.C.; Zhang, Y.X. New and bioactive natural products from an endophyte of *Panax notoginseng*. *RSC Adv.* **2017**, *7*, 38100–38109. [CrossRef]
62. Zhu, X.; Zhou, D.; Liang, F.; Wu, Z.; She, Z.; Li, C. Penochalasin K, a new unusual chaetoglobosin from the mangrove endophytic fungus *Penicillium chrysogenum* V11 and its effective semi-synthesis. *Fitoterapia* **2017**, *123*, 23–28. [CrossRef] [PubMed]
63. Zhu, M.; Yang, Z.; Feng, H.; Gan, Q.; Che, Q.; Zhu, T.; Gu, Q.; Han, B.; Li, D. Trichodermamides D-F, heterocyclic dipeptides with a highly functionalized 1,2-oxazadecaline core isolated from the endophytic fungus *Penicillium janthinellum* HDN13-309. *RSC Adv.* **2017**, *7*, 48019–48024. [CrossRef]
64. Huang, S.; Chen, H.; Li, W.; Zhu, X.; Ding, W.; Li, C. Bioactive chaetoglobosins from the mangrove endophytic fungus *Penicillium chrysogenum*. *Mar. Drugs* **2016**, *14*, 172. [CrossRef] [PubMed]
65. Meng, L.H.; Wang, C.Y.; Mandi, A.; Li, X.M.; Hu, X.Y.; Kassack, M.U.; Kurtan, T.; Wang, B.G. Three diketopiperazine alkaloids with spirocyclic skeletons and one bisthiodiketopiperazine derivative from the mangrove-derived endophytic fungus *Penicillium brocae* MA-231. *Org. Lett.* **2016**, *18*, 5304–5307. [CrossRef] [PubMed]
66. Meng, L.H.; Li, X.M.; Lv, C.T.; Huang, C.G.; Wang, B.G. Brocazines A-F, cytotoxic bisthiodiketopiperazine derivatives from *Penicillium brocae* MA-231, an endophytic fungus derived from the marine mangrove plant *Avicenniam arina*. *J. Nat. Prod.* **2014**, *77*, 1921–1927. [CrossRef] [PubMed]
67. Darsih, C.; Prachyawarakorn, V.; Wiyakrutta, S.; Mahidol, C.; Ruchirawat, S.; Kittakoop, P. Cytotoxic metabolites from the endophytic fungus *Penicillium chermesinum*: Discovery of a cysteine-targeted Michael acceptor as a pharmacophore for fragment-based drug discovery, bioconjugation and click reactions. *RSC Adv.* **2015**, *5*, 70595–70603. [CrossRef]

68. Guo, W.; Li, D.; Peng, J.; Zhu, T.; Gu, Q.; Li, D. Penicitols A-C and Penixanacid A from the mangrove-derived *Penicillium chrysogenum* HDN11-24. *J. Nat. Prod.* **2015**, *78*, 306–310. [CrossRef] [PubMed]
69. Zheng, C.; Chen, Y.; Jiang, L.L.; Shi, X.M. Antiproliferative metabolites from the endophytic fungus *Penicillium* sp. FJ-1 isolated from a mangrove *Avicennia marina*. *Phytochem. Lett.* **2014**, *10*, 272–275.
70. Zhou, Z.F.; Kurtan, T.; Yang, X.H.; Mandi, A.; Geng, M.Y.; Ye, B.P.; Taglialatela-Scafati, O.; Guo, Y.W. Penibruguieramine A, a novel pyrrolizidine alkaloid from the endophytic fungus *Penicillium* sp. GD6 associated with Chinese mangrove *Bruguiera gymnorrhiza*. *Org. Lett.* **2014**, *16*, 1390–1393. [CrossRef] [PubMed]
71. Li, J.; Yang, X.; Lin, Y.; Yuan, J.; Lu, Y.; Zhu, X.; Li, J.; Li, M.; Lin, Y.; He, J.; et al. Meroterpenes and azaphilones from marine mangrove endophytic fungus *Penicillium* 303#. *Fitoterapia* **2014**, *97*, 241–246. [PubMed]
72. Yang, J.; Huang, R.; Qiu, S.X.; She, Z.; Lin, Y. A new isobenzofuranone from the mangrove endophytic fungus *Penicillium* sp. (ZH58). *Nat. Prod. Res.* **2013**, *27*, 1902–1905. [CrossRef] [PubMed]
73. Meng, L.H.; Li, X.M.; Lv, C.T.; Li, C.S.; Xu, G.M.; Huang, C.G.; Wang, B.G. Sulfur-containing cytotoxic curvularin macrolides from *Penicillium sumatrense* MA-92, a fungus obtained from the rhizosphere of the mangrove *Lumnitzera racemose*. *J. Nat. Prod.* **2013**, *76*, 2145–2149. [CrossRef] [PubMed]
74. Huang, Z.; Yang, J.; Cai, X.; She, Z.; Lin, Y. A new furanocoumarin from the mangrove endophytic fungus *Penicillium* sp. ZH16. *Nat. Prod. Res.* **2012**, *26*, 1291–1295. [CrossRef] [PubMed]
75. Zhang, L.; Niaz, S.I.; Khan, D.; Wang, Z.; Zhu, Y.; Zhou, H.; Lin, Y.; Li, J.; Liu, L. Induction of diverse bioactive secondary metabolites from the mangrove endophytic fungus *Trichoderma* sp. (Strain 307) by Co-Cultivation with *Acinetobacter johnsonii* (Strain B2). *Mar. Drugs* **2017**, *15*, 35. [CrossRef] [PubMed]
76. Zhang, L.; Niaz, S.I.; Wang, Z.; Zhu, Y.; Lin, Y.; Li, J.; Liu, L. α-Glucosidase inhibitory and cytotoxic botryorhodines from mangrove endophytic fungus *Trichoderma* sp. 307. *Nat. Prod. Res.* **2017**. [CrossRef] [PubMed]
77. Zhang, M.; Liu, J.M.; Zhao, J.L.; Li, N.; Chen, R.D.; Xie, K.B.; Zhang, W.J.; Feng, K.P.; Yan, Z.; Wang, N.; et al. Two new diterpenoids from the endophytic fungus *Trichoderma* sp. Xy24 isolated from mangrove plant *Xylocarpus granatum*. *Chin. Chem. Lett.* **2016**, *27*, 957–960. [CrossRef]
78. Shang, Z.; Li, X.M.; Li, C.S.; Wang, B.G. Diverse secondary metabolites produced by marine-derived fungus *Nigrospora* sp. MA75 on various vulture media. *Chem. Biodivers.* **2012**, *9*, 1338–1348. [CrossRef] [PubMed]
79. Tao, Y.W.; Lin, Y.C.; She, Z.G.; Lin, M.T.; Chen, P.X.; Zhang, J.Y. Anticancer activity and mechanism investigation of Beauvericin isolated from secondary metabolites of the mangrove endophytic fungi. *Anticancer Agents Med. Chem.* **2015**, *15*, 258–266. [CrossRef] [PubMed]
80. Zhang, J.; Lai, Z.; Huang, W.; Ling, H.; Lin, M.; Tang, S.; Liu, Y.; Tao, Y. Apicidin inhibited proliferation and invasion and induced apoptosis via mitochondrial pathway in non-small cell lung cancer glc-82 cells. *Anticancer Agents Med. Chem.* **2017**, *17*, 1374–1382. [CrossRef] [PubMed]
81. Huang, Z.J.; Guo, Z.Y.; Yang, R.Y.; She, Z.G.; Lin, Y.C. Alkaloid metabolites of mangrove endophytic fungus ZZF42 from the South China Sea. *Zhong Yao Cai* **2007**, *30*, 939–941. [PubMed]
82. Ding, L.; Dahse, H.M.; Hertweck, C. Cytotoxic alkaloids from Fusarium incarnatum associated with the mangrove tree *Aegiceras corniculatum*. *J. Nat. Prod.* **2012**, *75*, 617–621. [CrossRef] [PubMed]
83. Hemphill, C.F.P.; Daletos, G.; Hamacher, A.; Kassack, M.U.; Lin, W.; Mandi, A.; Kurtan, T.; Proksch, P. Absolute configuration and antitumor activity of torrubiellin B. *Tetrahedron Lett.* **2015**, *56*, 4430–4433. [CrossRef]
84. Hammerschmidt, L.; Debbab, A.; Ngoc, T.D.; Wray, V.; Hemphil, C.P.; Lin, W.H.; Broetz-Oesterhelt, H.; Kassack, M.U.; Proksch, P.; Aly, A.H. Polyketides from the mangrove-derived endophytic fungus *Acremonium strictum*. *Tetrahedron Lett.* **2014**, *55*, 3463–3468. [CrossRef]
85. Zeng, Y.B.; Gu, H.G.; Zuo, W.J.; Zhang, L.L.; Bai, H.J.; Guo, Z.K.; Proksch, P.; Mei, W.L.; Dai, H.F. Two new sesquiterpenoids from endophytic fungus J3 isolated from Mangrove Plant *Ceriopstagal*. *Arch. Pharm. Res.* **2015**, *38*, 673–676. [CrossRef] [PubMed]
86. Yang, J.X.; Qiu, S.X.; She, Z.G.; Lin, Y.C. Metabolites of mangrove endophytic fungus ZH-3 from South China Sea. *Huagong Jishu Yu Kaifa* **2014**, *43*, 1–3.
87. Yang, J.X.; Qiu, S.X.; She, Z.G.; Lin, Y.C. Metabolites of mangrove endophytic fungus 5094 from the South China Sea. *Shizhen Guoyi Guoyao* **2013**, *24*, 1059–1061.
88. Zhu, F.; Chen, G.Y.; Wu, J.S.; Pan, J.H. Structure revision and cytotoxic activity of marinamide and its methyl ester, novel alkaloids produced by co-cultures of two marine-derived mangrove endophytic fungi. *Nat. Prod. Res.* **2013**, *27*, 1960–1964. [CrossRef] [PubMed]

89. Yang, J.X.; Qiu, S.X.; She, Z.G.; Lin, Y.C. Metabolites of mangrove endophytic fungus Gx-3a from the South China Sea. *Guangxi Kexue* **2013**, *20*, 168–170.
90. Yang, J.X.; Qiu, S.X.; She, Z.G.; Lin, Y.C. Metabolites of mangrove endophytic fungus SK7RN3G1 from South China Sea. *Zhongguo Shiyan Fangjixue Zazhi* **2012**, *18*, 95–98.
91. Choodej, S.; Teerawatananond, T.; Mitsunaga, T.; Pudhom, K. Chamigrane sesquiterpenes from a basidiomycetous endophytic fungus XG8D associated with Thai mangrove *Xylocarpus granatum*. *Mar. Drugs* **2016**, *14*, 132. [CrossRef] [PubMed]
92. Wibowo, M.; Prachyawarakorn, V.; Aree, T.; Mahidol, C.; Ruchirawat, S.; Kittakoop, P. Cytotoxic sesquiterpenes from the endophytic fungus *Pseudolagarobasidium acaciicola*. *Phytochemstry* **2016**, *122*, 126–138. [CrossRef] [PubMed]
93. Wibowo, M.; Prachyawarakorn, V.; Aree, T.; Wiyakrutta, S.; Mahidol, C.; Ruchirawat, S.; Kittakoop, P. Tricyclic and spirobicyclicnorsesquiterpenes from the endophytic fungus *Pseudolagarobasidium acaciicola*. *Eur. J. Org. Chem.* **2014**, *19*, 3976–3980. [CrossRef]
94. Gao, S.S.; Li, X.M.; Williams, K.; Proksch, P.; Ji, N.Y.; Wang, B.G. Rhizovarins A-F, indole-diterpenes from the mangrove-derived endophytic fungus *Mucor irregularis* QEN-189. *J. Nat. Prod.* **2016**, *79*, 2066–2074. [CrossRef] [PubMed]
95. Fenical, W.; Jensen, P.R.; Palladino, M.A.; Lam, K.S.; Lloyd, G.K.; Potts, B.C. Discovery and development of the anticancer agent salinosporamide A (NPI-0052). *Bioorg. Med. Chem.* **2009**, *17*, 2175–2180. [CrossRef] [PubMed]
96. Knight, V.; Sanglier, J.J.; DiTullio, D.; Braccili, S.; Bonner, P.; Waters, J.; Hughes, D.; Zhang, L. Diversifying microbial natural products for drug discovery. *Appl. Microbiol. Biotechnol.* **2003**, *62*, 446–458. [CrossRef] [PubMed]
97. Brakhage, A.A.; Schroeckh, V. Fungal secondary metabolites–strategies to activate silent gene clusters. *Fungal Genet. Biol.* **2011**, *48*, 15–22. [CrossRef] [PubMed]
98. Bode, H.B.; Bethe, B.; Höfs, R.; Zeeck, A. Big effects from small changes: Possible ways to explore nature's chemical diversity. *ChemBioChem* **2002**, *3*, 619–627. [CrossRef]
99. Cichewicz, R.H. Epigenome manipulation as a pathway to new natural product scaffolds and their congeners. *Nat. Prod. Rep.* **2010**, *27*, 11–22. [CrossRef] [PubMed]
100. Shwab, E.K.; Bok, J.W.; Tribus, M.; Galehr, J.; Graessle, S.; Keller, N.P. Histone deacetylase activity regulates chemical diversity in *Aspergillus*. *Eukaryot. Cell* **2007**, *6*, 1656–1664. [CrossRef] [PubMed]
101. Ola, A.R.B.; Thomy, D.; Lai, D.; Brötz-Oesterhelt, H.; Proksch, P. Inducing secondary metabolite production by the endophytic fungus *Fusarium tricinctum* through coculture with *Bacillus subtilis*. *J. Nat. Prod.* **2013**, *76*, 2094–2099. [CrossRef] [PubMed]
102. Zhu, F.; Chen, G.; Chen, X.; Huang, M.; Wan, X. Aspergicin, a new antibacterial alkaloid produced by mixed fermentation of two marine-derived mangrove epiphytic fungi. *Chem. Nat. Compd.* **2011**, *47*, 767–769. [CrossRef]
103. Li, C.; Zhang, J.; Shao, C.; Ding, W.; She, Z.; Lin, Y. A new xanthone derivative from the co-culture broth of two marine fungi (strain No. E33 and K38). *Chem. Nat. Compd.* **2011**, *47*, 382–384. [CrossRef]
104. Cueto, M.; Jensen, P.R.; Kauffman, C.; Fenical, W.; Lobkovsky, E.; Clardy, J. Pestalone, a new antibiotic produced by a marine fungus in response to bacterial challenge. *J. Nat. Prod.* **2001**, *64*, 1444–1446. [CrossRef] [PubMed]
105. Oh, D.C.; Jensen, P.R.; Kauffman, C.A.; Fenical, W. Libertellenones A–D: Induction of cytotoxic diterpenoid biosynthesis by marine microbial competition. *Bioorg. Med. Chem.* **2005**, *13*, 5267–5273. [CrossRef] [PubMed]
106. Park, H.B.; Kwon, H.C.; Lee, C.-H.; Yang, H.O. Glionitrin A, an antibiotic-antitumor metabolite derived from competitive interaction between abandoned mine microbes. *J. Nat. Prod.* **2009**, *72*, 248–252. [CrossRef] [PubMed]
107. Williams, R.B.; Henrikson, J.C.; Hoover, A.R.; Lee, A.E.; Cichewicz, R.H. Epigenetic remodeling of the fungal secondary metabolome. *Org. Biomol. Chem.* **2008**, *6*, 1895–1897. [CrossRef] [PubMed]
108. Asai, T.; Chung, Y.M.; Sakurai, H.; Ozeki, T.; Chang, F.R.; Yamashita, K.; Oshima, Y. Tenuipyrone, a novel skeletal polyketide from the entomopathogenic fungus, *Isaria tenuipes*, cultivated in the presence of epigenetic modifiers. *Org. Lett.* **2012**, *14*, 513–515. [CrossRef] [PubMed]

109. Wang, X.; Filho, J.G.S.; Hoover, A.R.; King, J.B.; Ellis, T.K.; Powell, D.R.; Cichewicz, R.H. Chemical epigenetics alters the secondary metabolite composition of guttate excreted by an atlantic-florest-soil-derived *Penicillium citreonigrum*. *J. Nat. Prod.* **2010**, *73*, 942–948. [CrossRef] [PubMed]
110. Ul-Hassan, S.R.; Strobel, G.A.; Booth, E.; Knighton, B.; Floerchinger, C.; Sear, J. Modulation of volatile organic compound formation in the mycodiesel-producing endophyte *Hypoxylon* sp. CI-4. *Microbiology* **2012**, *158*, 465–473. [CrossRef] [PubMed]
111. Beau, J.; Mahid, N.; Burda, W.N.; Harrington, L.; Shaw, L.N.; Mutka, T.; Kyle, D.E.; Barisic, B.; Olphen, A.; Baker, B.J. Epigenetic tailoring for the production of anti-infective cytosporones from the marine fungus *Leucostoma persoonii*. *Mar. Drugs* **2012**, *10*, 762–774. [CrossRef] [PubMed]

4

Restoring Waning Production of Volatile Organic Compounds in the Endophytic Fungus *Hypoxylon* sp. (BS15)

Yuemin Wang and James K. Harper *

Department of Chemistry, University of Central Florida, 4111 Libra Drive, Orlando, FL 32816, USA; Yueminwang@knights.ucf.edu

* Correspondence: James.Harper@ucf.edu

Abstract: Certain endophytic fungi belonging to the *Hypoxylon* genus have recently been found to produce volatile organic compounds (VOCs) that have potential relevance as hydrocarbon fuels. Here, a recently discovered *Hypoxylon* sp. (BS15) was demonstrated to also produce VOCs, but with diminished VOC production after an extended period of in vitro growth. Restoring VOC production was partially achieved by growing BS15 in growth media containing finely ground woody tissue from the original host plant (*Taxodium distichum*). In an effort to isolate VOC production modulators, extracts from this woody tissue were made by sequentially extracting with dichloromethane, methanol, and water. Both the dichloromethane and water extracts were found to modulate VOC production, while the methanol extract had no effect. Surprisingly, the woody tissue remaining after exhaustive extraction was also shown to act as a VOC production modulator when included in the growth media, with changes observed in the production of four compounds. This woody tissue also induced production of two compounds not observed in the original BS15 extract. Filter paper had the same modulating effect as exhaustively extracted woody tissue, suggesting the modulation was perhaps due to cellulose degradation products. Overall, this study demonstrated that VOC production in BS15 can be influenced by multiple compounds in the woody tissue rather than a single modulator.

Keywords: endophytic fungi; volatile organic compounds; *Hypoxylon*; production modulators

---

1. Introduction

Endophytes are microorganisms, usually fungi and bacteria, that live inside the host plant without showing signs of their presence or causing apparent disease symptoms. The relationship between endophytes and their host plants varies from symbiotic to pathogenic [1]. Typically, tropical areas and rainforests are presumed to have the greatest diversity and abundance of endophytes due to their vast plant diversity [2]. Investigations involving endophytes are of considerable interest, in part due to their production of a remarkable variety of natural products [3–6]. Although there has been significant focus on endophytes, they remain relatively understudied.

Recently some work has focused on fungi that produce volatile organic compounds (VOCs), including some with potential usefulness as fuels or antimicrobials. For the purposes of this manuscript, VOCs are defined as compounds having sufficient volatility to be separable/mobile on gas chromatography. Table 1 summarizes the presently known fungi that can produce VOCs. Currently, only fungi producing components similar to fossil fuels have been considered to have fuel potential. These compounds include branched alkanes and their derivatives, substituted cyclohexanes, benzenes, alkyl alcohols, aldehydes, and polycyclic aromatic hydrocarbons [7]. For instance, *Gliocladium roseum* produces more than 40 VOCs with fuel potential, such as pentyl, hexyl, heptyl, and octyl alcohols,

3,3,5-trimethyldecane, and other branched hydrocarbons [7]. Among all of the VOCs with fuel potential characterized and reported, 1,8-cineole is a compound of special interest because a 70/30 (*v*/*v*) mixture of petrol/1,8-cineole has performance characteristics similar to petrol with less carbon monoxide emissions [8–10]. In 2010, a *Hypoxylon* sp. designated CI-4 was reported as the first non-plant source to produce 1,8-cineole (hereinafter referred to as cineole). More recently, other *Hypoxylon* spp. have also been found to produce cineole [11,12].

Table 1. A list of fungi presently known to produce volatile organic compounds.

Species	Site Isolation	Extraction Method [a]	No. of VOCs Detected	Reference
Aspergillus fumigatus	New Jersey, USA	SPME	>10	[13]
A. niger	New Jersey, USA	SPME		[13]
A. tubingensis	New Jersey, USA	SPME		[13]
A. niger	Malaysia	LLE	>295	[14]
Fusarium armeniacum	New Jersey, USA	SPME	>10	[13]
F. graminearum	New Jersey, USA	SPME		[13]
F. oxysporum	New Jersey, USA	SPME		[13]
F. proliferatum	New Jersey, USA	SPME		[13]
F. culmorum	Belgium	SPME	>10	[15]
F. langsethiae	Russia	SPME	>40	[16]
F. sibiricum	Russia	SPME		[16]
F. poae	Russia	SPME		[16]
F. sporotrichioides	Russia	SPME		[16]
Metarhizium anisopliae	New Jersey, USA	SPME	>5	[13]
Mucor racemsus	New Jersey, USA	SPME	>10	[13]
Penicillium chrysogenum	New Jersey, USA	SPME	>10	[13]
P. citreonigrum	New Jersey, USA	SPME		[13]
P. commune	New Jersey, USA	SPME		[13]
P. corylophilum	New Jersey, USA	SPME	>10	[13]
P. crustosum	New Jersey, USA	SPME		[13]
P. glabrum	New Jersey, USA	SPME		[13]
P. pinophilum	New Jersey, USA	SPME		[13]
P. polonicum	New Jersey, USA	SPME		[13]
P. sclerotiorum	New Jersey, USA	SPME		[13]
P. steckii	New Jersey, USA	SPME		[13]
P. sumatrense	New Jersey, USA	SPME		[13]
Nodulisporium	Canary Islands, Ecuador, Thailand, Nicaragua, South Australia, Colombia, and Wetlands of Florida	SPME	>40	[17]
Muscodor albus	Honduras, Thailand, and Ecuador	SPME	>20	[18]
M. crispans	Bolivian Amazon basin	SPME	>15	[19]
M. kashayum	India	SPME	>20	[20]
M. strobelii	India	SPME	>14	[21]
M. darjeelingensis	India	SPME	>20	[22]
M. tigerii	India	SPME	>20	[23]

Table 1. *Cont.*

Species	Site Isolation	Extraction Method [a]	No. of VOCs Detected	Reference
M. suthepensis	Thailand	SPME	>25	[24]
M. musae	Thailand	SPME	>15	[24]
M. oryzae	Thailand	SPME	>15	[24]
M. equiseti	Thailand	SPME	>15	[24]
M. sutura	Colombia	SPME	>20	[25]
M. fengyangensis	China	SPME	>20	[26]
Myrothecium inunduatum	India	SPME	>30	[27]
Bionectria ochroleuca	India	LLE	>5	[28]
Ampelomyces	Japan	SPME	>5	[29]
Phoma	Japan	SPME	>5	[29]
Cladosporium	Japan	SPME	<5	[29]
Phomopsis	Ecuador	SPME	>10	[30]
Gliocladium roseum	Northern Patagonia	SPME	>40	[7]
Beauveria bassiana	Montana, USA	SPME	6	[31]
Ascocoryne sarcoides	Northern Patagonia, UK, Germany, France, Norway, and Canada	SPME	>100	[32]
A. cylichnium	Norway, Switzerland	SPME		[32]
A. solitaria	Netherlands	SPME		[32]
Schizophylum commune	Chile	SPME	10	[33]
Hypoxylon	Thailand, Spain	SPME	>15	[34]

[a] The abbreviations LLE and SPME denote liquid-liquid extraction and solid phase micro-extraction, respectively.

An unexpected challenge involving cineole production in CI-4 was the observation that the production gradually decreased over a period of months when the organism was removed from the plant host. This decrease suggested the presence of one of more cineole production modulator compounds in the host plant. Nigg et al. isolated and characterized a modulator in an endophytic *Nodulisporium* species, the imperfect stage of *Hypoxylon* [35]. This modulator was able to restore cineole biosynthesis, and it is likely that similar outcomes can be obtained in other endophytic fungi where production of valuable products decrease over time. In related work, Hassan et al. reported that the treatment of *Hypoxylon* sp. with known epigenetic modulators not only cause phenotypic changes, but also modifies the VOCs production and the bioactivity [34]. All of these prior studies provide new insight into why such a diverse range of VOCs are found in different isolates of *Hypoxylon* spp.

Recently, a *Hypoxylon* sp. not corresponding to any named species was isolated from a bald cypress tree (*Taxodium distichum*) near Orange City, Florida, USA. A complete phylogenetic characterization of this fungus, designated BS15, will be given elsewhere. BS15 produces a variety of VOCs with possible relevance as fuels or antimicrobials. As with other *Hypoxylon* spp., BS15 was found to exhibit a significant decrease in VOC production over time and, motivated by the work of Nigg et al. [35], techniques for restoring VOC production were evaluated. This manuscript describes a process in which extracts from woody tissue of the plant from which BS15 was originally isolated were added to the growth media in an effort to restore VOC production. Serial extractions of the woody tissue were performed with dichloromethane (DCM), methanol, and water. In the following, we describe the changes from each extract and show that the DCM and water extracts, as well as the exhaustively extracted wood tissue, induce production of compounds. These changes were found to be inheritable, and three of the products are shown to differ from those originally produced by BS15, suggesting that the modifications represent epigenetic changes.

2. Materials and Methods

Dichloromethane (DCM) and methanol were purchased from Fisher Scientific (Salt Lake City, UT, USA). SiliaPrep C-18 columns were purchased from Silicycle (Quebec City, Canada). Potato dextrose broth (PDB) and agar were purchased from Microtech Scientific (Vista, USA). All reagents were used as received.

The *Hypoxylon* sp. BS15 was isolated from a bald cypress tree (*Taxodium distichum*) near the Saint Johns river near Orange City, Florida, USA. Initially, isolation of the fungus followed the procedures of Tomsheck et al. [11]. This involved treating branches with 70% ethanol, further sterilizing the wood in a flame, and then drying in a sterile laminar-flow hood. Outer tissue was cut away using a sterile knife blade and a square section of inner tissue was then placed on water agar. Any fungal hyphae growing out from the sample were transferred onto different plates of potato dextrose agar (PDA). One such sampling resulted in isolation of BS15.

Growth of BS15 samples without added modulators (i.e., extracts from *T. distichum*) was accomplished in PDB prepared by adding 2.4 g of potato dextrose broth to 100 mL purified water in a 500 mL Erlenmeyer flask. The flask was sealed with aluminum foil and autoclaved for 15 min to sterilize. A culture of BS15 growing on PDA was then added to the sterile broth, and it was resealed with aluminum foil and left to grow for 30 days in the lab at room temperature without stirring. The resulting broth was then vacuum filtered twice with Whatman Grade 4 filter paper to remove all particulates.

Extractions of woody tissue involved taking approximately 210 g of shredded cypress wood and extracting with 400 mL DCM for 90 min with stirring. This process was repeated two times. The DCM was then removed on a rotatory evaporator (Buchi Rotavapor R-205, Buchi, New Castle, DE, USA) at reduced pressure. The cypress wood was further extracted with methanol, and finally with water, using 400 mL of each solvent. In order to simulate the extractable components in a wood matrix, all three fractions were dripped onto filter papers, and the filter paper was allowed to completely dry. This process was repeated several times until the filter paper contained the desired mass of the extracted components. These filter papers were cut into pieces and added to 100 mL of the PDB medium in a 500 mL Erlenmeyer flask, and this flask was inoculated with BS15. As a control, filter paper containing no extractables was added to the PDB/BS15 media. As a final test, the exhaustively extracted cypress wood was also added to PDB. All flasks were then sealed with aluminum foil and autoclaved for 15 min. After cooling, BS15 was added to the sterile broth, and the resulting solution was allowed to grow for 30 days at room temperature without stirring.

Media used to evaluate the influence of the DCM, methanol, and water extracts on BS15 were prepared by removing a sample of the fungi grown in a solution containing PDB plus *T. distichum* extracts after 30 days of growth and transferring it onto a petri dish. Serial weekly transfers onto PDA were then performed over a period of 4 weeks to ensure that all changes in VOC production ultimately observed were epigenetic changes, and that exogenous contaminants from the extractable components were rigorously removed.

Isolation of VOCs involved solid phase extraction of the growth media on a C-18 stationary phase (500 mg). First, a C-18 cartridge (particle size: 40–63 μm) was washed with 5 mL of methanol, and then with 5 mL of water (three times). A total of 100 mL of filtered fungal broth was then passed through the column under vacuum. The column was washed with 15 mL of water to remove polar components (e.g., salts), and the column was dried by drawing air through the column for 30 min. The column was then eluted by passing 1.5 mL of methanol through the column to yield a clear brown solution. The eluent was filtered using a 0.22 μm syringe filter prior to gas chromatography/mass spectrometry (GC/MS) analysis. This solid-phase extraction methodology differs from the solid-phase microextraction (SPME) methodology usually employed when evaluating fungal VOCs. The methodology was employed in order to more efficiently retain compounds having low vapor pressure, and which may be missed by SPME.

The GC/MS method used was similar to that of Strobel et al. [36]. The instrument used was a Finnigan TraceGC Ultra with Trace DSQ detector (Thermo Scientific) and a Restek Rtx-225 capillary column (cyanopropyl-methyl/phenyl-methyl polysiloxane, 50/50, 30 m × 0.25 mm, film thickness 0.25 μm) (Restek, Bellefonte, PA, USA). The carrier gas was ultra-high purity helium with a 1.5 cm^3/min constant flow rate and an initial column head pressure of 77 kPa. The injector was set to 250 °C with a 1 μL injection volume using splitless injection mode. The column oven temperature was initially 45 °C and held for one minute, followed by a 10 °C/min ramp to 100 °C, where the temperature was held for 5 min. Finally, the temperature was increased by 5 °C/min to 200 °C and held for 7 min. The detector was set at 280 °C and set to scan 50–650 m/z. Data acquisition and processing were performed on Xcalibur software. Identification of compounds was made via library comparison using National Institute of Standards and Technology (NIST) database. In all the GC/MS analyses describe herein, quantities of individual compounds detected are not reported because many of the compounds are unknowns. This ambiguity prevents the construction of calibration curves required for quantitation.

3. Results and Discussion

The endophytic fungus BS15 was selected for study based on the observation that some of the compounds produced had a distinctive odor, which indicated production of volatile compounds. A GC/MS analysis of the original BS15 revealed a number of VOCs (Table 2). Unfortunately, the production of several compounds decreased with time in the absence of the host plant (Figure 1). The nominal masses of all compounds were obtained and five compounds were tentatively identified. A more complete characterization of the compounds present will be given elsewhere. When BS15 with diminished VOC production was transferred back to PDA containing woody tissue from the host plant (i.e., finely ground *Taxodium distichum* tissue), production of most VOCs was restored, albeit to varying degrees (Figure 1, top plot). The ability to restore VOC production in BS15 suggested the presence of a modulator compound or multiple modulators in the host plant.

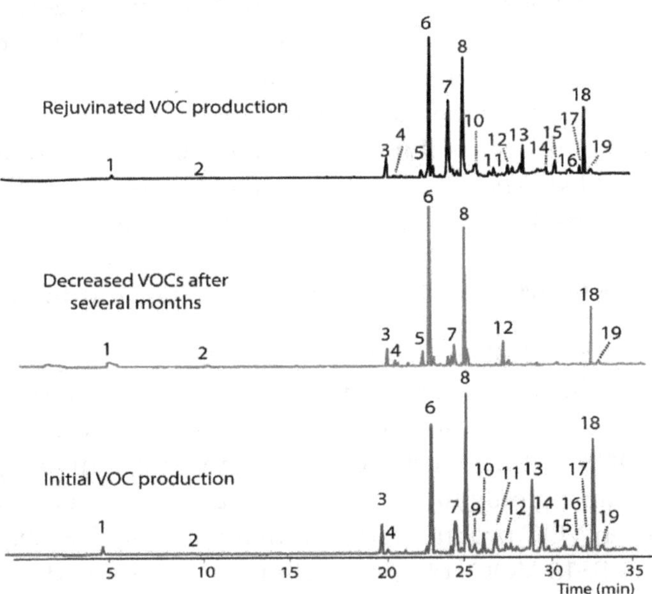

Figure 1. A gas chromatogram illustrating VOCs produced by BS15 showing the original production of VOCs immediately after isolation of BS15 (**bottom**) and decreased production after growing in the lab for several months (**middle**). Nominal mass of each peak and tentative identities are listed in Table 2. Production of VOCs was restored to varying extents (**top**) by growing BS15 on PDA containing finely ground woody tissue from the *Taxodium distichum* the fungus was originally isolated from.

Table 2. A GC/MS analysis of the VOCs produced by BS15.

Peak	R.T. (min)	Tentative Identify	Mol. Mass
1	4.75	unknown	70
2 [a]	9.41	1,8-cineole	154
3	19.15	unknown	142
4	19.78	unknown	120
5	21.84	unknown	126
6 [a]	22.35	Phenyl ethyl alcohol	122
7 [a]	23.85	2,3-naphthalenediamine	158
8	24.51	unknown	182
9	24.75	Unknown	184
10	25.87	unknown	220
11 [a]	26.58	Phenylacetic acid	136
12	27.62	Unknown	298
13 [a]	28.72	Diethyl phthalate	222
14	29.52	unknown	297
15	30.72	unknown	213
16	31.47	unknown	334
17	32.15	unknown	213
18	32.63	unknown	192
19	33.27	unknown	314

[a] Assignment confidence for peaks **2**, **6**, **7**, **11**, and **13** are, respectively, 86%, 89.3%, 82.7%, 76.4%, and 94%. No other peaks correspond to compounds in the NIST database.

4. Decreased VOC Production in BS15 after Extended In Vitro Growth

In order to investigate modulators from the host plant that restores production of VOCs in BS15, serial organic solvent extraction of woody tissue was made using DCM followed by methanol and then water. Each extract was then tested for its ability to restore VOC production in BS15. Since some of the extracted compounds were insoluble in the growth media, the extracted solutions were dripped onto filter paper and then air dried. Filter paper was employed to simulate the woody matrix of the original tissue. This process was repeated until the desired mass of extract had been loaded onto the filter paper (see Section 2). Growth media (PDB) was then prepared, and the filter paper impregnated with extractable compounds was included in the media. The filter paper was cut into strips of approximately $1'' \times \frac{1}{4}''$ to give a uniform distribution in solution. In each case, a control was also prepared containing filter paper with no extract added. The exhaustively extracted wood was also evaluated by including it in the growth media. The impact of each extract on production of volatiles is discussed below.

5. Assessing the Influence of DCM Extract/Filter Paper on VOC Production

A culture of BS15 grown in a PBD medium containing DCM extract/filter paper was found to alter the VOCs produced by inducing the production of three new compounds. Specifically, the peaks labeled **20**, **21**, and **22** in Figure 2, with respective nominal masses of 112, 216, and 154, were observed only after addition of the DCM extract, and thus appear to represent an epigenetic change to BS15. Surprisingly, the control containing only filter paper also induced production of compounds **21** and **22**. In both cases, these changes in VOC production were inheritable, and persist over several generations. Indeed, Figure 2 represents BS15 VOCs obtained from tissue removed from the DCM/filter paper media, then plated onto PDA, followed by weekly transfers onto PDA for one month, and finally regrown in PDB. In other words, Figure 2 represents a BS15 culture that was three generations removed from the initial DCM/filter paper treatment. Careful inspection of the chromatograms showed that the DCM extracts also increased production of peak **18** and decreased production of **6**. Overall, it appeared that DCM contained a modulator that altered production of peaks **6** and **18**, and created the ability to produce **20**. Remarkably, the filter paper appeared to be solely responsible for the production of compounds **21** and **22**, as discussed below. Isolation of individual modulator compounds from the DCM extract was not performed due to insufficient mass of DCM extract.

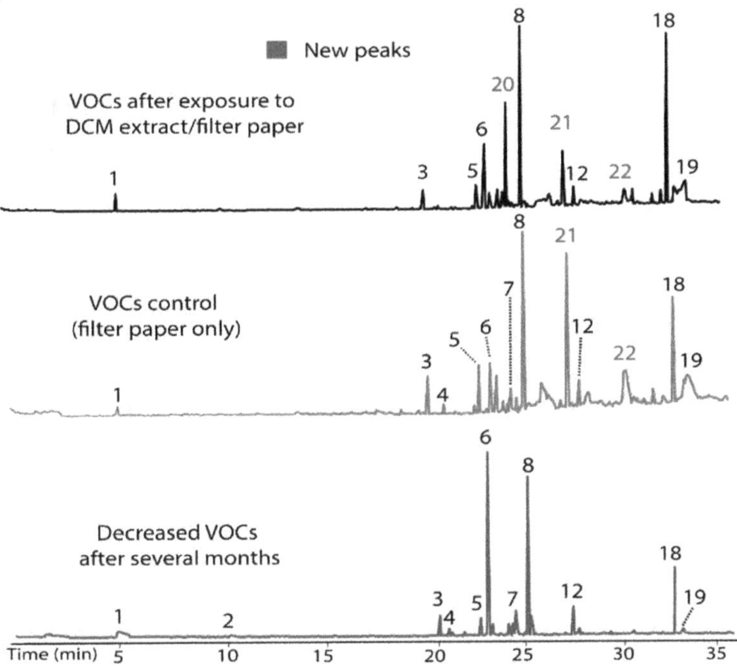

Figure 2. Gas chromatograms showing decreased VOCs production by BS15 after growing in vitro for several months (**bottom**). Adding BS15 to growth media (PDB) containing DCM extract/filter paper induced production of new compounds **20**, **21**, and **22** (**top**), and altered the production of compounds **6** and **18**. The **middle** plot demonstrates that filter paper containing no DCM extract also induces production of compounds **21** and **22** when included in growth media Nominal mass of each peak and their tentative identities are listed on Table 3.

Table 3. Compounds detected by GC/MS from BS15 modified by treatment with DCM extract/ filter paper.

Peak	R.T. (min)	Tentative Identify	Mol. Mass
1	4.75	Unknown	70
2	9.41	1,8-Cineole	154
3	19.15	Unknown	142
4	19.78	Unknown	120
5	21.84	Unknown	126
6	22.35	Phenyl ethyl alcohol	122
7	23.85	2,3-Naphthalenediamine	158
8	24.51	Unknown	182
9	24.75	Unknown	184
10	25.87	Unknown	220
11	26.58	Phenylacetic acid	136
12	27.62	Unknown	298
13	28.72	Diethyl phthalate	222
14	29.52	Unknown	297
15	30.72	Unknown	213
16	31.47	Unknown	334
17	32.15	Unknown	213
18	32.63	Unknown	192
19	33.27	Unknown	314
20 [a]	23.76	3-methyl-2,5-furandione	112
21	26.85	Unknown	216
22 [a]	29.82	4,4′-thiobis-benzeneamine	154

[a] Assignment confidence for peaks **20** and **22** are, respectively, 67% and 71%.

6. Evaluating the Influence of Methanol and Water Extracts on VOC Production in BS15

The influence of both the methanol and water extracts from *T. distichum* on BS15 were also evaluated using the process described above for the DCM extract. The methanol extract/filter paper produced almost no change, with the exception that peaks **21** and **22** were again observed. Chromatograms illustrating VOC production before and after addition of methanol extract/filter paper are shown in Figure 3. In this case, the control containing only filter paper also induced production of **21** and **22**. The production of these compounds also occurred in the DCM extract and their occurrence is thus attributed to the filter paper as a VOC production modifier rather than any compounds extracted by methanol.

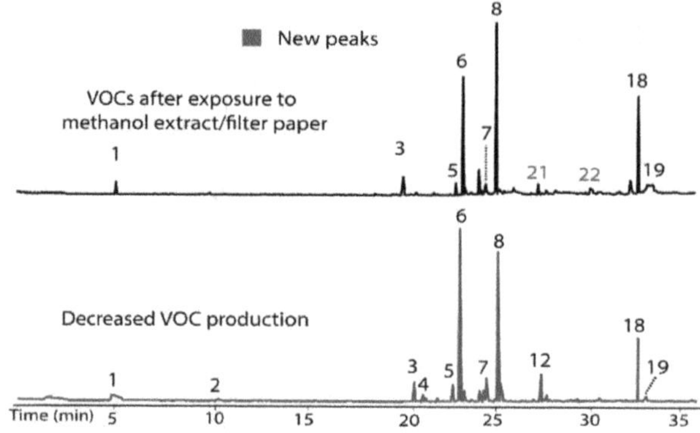

Figure 3. Chromatograms showing the negligible influence of the methanol extract/filter paper on VOC production in BS15. The new peaks (**21** and **22**) observed upon treatment (**top**) also occur in the control containing only filter paper. Their occurrence is therefore attributed to a change from the filter paper rather than the presence of VOC production modifiers extracted by methanol. The **bottom** plot illustrates BS15 after prolonged in vitro growth.

The water extract/filter paper was also evaluated (Figure 4) using the process described above. This extract increased production of compounds **7**, **12**, and **18**, while decreasing the quantity of **6**. The water extract thus likely contained a VOC production modulator. As in the other extracts, peaks **21** and **22** were again observed in both the extract/filter paper and in the control, strengthening the conclusion that filter paper induces their production. All changes from the water extract were inheritable.

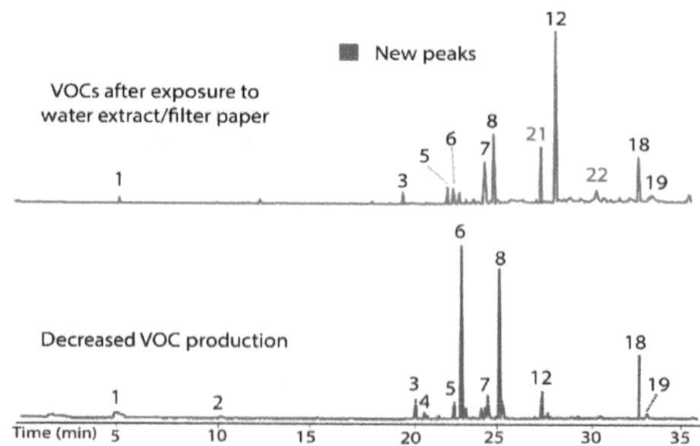

Figure 4. Chromatograms showing the influence of the water extract/filter paper of VOC production of BS15. Peaks **7**, **12**, and **18** increased upon exposure to the water extract while **6** decreased (**top**) versus BS15 with diminished VOC production (**bottom**). The water extract thus appeared to contain a modulator of VOC production. Peaks **21** and **22** were again observed to occur in both the water extract and in the control containing only filter paper.

7. The Influence of Exhaustively Extracted *T. distichum* Wood on VOC Production in BS15

As a final test of potential modulators of VOC production in BS15, the finely ground extracted woody tissue of *T. distichum* was evaluated. This unusual step was taken because the filter paper was repeatedly found to induce production of peaks **20** and **21** in prior extracts, and it was of interest to see if other cellulose contain materials could have the same effect. The extracted wood was observed to decrease production of peaks **7** and **12**, while it increased **5** and **18** (Figure 5). Of greatest interest was the observation that peaks **21** and **22** appeared as prominent peaks, supporting the contention that these peaks were induced by cellulose containing materials. A comparison of the woody tissue to the filter paper controls from each extract is illustrated in Figure 6. The remarkable similarity in the changes induced by filter paper and those induced by exhaustively extracted wood from *T. distichum* indicated that a common production modulator is involved.

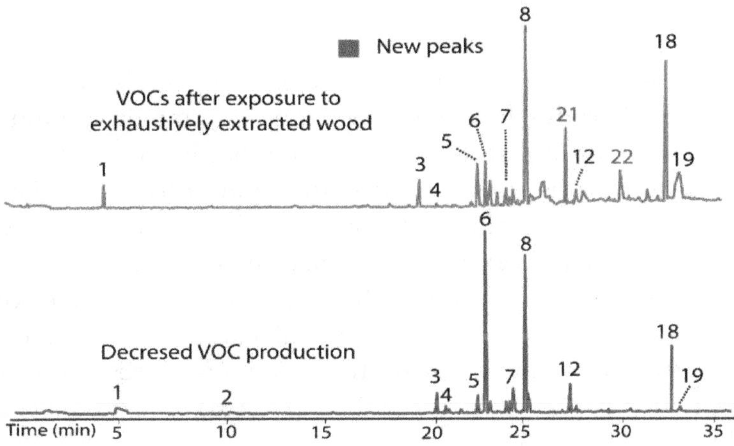

Figure 5. Chromatograms showing the influence on VOC production (**top**) of the addition of exhaustively extracted woody tissue of *T. distichum* to growth media (PDB). Peaks **5** and **18** increased, while **7** and **12** decreased versus BS15 after prolonged in vitro growth (**bottom**). Peaks **21** and **22** appeared as prominent components in the top plot, supporting the contention that these peaks were induced by cellulose containing materials.

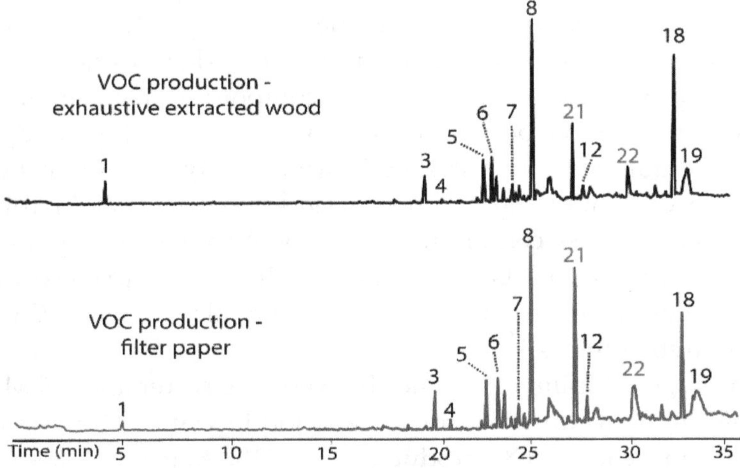

Figure 6. Chromatograms illustrating the changes in the production of VOCs by BS15 following exposure to filter paper (**bottom**) and to exhaustively extracted woody tissue from branches of *T. distichum* (**top**).

A direct comparison of each extract and its influence on VOC production is summarized in Table 4, and shows that the influence of DCM, water, and extracted wood (or equivalently, filter paper) clearly differed. This result indicated that multiple modulators were involved in VOC production as opposed to a single modulator.

Table 4. A summary of compounds influenced by different extracts of *T. disticum* [a].

Extract	Increased Production	Decreased Production	New Compounds Produced
DCM	18	6	20, 21, 22
Methanol	None	None	21, 22
Water	7, 12, 18	6	21, 22
Wood, extracted	5, 18	7, 12	21, 22
Filter paper	5, 18	7, 12	21, 22

[a] All analyses of VOC production were made on fungi from the fourth generation after initial exposure to the extract or woody tissue/filter paper.

An important question regarding the influence of modulators on VOC production is how long the fungus with compromised VOC production should be exposed to media containing ground plant tissue or extractable compounds in order to restore VOC production. In the case of BS15, we observed that continuous exposure to the woody tissue or extracts/filter paper for three generations was sufficient to restore production. For the purposes of this study, a transfer of the fungus was made each week to new media containing modulators (i.e., extracts). Thus, three generations correspond to three weeks. In all cases, the fungus was removed from the media containing extractables or wood and grown for a week to maturity before transfer to PDB to ensure that all exogenous contaminations were removed. Similarly, it was important to evaluate how long the fungus with restored production could grow in vitro before VOC production began to decrease. In BS15, a significant decrease in VOC production was observed after eight generations (8 weeks).

8. Conclusions

The research demonstrated that waning production of volatile organic compounds in an endophytic *Hypoxylon* sp. (BS15) can be partially restored by re-exposure to chemical constituents contained in the DCM and water extracts of the woody tissue of *T. distichum*. Surprisingly, the exhaustively extracted woody tissue also induced changes in VOC production from BS15 by causing production of two compounds never observed in the original growth media, and in altering production of four other compounds. Filter paper alone can also produce this change, suggesting that the differences are epigenetic changes, and that cellulose or its degradation products were the active component in altered VOC production rather than other compounds (e.g., lignans). The cellulose-based modification of production may have been caused by hydrolysis of the woody tissue or filter paper, which could create water soluble carbohydrates, and further study of this effect is needed. It is notable that carbohydrates are known to alter gene expression in some bacteria [37], thus there is a precedent for this change in BS15 from cellulose containing materials. An interesting parallel to this observation of activity in cellulose, a material that is essentially insoluble, is a prior study [35] reporting that breakdown products from lignin (i.e., ferulic acid and vanillin) also act as VOC production modulators in a closely related endophytic fungus.

It is notable that the large solubility difference between the water and DCM extracts suggests that the production modulator in DCM differed from that found in water. Thus, it is likely that more than one modulator effectively influenced VOC production in BS15. At present, insufficient amounts of the DCM and water extracts were available to allow isolation of individual compounds, and future work will focus on identifying compounds in these extracts involved in VOC modulation.

Solid-phase extraction was used in this study, rather than the more commonly employed approach of solid-phase microextraction, in order to more effectively include compounds having lower vapor pressures. One possible limitation to this approach is that compounds having high vapor pressures may be underrepresented. Further study is underway to directly compare these two techniques.

Author Contributions: Y.W. performed the growth of all BS15 cultures, performed all gas chromatography/mass spectrometry analyses, provided a comprehensive literature review of prior work on endophytic fungi producing volatile organic compounds, and assisted in manuscript preparation. The idea for this work was conceived of by J.K.H., and he wrote much of the manuscript.

Acknowledgments: Domenic Valenti is acknowledged for initially isolating the fungi BS15 from an area along the Saint Johns River and near Orange City, FL, USA.

References

1. Wilson, D. Endophyte: The evolution of a term, and clarification of its use and definition. *Oikos* **1995**, *73*, 274–276. [CrossRef]
2. Strobel, G.; Daisy, B. Bioprospecting for microbial endophytes and their natural products. *Microbiol. Mol. Biol. Rev.* **2003**, *67*, 491–502. [CrossRef] [PubMed]
3. Azevedo, J.L.; Maccheroni, W., Jr.; Pereira, J.O.; Araújo, W.L.D. Endophytic microorganisms: A review on insect control and recent advances on tropical plants. *Electron. J. Biotechnol.* **2000**, *3*, 40–65. [CrossRef]
4. Gao, F.K.; Dai, C.C.; Liu, X.Z. Mechanisms of fungal endophytes in plant protection against pathogens. *Afr. J. Microbiol. Res.* **2010**, *4*, 1346–1351.
5. Li, J.Y.; Harper, J.K.; Grant, D.M.; Tombe, B.O.; Bashyal, B.; Hess, W.M.; Strobel, G.A. Ambuic acid, a highly functionalized cyclohexenone with antifungal activity from *Pestalotiopsis* spp. and *Monochaetia* sp. *Phytochemistry* **2001**, *56*, 463–468. [CrossRef]
6. Castillo, U.; Harper, J.K.; Strobel, G.A.; Sears, J.; Alesi, K.; Ford, E.; Lin, J.; Hunter, M.; Maranta, M.; Ge, H.; et al. Kakadumycins, novel antibiotics from *Streptomyces* sp. NRRL 30566, an endophyte of *Grevillea pteridifolia*. *FEMS Microbiol. Lett.* **2003**, *224*, 183–190. [CrossRef]
7. Strobel, G.A.; Knighton, B.; Kluck, K.; Ren, Y.; Livinghouse, T.; Griffin, M.; Spakowicz, D.; Sears, J. The production of myco-diesel hydrocarbons and their derivatives by the endophytic fungus *Gliocladium roseum* (NRRL 50072). *Microbiology* **2008**, *154*, 3319–3328. [CrossRef] [PubMed]
8. Tamilvendhan, D.; Ilangovan, V.; Karthikeyan, R. Optimisation of engine operating parameters for eucalyptus oil mixed diesel fueled diesel engine using Taguchi method. *ARPN J. Eng. Appl. Sci.* **2011**, *6*, 14–22.
9. Tarabet, L.; Loubar, K.; Lounici, M.S.; Hanchi, S.; Tazerout, M. Eucalyptus biodiesel as an alternative to diesel fuel: Preparation and tests on diesel engine. *J. Biomed. Biotechnol.* **2012**, *2012*, 235485. [CrossRef] [PubMed]
10. Kazuo Sugito, K.S.T. Fuel Composition. U.S. Patent No. 4,297,109, 27 October 1981.
11. Tomsheck, A.R.; Strobel, G.A.; Booth, E.; Geary, B.; Spakowicz, D.; Knighton, B.; Floerchinger, C.; Sears, J.; Liarzi, O.; Ezra, D. *Hypoxylon* sp., an endophyte of *Persea indica*, producing 1,8-cineole and other bioactive volatiles with fuel potential. *Microb. Ecol.* **2010**, *60*, 903–914. [CrossRef] [PubMed]
12. Strobel, G. Methods of discovery and techniques to study endophytic fungi producing fuel-related hydrocarbons. *Nat. Prod. Rep.* **2014**, *31*, 259–272. [CrossRef] [PubMed]
13. Zhao, G.; Yin, G.; Inamdar, A.A.; Luo, J.; Zhang, N.; Yang, I.; Buckley, B.; Bennett, J.W. Volatile organic compounds emitted by filamentous fungi isolated from flooded homes after Hurricane Sandy show toxicity in a *Drosophila* bioassay. *Indoor Air* **2017**, *27*, 518–528. [CrossRef] [PubMed]
14. Siddiquee, S.; Azad, S.A.; Abu Bakar, F.; Naher, L.; Vijay Kumar, S. Separation and identification of hydrocarbons and other volatile compounds from cultures of *Aspergillus niger* by GC–MS using two different capillary columns and solvents. *J. Saudi Chem. Soc.* **2015**, *19*, 243–256. [CrossRef]
15. Fiers, M.; Lognay, G.; Fauconnier, M.-L.; Jijakli, M.H. Volatile compound-mediated interactions between barley and pathogenic fungi in the soil. *PLoS ONE* **2013**, *8*, e66805. [CrossRef] [PubMed]
16. Sav Savelieva, E.I.; Gustyleva, L.K.; Kessenikh, E.D.; Khlebnikova, N.S.; Leffingwell, J.; Gavrilova, O.P.; Gagkaeva, T.Y. Study of the vapor phase over *Fusarium* fungi cultured on various substrates. *Chem. Biodivers.* **2016**, *13*, 891–903. [CrossRef] [PubMed]
17. Strobel, G.A. Bioprospecting—Fuels from fungi. *Biotechnol. Lett.* **2015**, *37*, 973–982. [CrossRef] [PubMed]
18. Strobel, G. *Muscodor* species—Endophytes with biological promise. *Phytochem. Rev.* **2011**, *10*, 165–172. [CrossRef]

19. Mitchell, A.M.; Strobel, G.A.; Moore, E.; Robison, R.; Sears, J. Volatile antimicrobials from *Muscodor crispans*, a novel endophytic fungus. *Microbiology* **2010**, *156*, 270–277. [CrossRef] [PubMed]
20. Meshram, V.; Kapoor, N.; Saxena, S. *Muscodor kashayum* sp. nov.—A new volatile anti-microbial producing endophytic fungus. *Mycology* **2013**, *4*, 196–204. [CrossRef] [PubMed]
21. Meshram, V.; Saxena, S.; Kapoor, N. *Muscodor strobelii*, a new endophytic species from South India. *Mycotaxon* **2014**, *128*, 93–104. [CrossRef]
22. Saxena, S.; Meshram, V.; Kapoor, N. *Muscodor darjeelingensis*, a new endophytic fungus of *Cinnamomum camphora* collected from northeastern Himalayas. *Sydowia* **2014**, *66*, 55–67.
23. Saxena, S.; Meshram, V.; Kapoor, N. *Muscodor tigerii* sp. nov.-volatile antibiotic producing endophytic fungus from the Northeastern Himalayas. *Ann. Microbiol.* **2015**, *65*, 47–57. [CrossRef]
24. Suwannarach, N.; Kumla, J.; Bussaban, B.; Hyde, K.D.; Matsui, K.; Lumyong, S. Molecular and morphological evidence support four new species in the genus *Muscodor* from northern Thailand. *Ann. Microbiol.* **2013**, *63*, 1341–1351. [CrossRef]
25. Kudalkar, P.; Strobel, G.; Riyaz-Ul-Hassan, S.; Geary, B.; Sears, J. *Muscodor sutura*, a novel endophytic fungus with volatile antibiotic activities. *Mycoscience* **2012**, *53*, 319–325. [CrossRef]
26. Zhang, C.L.; Wang, G.P.; Mao, L.J.; Komon-Zelazowska, M.; Yuan, Z.L.; Lin, F.C.; Druzhinina, I.S.; Kubicek, C.P. *Muscodor fengyangensis* sp. nov. from southeast China: Morphology, physiology and production of volatile compounds. *Fungal Biol.* **2010**, *114*, 797–808. [CrossRef] [PubMed]
27. Banerjee, D.; Strobel, G.A.; Booth, E.; Geary, B.; Sears, J.; Spakowicz, D.; Busse, S. An endophytic *Myrothecium inundatum* producing volatile organic compounds. *Mycosphere* **2010**, *1*, 229–240.
28. Samaga, P.V.; Rai, V.R.; Rai, K.M.L. *Bionectria ochroleuca* NOTL33—An endophytic fungus from *Nothapodytes foetida* producing antimicrobial and free radical scavenging metabolites. *Ann. Microbiol.* **2014**, *64*, 275–285. [CrossRef]
29. Naznin, H.A.; Kiyohara, D.; Kimura, M.; Miyazawa, M.; Shimizu, M.; Hyakumachi, M. Systemic resistance induced by volatile organic compounds emitted by plant growth-promoting fungi in *Arabidopsis thaliana*. *PLoS ONE* **2014**, *9*, e86882. [CrossRef] [PubMed]
30. Singh, S.K.; Strobel, G.A.; Knighton, B.; Geary, B.; Sears, J.; Ezra, D. An endophytic *Phomopsis* sp. possessing bioactivity and fuel potential with its volatile organic compounds. *Microb. Ecol.* **2011**, *61*, 729–739. [CrossRef] [PubMed]
31. Crespo, R.; Pedrini, N.; Juárez, M.P.; Dal Bello, G.M. Volatile organic compounds released by the entomopathogenic fungus *Beauveria* bassiana. *Microbiol. Res.* **2008**, *163*, 148–151. [CrossRef] [PubMed]
32. Griffin, M.A.; Spakowicz, D.J.; Gianoulis, T.A.; Strobel, S.A. Volatile organic compound production by organisms in the genus *Ascocoryne* and a re-evaluation of myco-diesel production by NRRL 50072. *Microbiology* **2010**, *156*, 3814–3829. [CrossRef] [PubMed]
33. Strobel, G.; Singh, S.K.; Riyaz-Ul-Hassan, S.; Mitchell, A.M.; Geary, B.; Sears, J. An endophytic/pathogenic *Phoma* sp. from creosote bush producing biologically active volatile compounds having fuel potential. *FEMS Microbiol. Lett.* **2011**, *320*, 87–94. [CrossRef] [PubMed]
34. Ul-Hassan, S.R.; Strobel, G.A.; Booth, E.; Knighton, B.; Floerchinger, C.; Sears, J. Modulation of volatile organic compound formation in the Mycodiesel-producing endophyte *Hypoxylon* sp CI-4. *Microbiology* **2012**, *158*, 465–473. [CrossRef] [PubMed]
35. Nigg, J.; Strobel, G.; Knighton, W.B.; Hilmer, J.; Geary, B.; Riyaz-Ul-Hassan, S.; Harper, J.K.; Valenti, D.; Wang, Y. Functionalized para-substituted benzenes as 1,8-cineole production modulators in an endophytic *Nodulisporium* sp. *Microbiology* **2014**, *160*, 1772–1782. [CrossRef] [PubMed]
36. Strobel, G.A.; Dirkse, E.; Sears, J.; Markworth, C. Volatile antimicrobials from *Muscodor albus*, a novel endophytic fungus. *Microbiology* **2001**, *147*, 2943–2950. [CrossRef] [PubMed]
37. Chhabra, S.R.; Shockley, K.R.; Conners, S.B.; Scott, K.L.; Wolfinger, R.D.; Kelley, R.M. Carbohydrate-induced differential gene expression patterns in the hyperthermophilic bacterium *Thremotoga maritima*. *J. Biol. Chem.* **2003**, *278*, 7540–7552. [CrossRef] [PubMed]

5

From Concept to Commerce: Developing a Successful Fungal Endophyte Inoculant for Agricultural Crops

Brian R. Murphy [1,*], Fiona M. Doohan [2] and Trevor R. Hodkinson [1]

1 School of Natural Sciences & Trinity Centre for Biodiversity Research, Trinity College Dublin, The University of Dublin, College Green, Dublin 2, Ireland; hodkinst@tcd.ie
2 UCD School of Biology & Environmental Science and UCD Earth Institute, University College Dublin, Dublin 4, Ireland; fiona.doohan@ucd.ie
* Correspondence: murphb16@tcd.ie

Abstract: The development of endophyte inoculants for agricultural crops has been bedevilled by the twin problems of a lack of reliability and consistency, with a consequent lack of belief among end users in the efficacy of such treatments. We have developed a successful research pipeline for the production of a reliable, consistent and environmentally targeted fungal endophyte seed-delivered inoculant for barley cultivars. Our approach was developed de novo from an initial concept to source candidate endophyte inoculants from a wild relative of barley, *Hordeum murinum* (wall barley). A careful screening and selection procedure and extensive controlled environment testing of fungal endophyte strains, followed by multi-year field trials has resulted in the validation of an endophyte consortium suitable for barley crops grown on relatively dry sites. Our approach can be adapted for any crop or environment, provided that the set of first principles we have developed is followed. Here, we report how we developed the successful pipeline for the production of an economically viable fungal endophyte inoculant for barley cultivars.

Keywords: barley; commercialisation; crop wild relatives; endophytic fungi; yield enhancement

---

1. Introduction

The use of microorganisms to improve agricultural crop performance has a long history, but only in the last few decades have we been able to describe in any detail how bacteria, fungi, algae and protozoans interact with plants to enhance desired traits [1–3]. Mankind has understood the benefits of various practices, often without conscious intent, that directly or indirectly involve the manipulation of the soil and plant microbiota [4,5]. Only in the last century have we incorporated the use of microorganisms in agriculture in a deliberate, focused and knowledgeable way [6,7]. However, it is the plant that is the ultimate arbiter of how and when these microorganisms are incorporated into the functional plant microbiome, often in an unpredictable fashion [8,9]. It is the goal of all plant-microbiome research to understand and use these relationships in the most effective way, but we are still some way from obtaining this full understanding [10,11]. Such a lack of informed and validated knowledge has resulted in inconsistent results when using microorganisms in the field situation, resulting in a reduced end-user confidence and uptake of the technology [12,13].

Endophytes are a class of plant-associated microorganisms that have shown particular promise in agriculture [14–21]. Endophytes (bacteria, fungi and unicellular eukaryotes) live at least part of their life cycle inter- or intracellularly inside plants, usually without inducing pathogenic symptoms. This can include competent, facultative, obligate, opportunistic and passenger endophytes (a passenger endophyte enters the plant by accident in the absence of selective forces maintaining it in the internal tissue of the plant [22]). Endophytes can have several effects on plants and/or may change function during their life-cycle [14]. Bacterial and fungal endophytes have shown promise as beneficial crop inoculants, and many are known to enhance abiotic and biotic plant stress tolerance [23–28]. But a large proportion of the associated studies have been conducted in a controlled environment and do not translate successfully to the field [29].

The approach adopted by our research group has consciously focused on making this transition from 'pot to plot' more successful. We aimed to remove the inconsistency from endophyte application in field crops by targeting a single crop, barley, as our model plant, and testing the efficacy of endophyte strains under various environmental stresses, culminating in extensive field trials [30–35]. Only with this field validation can we say with any confidence that a microbial inoculant really "works" [36]. We have demonstrated the effectiveness of this approach by developing a fungal endophyte inoculum that consistently increases barley grain yield over several seasons under a variety of chemical fertilizer inputs in dry growing environments [31]. The outcomes from our research and field application have been so successful that we have been able to bring our endophyte technology into commerce, which will result in significant reductions in economic and environmental costs.

This article will review the development of this technology, outlining the concepts involved and the methods employed to create a viable, consistent and environmentally targeted commercial endophyte inoculant for barley crops. We will discuss a set of first principles that are important for success, and we will outline future research directions that will increase the portfolio of endophyte inoculants and expand the scope of target crop species.

2. Concept

Before any resources are allocated to an endophyte research project, it is important to understand the problem that is being addressed and the solution that is proposed. The problem must be a real commercial and/or functional need of the eventual user of the solution. Nearly half of all company start-ups fail because there was no market need for the product [37]. So it is essential that the proposed concept is going to provide a viable and needed solution. This solution should also attract potential commercial partners and collaborators on the project, thus increasing the resources available for successful implementation. For illustration, we focused on the need to reduce the high economic cost of chemical crop inputs; any product that successfully addresses this need would provide real and tangible benefits for the crop growers and also reduce environmental impact of the chemical inputs.

Crop wild relatives are an important genetic resource for breeding desired plant traits into related crop cultivars [38]. We suspected that the same was true for microbial resources, and particularly for fungal endophytes. Our hypothesis was that endophytes that were recoverable from crop wild relatives would be more compatible with the related crop, both for inoculation and colonization purposes and for *in planta* behaviour. We further hypothesized that endophytes recovered from a particular type of environment would be more compatible and effective on crops growing in a similar environment, and that endophyte crop wild relative interactions were more likely to have been maintained through time than endophyte-free elite crop cultivars.

It was important to retain this focus in all subsequent experiments so that we could produce a set of results that may or may not support our hypotheses. We were also aware that to produce a comprehensive and useful dataset, with the resources available, we would need to target only one important crop species. For our purposes, barley was the ideal crop as it is the most important arable crop grown in Ireland, where it is grown on over 200,000 ha, and there are several local wild barley relatives available for sampling.

The endophyte source we selected was the wild barley species *Hordeum murinum* ssp. *murinum* L. (wall barley), as it is relatively common and occurs in a wide variety of habitats on the east coast of Ireland, especially those subject to human disturbance. This enabled us to sample from diverse environments and to establish any significant correlations between endophyte recovery and alpha/beta diversity. It also gave us the largest possible set of endophyte strains from which to screen potential study taxa. The only other naturally occurring wild barley species in Ireland are *Hordeum jubatum* and *Hordeum secalinum*, and these only occur in rare and scattered populations [39].

By targeting endophytes recovered from natural and native plant populations we avoid introducing exotic microorganisms into the environment which may face regulatory and safety issues. They would also likely be more suited to local barley crop growing conditions, as they would have been selected by historical climatic conditions. Finally, as these host plants and geographical locations have never previously been sampled for fungal endophytes, we expected that the strains would be novel and previously uncharacterized, a real advantage when there is a need to protect intellectual property.

3. Developing the Concept

3.1. Endophyte Recovery

Having selected the barley wild relative to sample, it was important to ensure that we could recover the greatest variety of endophytes for subsequent biofertilisation and biocontrol screening in agricultural applications. Here, we followed a set of general principles derived from years of experience, and which we recommend for similar research.

The most important of these was the selection of the sampling sites, as this will determine the direction and success of all subsequent work:

- Sampling site should contain a minimum population of 10 genetically individual plants
- Plants on sampling site should be healthy and free of disease
- If selected for biotic tolerance, the site should preferably be in an area of potentially high disease pressure for the species
- There should be some form of abiotic stress present at the site, such as a very dry or nutrient-poor soil, which must be measureable
- The site should be relatively undisturbed and natural, i.e., not recently worked
- The site should not be exposed to contaminants, such as close to a busy roadside
- The site should preferably contain no alien plant species
- The host plant species should be identifiable using a recognised and reliable key
- The plants should be actively growing and not in a senescent phase
- The plants should be sampled when the site is in its 'normal' condition, i.e., not after any heavy rain, frost or unusual weather event, nor after any disturbance

While collecting the plant material it is vital to obtain as much environmental data as possible, as this information will be the major deciding factor when selecting individual strains for consortia application at the crop growing sites. For example, endophyte strains that are sourced from a particularly dry site may be more suitable for a relatively drought-stressed crop growing environment. Recording of environmental data should include variables such as GPS location, soil pH, soil moisture, soil nutrients, soil salinity, soil type, accompanying ground vegetation, tree cover, and exposure. We also recommend collecting a soil sample onto dry ice for later analysis of soil microbial community and elemental constituents. It is also important to ensure that plant sampling is aligned with principles outlined in The Nagoya Protocol on Access to Genetic Resources and the Fair and Equitable Sharing of Benefits Arising from their Utilization to the Convention on Biological Diversity, also known as the Nagoya Protocol on Access and Benefit Sharing (ABS) [40].

For our experimental work, we chose endophyte source sites that were particularly dry, with thin sandy soils and relatively high soil pH and salinity (Table 1). These sites were chosen as we ultimately wanted to test the effects of the endophytes on field barley crops growing in similar environments; so we had the experimental goal in mind at all times from initial concept to pre-planning to field sampling. This goal-oriented approach is at the heart of our methodology, and can be applied to the selection of endophytes for other environmental conditions.

Table 1. Soil variables for endophyte recovery sites.

Location	pH	Moisture Content %	Salinity *	Soil Type
1	7.8 ± 0.0	9.1 ± 0.0	1.18 ± 0.0	Loamy silt, stony
2	8.0 ± 0.0	13.4 ± 0.0	1.41 ± 0.0	Loamy silt, shallow
3	7.6 ± 0.0	0 ± 0.0	1.22 ± 0.0	Sandy silt, shallow
4	7.9 ± 0.0	4.4 ± 0.0	1.39 ± 0.0	Sandy silt, stony
5	7.9 ± 0.0	0 ± 0.0	1.45 ± 0.0	Sandy silt, shallow
6	7.7 ± 0.0	19.5 ± 0.0	1.26 ± 0.0	Clay loam, shallow
7	7.7 ± 0.0	0 ± 0.0	1.49 ± 0.0	Sandy silt, shallow
8	7.7 ± 0.0	0 ± 0.0	1.51 ± 0.0	Sandy silt, shallow
MEANS	7.8	8.3	1.37	

* Salinity is Osmotic Pressure in bars. pH, Moisture content % and Salinity are mean values ± standard error (n = 10).

The plant tissue that is chosen for sampling must be disease- and blemish-free, especially when sampling roots. The sampled tissue should be placed in sealed plastic bags immediately and processed in the laboratory as soon as possible, preferably within one day of sampling. Well-established protocols can then be followed for inoculation and incubation of the plant tissue to enable maximum recovery of endophyte isolates. A general principle to follow when preparing the culture media is to use the medium concentrate at only 50% of the manufacturers' recommendation, i.e., double the proportion of water. This will reduce the chances that the endophyte will experience any osmotic shock when emerging from the plant tissue and also make the culture medium more open and accessible for explorative hyphae. We also found that adding a high proportion (50%) of autoclaved plant extract increased the number of endophyte isolates recovered [41], an effect also reported by Prior et al., 2014 [42]. This plant extract should ideally be obtained from the same source species that is being sampled for endophytes. The endophytes are thus emerging into a chemically similar environment to that of the plant interior. Emerging hyphae need to be subcultured immediately, and sometimes single spore or hyphal culturing is required.

Identification of the isolates can be carried out using a combination of standard morphological and genetic barcoding techniques with analytical tools [43]. Establishing the endophyte strain identity is the most important process in screening the isolates for future experimental work.

3.2. *Endophyte Screening*

The initial recovery and isolation of endophytes from the host populations is only the first step in a comprehensive screening process to select the isolates with the greatest potential as crop inoculants. We took several factors into account when selecting strains for efficacy testing with barley crops:

1. The isolate should not be related to any known human or plant pathogen. This is a crucial factor which needs careful qualifying. The degree of DNA similarity between isolate and closest match is important in deciding the identity and whether to proceed with the isolate. Genetic similarity can best be judged by comparing a standard barcoding gene from the isolate with known accessions deposited in a genetic database such as GenBank (NCBI). For our purposes we compared the nuclear ribosomal Internal Transcribed Spacer gene (ITS), as it is sufficiently discriminatory to the species level when assigning taxonomy [44–49]. If the isolate is only distantly related to a known pathogenic strain, then it may be worth pursuing but is a personal judgement call depending on taxonomic group. Fungal isolate growth and proliferation at 37 °C would also be a contra-indicator for selection.
2. The isolate must show vigorous growth on a range of substrates.
3. The isolate should produce early and copious spores (endophyte cultures that do not easily produce spores are of limited use in large-scale agricultural applications).
4. The isolate should be pure; i.e., subcultured from a single spore.

It is likely that the sampling of host populations will result in a large collection of endophyte isolates and final selection of which isolates to use in subsequent experiments will depend largely on the resources available—it may be unrealistic to test 100 s or 1000 s of individual isolates. Careful screening and selection at this point will enable a more focused approach, and the controlled environment experiments will enable a further reduction of isolates selected for field trials. So we recommend that a set of criteria based on desired strain characteristics be used to select the final experimental subjects. For example, fungal endophytes should grow readily on a broad range of substrates, be of easy culture, retain competence over many cycles of sub-culturing, and should produce copious spores/conidia at a relatively early stage after initial recovery and culture. We have found that relatively few strains pass these criteria; from over 100 recovered strains, only 12 were finally selected for further experimentation [50].

4. Proof of Concept

With the basic ideas now in place we proceeded to obtain a proof of concept for our approach. At this stage, we have a selection of identified endophyte strains collected from a particular environment. Before using these strains in experimental work, the target agronomic traits must be clear from the very start, and the statistical analyses that will be used should be established; as the eminent statistician Sir Ronald Fisher once said: 'To consult the statistician after an experiment is finished is often merely to ask him to conduct a post mortem examination. He can perhaps say what the experiment died of [51]. There is always a danger that any post-hoc selection of statistical methods may bias the analyses in favour of the desired results [52].

For many agricultural crops, yield is often the most important trait of interest and is related to a whole plethora of influences: pathogen and disease resistance, nutrient use efficiency, photosynthetic efficiency, hormone and metabolite balance, abiotic stress resistance, etc. So it makes sense to test the effect of the endophyte inoculant on as many of these factors as possible; there is no short cut to developing an effective endophyte inoculant. The crucial point here is to focus on the main influences on crop performance that will be experienced by the target crop species in a particular environment.

To achieve a comprehensive set of data for barley-endophyte responses under different conditions, we decided to carry out extensive controlled environment experiments under a variety of stresses using the fungal endophytes recovered from the sampled populations of *H. murinum*.

Prior to the experimental cycle, a range of commercial culture media combined with whole plant extract of *H. murinum* were tested for their effect on endophyte recovery from the roots of their host (*H. murinum*), and for their subsequent growth and sporulation [41]. It was found that there were significant differences between the media in endophyte recovery, endophyte mycelial growth and time to sporulation. A significantly greater number of different endophytes were recovered from roots on the malt extract plus whole plant extract medium than any of the other media tested. These results indicated that different media are suitable either for the initial recovery and isolation of fungal root endophytes or for increasing fungal biomass and inducing earlier sporulation.

The first controlled-environment experiment with barley examined the effects of ten endophyte strains on disease development on seeds of a barley cultivar using five artificial and one soil-based growth media [30]. A co-inoculant of all ten isolates, as well as two individual isolates, successfully suppressed the development of seed-borne fungal infections on germinated and ungerminated seed. We further reported that the ability to suppress the seed-borne infections was related to the soil properties of the isolate origin [33].

Secondly, we found that inoculation with six different individual endophytes increased grain yield in a nutrient-starved barley cultivar by up to 29% [32]. Furthermore, we also showed that inoculation with the isolates induced increases of up to 70% in shoot dry weight in the nutrient-starved spring barley; the nutrient input was only 10–15% of that recommended.

Thirdly, five individual endophyte strains induced significant improvements in agronomic traits for a severely drought-stressed barley cultivar, including the number of tillers, grain yield and shoot biomass [35]. Soil moisture content was allowed to reach 10–15% of field capacity before watering. The trait that showed the greatest significant difference in the drought-stressed plants was the number of tillers, where all of the endophyte treatments induced a greater number of tillers per plant. However, except in one case, the mean dry root weight for all plants was greater in the control plants, indicating preferential allocation of resources to above-ground parts in the endophyte treatments.

Fourthly, five fungal root endophytes, either individually or combined, were inoculated onto seeds of a barley cultivar grown in optimal conditions and under a combined drought (10–15% soil moisture), heat (33 °C), nutrient (15% of recommended input) and pathogen (*Gaeumannomyces graminis*) stress (multiply-stressed). We found a greater endophyte-induced improvement in important agronomic traits in the multiply-stressed plants compared with the plants grown in optimal conditions [53]. For the multiply-stressed plants, only 13% of the controls survived to the end of the experiment compared with 80% of the endophyte treatments. In multiply-stressed plants, the endophytes induced increases in the number of tillers and root and shoot biomass. The improvements were most significant for barley inoculated with a combination of all five endophytes.

Finally, we tested a range of seed dressings to evaluate the effects of the endophytes on germination and early growth of barley. Seeds were either untreated, dressed with a fungicide (triticonazole and prochloraz), an endophyte spore solution or a combination of endophyte and fungicide. We found significant increases in mean barley seedling length induced by the endophytes after 28 days of seedling growth at time points of up to 6 months from seed dressing. The increases in mean seedling length were greatest for the combined fungal endophyte and fungicide treatment. These results indicated that the endophytes tested were persistent in a seed dressing, enhanced early seedling growth and were fungicide tolerant [34].

Results from these experiments showed that novel symbiotic associations between barley and fungal root endophytes significantly increased yield and biomass in barley grown under nutrient, drought, heat and pathogen stress, and also suppressed the development of seed-borne pathogenic infections.

5. Translation

Isolating, selecting and testing the endophytes in controlled environments to determine their potential as crop inoculants is only the first step towards future agricultural application. We now needed to see whether the positive results translated to field grown barley. The four endophyte strains that induced the greatest significant yield increase in nutrient- and drought-stressed barley were selected as the consortium members. We reasoned that the different modes of action associated with each of the two stresses would allow a degree of compatibility as a consortium, with each endophyte bringing different functional mechanisms to the plant in a particular ecological niche space [54–58].

To obtain the maximum benefit from the endophytes as a crop treatment, it is essential to select the most effective inoculant delivery formulation. The formulation will depend to a large extent on the type of fungal material that is to be used as the active ingredient. The method of applying the treatment should be economically and ergonomically viable, and be easy to apply. The inoculant carrier substrate should provide a stable environment for the microbial fractions and prolong product shelf-life [59]. Some microbial products have a very short shelf-life [12], with a severely reduced window of effectiveness. Our decision to dress the barley seeds with environmentally stable fungal endophyte spores helps to provide robustness and preserve vitality of the treatment, even several months after dressing [34]. Dressing the seeds using standard seed dressing equipment means that the seed supplier does not need to purchase extra and possibly expensive equipment.

Dressing the seed also ensures that the endophyte inoculant will be deployed in the most efficient way, as the inoculant is delivered precisely to where it will be most effective, with no wastage. The germinated seedling root(s) will come into immediate contact with the fungal spores, which will then colonise the plant at the earliest possible stage. This will also have the added benefit of reducing the release of the strains into the environment. The endophyte may be more effective if it colonises the plant tissue before any competitors, as the order of arrival of endophyte and pathogen may even change the effect on the plant from beneficial to detrimental [60].

This scenario depends on a productive relationship with industry partners, and this is often the point at which transition from promising laboratory results to real-world applications fails. Unless specific funding is available to support this process then the researcher must rely on the good will of the industry partner(s). Getting the right partners in the right place at the right time will have a big effect on the success of the translational research effort. We have found it helpful to include such funding in the associated research proposal. This is particularly important when trialing field crops, as multi-year validation is required before any robust claims can be made for the effectiveness or otherwise of the endophyte treatment.

The approach we have adopted in developing a crop inoculant, as outlined in this paper, has resulted in the successful translation of promising laboratory-based research to agriculture. In field trials with barley, we found that for an endophyte-associated increase in grain yield, a strong correlation was found between increased yield and low seasonal rainfall [31]. Furthermore, the endophytes were just as effective with regular foliar fungicidal treatment. Another recently completed field trial on dry sites also returned significant endophyte-induced yield increases (manuscript in preparation), providing strong validation for the value of our methodology.

6. Discussion

Many promising scientific discoveries do not make a successful transition to commerce, for any number of reasons, while some discoveries happen almost by accident [61]. The most common point of failure may be in the initial research project design, where a focused, achievable and outcome-based plan is not in place [62]. This is especially true of research into the potential of endophytes in agricultural crops, where many valuable discoveries languish in the limbo between the laboratory bench and the farmer's field. In this paper, we have shown that a properly focused endophyte project with a continuous emphasis on one particular outcome can result in a more successful transition from idea to application. We have described a set of best practices leading to the development of a successful endophyte field crop inoculant, and this approach can, in principle, be adapted for any crop or agricultural environment (including marginal lands) provided that a similar methodology is followed. The flowchart (Figure 1) shows the basic steps and decision points of our approach.

The prospecting and discovery of endophytes that have beneficial application to agricultural crops is an ongoing and potentially open-ended enterprise, with rates of successful application only limited by available resources and unproven methodologies. Resources are finite and competitive, but ensuring that the methods and principles applied are sound will maximize the chance of success. While the use of endophytes as beneficial crop inoculants has potential, the solution to a defined agricultural need is often complex and multi-factorial and it is unlikely that a single approach will provide the full answer. Future research directions should focus on combining different approaches that will provide more robust and persistent solutions for agriculture. For instance, an endophyte crop treatment combined with a change in agricultural practice can give extra benefits. For example, moving towards a 'no-till' method of cultivation along with longer rotations and the use of cover crops can drastically reduce chemical inputs and labour costs [63] and, if combined with an endophyte inoculant, may produce surprising synergistic results.

Global climate change will bring new and greater stresses to bear on crops, and future research efforts should focus on the solutions to the expected changes. It may be best to target expected conditions in particular geographic areas experiencing particular stresses, especially in poorer areas where farmers cannot afford to use expensive chemicals. Endophyte-based solutions to stress-related problems that minimize economic and environmental costs have the potential to bring real benefit to those who need it the most.

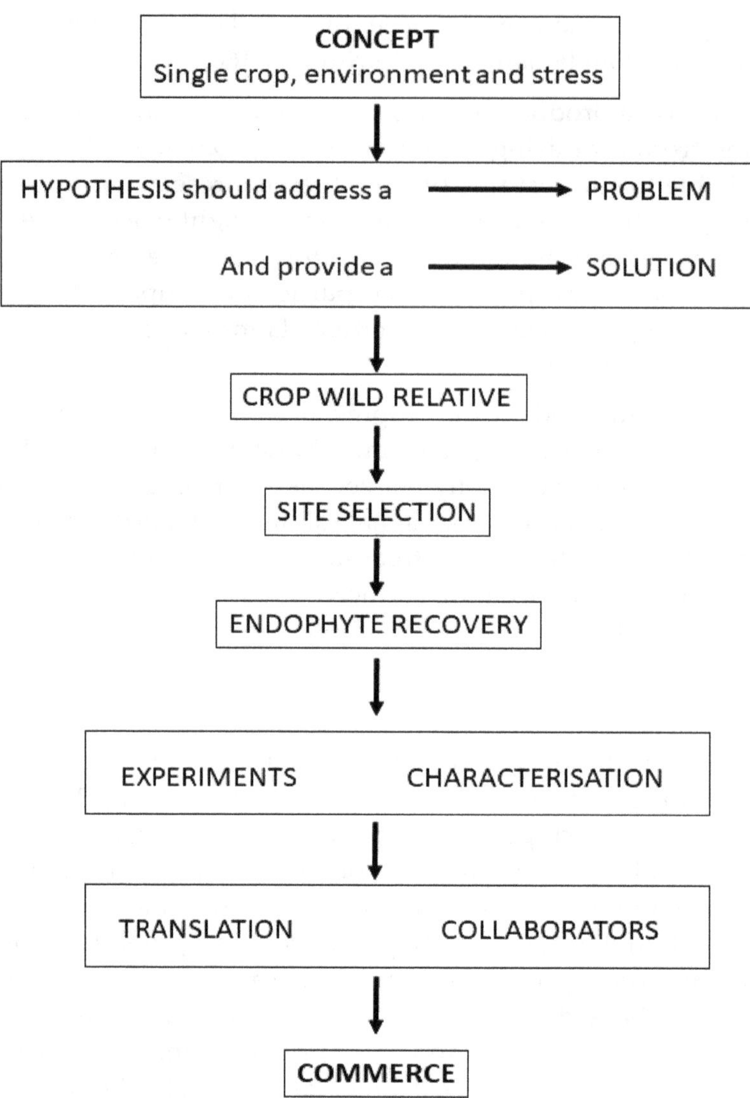

Figure 1. From Concept to Commerce: a schematic flowchart for the prospecting of endophytes for use in agriculture.

Acknowledgments: We thank: Goldcrop Limited, Cork, Ireland for the generous supply and dressing of barley seeds, and for advice on suitable cultivars to use; Crop Plot Trials and the Irish Department of Agriculture, Food and the Marine for carrying out the field trials; Helena Murphy for proof reading and the de-cluttering of technical terms; laboratory technicians at Trinity College Dublin and University College Dublin for providing supplies and technical support. Science Foundation Ireland funded the 2016 field trials through a Technology Innovation Development Award (Grant Number R15449). Enterprise Ireland are supporting further research and commercialisation through a Commercialisation Fund (Grant number CF 2017 0625-P). Trinity College Dublin provided financial support for the 2015 field trial through a PhD studentship grant.

Author Contributions: Brian R. Murphy, Fiona M. Doohan and Trevor R. Hodkinson conceived and designed the experiments; Brian R. Murphy performed the experiments; Brian R. Murphy analyzed the data; Brian R. Murphy wrote the paper.

References

1. Van der Heijden, M.G.A.; Hartmann, M. Networking in the Plant Microbiome. *PLoS Biol.* **2016**, *14*, 1–9. [CrossRef] [PubMed]
2. Kemen, E.; Kemen, A.C.; Agler, M.T.; Kemen, E. Tansley review Host—microbe and microbe—microbe interactions in the evolution of obligate plant parasitism. *New Phytol.* **2015**, *206*, 1–18. [CrossRef] [PubMed]

3. Müller, D.B.; Vogel, C.; Bai, Y.; Vorholt, J.A. The Plant Microbiota: Systems-Level Insights and Perspectives. *Annu. Rev. Genet.* **2016**, *50*, 211–234. [CrossRef] [PubMed]
4. Behera, K.K.; Alam, A.; Vats, S.; Sharma, H.P.; Sharma, V. *Agroecology and Strategies for Climate Change*; Springer: Dordrecht, The Netherlands, 2012; pp. 287–328.
5. Thibodeau, P. *A Companion to Science, Technology, and Medicine in Ancient Greece and Rome: Volume II*; Irby-Massie, G.L., Georgia, L., Eds.; John Wiley and Sons: Chichester, UK, 2016; pp. 519–533.
6. Schlaeppi, K.; Bulgarelli, D. The Plant Microbiome at Work. *Mol. Plant-Microbe Interact.* **2015**, *28*, 212–217. [CrossRef] [PubMed]
7. Schneider, A. Observations on Some American Rhizobia. *Bull. Torrey Bot. Club* **1892**, *19*, 203–218. [CrossRef]
8. Lareen, A.; Burton, F.; Schäfer, P. Plant root-microbe communication in shaping root microbiomes. *Plant Mol. Biol.* **2016**, *90*, 575–587. [CrossRef] [PubMed]
9. Kogel, K.-H.; Franken, P.; Hückelhoven, R. Endophyte or parasite—what decides? *Curr. Opin. Plant Biol.* **2006**, *9*, 358–363. [CrossRef] [PubMed]
10. Berendsen, R.L.; Pieterse, C.M.J.; Bakker, P.A.H.M. The rhizosphere microbiome and plant health. *Trends Plant Sci.* **2012**, *17*, 478–486. [CrossRef] [PubMed]
11. Bisseling, T.; Dangl, J.L.; Schulze-Lefert, P. Next-generation communication. *Science* **2009**, *324*, 691. [CrossRef] [PubMed]
12. Owen, D.; Williams, A.P.; Griffith, G.W.; Withers, P.J.A. Use of commercial bio-inoculants to increase agricultural production through improved phosphrous acquisition. *Appl. Soil Ecol.* **2015**, *86*, 41–54. [CrossRef]
13. Nadeem, S.M.; Ahmad, M.; Zahir, Z.A.; Javaid, A.; Ashraf, M. The role of mycorrhizae and plant growth promoting rhizobacteria (PGPR) in improving crop productivity under stressful environments. *Biotechnol. Adv.* **2014**, *32*, 429–448. [CrossRef] [PubMed]
14. Murphy, B.R.; Doohan, F.M.; Hodkinson, T.R. Fungal endophytes of barley roots. *J. Agric. Sci.* **2013**, *152*, 1–14. [CrossRef]
15. Achatz, B.; Rüden, S.; Andrade, D.; Neumann, E.; Pons-Kühnemann, J.; Kogel, K.-H.; Franken, P.; Waller, F. Root colonization by Piriformospora indica enhances grain yield in barley under diverse nutrient regimes by accelerating plant development. *Plant Soil* **2010**, *333*, 59–70. [CrossRef]
16. Schulz, B.; Boyle, C. What are Endophytes? In *Microbial Root Endophyte*; Schulz, B., Boyle, C., Sieber, T.N., Eds.; Springer: New York, NY, USA, 2006; Volume 9, pp. 1–13.
17. Kuldau, G.; Bacon, C. Clavicipitaceous endophytes: Their ability to enhance resistance of grasses to multiple stresses. *Biol. Control* **2008**, *46*, 57–71. [CrossRef]
18. Saikkonen, K.; Gundel, P.E.; Helander, M. Chemical Ecology Mediated by Fungal Endophytes in Grasses. *J. Chem. Ecol.* **2013**, *39*, 962–968. [CrossRef] [PubMed]
19. O'Hanlon, K.A.; Knorr, K.; Jørgensen, L.N.; Nicolaisen, M.; Boelt, B. Exploring the potential of symbiotic fungal endophytes in cereal disease suppression. *Biol. Control* **2012**, *63*, 69–78.
20. Busby, P.E.; Ridout, M.; Newcombe, G. Fungal endophytes: modifiers of plant disease. *Plant Mol. Biol.* **2015**, *90*, 645–655. [CrossRef] [PubMed]
21. Rodriguez, R.J.; White, J.F.; Arnold, A.E.; Redman, R.S. Fungal endophytes: diversity and functional roles. *New Phytol.* **2009**, *182*, 314–330. [CrossRef] [PubMed]
22. Hardoim, P.R.; van Overbeek, L.S.; van Elsas, J.D. Properties of bacterial endophytes and their proposed role in plant growth. *Trends Microbiol.* **2008**, *16*, 463–471. [CrossRef] [PubMed]
23. Belesky, D.P.; West, C.P. Abiotic Stresses and Endophyte Effects. In *Tall Fescue for the 21st Century*; Fribourg, H., Hannaway, D., West, C., Eds.; American Society of Agronomy: Madison, WI, USA, 2009; pp. 49–64.
24. Waller, F.; Achatz, B.; Baltruschat, H.; Fodor, J.; Becker, K.; Fischer, M.; Heier, T.; Hückelhoven, R.; Neumann, C.; von Wettstein, D.; et al. The endophytic fungus Piriformospora indica reprograms barley to salt-stress tolerance, disease resistance, and higher yield. *Proc. Natl. Acad. Sci. USA* **2005**, *102*, 13386–13391. [CrossRef] [PubMed]
25. Rodriguez, R.J.; Henson, J.; Van Volkenburgh, E.; Hoy, M.; Wright, L.; Beckwith, F.; Kim, Y.-O.; Redman, R.S. Stress tolerance in plants via habitat-adapted symbiosis. *ISME J.* **2008**, *2*, 404–416. [CrossRef] [PubMed]
26. Gill, S.S.; Gill, R.; Trivedi, D.K.; Anjum, N.A.; Sharma, K.K.; Ansari, M.W.; Ansari, A.A.; Johri, A.K.; Prasad, R.; Pereira, E.; et al. Piriformospora indica: Potential and Significance in Plant Stress Tolerance. *Front. Microbiol.* **2016**, *7*, 1–20. [CrossRef] [PubMed]

27. Johnson, J.M.; Alex, T.; Oelmüller, R. Piriformospora indica: The versatile and multifunctional root endophytic fungus for enhanced yield and tolerance to biotic and abiotic stress in crop plants. *J. Trop. Agric.* **2014**, *52*, 103–122.
28. Singh, L.P.; Gill, S.S.; Tuteja, N. Unraveling the role of fungal symbionts in plant abiotic stress tolerance. *Plant Signal. Behav.* **2011**, *6*, 175–191. [CrossRef] [PubMed]
29. Nelissen, H.; Moloney, M.; Inzé, D. Translational research: from pot to plot. *Plant Biotechnol. J.* **2014**, *12*, 277–285. [CrossRef] [PubMed]
30. Murphy, B.R.; Doohan, F.M.; Hodkinson, T.R. Persistent fungal root endophytes isolated from a wild barley species suppress seed- borne infections in a barley cultiva. *Biocontrol* **2015**, *60*, 281–292. [CrossRef]
31. Murphy, B.R.; Hodkinson, T.R.; Doohan, F.M. A fungal endophyte consortium counterbalances the negative effects of reduced nitrogen input on the yield of field-grown spring barley. *J. Agric. Sci.* **2017**, *155*, 1324–1331. [CrossRef]
32. Murphy, B.R.; Doohan, F.M.; Hodkinson, T.R. Fungal root endophytes of a wild barley species increase yield in a nutrient-stressed barley cultivar. *Symbiosis* **2015**, *65*, 1–7. [CrossRef]
33. Murphy, B.R.; Doohan, F.M.; Hodkinson, T.R. Suppression of vertically transmitted infections of barley by fungal root endophytes is linked to the soil properties of the isolate origin. *IOBC-WPRS Bull.* **2014**, *115*, 31–36.
34. Murphy, B.R.; Doohan, F.M.; Hodkinson, T.R. A seed dressing combining fungal endophyte spores and fungicides improves seedling survival and early growth in barley and oat. *Symbiosis* **2017**, *71*, 69–76. [CrossRef]
35. Murphy, B.R.; Martin Nieto, L.; Doohan, F.M.; Hodkinson, T.R. Fungal Endophytes Enhance Agronomically Important Traits in Severely Drought-Stressed Barley. *J. Agron. Crop Sci.* **2015**, *201*, 419–427. [CrossRef]
36. Rosier, A.; Bishnoi, U.; Lakshmanan, V.; Sherrier, D.J.; Bais, H.P. A perspective on inter-kingdom signaling in plant-beneficial microbe interactions. *Plant Mol. Biol.* **2016**, *90*, 537–548. [CrossRef] [PubMed]
37. CB Insights. The Top 20 Reasons Startups Fail. Available online: https://www.cbinsights.com/research/startup-failure-reasons-top/?utm_source=CB+Insights+Newsletter&utm_campaign=db99525228-TuesNL_10_31_2017&utm_medium=email&utm_term=0_9dc0513989-db99525228-89648729 (accessed on 2 November 2017).
38. Zhang, H.; Mittal, N.; Leamy, L.J.; Barazani, O.; Song, B.-H. Back into the wild-Apply untapped genetic diversity of wild relatives for crop improvement. *Evol. Appl.* **2017**, *10*, 5–24. [CrossRef] [PubMed]
39. Botanical Society of Britain & Ireland (BSBI). Available online: www.brc.ac.uk/plantatlas/finder/plant/Hordeum (accessed on 1 November 2017).
40. Secretariat of CBD. *Nagoya Protocol on Access to Genetic Resources and the Fair and Equitable Sharing of Benefits Arising Convention on Biological Diversity*; Convention on Biological Diversity United Nations: Montreal, QC, Canada, 2011; p. 15.
41. Murphy, B.R.; Batke, S.P.; Doohan, F.M.; Hodkinson, T.R. Media Manipulations and the Culture of Beneficial Fungal Root Endophytes. *Int. J. Biol.* **2015**, *7*, 94. [CrossRef]
42. Prior, R.; Görges, K.; Yurkov, A.; Begerow, D. New isolation method for endophytes based on enzyme digestion. *Mycol. Prog.* **2014**, *13*, 849–856. [CrossRef]
43. Tamura, K.; Stecher, G.; Peterson, D.; Filipski, A.; Kumar, S. MEGA6: Molecular Evolutionary Genetics Analysis version 6.0. *Mol. Biol. Evol.* **2013**, *30*, 2725–2729. [CrossRef] [PubMed]
44. Bellemain, E.; Carlsen, T.; Brochmann, C.; Coissac, E.; Taberlet, P.; Kauserud, H. ITS as an environmental DNA barcode for fungi: an in silico approach reveals potential PCR biases. *BMC Microbiol.* **2010**, *10*, 189. [CrossRef] [PubMed]
45. Dentinger, B.T.M.; Didukh, M.Y.; Moncalvo, J.-M. Comparing COI and ITS as DNA barcode markers for mushrooms and allies (Agaricomycotina). *PLoS ONE* **2011**, *6*, e25081. [CrossRef] [PubMed]
46. Korabecna, M. The Variability in the Fungal Ribosomal DNA (ITS1, ITS2, and 5.8 S rRNA Gene): Its Biological Meaning and Application in Medical Mycology. In *Communicating Current Research and Educational Topics and Trends in Applied Microbiology*; Mendez-Vilas, A., Ed.; Formatex: Badajoz, Spain, 2007; pp. 783–787.
47. Blattner, F.R. Phylogenetic analysis of Hordeum (Poaceae) as inferred by nuclear rDNA ITS sequences. *Mol. Phylogenet. Evol.* **2004**, *33*, 289–299. [CrossRef] [PubMed]
48. Blaalid, R.; Kumar, S.; Nilsson, R.H.; Abarenkov, K.; Kirk, P.M.; Kauserud, H. ITS1 versus ITS2 as DNA metabarcodes for fungi. *Mol. Ecol. Resour.* **2013**, *13*, 218–224. [CrossRef] [PubMed]

49. Schoch, C.L.; Seifert, K.A.; Huhndorf, S.; Robert, V.; Spouge, J.L.; Levesque, C.A.; Chen, W. Nuclear ribosomal internal transcribed spacer (ITS) region as a universal DNA barcode marker for Fungi. *Proc. Natl. Acad. Sci. USA* **2012**, *109*, 6241–6246. [CrossRef] [PubMed]
50. Murphy, B.R.; Martin Nieto, L.; Doohan, F.M.; Hodkinson, T.R. Profundae diversitas: The uncharted genetic diversity in a newly studied group of fungal root endophytes. *Mycology* **2015**, *6*, 139–150. [CrossRef]
51. Fisher, R.A. Presidential Address to the First Indian Statistical Congress. *Sankhya Indian J. Stat.* **1938**, *4*, 14–17.
52. Elliott, H.L. Post hoc analysis: use and dangers in perspective. *J. Hypertens. Suppl.* **1996**, *14*, S21–S25. [CrossRef] [PubMed]
53. Murphy, B.R.; Martin Nieto, L.; Doohan, F.M.; Hodkinson, T.R. The difference between life and death. *Glob. J. Sci. Front. Res.* **2015**, *15*, 1–9.
54. Kearney, M.; Simpson, S.J.; Raubenheimer, D.; Helmuth, B. Modelling the ecological niche from functional traits. *Philos. Trans. R. Soc. Lond. Ser. B Biol. Sci.* **2010**, *365*, 3469–3483. [CrossRef] [PubMed]
55. Gopal, M.; Gupta, A.; Thomas, G.V. Bespoke microbiome therapy to manage plant diseases. *Front. Microbiol.* **2013**, *4*. [CrossRef] [PubMed]
56. Vandenkoornhuyse, P.; Quaiser, A.; Duhamel, M.; Le Van, A.; Dufresne, A. Tansley review The importance of the microbiome of the plant holobiont. *New Phytol.* **2015**, *206*, 1196–1206. [CrossRef] [PubMed]
57. De Souza, R.S.C.; Okura, V.K.; Armanhi, J.S.L.; Jorrín, B.; Lozano, N.; da Silva, M.J.; González-Guerrero, M.; de Araújo, L.M.; Ferreira, N.V.; Bagheri, H.C.; et al. Unlocking the bacterial and fungal communities assemblages of sugarcane microbiome. *Nat. Sci. Rep.* **2016**, *6*, 28774. [CrossRef] [PubMed]
58. Busby, P.E.; Soman, C.; Wagner, M.R.; Friesen, M.L.; Kremer, J.; Bennett, A.; Morsy, M.; Eisen, J.A.; Leach, J.E.; Dangl, J.L. Research priorities for harnessing plant microbiomes in sustainable agriculture. *PLoS Biol.* **2017**, *15*, 1–14. [CrossRef] [PubMed]
59. Malusá, E.; Sas-Paszt, L.; Ciesielska, J. Technologies for beneficial microorganisms inocula used as biofertilizers. *Sci. World J.* **2012**, *2012*, 491206. [CrossRef] [PubMed]
60. Adame-Álvarez, R.-M.; Mendiola-Soto, J.; Heil, M. Order of arrival shifts endophyte-pathogen interactions in bean from resistance induction to disease facilitation. *FEMS Microbiol. Lett.* **2014**, *355*, 100–107. [CrossRef] [PubMed]
61. Ward, A. 24 Important Scientific Discoveries That Happened by Accident | Mental Floss. Available online: http://mentalfloss.com/article/53646/24-important-scientific-discoveries-happened-accident (accessed on 31 October 2017).
62. Armstrong, J.S. The value of formal planning for strategic decisions: Review of empirical research. *Strateg. Manag. J.* **1982**, *3*, 197–211. [CrossRef]
63. Montgomery, D.R. *Growing a Revolution: Bringing Our Soil Back to Life*; W.W. Norton and Company: New York, NY, USA, 2017.

6

A Solvent-Free Approach for Converting Cellulose Waste into Volatile Organic Compounds with Endophytic Fungi

Tyler Maxwell [1], Richard G. Blair [2], Yuemin Wang [1], Andrew H. Kettring [3], Sean D. Moore [3], Matthew Rex [1] and James K. Harper [1,*]

1 Department of Chemistry, University of Central Florida, 4111 Libra Drive, Orlando, FL 32816, USA; Tyler.Maxwell@knights.ucf.edu (T.M.); Yueminwang@knights.ucfe.edu (Y.W.); Matthew.Rex@ucf.edu (M.R.)
2 Florida Space Institute, University of Central Florida, 12354 Research Parkway, Suite 214, Orlando, FL 32826, USA; Richard.Blair@ucf.edu
3 Burnett School of Biomedical Sciences, University of Central Florida, 4110 Libra Dr., Orlando, FL 32816, USA; Akettring@knights.ucf.edu (A.H.K.); Sean.Moore@ucf.edu (S.D.M.)
* Correspondence: James.Harper@ucf.edu

Abstract: Simple sugars produced from a solvent-free mechanocatalytic degradation of cellulose were evaluated for suitability as a growth medium carbon source for fungi that produce volatile organic compounds. An endophytic *Hypoxylon* sp. (CI-4) known to produce volatiles having potential value as fuels was initially evaluated. The growth was obtained on a medium containing the degraded cellulose as the sole carbon source, and the volatile compounds produced were largely the same as those produced from a conventional dextrose/starch diet. A second *Hypoxylon* sp. (BS15) was also characterized and shown to be phylogenetically divergent from any other named species. The degraded cellulose medium supported the growth of BS15, and approximately the same quantity of the volatile compounds was produced as from conventional diets. Although the major products from BS15 grown on the degraded cellulose were identical to those from dextrose, the minor products differed. Neither CI-4 or BS15 exhibited growth on cellulose that had not been degraded. The extraction of volatiles from the growth media was achieved using solid-phase extraction in order to reduce the solvent waste and more efficiently retain compounds having low vapor pressures. A comparison to more conventional liquid–liquid extraction demonstrated that, for CI-4, both methods gave similar results. The solid-phase extraction of BS15 retained a significantly larger variety of the volatile compounds than did the liquid–liquid extraction. These advances position the coupling of solvent-free cellulose conversion and endophyte metabolism as a viable strategy for the production of important hydrocarbons.

Keywords: endophytic fungi; Mechanocatalysis; cellulose degradation; volatile organic compounds; myco-diesel; *Hypoxylon*

---

1. Introduction

The endophytic fungi are organisms that colonize the tissue of living plants. In most cases, this relationship is asymptomatic and may even provide benefits to plants [1]. Endophytes have been studied extensively and found to produce a remarkable variety of natural chemical products [2]. While much of the interest has focused on bioactive compounds, the production of other important compounds has also been reported. A recent noteworthy discovery is that certain endophytes can produce hydrocarbons that have the potential to be used as fuels or fuel additives [3]. These products have been compared to diesel fuel and even described as "myco-diesel", because they include compounds normally associated with diesel fuel. Over the past decade, interest in fungi producing volatile organic products with the potential for use as fuels has increased, and several studies have identified potentially useful fungi [4–10]. Related work has also identified fungi producing volatile products but has not focused on their potential usefulness as fuels [11–25].

The availability of hydrocarbon fuels from fungi complements fuel products produced by other organisms. For example, certain algae produce aliphatic fatty acids and considerable effort has been expended into developing these into viable biofuels [26]. Likewise, yeast fermentation has been prominently utilized to convert carbohydrates from corn into ethanol for fuel [27]. In general, fungal products contain a more complex variety of volatile compounds than either algae or yeast, including ketones, esters, alcohols, and a remarkable variety of hydrocarbon products. All of these biofuels complement more conventional fuels and thus represent important pathways worthy of exploration given the current interest in developing alternative fuels. However, one of the concerns that exists when producing hydrocarbon fuel from fungi is that they require a refined carbohydrate source (e.g., sucrose) in their diet. There has been debate regarding the suitability of devoting carbohydrates to fuel production. A solution to this dilemma would be to find an alternative food source for the fungi.

Recently, a "green chemistry" mechanocatalytic method has been reported that allows cellulose waste products to be converted into simple carbohydrates [28]. This process involves ball milling performed in the solid state and is thus entirely solvent free and capable of rapidly producing large quantities of carbohydrates (see experimental). The major water-soluble products from this process have been shown to be glucose, fructose, and levoglucosan. No oligosaccharides larger than dimers survive the milling, even after short processing times (e.g., 30 min) [28]. This process has been successfully demonstrated using a remarkable variety of cellulose-based feedstock materials from plants (Table 1) and includes numerous materials normally regarded as unusable waste such as orange peels, cherry pits, coffee grounds, and discarded newspaper.

At the present time, however, it has not been demonstrated that fungi can actually grow on the carbohydrates created from the cellulose breakdown process. The aim of this manuscript is to demonstrate that carbohydrates produced from this solvent-free degradation process are a suitable carbon source for fungal growth and that the volatile products produced from the degraded cellulose closely match compounds produced from a more conventional diet. In the following, sugars from only one cellulose source (oak) are tested. The results from other materials in Table 1 are expected to give similar results, as it has been demonstrated that cellulose from various sources consistently breaks down into simple sugars [28].

Table 1. Waste materials containing cellulose that can be converted into simple sugars.

Feedstock	Percent Hydrolyzed [a]
Cherry pit	95.7
Flint corn kernals	93.4
St. Augustine grass	92.5
Oat	90.3
Orange peel	85.0
Corn cobs	81.5
Bamboo	75.1
Cedar	74.0
Red Oak	72.4
Maple	72.0
Douglas Fir	71.1
Nannochloropsis	69.2
Aspen	68.0
Poplar	66.9
Yellow pine	65.3

Table 1. *Cont.*

Feedstock	Percent Hydrolyzed [a]
Wheat	65.0
Miscanthus grass	64.7
White pine	64.4
Mixed yard waste	58.1
Switch grass	57.9
Hickory	55.9
Paper, newsprint	54.7
Flint corn stover	52.1
Banana leaf	52.0
Big blue stem grass	50.1
Little blue stem grass	48.9
Coffee grounds	45.2

[a] This value represents hydrolysis of the holocellulose present in the material.

2. Materials and Methods

The cellulose employed in this study to create the simple sugars was obtained from water oak (*Quercus nigra*) sawdust sourced from a local sawmill. The oak was dried at room temperature to a moisture content of <10% and cut into 2 cm or smaller pieces. Delaminated kaolinite (Kaopaque 10, IMERYS) was used as received.

The mechanical processing of cellulose employed 8000M and 8000D mixer mills (SPEX Certiprep, Metuchen, NJ, USA). Two grams of a 1:1 mixture of the kaolinite clay catalyst and biomass source were processed for two hours in 65-mL vials (1.5″ ID × 2.25″ deep) made of 440C steel, utilizing three 0.5-inch diameter balls composed of the same material as the milling vial. Energy was applied in 30-min intervals with 30 min of cooling time to minimize the effects of frictional heating. Hydrolysis of hemicellulose and cellulose (holocellulose) was monitored gravimetrically. Conversion of holocellulose to water-soluble oligosaccharides was determined by stirring 0.1 g of the reaction mixture in 30 mL of water. The production of water-soluble products was measured by filtration through a 47-mm diameter Whatman Nuclepore® track etched polycarbonate membrane filter with a pore size of 0.220 μm. The residue was dried in a 60 °C oven for 12 h and then weighed.

The potato dextrose broth and agar were purchased from Becton Dickinson. Ammonium sulfate, acetonitrile, ethyl acetate, anhydrous magnesium sulfate, and methanol were purchased from Fisher Scientific. A sample of 1,8-cineole was obtained from TCI chemicals (Portland, OR, USA). The yeast nitrogen base was purchased from Sigma Aldrich. HyperSep C-18 solid-phase extraction columns (1 g bed weight) were purchased from Thermo Scientific. Potato dextrose agar was purchased from Microtech Scientific. All the reagents were used as received.

The isolation of the *Hypoxylon* sp., BS15, was from branch clippings of a *Taxodium distichum* (Bald Cyprus) gathered near Orange City, Florida, USA. The branches were treated with 70% ethanol, flame sterilized, and then dried in a sterile laminar-flow hood. A sterile knife blade was then used to cut away the outer tissue from the clipping, and a square wedge of the inner tissue was placed on water agar. This dish was incubated, and any fungal hyphae observed growing from the sample were transferred onto separate plates of potato dextrose agar.

The potato dextrose broth was prepared by adding 2.4 g of the potato dextrose broth to 100 mL of purified water in a 500-mL Erlenmeyer flask. The flask was then sealed with aluminum foil and autoclaved for 15 min for sterilization. The fungi of interest (CI-4 or BS15) were then added to the sterile broth, and it was resealed with foil. Cellulose broth was prepared using 250 mL of purified water, 5 g of degraded cellulose, 1.5 g of ammonia sulfate, and 1.7 g of the yeast nitrogen base without amino acids. In both growth media, the fungi were then left to grow for 25 days in the lab at 20–25 °C without stirring. Each broth was then vacuum filtered twice through Whatman Grade 4 filter paper to remove all particulates.

A control sample containing cellulose not subjected to the mechanocatalytic degradation process was prepared by adding 2 g of finely ground cellulose powder and 0.5 g of ammonium sulfate to 250 mL of distilled water. This medium was autoclaved for 15 min, and, after cooling, two separate solutions were prepared by adding CI-4 or BS15 to the liquid. This culture was allowed to grow for 2 weeks at 20–25 °C without stirring.

For the solid-phase extraction of the fungal volatile compounds, a C-18 cartridge was first washed with 4 mL of methanol and then with 4 mL of water. Filtered fungal broth (50 mL) was then passed through the column slowly under vacuum. The column was washed again with 4 mL of water to remove any contaminants and then dried by drawing air through the column for 15 min. The retained compounds were then eluted by passing acetonitrile through the column. A clear brown solution was typically recovered from this process. The eluent was then filtered with a 0.22-μm syringe filter prior to analysis.

For the liquid–liquid extractions, a total of 300 mL of the filtered fungal broth was shaken in a separatory funnel with 50 mL of ethyl acetate. The ethyl acetate was then separated from the water and dried over anhydrous magnesium sulfate. The solution was then filtered with a 0.22-μm syringe filter prior to analysis.

The gas chromatography/mass spectrometry (GC/MS) analysis for the volatile compounds was performed using a method described previously with slight modification [29,30]. An Agilent 6850 was used with a 5975CVC MS detector and a Restek Rxi-5HT capillary column (30 m × 0.25 mm, film thickness 0.25 μm). The carrier gas was ultrahigh purity helium with a one cm^3/min constant flow rate and an initial column head pressure of 77 kPa. The injector split was set to 250 °C at a 20:1 split ratio with 1-μL volume per injection. The column oven temperature was programmed to 45 °C with an initial temperature hold for 1 min with a 10 °C/min ramp to 100 °C and hold for 5 min, followed by a 5 °C/min ramp to 200 °C and a hold for 5 min. The detector was set at a constant 280 °C and set to scan 30–350 m/z. The data acquisition and processing were performed on Agilent MSD ChemStation software. The identification of the compounds was made via library comparison using the National Institute of Standards and Technology (NIST, Gaithersburg, MD, USA) database.

For DNA extraction from BS15, a small sample of the fungal tissue (50–100 mg) was collected into a microcentrifuge tube from the surface of a potato dextrose agar plate after 1 week of growth at room temperature. The tissue was lysed using a FastPrep Homogenizer (MP Biomedicals, Santa Ana, CA, USA) by zirconia–silica bead beating in 1 mM of sodium dodecyl sulfate, 5 mM of EDTA, and 10 mM of Tris-HCl, pH 8.0 with 10 μg/mL RNase A. The lysate was centrifuged, and then, the DNA was purified from the supernatant by silica column binding in guanidinium thiocyanate [31].

Diagnostic gene sequences used for identification by genetic barcoding were amplified by a routine polymerase chain reaction (PCR) with Taq [32,33]. Primers ITS1-F_KYO1 and ITS4_KYO1 were used to target the internally transcribed spacers (ITS1 and ITS2) and the flanking portions of the ribosomal RNA encoding genes (SSU, 5.8S, and LSU) [34]. The protein-coding genes α-actin and β-tubulin were amplified by primers ACT-512F/ACT-783R and T1/T22, respectively [35,36]. The PCR products were visualized by agarose gel electrophoresis, similarly purified by silica column binding, and then sequenced commercially (GENEWIZ, Plainfield, NJ, USA). The sequences were deposited in GenBank under accession number MH223406 (ITS), MH465497 (actin), and MH465498 (tubulin).

The ribosomal gene sequences were analyzed with a series of BioPython-based scripts [37]. First, full-length ITS sequences were extracted via ITSx and used to locally query the UNITE+INSDC fungal database by BLAST search [38–40]. Based on these search results, relevant taxa were selected, and a list was compiled of all the unambiguous binomial species within these taxa. The corresponding UNITE records were pooled and analyzed by ITSx. For each species, a single representative full-length ITS2 record was chosen for alignment. Relevant α-actin and β-tubulin records used in alignment were retrieved from GenBank [41].

The phylograms were generated using MEGA software [42]. The sequences were aligned by the MUSCLE algorithm and then clustered by the maximum likelihood method with 1000 bootstrap replicates [43,44]. Both the α-actin and β-tubulin sequences were treated as protein-coding during the phylogenetic analyses, while ITS sequences were not. All other settings in MEGA were unchanged and no manual modifications were made during the alignment or clustering. The resultant phylograms were exported and visualized via Interactive Tree of Life (iTOL) web software [45]. The nodes were pruned on the basis of relatedness to BS15 and intra-generic species richness. The alignments and phylograms were deposited in TreeBase under submission number 23089.

3. Results and Discussion

3.1. Growth of the *Hypoxylon* CI-4 on Degraded Cellulose

As an initial test of the feasibility of using carbohydrates from mechanocatalytic cellulose degradation as a fungal diet, a *Hypoxylon* sp. was added to a growth medium consisting of the degraded cellulose as the sole carbon source (see experimental). A control sample was also prepared, having the fungus on a conventional diet of potato dextrose broth. The particular *Hypoxylon* fungus used for this study (designated CI-4) was selected because it has been previously shown to produce a diverse variety of volatile organic hydrocarbons [46,47]. Both cultures exhibited similar fungal growth and were incubated for three weeks. The hydrocarbon fraction was extracted from the growth media using a solid-phase extraction process (see experimental). A gas chromatography/mass spectrometry (GC/MS) analysis exhibited a diverse range of volatile products, as expected from the previous study [46]. A comparison of the volatile compounds produced from each growth condition is illustrated in Figure 1 and demonstrates that the degraded cellulose material produces the same major products as a conventional carbohydrate-rich diet. A notable difference, however, between the growth media is that the amounts of the volatile compounds produced from the cellulose degradation products were roughly two–five times less than the same products produced from the potato dextrose broth.

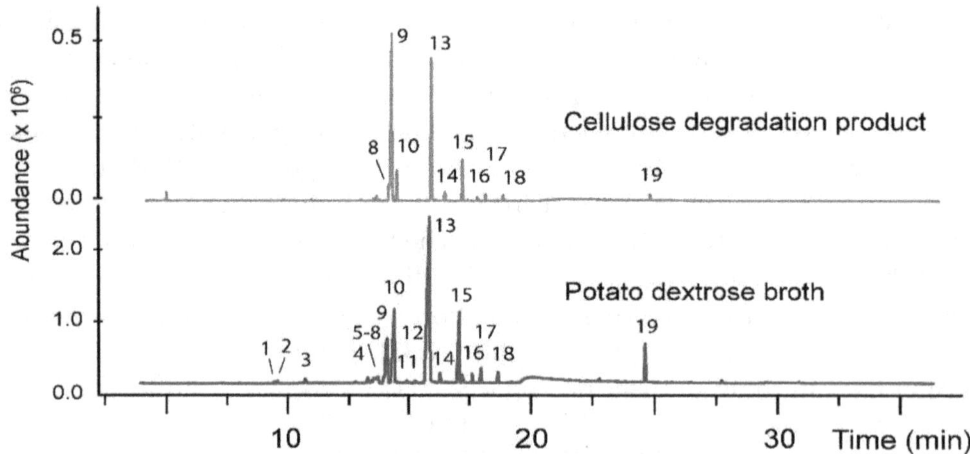

Figure 1. A gas chromatogram showing the volatile organic products produced by the fungus CI-4 growth on a conventional media (**bottom**) versus a diet containing carbohydrates produced from cellulose degradation (**top**). The nominal masses for each numbered peak are given in Table 2. In each case a control sample was also analyzed consisting of the growth medium without fungi added. This solution was processed identically to the fungi-containing samples. In each case, no peaks from the control samples corresponded to any of the peaks shown above.

The molecular masses and tentative identification of individual compounds from CI-4 were made by comparing the mass spectrum of each peak against the data in the NIST database. Although the nominal masses were obtained in all cases, most compounds were not identifiable. All the results are summarized in Table 2. Also included in Table 2 are the results from the analysis of a second fungus (BS15, described below).

Table 2. A list of the volatile compounds produced by CI-4 or BS15 grown on either the potato dextrose broth (PD) or the degraded cellulose (DC), showing tentative compound identification where possible.

Fungus	Peak # [a]	R.T. (min)	Area (%) [b] PD, DC	Tentative Identity [c]	Mol. Mass (Da)	Qual. [d]
CI-4	1	9.58	0.1, -	Unknown	126	-
CI-4	2	9.72	0.3, -	Unknown	138	-
CI-4	3	10.88	0.5, -	Unknown	124	-
CI-4	4	12.93	0.2, -	Unknown	152	-
CI-4	5	13.43	0.8, -	Unknown	122	-
CI-4	6	13.57	0.3, -	Unknown	154	-
CI-4	7	13.68	1.4, -	3-Ethenyl-2-methylene cyclopentanecarboxylic acid	152	50
CI-4	8	14.11	0.6, 2.8	Unknown	152	-
CI-4	9	14.25	10.6, 45.7	Unknown	150	-
CI-4	10	14.57	15.5, 6.0	Unknown	154	-
CI-4	11	15.08	0.3, -	Unknown	154	-
CI-4	12	15.44	0.3, -	Unknown	152	-
CI-4	13	16.02	43.8, 2.0	Unknown	168	-
CI-4	14	16.45	1.1, 2.0	1-Acetyl-2-(1-hydroxyethyl)-cyclohexene	168	50
CI-4	15	17.24	10.4, 7.6	Unknown	150	-
CI-4	16	17.79	1.0, 0.8	Unknown	170	-
CI-4	17	18.14	1.5, 1.2	Unknown	170	-
CI-4	18	18.84	1.1, 1.1	3-Isopropoxy 5-methyl-phenol	166	61
CI-4	19	24.78	4.0, 1.0	2,3-Dimethoxy-naphthalene	188	85
BS15	1	4.82	1.1, 2.7	Furfuryl alcohol	98	72
BS15	2	5.07	9.9, 2.4	Methyl 4-oxo-2-butenoate	114	94
BS15	3	7.85	9.3, 1.7	Benzeneacetaldehyde	120	70
BS15	4	8.22	1.6, -	4-methoxy-2,5-dimethyl-3 (2H)-furanone	142	77
BS15	5	8.45	36.7, 26.1	2,5-furandione dihydro-3-methylene	112	55
BS15	6	8.65	-, 21.5	Levoglucosenone	126	78
BS15	7	9.65	20.9, 12.8	2-Phenyethanol	122	86
BS15	8	12.24	3.0, -	Unknown	158	-
BS15	9	12.34	-, 20.3	5-(Hydroxymethyl)furfural	126	91
BS15	10	13.84	17.8	Unknown	86	–
BS15	11	14.56	2.5, -	Phenylacetic acid	136	75
BS15	12	16.23	3.0, -	Unknown	138	-
BS15	13	16.45	-, 3.2	Unknown	142	-
BS15	14	18.20	3.7, –	Unknown	154	-
BS15	15	20.47	-, 8.1	Unknown	162	-
BS15	16	22.73	-, 3.2	2,4-dihydroxy-3,6-dimethyl Benzoic acid, methyl ester	196	72
BS15	17	25.58	-, 2.1	Dihydro-5-(2-oxocyclohexylidene) 2(3H)-furanone	180	70
BS15	18	26.63	1.7, -	Furo [3, 4-f][1,3] benzodioxole-5,7-dione	192	65
BS15	19	26.78	2.9, -	Unknown	97	-
BS15	20	27.06	8.7, -	Unknown	127	-
BS15	21	28.04	1.8, -	Unknown	127	-

[a] Peak numbers correspond to the numbering shown in Figure 1 (CI-4) or Figure 4 (BS15). [b] The labels PD and DC refer, respectively, to the potato dextrose broth and the degraded cellulose. The areas listed are the relative peak areas. [c] All the assignments of structure were made on the basis of the match to the National Institute of Standards and Technology (NIST) database. [d] Qual. refers to the highest listed quality value for the peaks that occur in both the growth media ort, for the peaks that occur only in a single medium, to the value from that solution.

To verify that the volatile compounds produced from CI-4 and grown on the degraded cellulose are the result of the presence of simple sugars rather than residual cellulose, a control containing cellulose not degraded by the mechanocatalytic process was also prepared for comparison (see Materials and Methods). After two weeks of incubation on this medium, CI-4 showed no growth.

A notable difference between the compounds extracted here by the solid-phase extraction and the previous study of CI-4 [46,47] is that the solid-phase extraction failed to recover some of the early eluting peaks. As discussed below, a liquid–liquid extraction demonstrated that these compounds are, in fact, present, and the cause of their omission from the solid-phase extraction sample is currently under investigation.

3.2. *Phylogenetic Characterization of a New Hypoxylon sp., BS15*

Recently a second fungus producing volatile organic products was isolated from a Bald Cyprus (*Taxodium distichum*) near Orange City, FL (USA). This fungus, designated BS15, was selected for study based on the serendipitous observation that compounds having a distinctive odor were produced.

The identification of BS15 involved extracting genomic DNA, amplifying and sequencing its ribosomal internally transcribed spacer regions (ITS), and then applying an improved bioinformatics analysis based on existing methods. The detection of flanking ribosomal genes in the BS15 sequence by ITSx allowed for the extraction of full-length ITS1 and ITS2 sub-sequences, a critical factor for producing alignments where gap site data is utilized in the phylogenetic analyses [48]. Independent BLAST searches using these sub-sequences to query the UNITE+INSDC database returned alignments with species exclusively of the taxonomic family *Xylariaceae*. Therefore, all public sequence records pertaining to the family *Xylaraiaceae* were comprehensively screened. The ITS sequences were detected by ITSx in 3443 of 3470 records from 394 unique binomial species.

A notable discrepancy regarding the naming and classification of organisms described in this work is the recent recognition of the family *Hyopxylaceae* by INSDC, whose members were previously included within *Xylariaceae* [49]. However, these records have not yet been updated in UNITE at this time. For the present work, non-*Hypoxylaceae* species were included in the alignment and clustering but pruned from the ITS phylogram with the exception of *Xylaria hypoxylon*, presented as a rooted out-group (Figure 2).

The relative richness of the full-length sequence records and the consistency in the sequence length made ITS2 a more favorable target for multiple alignment than ITS1 for the family *Xylariaceae*. An analysis of ITSx outputs revealed a bias for sequences containing the large ribosomal subunit sequence (LSU) compared with the small subunit (SSU) sequences among the UNITE records for the family *Xylariaceae*. Because the detection of these flanking ribosomal sequences is required for the full-length extraction of ITS sequences by ITSx, there were nearly twice the number of full-length ITS2 sequences (n = 2165) available for alignment compared with ITS1 (n = 1212). The sequence lengths were considerably less variable for ITS2 (SD = 5) than ITS1 (SD = 51).

Our taxonomic evaluations are consistent with other authors who found protein-coding genes more congruent with phenotypic observations than non-coding ITS sequences for *Hypoxylon* and related genera [41]. The phylograms generated from the ITS2 sequences were remarkably unresolved regardless of the alignment and clustering methods, with several genera not clustered into the monophyletic groups (e.g., *Annulohypoxylon* spp., *Daldinia* spp.) (Figure 2). Both the α-actin and β-tubulin genetic analyses were able to fully resolve these taxa, albeit with fewer specimens (n = 78) than ITS (Figure 3). For all three genetic markers, the fungal strain BS15 was consistently clustered among *Hypoxylon* spp. and most closely associated with *H. investiens*.

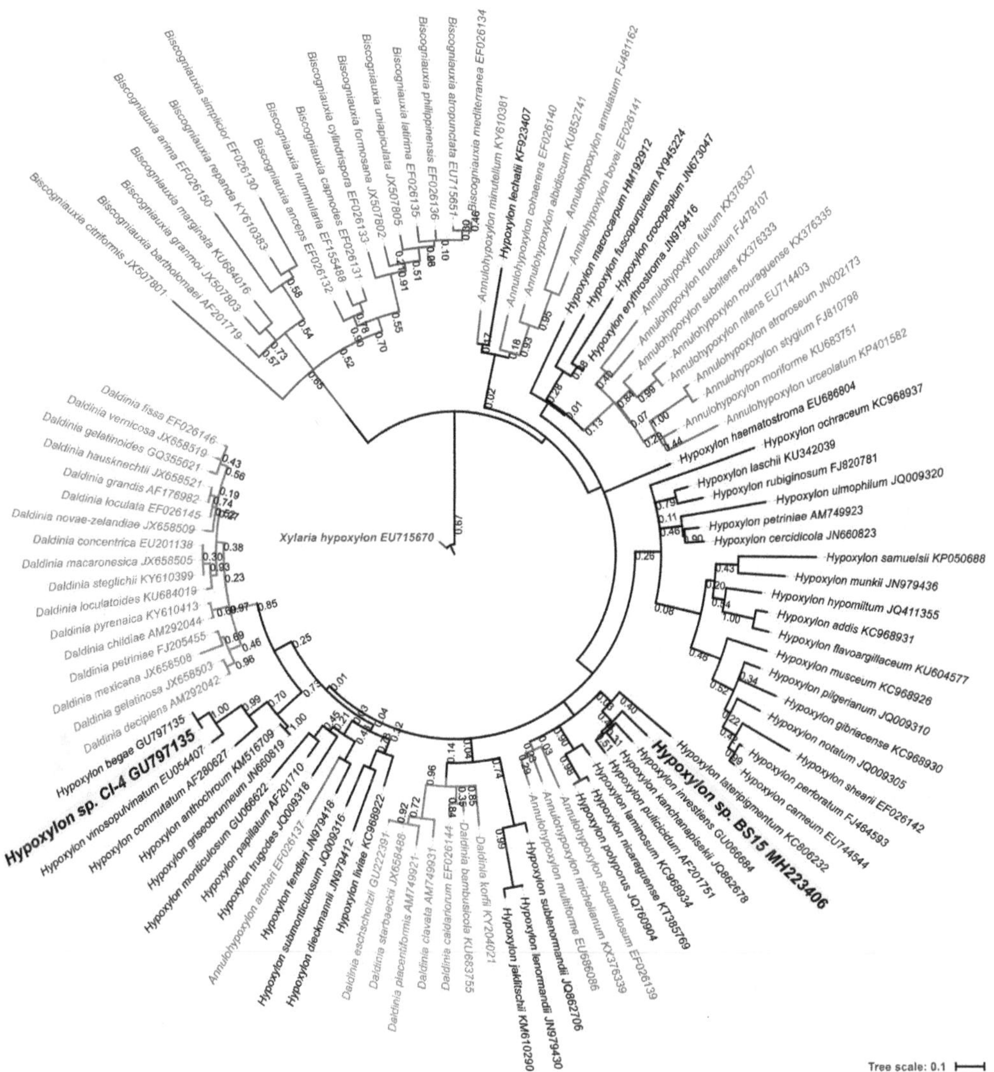

Figure 2. Phylogenetic reconstruction of *Hypoxylon* sp. BS15 and related organisms generated from maximum likelihood clustering of MUSCLE-aligned ITS2 sequences. Branch lengths are drawn to scale, representing the average number of nucleotide substitutions per site between the sequences. 121 nodes were selected for inclusion in the present figure from 331 nodes in the original phylogram. The bootstrap values at the nodes are from 1000 bootstrap iterations.

3.3. *Growth of BS15 on Degraded Cellulose and Analysis of Volatile Hydrocarbons*

In order to more generally evaluate the suitability of the degraded cellulose as a carbon source for fungi, BS15 was also evaluated for its ability to grow on the material. The procedure described above using two separate diets was employed with BS15 growth. The first included the degraded cellulose as the sole carbon source (see experimental), and the second contained potato dextrose broth. Both cultures exhibited strong fungal growth with mycelium covering the entire surface of the liquid media in approximately two weeks. The hydrocarbon fraction was extracted after three weeks using the solid-phase extraction process described above. A GC/MS analysis exhibited a large number of volatile products. A chromatographic comparison of the volatile compounds produced from each growth condition is illustrated in Figure 4, with tentative structural assignments and molecular weights listed in Table 2. The structures of the compounds listed in Table 2 are illustrated in Figure 5. In the case of BS15, both diets produced compounds **1**, **2**, **3**, **5**, and **7** but all the other products differed depending on the diet employed. Another notable difference in comparison with CI-4 is that BS15 on the degraded cellulose diet produced approximately the same amounts of volatile products as the potato dextrose diet.

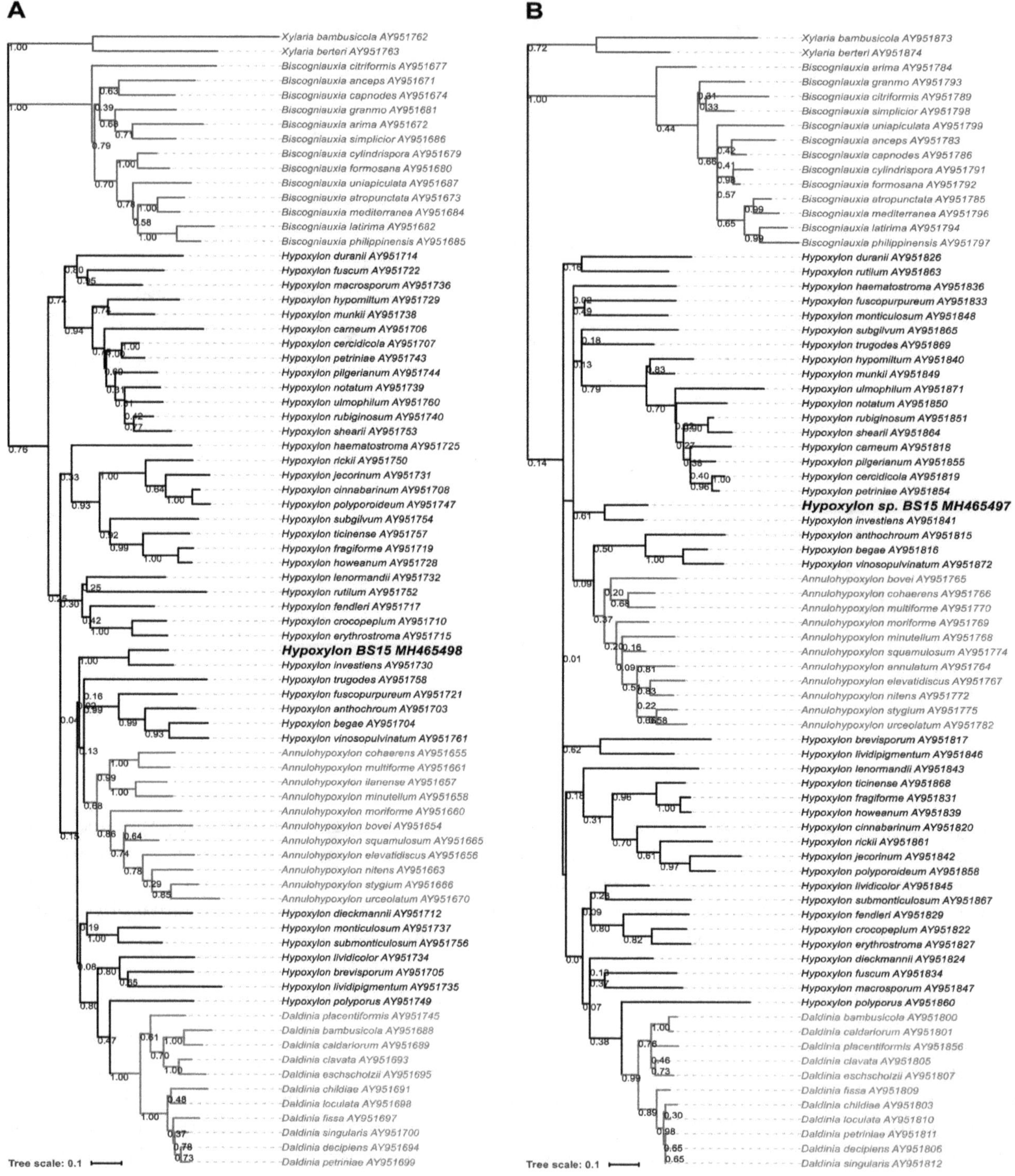

Figure 3. Phylogenetic reconstruction of *Hypoxylon* sp. BS15 and related organisms generated from maximum likelihood clustering of MUSCLE-aligned protein-coding gene sequences. A total of 78 sequences were analyzed for α-actin (**A**) and β-tubulin (**B**). The branch lengths are shown to scale, representing the average number of nucleotide substitutions per site between sequences. The bootstrap values at the nodes are from 1000 bootstrap iterations.

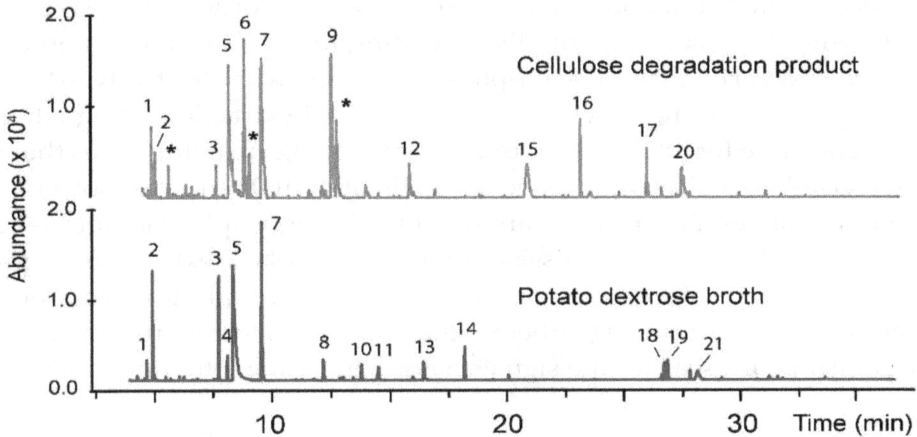

Figure 4. A gas chromatogram showing the volatile organic products produced by the fungus BS15 growth on a conventional media (**bottom**) and on carbohydrates produced from the degraded cellulose (**top**). The asterisks (*) denote volatile contaminants occurring in the degraded cellulose media as determined by analyzing a control sample with no BS15 added. The tentative identification and nominal masses of the individual peaks numbered in the chromatogram are given in Table 2.

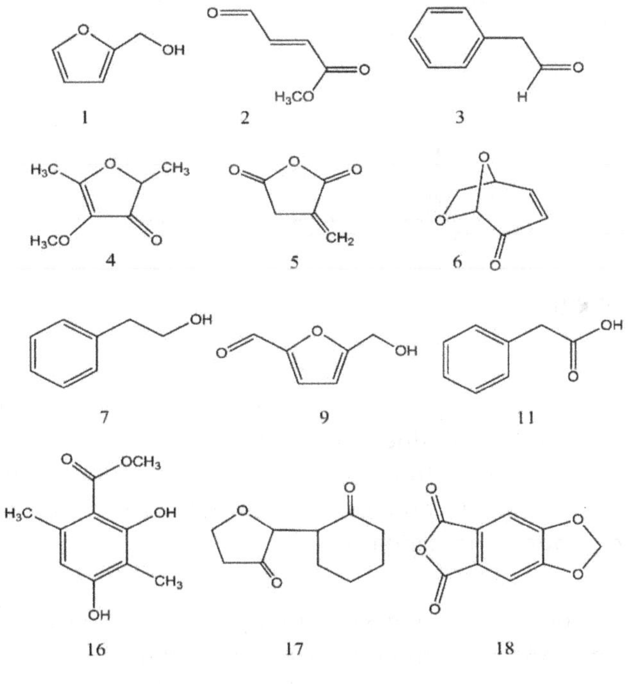

Figure 5. Structures of the volatile compounds tentatively identified from BS15 by comparison with the mass spectra in the NIST database. The compound numbers correspond to the peak numbers listed in Table 2 and in Figure 4.

In order to verify that the volatile compounds resulting from growth of BS15 on the degraded cellulose were being generated from simple sugars rather than from residual cellulose, a control was prepared containing non-degraded cellulose as described above for CI-4. After two weeks of growth on this cellulose medium, no growth was observed.

3.4. Comparing Solid-Phase and Liquid–Liquid Extraction Methods

All extractions of the volatile compounds in this study were performed using a solid-phase extraction with a C-18 stationary phase. This method is widely viewed as a "green chemistry" alternative to liquid–liquid extractions that require two–three orders of magnitude less solvent. However, it comes with the risk of potentially extracting fewer or different compounds from the growth media. In order to verify that the solid-phase extraction was effectively extracting the growth media, a direct comparison was made versus a liquid–liquid extraction using ethyl acetate/water. The comparisons were made for CI-4 and for BS15. For CI-4, Figure 6 illustrates that the solid-phase extraction gave very similar results, while using approximately 100 times less solvent. The differences in the relative amounts of certain products are notable. For example, the liquid–liquid extraction includes 1,8-cineole, while this product is missing from the solid-phase extraction. At present, the cause of this difference is unknown and further study is needed. Overall, however, the majority of the compounds extracted are the same using either method, and each technique extracts similar amounts as judged by the similar peak areas for the signals having the same retention times.

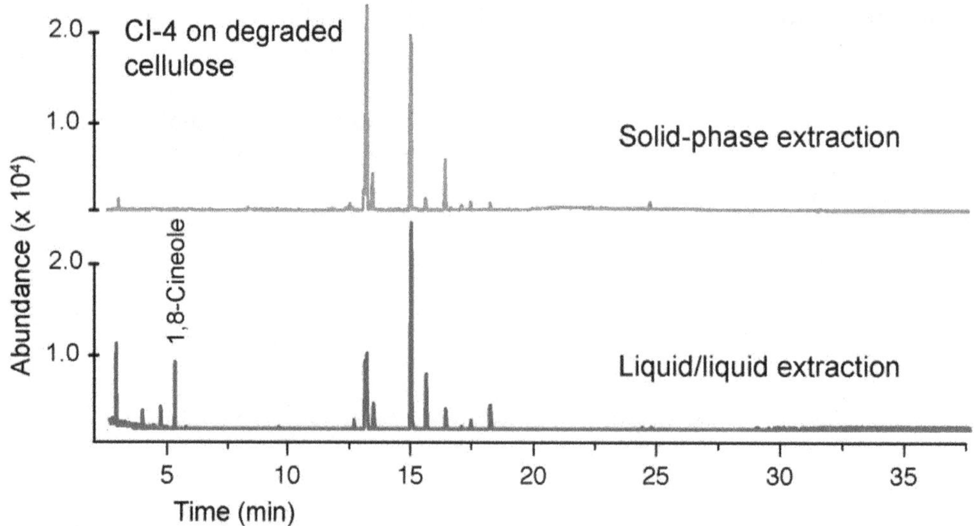

Figure 6. A comparison of the solid-phase extraction versus the liquid–liquid extraction (ethyl acetate/water). The growth media contained the degraded cellulose as a carbon source. A notable difference is the presence of 1,8-cineole in the liquid–liquid extraction. A comparison of the potato dextrose broth gave very similar results and, therefore, is not shown. The identity of 1,8-cineole was verified by comparison to an authentic standard.

A similar comparison of the effectiveness of the liquid–liquid extraction versus the solid-phase extraction was performed using BS15 grown on the degraded cellulose (Figure 7). Here, the differences between the methods were more pronounced with the solid-phase process extracting a more diverse range of products than did the liquid–liquid extraction. The absence of cineole in Figure 7 is a notable omission.

The solid-phase extraction process used herein differs from the vast majority of prior studies on volatile products from fungi, which have utilized solid-phase micro-extraction (SPME). This choice was made because SPME preferentially measures compounds having significant populations in the vapor phase and thus biases analysis against materials having low vapor pressures. Because compounds having potential use as fuels may have low vapor pressure, a methodology was employed here that includes these compounds. Admittedly, the use of a solid-phase extraction (SPE) represents a significant deviation from the common practice in the analysis of volatiles, and future work is needed to directly compare SPE and SPME to clearly identify the advantages and limitations.

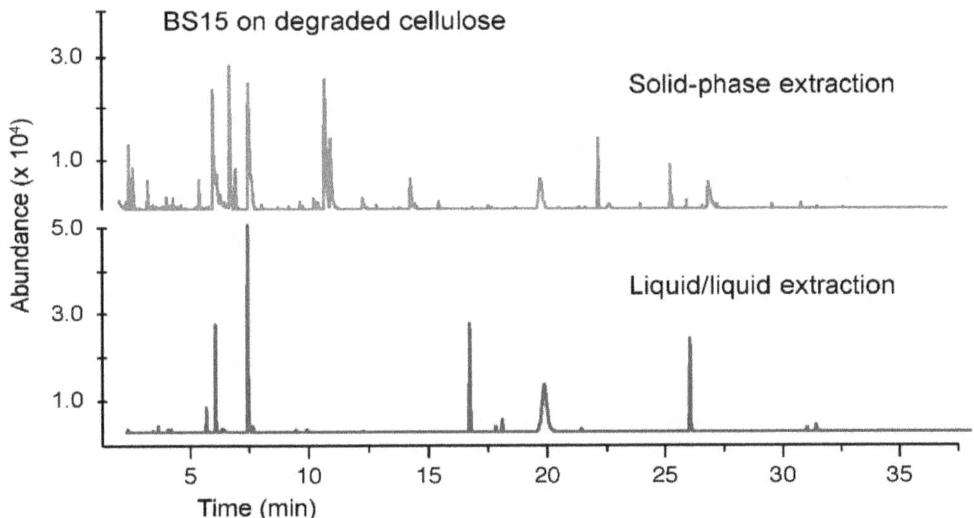

Figure 7. A comparison of the solid-phase extraction versus the liquid–liquid extraction (ethyl acetate/water). The growth media contained BS15 grown on the degraded cellulose as a carbon source.

4. Conclusions

The work described herein establishes the ability of certain endophytic fungi to convert a mechanochemically degraded cellulose product into volatile organic products with potential relevance as fuels. This process has been shown to be feasible with two different *Hypoxylon* sp. to demonstrate that the results are not limited to a single organism. One of the fungi employed (BS15) is described here for the first time, and the phylogenetic analysis demonstrated that it is substantially divergent from any other named species. BS15 produced a range of volatile products that differed significantly from those previously described from CI-4, emphasizing the importance of intra-generic species variation in metabolic studies. Here, the measurement of the total concentration of volatile products is not reported, because the quantification of the individual peaks in the chromatograms is not possible without standards and, at present, several products remain unknown. Our future work will focus on a more complete characterization of the individual compounds and the measurement of the total production.

Author Contributions: Conceptualization, T.M. and R.G.B.; Data curation, M.R.; Formal analysis, T.M., Y.W., and A.H.K.; Methodology, A.H.K. and S.D.M.; Supervision, S.D.M. and J.K.H.; Writing of original draft, J.K.H.

Acknowledgments: The authors would like to thank Gary Strobel (Montana State University) for providing a sample of the fungus CI-4 for the work described herein. Domenic Valenti is acknowledged for initially isolating the fungus BS15 from an area along the Saint Johns River and near Orange City, FL, USA.

References

1. Stone, J.K.; Bacon, C.W.; White, J.F., Jr. An overview of endophytic microbes: Endophytism defined. In *Microbial Endophytes*; Bacon, C.W., White, J.F., Jr., Eds.; Marcel Dekker: New York, NY, USA, 2000; pp. 29–33. ISBN 0.8247-8831-1.
2. Strobel, G.A.; Daisy, B.; Castillo, U.C.; Harper, J.K. Natural products from endophytic microorganisms. *J. Nat. Prod.* **2004**, *67*, 257–268. [CrossRef] [PubMed]
3. Strobel, G.A.; Knighton, B.; Kluck, K.; Ren, Y.; Livinghouse, T.; Griffin, M.; Spakowicz, D.; Sears, J. The production of myco-diesel hydrocarbons and their derivatives by the endophytic fungus Gliocladium roseum (NRRL 50072). *Microbiology* **2008**, *154*, 3319–3328. [CrossRef] [PubMed]
4. Strobel, G.A. Bioprospecting—Fuels from fungi. *Biotechnol. Lett.* **2015**, *37*, 973–982. [CrossRef] [PubMed]

5. Strobel, G. Muscodor species-endophytes with biological promise. *Phytochem. Rev.* **2011**, *10*, 165–172. [CrossRef]
6. Banerjee, D.; Strobel, G.A.; Booth, E.; Geary, B.; Sears, J.; Spakowicz, D.; Busse, S. An endophytic *Myrothecium inundatum* producing volatile organic compounds. *Mycosphere* **2010**, *1*, 229–240.
7. Ul-Hassan, S.R.; Strobel, G.A.; Booth, E.; Knighton, B.; Floerchinger, C.; Sears, J. Modulation of volatile organic compound formation in the Mycodiesel-producing endophyte *Hypoxylon* sp. CI-4. *Microbiology* **2012**, *158*, 465–473. [CrossRef] [PubMed]
8. Singh, S.K.; Strobel, G.A.; Knighton, B.; Geary, B.; Sears, J.; Ezra, D. An Endophytic *Phomopsis* sp. possessing bioactivity and fuel potential with its volatile organic compounds. *Microb. Ecol.* **2011**, *61*, 729–739. [CrossRef] [PubMed]
9. Griffin, M.A.; Spakowicz, D.J.; Gianoulis, T.A.; Strobel, S.A. Volatile organic compound production by organisms in the genus *Ascocoryne* and a re-evaluation of myco-diesel production by NRRL 50072. *Microbiology* **2010**, *156*, 3814–3829. [CrossRef] [PubMed]
10. Strobel, G.; Singh, S.K.; Riyaz-Ul-Hassan, S.; Mitchell, A.M.; Geary, B.; Sears, J. An endophytic/pathogenic *Phoma* sp. from creosote bush producing biologically active volatile compounds having fuel potential. *FEMS Microbiol. Lett.* **2011**, *320*, 87–94. [CrossRef] [PubMed]
11. Zhao, G.; Yin, G.; Inamdar, A.A.; Luo, J.; Zhang, N.; Yang, I.; Buckley, B.; Bennett, J.W. Volatile organic compounds emitted by filamentous fungi isolated from flooded homes after Hurricane Sandy show toxicity in a *Drosophila* bioassay. *Indoor Air* **2017**, *27*, 518–528. [CrossRef] [PubMed]
12. Siddiquee, S.; Azad, S.A.; Abu Bakar, F.; Naher, L.; Vijay Kumar, S. Separation and identification of hydrocarbons and other volatile compounds from cultures of *Aspergillus niger* by GC–MS using two different capillary columns and solvents. *J. Saudi Chem. Soc.* **2015**, *19*, 243–256. [CrossRef]
13. Fiers, M.; Lognay, G.; Fauconnier, M.-L.; Jijakli, M.H. Volatile Compound-Mediated Interactions between Barley and Pathogenic Fungi in the Soil. *PLoS ONE* **2013**, *8*, e66805. [CrossRef] [PubMed]
14. Savelieva, E.I.; Gustyleva, L.K.; Kessenikh, E.D.; Khlebnikova, N.S.; Leffingwell, J.; Gavrilova, O.P.; Gagkaeva, T.Y. Study of the vapor phase over *Fusarium* fungi cultured on various substrates. *Chem. Biodivers.* **2016**, *13*, 891–903. [CrossRef] [PubMed]
15. Mitchell, A.M.; Strobel, G.A.; Moore, E.; Robison, R.; Sears, J. Volatile antimicrobials from *Muscodor crispans*, a novel endophytic fungus. *Microbiology* **2010**, *156*, 270–277. [CrossRef] [PubMed]
16. Meshram, V.; Kapoor, N.; Saxena, S. *Muscodor kashayum* sp. nov.—A new volatile anti-microbial producing endophytic fungus. *Mycology* **2013**, *4*, 196–204. [CrossRef] [PubMed]
17. Meshram, V.; Saxena, S.; Kapoor, N. *Muscodor strobelii*, a new endophytic species from South India. *Mycotaxon* **2014**, *128*, 93–104. [CrossRef]
18. Saxena, S.; Meshram, V.; Kapoor, N. *Muscodor darjeelingensis*, a new endophytic fungus of Cinnamomum camphora collected from northeastern Himalayas. *Sydowia* **2014**, *66*, 55–67.
19. Saxena, S.; Meshram, V.; Kapoor, N. *Muscodor tigerii* sp. nov.-Volatile antibiotic producing endophytic fungus from the Northeastern Himalayas. *Ann. Microbiol.* **2015**, *65*, 47–57. [CrossRef]
20. Suwannarach, N.; Kumla, J.; Bussaban, B.; Hyde, K.D.; Matsui, K.; Lumyong, S. Molecular and morphological evidence support four new species in the genus *Muscodor* from northern Thailand. *Ann. Microbiol.* **2013**, *63*, 1341–1351. [CrossRef]
21. Kudalkar, P.; Strobel, G.; Riyaz-Ul-Hassan, S.; Geary, B.; Sears, J. *Muscodor sutura*, a novel endophytic fungus with volatile antibiotic activities. *Mycoscience* **2012**, *53*, 319–325. [CrossRef]
22. Zhang, C.L.; Wang, G.P.; Mao, L.J.; Komon-Zelazowska, M.; Yuan, Z.L.; Lin, F.C.; Druzhinina, I.S.; Kubicek, C.P. *Muscodor fengyangensis* sp. nov. from southeast China: Morphology, physiology and production of volatile compounds. *Fungal Biol.* **2010**, *114*, 797–808. [CrossRef] [PubMed]
23. Samaga, P.V.; Rai, V.R.; Rai, K.M.L. *Bionectria ochroleuca* NOTL33-an endophytic fungus from *Nothapodytes foetida* producing antimicrobial and free radical scavenging metabolites. *Ann. Microbiol.* **2014**, *64*, 275–285. [CrossRef]
24. Naznin, H.A.; Kiyohara, D.; Kimura, M.; Miyazawa, M.; Shimizu, M.; Hyakumachi, M. Systemic Resistance Induced by Volatile Organic Compounds Emitted by Plant Growth-Promoting Fungi in *Arabidopsis thaliana*. *PLoS ONE* **2014**, *9*, e86882. [CrossRef] [PubMed]
25. Crespo, R.; Pedrini, N.; Juárez, M.P.; Dal Bello, G.M. Volatile organic compounds released by the entomopathogenic fungus *Beauveria bassiana*. *Microbiol. Res.* **2008**, *163*, 148–151. [CrossRef] [PubMed]

26. Shurin, J.B.; Burkart, M.D.; Mayfield, S.P.; Smith, V.H. Recent progress and future challenges in algal biofuel production. *F1000Research* **2016**, *5*, 2434. [CrossRef] [PubMed]
27. Nass, L.L.; Pereira, P.A.A.; Ellis, D. Biofuels in Brazil: An overview. *Crop Sci.* **2007**, *47*, 2228–2237. [CrossRef]
28. Hicks, S.M.; Griebel, C.; Restrepo, D.T.; Truitt, J.H.; Baker, E.J.; Bylda, C.; Blair, R.G. Mechanocatalysis for biomass-derived chemicals and fuels. *Green Chem.* **2010**, *12*, 468–474. [CrossRef]
29. Strobel, G.A.; Dirkse, E.; Sears, J.; Markworth, C. Volatile antimicrobials from *Muscodor albus*, a novel endophytic fungus. *Microbiology* **2001**, *147*, 2943–2950. [CrossRef] [PubMed]
30. Riyaz-Ul-Hassan, S.; Strobel, G.A.; Geary, B.; Sears, J. An endophytic *Nodulisporium* sp. from Central America producing volatile organic compounds with both biological and fuel potential. *J. Microbiol. Biotechnol.* **2013**, *23*, 29–35. [CrossRef] [PubMed]
31. Boom, R.; Sol, C.J.A.; Salimans, M.M.M.; Jansen, C.L.; Wertheim-van Dillen, P.M.E.; van der Noordaa, J. Rapid and simple method for purification of nucleic acids. *J. Clin. Microbiol.* **1990**, *28*, 495–503. [PubMed]
32. Schoch, C.L.; Seifert, K.A.; Huhndorf, S.; Robert, V.; Spouge, J.L.; Levesque, C.; André; Chen, W. Nuclear ribosomal internal transcribed spacer (ITS) region as a universal DNA barcode marker for Fungi. *Proc. Natl. Acad. Sci. USA* **2012**, *109*, 6241–6246. [CrossRef] [PubMed]
33. Saiki, R.K.; Scharf, S.; Faloona, F.; Mullis, K.B.; Horn, G.T.; Erlich, H.A.; Arnheim, N. Enzymatic amplification of beta-globin genomic sequences and restriction site analysis for diagnosis of sickle cell anemia. *Science* **1985**, *230*, 1350–1354. [CrossRef] [PubMed]
34. Toju, H.; Tanabe, A.S.; Yamamoto, S.; Sato, H. High-Coverage ITS Primers for the DNA-Based Identification of Ascomycetes and Basidiomycetes in Environmental Samples. *PLoS ONE* **2012**, *7*, e40863. [CrossRef] [PubMed]
35. Carbone, I.; Kohn, L.M. A method for designing primer sets for speciation studies in filamentous ascomycetes. *Mycologia* **1999**, *91*, 553–556. [CrossRef]
36. O'Donnell, K.; Cigelnik, E. Two divergent intragenomic rDNA ITS2 types within a monophyletic lineage of the fungus Fusarium are nonorthologous. *Mol. Phylogenet Evol.* **1997**, *7*, 103–116. [CrossRef] [PubMed]
37. Cock, P.J.A.; Antao, T.; Chang, J.T.; Chapman, B.A.; Cox, C.J.; Dalke, A.; Friedberg, I.; Hamelryck, T.; Kauff, F.; Wilczynski, B.; et al. Biopython: Freely available Python tools for computational molecular biology and bioinformatics. *Bioinformatics* **2009**, *25*, 1422–1423. [CrossRef] [PubMed]
38. Bengtsson-Palme, J.; Ryberg, M.; Hartmann, M.; Branco, S.; Wang, Z.; Godhe, A.; De Wit, P.; Sánchez-García, M.; Ebersberger, I.; de Sousa, F.; et al. ITSx: Improved software detection and extraction of ITS1 and ITS2 from ribosomal ITS sequences of fungi and other eukaryotes for use in environmental sequencing. *Methods Ecol. Evol.* **2013**, *4*, 914–919. [CrossRef]
39. Kõljalg, U.; Larsson, K.-H.; Abarenkov, K.; Nilsson, R.H.; Alexander, I.J.; Eberhardt, U.; Erland, S.; Høiland, K.; Kjøller, R.; Larsson, E.; et al. UNITE: A database providing web-based methods for the molecular identification of ectomycorrhizal fungi. *New Phytol.* **2005**, *166*, 1063–1068. [CrossRef]
40. Boratyn, G.M.; Camacho, C.; Cooper, P.S.; Coulouris, G.; Fong, A.; Ma, N.; Madden, T.L.; Matten, W.T.; McGinnis, S.D.; Merezhuk, Y.; et al. BLAST: A more efficient report with usability improvements. *Nucleic Acids Res.* **2013**, W29–W33. [CrossRef] [PubMed]
41. Hsieh, H.-M.; Ju, Y.-M.; Rogers, J.D. Molecular Phylogeny of *Hypoxylon* and Closely Related Genera. *Mycologia* **2005**, *97*, 844–865. [CrossRef] [PubMed]
42. Kumar, S.; Stecher, G.; Tamura, K. MEGA7: Molecular Evolutionary Genetics Analysis Version 7.0 for Bigger Datasets. *Mol. Biol. Evol.* **2016**, *33*, 1870–1874. [CrossRef] [PubMed]
43. Edgar, R.C. MUSCLE: Multiple sequence alignment with high accuracy and high throughput. *Nucleic Acids Res.* **2004**, *32*, 1792–1797. [CrossRef] [PubMed]
44. Holder, M.; Lewis, P.O. Phylogeny estimation: Traditional and Bayesian approaches. *Nat. Rev. Genet.* **2003**, *4*, 275–284. [CrossRef] [PubMed]
45. Letunic, I.; Bork, P. Interactive tree of life (iTOL) v3: An online tool for the display and annotation of phylogenetic and other trees. *Nucleic Acids Res.* **2016**, *44*, W242–W245. [CrossRef] [PubMed]
46. Tomsheck, A.R.; Strobel, G.A.; Booth, E.; Geary, B.; Spakowicz, D.; Knighton, B.; Floerchinger, C.; Sears, J.; Liarzi, O.; Ezra, D. *Hypoxylon* sp., an endophyte of *Persea indica*, producing 1,8-cineole and other bioactive volatiles with fuel potential. *Microb. Ecol.* **2010**, *60*, 903–914. [CrossRef] [PubMed]

47. Shaw, J.J.; Berbasova, T.; Sasaki, T.; Jefferson-George, K.; Spakowicz, D.J.; Dunican, B.F.; Portero, C.E.; Narváez-Trujillo, A.; Strobel, S.A. Identification of a fungal 1,8-cineole synthase from *Hypoxylon* sp. with common specificity determinants to the plant synthases. *J. Biol. Chem.* **2015**, *290*, 8511–8526. [CrossRef] [PubMed]
48. Warnow, T. Standard maximum likelihood analyses of alignments with gaps can be statistically inconsistent. *PLoS Curr.* **2012**, *4*, RRN1308. [CrossRef] [PubMed]
49. Wendt, L.; Sir, E.B.; Kuhnert, E.; Heitkämper, S.; Lambert, C.; Hladki, A.I.; Romero, A.I.; Luangsa-ard, J.J.; Srikitikulchai, P.; Peršoh, D.; et al. Resurrection and emendation of the Hypoxylaceae, recognised from a multigene phylogeny of the Xylariales. *Mycol. Prog.* **2018**, *17*, 115–154. [CrossRef]

7

Endophytic Fungi: A Source of Potential Antifungal Compounds

Sunil K. Deshmukh [1,*], Manish K. Gupta [1], Ved Prakash [2] and Sanjai Saxena [3]

1 TERI-Deakin Nano Biotechnology Centre, The Energy and Resources Institute (TERI), Darbari Seth Block, IHC Complex, Lodhi Road, New Delhi 110003, India; manish.gupta@teri.res.in
2 Department of Biotechnology, Motilal Nehru National Institute of Technology, Allahabad 211004, India; ved.prakash2012@vitalum.ac.in
3 Department of Biotechnology, Thapar Institute of Engineering & Technology, Deemed to be a University, Patiala, Punjab 147004, India; sanjaibiotech@yahoo.com
* Correspondence: sunil.deshmukh@teri.res.in or sunil.deshmukh1958@gmail.com

Abstract: The emerging and reemerging forms of fungal infections encountered in the course of allogeneic bone marrow transplantations, cancer therapy, and organ transplants have necessitated the discovery of antifungal compounds with enhanced efficacy and better compatibility. A very limited number of antifungal compounds are in practice against the various forms of topical and systemic fungal infections. The trends of new antifungals being introduced into the market have remained insignificant while resistance towards the introduced drug has apparently increased, specifically in patients undergoing long-term treatment. Considering the immense potential of natural microbial products for the isolation and screening of novel antibiotics for different pharmaceutical applications as an alternative source has remained largely unexplored. Endophytes are one such microbial community that resides inside all plants without showing any symptoms with the promise of producing diverse bioactive molecules and novel metabolites which have application in medicine, agriculture, and industrial set ups. This review substantially covers the antifungal compounds, including volatile organic compounds, isolated from fungal endophytes of medicinal plants during 2013–2018. Some of the methods for the activation of silent biosynthetic genes are also covered. As such, the compounds described here possess diverse configurations which can be a step towards the development of new antifungal agents directly or precursor molecules after the required modification.

Keywords: endophytic fungi; antifungal compounds; medicinal plants; co-culture; epigenetic modification

---

1. Introduction

The undisputed potential of fungi to produce bioactive secondary metabolites has long been established. To date, merely 5% of the entire number of species has been elaborated of an estimated 1.5 million. Out of these (69,000 species), merely 16% (11,500) have been cultured [1]. Amongst the fungi, endophytes represent a wide source of unexplored and uncharacterized microorganisms capable of producing novel metabolites. Endophytes generally exist asymptomatically, coexisting with their hosts and representing an underutilized group of microorganisms for the discovery of new compounds. Endophytes produce diverse metabolites and have the ability to synthesize compounds which are solely produced and isolated from higher plants [2,3]. Strobel and Daisy [4] commented that endophytes could be a goldmine of secondary metabolites. *Pestalotiopsis* sp. can be considered as "the *E. coli* of the rain forests" and *P. microspora*, a "microbial factory" of bioactive secondary metabolites. As per them, numerous chemical structures such as Ambuic acid, Cryptocandin, Taxol, Torreyanic acid, Subglutinol A and B, and many others have been identified. An array of metabolites of different chemical classes profiles have been deciphered, such as alkaloids, cytochalasines, flavonoids, furandiones, phenylpropanoids, lignans, peptides, phenol, phenolic acids, steroids, terpenoids, quinones, aliphatic acid, and chlorinated compounds. Secondary metabolites derived from endophytes comprise classes of compounds such as steroids, xanthones, phenols, isocoumarins, perylene derivatives, quinines, furandiones, terpenoids, depsipeptides, and cytochalasins, which are identified to possess biological activities with antibiotic, antiviral, volatile antibiotic, anticancer, antioxidant, insecticidal, antidiabetic, and immunosuppressive properties [5–9]. Endophytes play a major role in the physiological activities of host plants, influencing the enhancement of stress, insects, nematodes, and disease resistance [10–13].

This review covers the antifungal fungal metabolites reported from endophytic fungi from medicinal plants during 2013–2018 and their potential as antifungal agents. The antifungal activity of these compounds against the selected fungal pathogens are described briefly and some details such as producing organisms, plant sources, place of collections, and the antifungal properties of many of these compounds are shown in Tables 1 and 2.

2. Medicinal Plants

2.1. Compounds Produced by Coelomycetes

Pestalotiopsis is an important genus of coelomycetes and different species of this genus have been identified for production of bioactive compounds for various biological properties which include antimicrobial, antifungal, antiviral, antineoplastic, and antioxidant activities [14]. Some of the antifungal reported from this genus includes a new monoterpene lactone, (3R,4R,6R,7S)-7-hydroxyl-3,7-dimethyl-oxabicyclo[3.3.1]nonan-2-one (**1**) (Figure 1), along with one related known compound, (3R,4R)-3-(7-methylcyclohexenyl)-propanoic acid (**2**) (Figure 1), were discovered from endophytic fungus *Pestalotiopsis foedan* obtained from the branch of *Bruguiera sexangula* in Hainan, China. Compounds **1–2** showed antifungal activity against *Botrytis cinerea* and *Phytophthora nicotianae* with MIC values of 3.1 and 6.3 μg/mL, respectively, while the known antifungal drug ketoconazole showed comparable activity (MIC 3.1 μg/mL each) Compound **2** also exhibited satisfactory activity against *Candida albicans* (MIC value of 50 μg/mL) while ketoconazole showed MIC of 6.3 μg/mL [15].

Plants of *Dendrobium officinale* were collected in Yandang Mountain, Zhejiang Province, China. *Pestalotiopsis* sp. DO14 was obtained from the shoots of *D. officinale* endophytic fungus which yielded two novel antifungal constituents, (4S,6S)-6-[(1S,2R)-1,2-dihydroxybutyl]-4-hydroxy-4-methoxytetrahydro-2H-pyran-2-one (**3**) and (6S,2E)-6-hydroxy-3-methoxy-5-oxodec -2-enoic acid (**4**), and two known compounds, LL-P880γ (**5**) and LL-P880α (**6**) (Figure 1) were isolated. Compounds **3–6** exhibited good anti-fungal activities (MIC $\leq$ 50 μg/mL) against *C. albicans*, *Cryptococcus neoformans*, *Trichophyton rubrum*, and *Aspergillus fumigatus*. Compounds **3** and **4** possess the strong activities with the MIC values $\leq$25 μg/mL against tested strains [16].

Endophytic fungus *Pestalotiopsis fici* obtained from the branches of *Camellia sinensis* collected from the suburb of Hangzhou, China was the source of a new a-pyrone derivative ficipyrone A (**7**) (Figure 1). Compound **7** showed antifungal activity against the plant pathogen *Gibberella zeae* with an IC_{50} value of 15.9 μM (the positive control ketoconazole showed an IC_{50} value of 6.02 μM) [17].

Endophytic fungus *Pestalotiopsis mangiferae* associated with *Mangifera indica* Linn collected from Maduravoyal, Tamil Nadu Province, India was the source of a new phenolic compound 4-(2,4,7-trioxa-bicyclo[4.1.0]heptan-3-yl) phenol (**8**) (Figure 1). Compound **8** exhibited strong antifungal activity against *C. albicans* with MIC value of 0.039 μg/mL, while nystatin showed MIC 10.0 μg/mL [18].

Phomopsis is an important genus that is a prolific producer of bioactive compounds including Cytochalasin H (**9**) (Figure 1), which was isolated from the endophytic fungus *Phomopsis* sp. of *Senna spectabilis* (Fabaceae) collected from São Paulo, Brazil. Compound **9** exhibited activity against *Cladosporium cladosporioides* and *C. sphaerosphermum* with MIC values of 10.0 and 25.0 μg, respectively, while the MIC of nystatin the reference compound was 1.0 μg [19].

Endophytic fungus *Phomopsis* sp. isolated from *Aconitum carmichaeli* collected in Huize County, Yunnan Province, China yielded (14β,22E)-9,14-dihydroxyergosta-4,7,22-triene-3,6-dione (**10**) and (5α,6β,15β,22E)-6-ethoxy-5,15 -dihydroxyergosta-7,22-dien-3-one (**11**), calvasterol A (**12**), and ganodermaside D (**13**) (Figure 1). All compounds were evaluated for their antifungal activities against *Candida albicans, Aspergillus niger, Pyricularia oryzae, Fusarium avenaceum, Hormodendrum compactum*, and *Trichophyton gypseum*. Compound **10** exhibited average antifungal activities against *C. albicans, H. compactum*, and *A. niger*, with MIC values of 64, 64, and 128 μg/mL, respectively. Compound **11** showed poor inhibitory activity against *C. albicans* and *F. avenaceum* with MIC values of 128 μg/mL. Compounds **12** and **13** showed average inhibitory activities against *F. avenaceum* (MIC 64 μg/mL for both compounds). Only compound **12** exhibited weak antifungal activities against *P. oryzae* and *T. gypseum* (MIC of 128 and 256 μg/mL, respectively) [20].

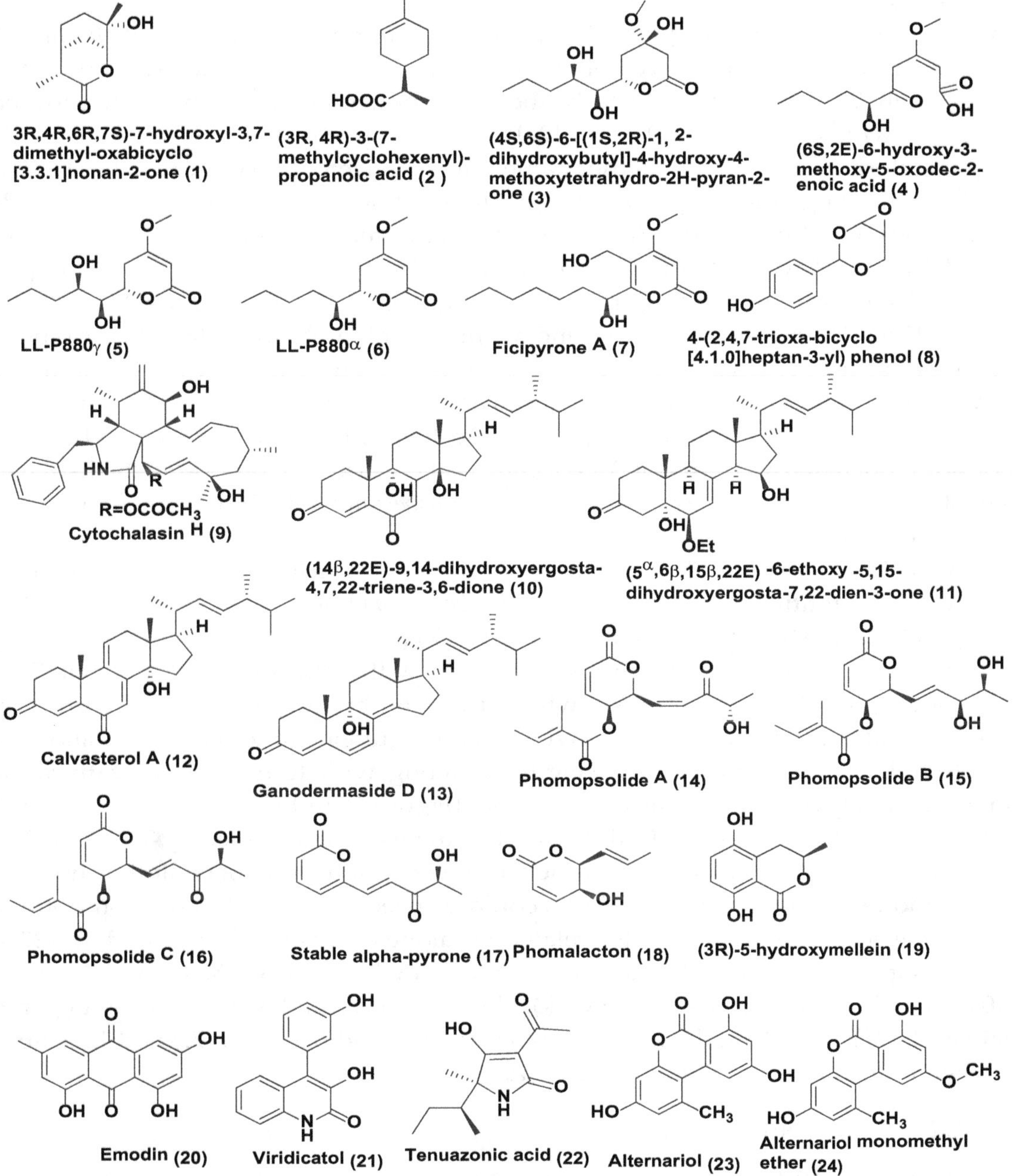

Figure 1. Structures of metabolites isolated from Coelomycetes (**1–24**).

Diaporthe maritima an endophytic fungus obtained from needles of *Picea* sp. in the Acadian forest of Eastern Canada yielded three dihydropyrones, phomopsolides A (**14**), B (**15**), and C (**16**), and a stable alpha-pyrone (**17**) (Figure 1). Compound **14** demonstrated growth inhibition at 25 μM against *Microbotryum violaceum* and *Saccharomyces cerevisiae* whereas Compounds **15–17** were active at 250 μM [21].

Another coelomycete *Phoma* is known to produce diverse compounds [22]. From *Phoma* sp. an endophytic fungus of the plant *Fucus serratus* yielded phomalacton (**18**), (3R)-5-hydroxymellein (**19**) and emodin (**20**) (Figure 1). Phomalactone (**18**), (3R)-5-hydroxymellein (**19**) and emodin (**20**) Compounds **18–20** exhibited antifungal activity against *Microbotryum violaceum* with 5, 6 and 5 mm zone of inhibition (0.05 mg was pipetted onto 9 mm sterile filter disk) [23].

Viridicatol (**21**) (Figure 1), tenuazonic acid (**22**), alternariol (**23**), and alternariol monomethyl ether (**24**) (Figure 1) were isolated from endophytic fungi *Phoma* sp. WF4 of *Eleusine coracana* grown under semi-hydroponic conditions Arkell Field Station, Arkell, ON, Canada. Compounds **21–24** caused reasonable breakage of *Fusarium graminearum* hyphae in vitro [24].

Endophytic fungus *Rhizopycnis vagum* Nitaf 22 obtained from the healthy root of *Nicotiana tabacum* grown at China Agricultural University Beijing, China was the source of Rhizopycnin D (**25**) and TMC-264 (**26**) (Figure 2). Compounds **25** and **26** showed strong inhibition of the spore germination of *Magnaporthe oryzae* with IC_{50} values of 9.9 and 12.0 μg/mL, respectively [25].

A new polychlorinated triphenyl diether named microsphaerol (**27**) (Figure 2) has been isolated from the endophytic fungus *Microsphaeropsis* sp. (internal strain No. 8883), and phytochemical investigation of the endophytic fungus *Seimatosporium* sp. (internal strain No. 8883) associated with *Salsola oppositifolia* from Playa del Ingles (Gomera, Spain) led to the isolation of a new naphthalene derivative named seimatorone (**28**) (Figure 2). In antifungal assay, compounds **27** and **28** showed activity against *Microbotryum violaceum* with 9 and 5 mm zone of inhibition (0.05 mg was pipetted onto 9 mm sterile filter paper disk). In addition, there was some growth within the zone of inhibition [26].

Endophytic fungus *Colletotrichum gloeosporioides* associated with *Michelia champaca* isolated from São Paulo State University (UNESP), Araraquara, São Paulo, Brazil was found to be source of a new compound, 2-phenylethyl 1H-indol-3-yl-acetate (**29**) (Figure 2). Compound **29** displayed good activity against *Cladosporium cladosporioides* and *C. sphaerospermum* which was analogous to nystatin, the positive control [27].

Colletonoic acid (**30**) (Figure 2) was isolated from *Colletotrichum* sp. from Gomera (Spain). Colletonoic acid exhibit antifungal activity against *Microbotryum violaceum* with 7 mm zone of inhibition (0.05 mg was pipetted onto 9 mm a sterile filter paper disk) [28].

Coniothyrium sp., an endophytic fungus associated with *Salsola oppostifolia* from Gomera in the Canary Islands, was the source of known hydroxy anthraquinones 1,7-dihydroxy3-methyl-9,10-anthraquinone (**31**), 1,6-dihydroxy-3-methyl-9,10-anthraquinone (phomarin) (**32**), and 1-hydroxy-3-hydroxymethyl-9,10-anthraquinone (**33**) (Figure 2) along with four new derivatives having a tetralone moiety, namely coniothyrinones A–C (**34–36**) (Figure 6) and D (**37**) (Figure 2). The absolute configurations of coniothyrinones A (**34**), B (**35**), and D (**37**) were determined by TDDFT calculations of CD spectra, allowing the determination of the absolute configuration of coniothyrinone C (**36**) as well. Coniothyrinones A (**34**), B (**35**), and D (**37**) could be used as ECD reference compounds in the determination of absolute configuration for related tetralone derivates. Compounds **31–37** showed inhibitory effects against the fungus *Microbotryum violaceum* with 7, 10, 8, 7.5, 6, 8 and 7.5 mm zone of inhibition (0.05 mg/9-mm sterile filter paper disk). Compounds **32** and **34** exhibited strong antifungal activity against *M. violaceum* (10 and 9 mm zone of inhibition) and *B. cinerea* (7.5 and 12.5 mm zone of inhibition) when tested under similar conditions [29].

2.2. Compounds Produced by Ascomycetes

Xylaria is very important ascomycetous genus and a good sources of novel bioactive compounds, and some of the compounds reported to have drug-able properties relevant for drug discovery [30,31]. Nine oxygenated guaiane-type sesquiterpenes (**38–46**) and three isopimarane diterpenes (**47–49**), (Figure 2) were obtained from *Xylaria* sp. YM 311647, an endophytic fungus associated with *Azadirachta indica* collected from Yuanjiang County, Yunnan Province, China. All compounds were evaluated for their antifungal activities against *Candida albicans, Aspergillus niger, Pyricularia oryzae, Fusarium avenaceum,* and *Hormodendrum compactum*. Compounds **38–46** were moderately active against *C. albicans* and *H. compactum* (MIC values ranging from 32 to 256 µg/mL), while compound **47–49** were more active against all the tested strains (MIC values ranging from 16 to 256 µg/mL). Compound **49** exhibited the most promising activity against *C. albicans* and *P. oryzae* with MIC values of 16 µg/mL [32].

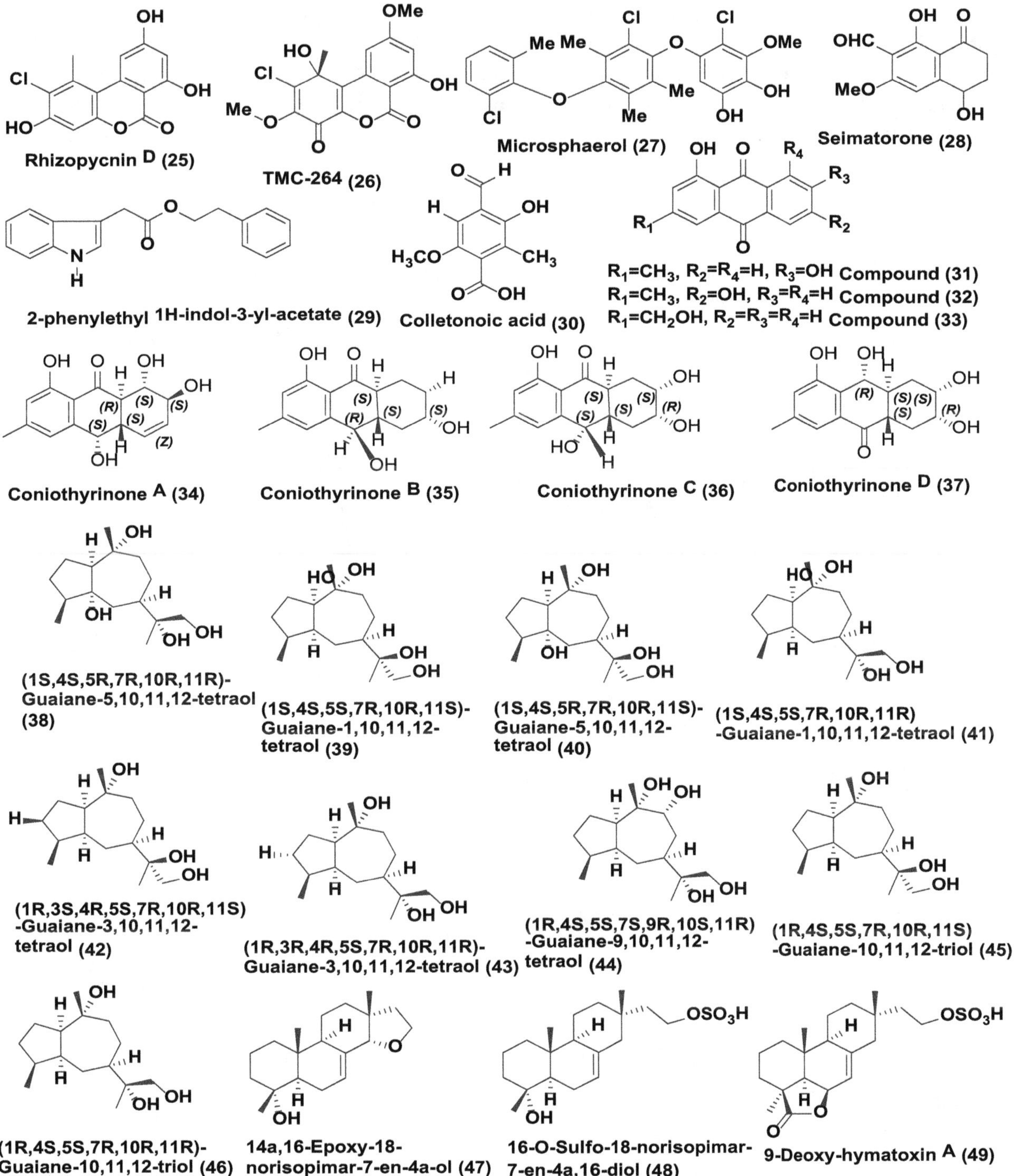

Figure 2. Structures of metabolites isolated from Coelomycetes (**25–37**) and Ascomycetes (**38–49**).

Endophytic fungus *Xylaria* sp. YM 311647 associated with *Azadirachta indica* from Yuanjiang County, China was also reported to produce five new guaiane sesquiterpenes, (**50–54**) (Figure 3). The antifungal activities of **50–54** were evaluated by means of the broth microdilution method against *C. albicans, A. niger, P. oryzae, F. avenaceum* and *H. compactum* Compounds **50–54** exhibited average or poor antifungal activities against *P. oryzae* and *H. compactum* (MIC values in the range of 32–256 μg/mL). Among them, **53** exhibited the most promising inhibitory activity against *P. oryzae* with a MIC value of 32 μg/mL. Compounds **52** and **53** showed average antifungal activities against *H. compactum* with MIC values of 64 μg/mL. In addition, **53** and **54** exhibited the most promising antifungal activities against *C. albicans* with MIC values of 32 μg/mL. Compound **52** showed average inhibitory activities against *C. albicans, A. niger*, and *H. compactum* with MIC values of 64 μg/mL. All compounds showed no notable inhibitory activities against *Fusarium avenaceum* [33]. Amazonian endophytic fungus *X. feejeensis* residing in *Croton lechleri* yielded nonenolide, xyolide (**55**) (Figure 3). Compound **55** exhibited antifungal activity against oomycetes *Pythium ultimum* with a MIC value of 425 μM [34].

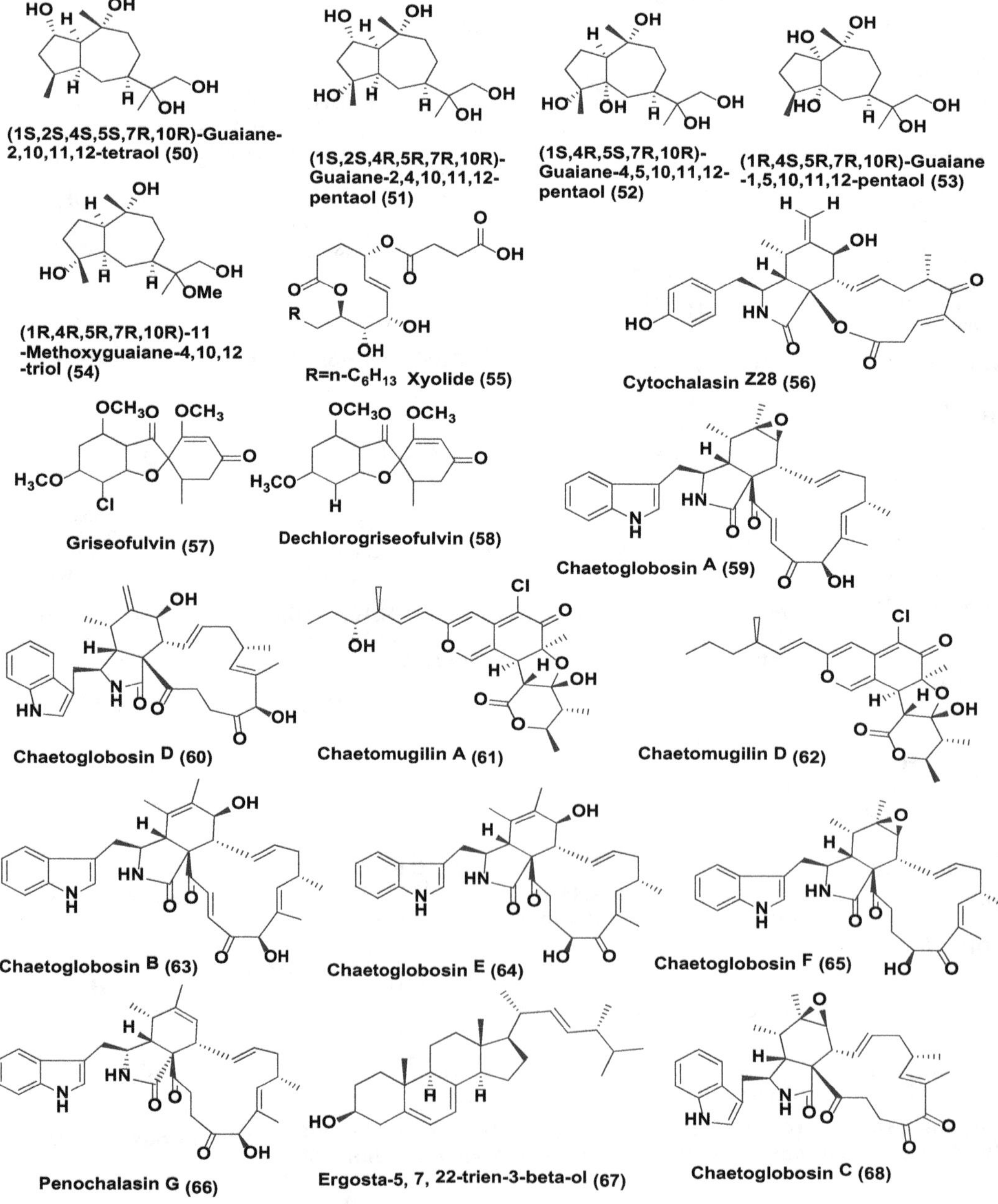

Figure 3. Structures of metabolites isolated from Ascomycetes (**50–68**).

Endophytic fungus *Xylaria sp.* XC-16 associated with *Toona sinensis* was isolated from Yangling, Shaanxi Province, China and was observed to produce a potent antifungal compound Cytochalasin Z28 (**56**) (Figure 3), displaying enhanced activity with an MIC of 12.5 μM as opposed to the antifungal activity possessed by hymeaxszol possessing an MIC value of 25 μM against the plant pathogen *Gibberella saubinetti* [35].

Various isolates of *Xylaria* produce griseofulvin (**57**) (Figure 3); *Xylaria sp.* PSU G12 associated with *Garcinia hombroniana* [36], and *X. cubiensis* residing in *Asimina triloba* [37], along with 13 strains of *Xylaria* sp. inhabiting *Pinus strobus* and six strains associated with *Vaccinum augustifolium* found in the Acadian forest of New Brunswick and Nova Scotia Canada [38] are known to produce griseofulvin and a few of them can also produce dechlorogriseofulvin (**58**) (Figure 3) [36,38].

Griseofulvin (**57**) is very potent against the phytopathogenic fungi, but not against oomycetes [38]. Griseofulvin has been validated to possess antifungal activity against *Alternaria mali, B. cinerea, C. gloeosporioides, Corticium sasaki, Fusarium oxysporum* and *Magnaporthe grisea* in vitro with IC_{50} values of 18.0, 5.0, 1.7, 11.0, 30.0, and 1.7 μg/mL, respectively. Dechlorogriseofulvin (**58**) demonstrated poor activity, with an IC_{50} value of 200 μg/mL for each fungus. Griseofulvin (**57**) also hinders the growth of *M. grisea, C. sasaki, B. cinerea, Puccinia recondite* and *Blumeria graminis* f. sp. *hordei* in vivo, with a percentage of fungal control of 95, 100, 60, 90 and 90, respectively, at 150 μg/mL. Griseofulvin (**57**) is used to cure dermatophytic infections caused by fungi such as *Epidermophyton* and *Trichophyton* species [38].

Chaetomium is another genus of ascomycete and prolific producer of bioactive compounds [39,40]. Chaetoglobosin A (**59**) and D (**60**) (Figure 3) were isolated *Chaetomium globosum* CDW7, an endophyte from *Ginkgo biloba* located in Taixing and Nanjing in Jiangsu Province and Chengdu in Sichuan Province, China. Compounds **59** and **60** showed antifungal activity against *Sclerotinia sclerotiorum* with IC_{50} values of 0.35 and 0.62 μg/mL, respectively, compared with carbendazim (0.17 μg/mL) [41].

Compounds Chaetomugilin A (**61**), Chaetomugilin D (**62**) Chaetoglobosin A (**59**), Chaetoglobosin B (**63**), Chaetoglobosin E (**64**), Chaetoglobosin F (**65**), and Penochalasin G (**66**) (Figure 3) were obtained from *C. globosum* endophyte obtained from seeds of *Panax notoginseng* collected at the Wenshan, Yunnan, China. Compounds **59** and **61–66** exhibited antifungal activity against *Phoma herbarum* (MIC in the range of 16–128 μg/mL) and *Epicoccum nigrum* (MIC in the range of <1–16 μg/mL). Both fungi are phytopathogenic fungi causing root rot of *Panax notoginseng* [42].

Ergosta-5,7,22-trien-3beta-ol (**67**) (Figure 3) was isolated from *Chaetomium cupreum* ZJWCF079 of *Macleaya cordata.* It exhibited antifungal activity against *Sclerotinia sclerotiorum* and *B. cinereal*, plant pathogenic fungi with EC_{50} values of 125 μg/mL and 190 μg/mL respectively, but had no effects on *Pythium ultimum, Rhizoctonia solani* and *F. oxysporum* [43].

Chaetoglobosin A (**59**) (Figure 3), D (**60**), E (**64**), C (**68**), (Figure 3) G (**69**), and R (**70**) (Figure 4) were isolated from *Chaetomium globosum* No.04 obtained from barks of *Ginkgo biloba*, growing in Linyi, Shandong Province, China. Compounds **59**, **60**, **64**, and **68–70** showed good growth inhibitory activity at a concentration of 20 μg/disk, against *Rhizopus stolonifer* and *Coniothyrium diplodiella* [44].

A new tetranorlabdane diterpenoids botryosphaerin H (**71**) and a known tetranorlabdane diterpenes 13,14,15,16-tetranorlabd-7-en-19,6β:12,17-diolide (**72**) (Figure 4) were obtained from *Botryosphaeria* sp. P483, an endophyte of *Huperzia serrata* collected in Xichou County, Yunnan Province, China. When tested at 100 μg/disk, compound **71** showed zone of inhibition of 9, 7, 7, 8, and 8 mm, against *Gaeumannomyces graminis, Fusarium solani, Pyricularia oryzae, Fusarium moniliforme,* and *F. oxysporum* while compound **72** showed zone of inhibition of 12, 10, 10, 11, 13 mm against *G. graminis, F. solani, P. oryzae, F. moniliforme,* and *F. oxysporum.* The standard Carbendazim (50 μg/disk) exhibited activity against *G. graminis, F. solani, P. oryzae, F. moniliforme,* and *F. oxysporum*, with the zone of inhibition of 14, 18, 15, 17 and 15, mm respectively [45].

Endophytic fungus *Botryosphaeria dothidea* KJ-1 associated with the stems of *Melia azedarach* collected at Yangling, Shaanxi Province, China was the source of pycnophorin (**73**), stemphyperylenol (**74**), chaetoglobosin C (**68**), djalonensone (**75**) (Figure 4), alternariol (**76**), β-sitosterol glucoside (**77**), and 5-hydroxymethylfurfural (**78**) (Figure 4). Stemphyperylenol (**74**) exhibited good antifungal activity against *Alternaria solani* the plant pathogen with the MIC value of 1.57 μM comparable to commonly used fungicide, the carbendazim. Compounds **68, 73,** and **75–78** showed good to average antifungal activities against *A. solani* (MICs of 6.25−25 μM) [46].

Two eicosanoic acids, 2-amino-3,4-dihydroxy-2-25-(hydroxymethyl)-14-oxo-6,12-eicosenoic acid (**79**) and myriocin (**80**) (Figure 4), were isolated from *Mycosphaerella* sp. an endophytic fungus of *Eugenia bimarginata* DC. (Myrtaceae) collected in Brazil (Savannah). These compounds displayed antifungal activities against several isolates of *C. neoformans* and *C. gattii,* with MIC values for compound **79** ranging from 1.3 to 2.50 μg/mL and for compound **80** was 0.5 μg/mL [47]. Both compounds exhibited antifungal activities against several isolates of *C. neoformans* and *C. gattii,* with MIC values ranging from 0.49 to 7.82 μM for compound **79** and 0.48–1.95 μM for compound **80** in another study. When checked by the checkerboard microtiter assay, both compounds exhibited synergistic activity against *C. gattii* with amphotericin B. Ultrastructural analysis divulges various signs of damage in *C. gattii* and *C. neoformans* cells treated with compounds **79** and **80**, including deformities in cell shape, depressions on the surface, and withered cells. Compounds **79** and **80** showed less loss of cellular material in cells of *C. gattii* compared to those treated with amphotericin B.

The difference in cellular material loss increased in a test compound concentration-dependent manner. Compound **80** also induced the formation of several pseudohyphae, suggesting that it could reduce virulence in *C. gattii* cells [48].

Endophytic fungus *Guignardia* sp., associated with *Euphorbia sieboldiana* collected from Nanjing, Jiangsu, China was the source of guignardone N (**81**) and guignardic acid (**82**) (Figure 4). Both compounds were evaluated for their inhibitory effects alone and with fluconazole on the growth and biofilms of *Candida albicans*. At 6.3 μg/mL combined with 0.031 μg/mL of fluconazole, compounds **81** and **82** were found to have prominent inhibition on the growth of *C. albicans* with fractional inhibitory concentration (FIC) index values of 0.23 and 0.19, respectively. Combined with fluconazole, both (40 μg/mL for (**81**) and 20 μg/mL for (**82**) could also inhibit *C. albicans* biofilms and reverse the tolerance of *C. albicans* biofilms to fluconazole [49].

Antifungal hyalodendriol C (**83**), rhizopycnin D (**84**), palmariol B (**85**), TMC-264 (**86**), penicilliumolide B (**87**) and alternariol 9-methyl ether (**88**) (Figure 4) were obtained from the endophytic fungus *Hyalodendriella* sp. Ponipodef 12 associated with the healthy stems of the "Neva" hybrid of *Populus deltoides* Marsh × *P. nigra* L. were collected from Longhua in Hebei Province of China. Compound **83** displayed antifungal effects against the spore germination of *M. oryzae* with potent inhibition with the IC_{50} value of 11.6 μg/mL, which was comparable with the positive control, carbendazim (IC_{50} 6.9 μg/mL) [50]. Previously, it found that rhizopycnin D (**84**), palmariol B (**85**), TMC-264 (**86**), penicilliumolide B (**87**), and alternariol 9-methyl ether (**88**) exhibits antifungal activity against the spore germination of *M. oryzae* [51,52].

Mellein (**89**) (Figure 4), was isolated from *Pezicula* sp. associated with the twigs of *Forsythia viridissima*, Zhejiang Province, Southeast China. Antifungal activity of this compound was tested against *B. cinerea, Pythium ultimum, Fusarium oxysporium* f. sp. *cucumerinum, Colletotrichum orbiculare, Verticillium dahliae, Pyricularia oryzae, Pestalotia diospyri, Sclerotinia sclerotiorum* and *Fulvia fulva*. Compound **89** displayed antifungal activity against 9 plant pathogenic fungi, esp. *B. cinerea* and *F. fulva* with EC_{50} values below 50 μg/mL [53].

Endophytic fungus *Nodulisporium* sp. A21 associated with the leaves of *Ginkgo biloba* collected from Nanjing in Jiangsu Province, China was a source of anti-phytopathogenic sporothriolide (**90**) (Figure 4). In mycelia growth inhibition method, sporothriolide (**90**) showed antifungal activity against *Rhizoctonia solani* with the EC_{50} value of 3.04 μg/mL (11.6 μM) while the EC_{50} of positive control carbendazim was 1.84 μg/mL (9.6 μM). Sporothriolide (**90**) at 200 μg/mL had a protective efficacy of 71.7% against Rice Sheath Blight in comparison with the protective efficacy 90.1% of the positive control of validamycin A at 200 μg/mL. Conidia of *Magnaporthe oryzae* could not form the germ tube and appressorium germinate in the sporothriolide with solution at the concentration of 1.5 μg/mL (5%

DMSO). In vivo, sporothriolide at 50 μg/mL, compared to tricyclazole of 2.5 μg/mL, could control the developing of Rice Blast [54]. Sporothriolide (**90**) also showed antifungal activity against *Sclerotinia sclerotiorum* with EC_{50} of sporothriolide against was 2.78 μg/mL (10.7 μM) while the EC_{50} of positive control carbendazim was 0.17 μg/mL (0.89 μM). When it comes to the protective activity on rape leaves, the positive control carbendazim of 250 μg/mL was 57.6% and sporothriolide of 250 μg/mL was 41.5% [54].

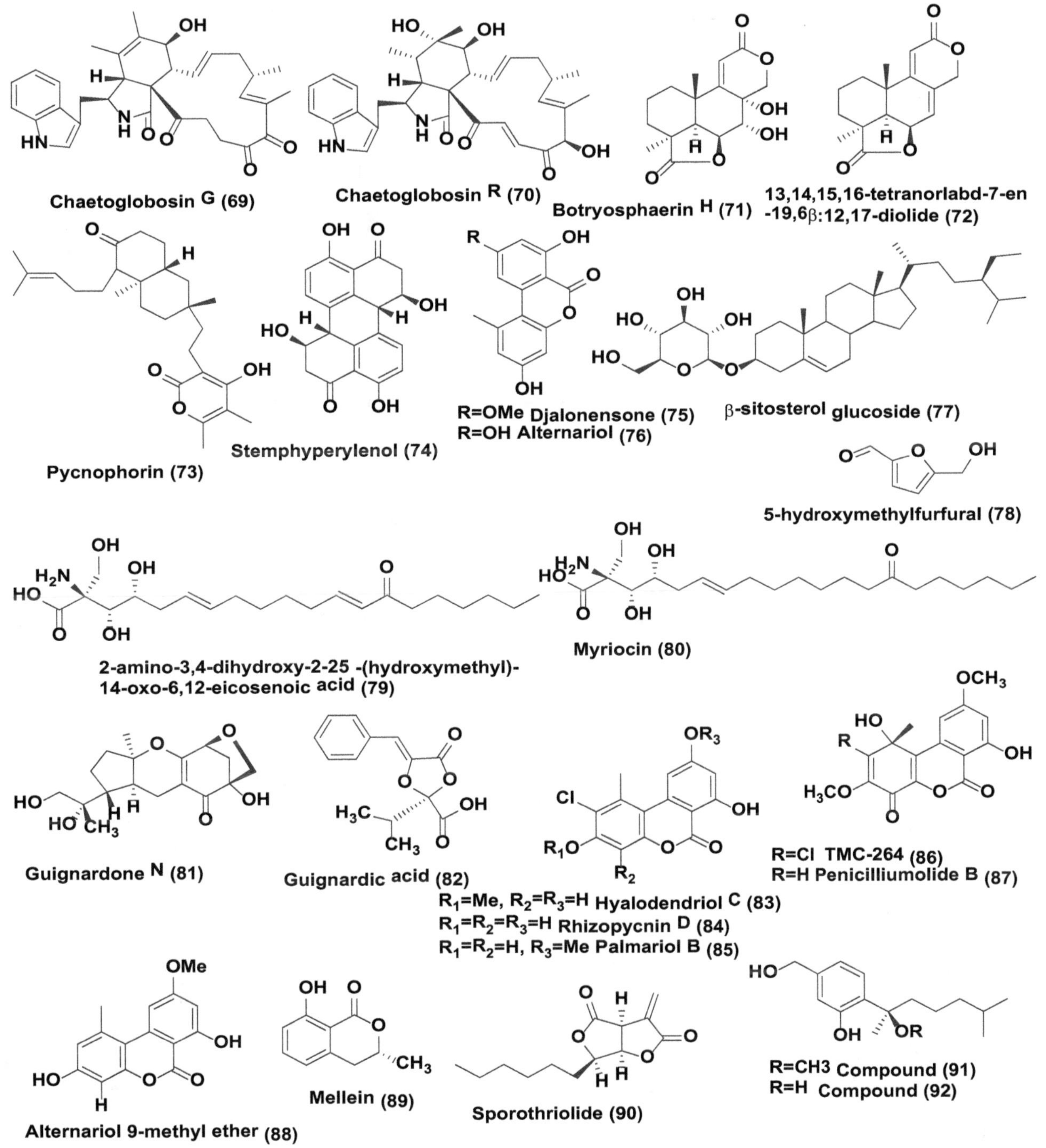

Chaetoglobosin G (69)
Chaetoglobosin R (70)
Botryosphaerin H (71)
13,14,15,16-tetranorlabd-7-en -19,6β:12,17-diolide (72)
Pycnophorin (73)
Stemphyperylenol (74)
R=OMe Djalonensone (75)
R=OH Alternariol (76)
β-sitosterol glucoside (77)
5-hydroxymethylfurfural (78)
2-amino-3,4-dihydroxy-2-25 -(hydroxymethyl)-14-oxo-6,12-eicosenoic acid (79)
Myriocin (80)
Guignardone N (81)
Guignardic acid (82)
R_1=Me, R_2=R_3=H Hyalodendriol C (83)
R_1=R_2=R_3=H Rhizopycnin D (84)
R_1=R_2=H, R_3=Me Palmariol B (85)
R=Cl TMC-264 (86)
R=H Penicilliumolide B (87)
Alternariol 9-methyl ether (88)
Mellein (89)
Sporothriolide (90)
R=CH3 Compound (91)
R=H Compound (92)

Figure 4. Structures of metabolites isolated from Ascomycetes (**69**–**92**).

Six phenolic bisabolane-type sesquiterpenoids (**91**, **92**) (Figure 4) (**93**–**96**), along with a macrolide, pyrenophorin (**97**) (Figure 5) were isolated from *Lopherdermium nitens* DAOM 250027 endophyte of

Pinus strobus (eastern white pine) near Sussex, NB, Canada. These compounds were characterized based on interpretation of spectroscopic data (NMR, OR, UV) and HRMS. All compounds were tested for antifungal activity. Pyrenophorin (**97**) significantly reduced the growth of *Microbotryum violaceum* and *Saccharomyces cerevisiae* at 5 μM, whereas sesquiterpenoids (**91–96**) were antifungal at 50 μM to both species tested [55]. Isocoumarin derivative exserolide C (**98**) and (12R)-12-hydroxymonocerin (**99**) (Figure 5) were isolated from endophyte *Exserohilum* sp. associated with *Acer truncatum* collected from Beijing, China. Compounds **98** and **99** displayed antifungal activity against *Fusarium oxysporum*, with MIC value of 20 μg/mL for both compounds, while Amphotericin B the positive control showed the MIC value of 0.63 μg/mL [56].

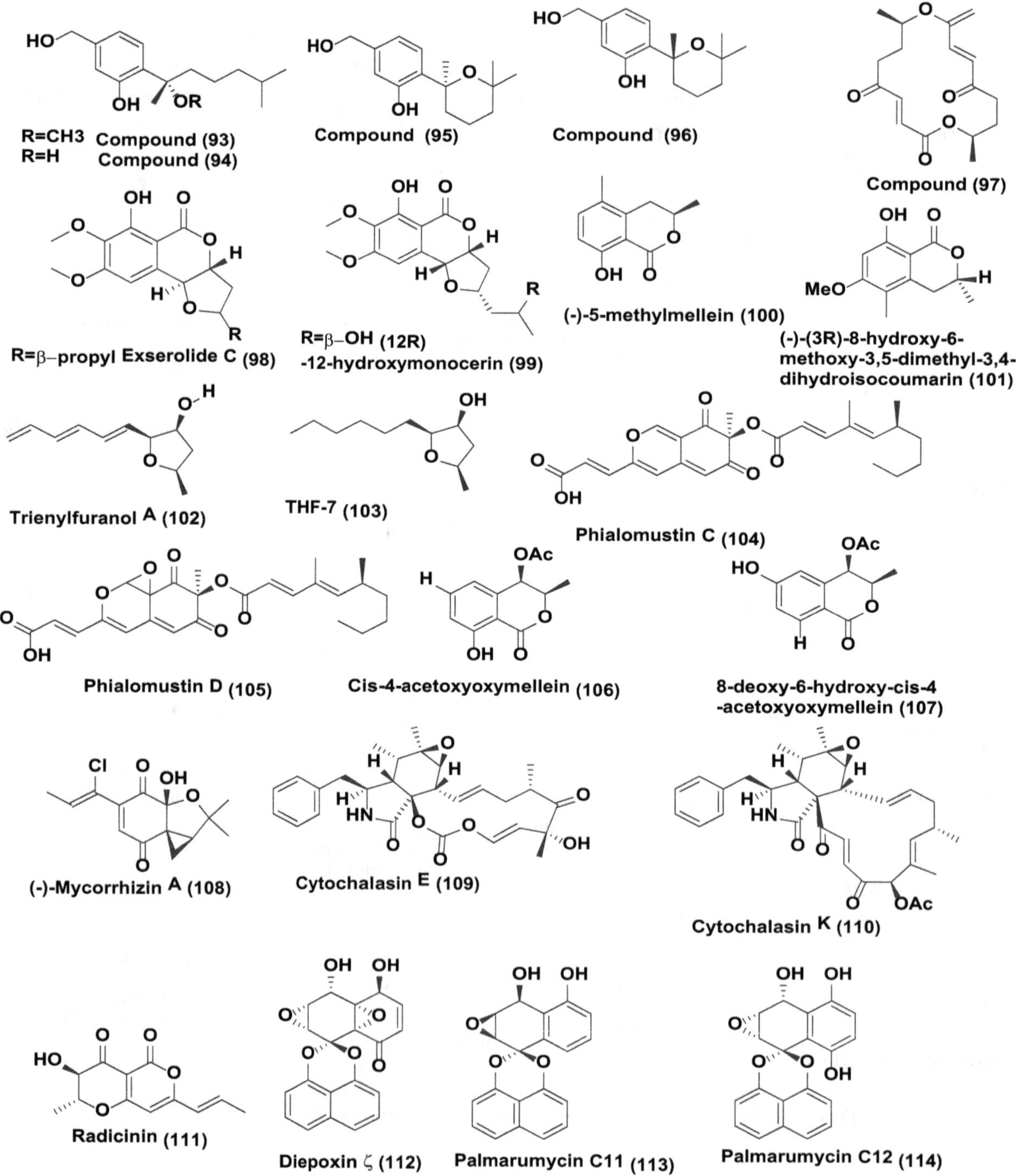

Figure 5. Structures of metabolites isolated from Ascomycetes (**93–114**).

Endophytic fungus of *Echinacea purpurea* associated with *Biscogniauxia mediterranea* EPU38CA from the wild in Missouri, USA was the source of (−)-5-methylmellein (**100**) and (−)-(3R)-8-hydroxy-6-methoxy-3,5-dimethyl-3, 4-dihydroisocoumarin (**101**) (Figure 5). Compound **100** exhibited poor activity against *Phomopsis obscurans, P. viticola*, and *Fusarium oxysporum*, and stimulated the growth of *Colletotrichum fragariae, C. acutatum, C. gloeosporioides*, and *B. cinerea*. Compound **101** was acknowledged to be marginally more active in the microtiter method than 5-methylmellein [57]. Compound **101** was found to be slightly more active in the microtiter method than 5-methylmellein [57]. Trienylfuranol A (**102**) (Figure 5) was isolated from isolated from *Hypoxylon submonticulosum* the endophyte of *Rubus idaeus* collected from Jordan Station, ON, Canada. It was identified based on high-resolution LC-MS and 1- and 2-D NMR spectroscopy. Absolute stereochemical configurations of the compounds were confirmed by NOE NMR experiments and by the preparation of Mosher esters. Complete hydrogenation of I yielded THF 7 (**103**) (Figure 5) that was used for stereochemical characterization and assessment of antifungal activity. Compound THF 7 (**103**) significantly inhibited the growth of *Saccharomyces cerevisiae* (74 $\pm$ 4% inhibition) at a concentration of 250 μg/mL as compared with complete inhibition by nystatin at 10 μg/mL [58].

Endophytic fungus *Phialophora mustea* associated with *Crocus sativus* was the source of an unprecedented azaphilone derived skeleton, Phialomustin C (**104**) and D (**105**) (Figure 5). Compounds **104** and **105** showed potent activities against *Candida albicans*, with IC_{50} values of 14.3 and 73.6 μM respectively [59].

An unidentified ascomycete, associated with *Melilotus dentatus* was the source of two new polyketide metabolites *cis*-4-acetoxyoxymellein (**106**) and 8-deoxy-6-hydroxy-*cis*-4-acetoxyoxymellein (**107**) (Figure 5). Compounds **106** and **107** displayed potent antifungal activities toward *Microbotryum violaceum* and *B. cinerea*, with 8 mm zone of inhibition for both fungi tested (0.05 mg was pipetted onto 9 mm sterile filter paper disk). In the case of *B. cinereal*, there was some growth within the zone of inhibition [60].

(−)-Mycorrhizin A (**108**) (Figure 6) was isolated from *Plectophomella* sp. while cytochalasins E (**109**) and K (**110**) (Figure 5) were isolated from *Physalospora* sp. Similarly, radicinin (**111**) (Figure 5) was purified from the endophytic fungus *Crataegus monogyna*. (-)-Mycorrhizin A showed good antifungal activity towards *Ustilago violacea* and *Eurotium repens*. Cytochalasins E (**109**) and K (**110**) showed potent activity against *E. repens* and *Mycotypha microspora*. Radicinin (**111**) (Figure 5) showed good activity against *E. repens* and *M. microspore* [61].

Diepoxin ζ (**112**), palmarumycin C11 (**113**), palmarumycin C12 (**114**) (Figure 5), cladospirone B (**115**), palmarumycin C6 (**116**), 1,4,7β-trihydroxy-8-(spirodioxy-1′,8′-naphthyl) -7,8-dihydronaphthalene (**117**), and palmarumycin C8 (**118**) (Figure 6) were obtained from *Berkleasmium* sp., an endophyte associated with *Dioscorea zingiberensis* from Hubei Province, China. Compounds **112–118** were evaluated for their antifungal activity against the spore germination of *M. oryzae*. Compounds **112–118** inhibited spore germination of *M. oryzae* with IC_{50} values in the range 9.1–124.5 μg/mL. Palmarumycin C8 (**118**) showed the best inhibitory activity (IC_{50} 9.1 μg/mL) among the compounds tested, although not as active as the positive control carbendazim (IC_{50} 6.3 μg/mL) [62].

Bipolamide B (**119**) (Figure 6) was isolated from *Bipolaris* sp. MU34, the endophytic fungus associated with the leaves of *Gynura hispida* Thwaites collected from Mahidol University, Bangkok, Thailand. The compounds were characterized based on NMR and MS experiments. Bipolamide B (**119**) exhibited average antifungal activity with MIC values of 16, 32, 32, 64 and 64 μg/mL, against *Cladosporium cladosporioides, C. cucumerinum, Saccharomyces cerevisiae, Aspergillus niger* and *Rhizopus oryzae* respectively [63].

Altenusin (**120**) (Figure 6), a biphenyl derivative, was isolated from an endophytic fungus, *Alternaria alternata* Tche-153 of *Terminalia chebula*, collected from Bangkok, Thailand. Employing disk diffusion method and the microdilution checkerboard technique, altenusin (**120**) in amalgamation with each of three azole drugs, ketoconazole, fluconazole or itraconazole at their low sub-inhibitory concentrations displayed potent synergistic activity against *C. albicans* with the fractional inhibitory concentration index range of 0.078 to 0.188 [64]. It is reported that *Schizosaccharomyces pombe* cells treated with altenusin were more rounded in shape than untreated cells which suggest that altenusin could act through the inhibition of cell wall synthesis or assembly in *S. pombe* [65].

Cladosporin (**121**) and isocladosporin (**122**) (Figure 6) were isolated from endophytic fungus *Cladosporium cladosporioides*. Compound **121** exhibited growth inhibition against *Colletotrichum acutatum*, *C. fragariae*, *C. gloeosporioides* and *Phomopsis viticola* at 30 μM with 92.7%, 90.1%, 95.4%, and 79.9%, respectively. Similarly, compound **122** showed 50.4%, 60.2%, and 83.0% growth inhibition against *C. fragariae*, *C. gloeosporioides*, and *P. viticola*, respectively, at 30 μM [66].

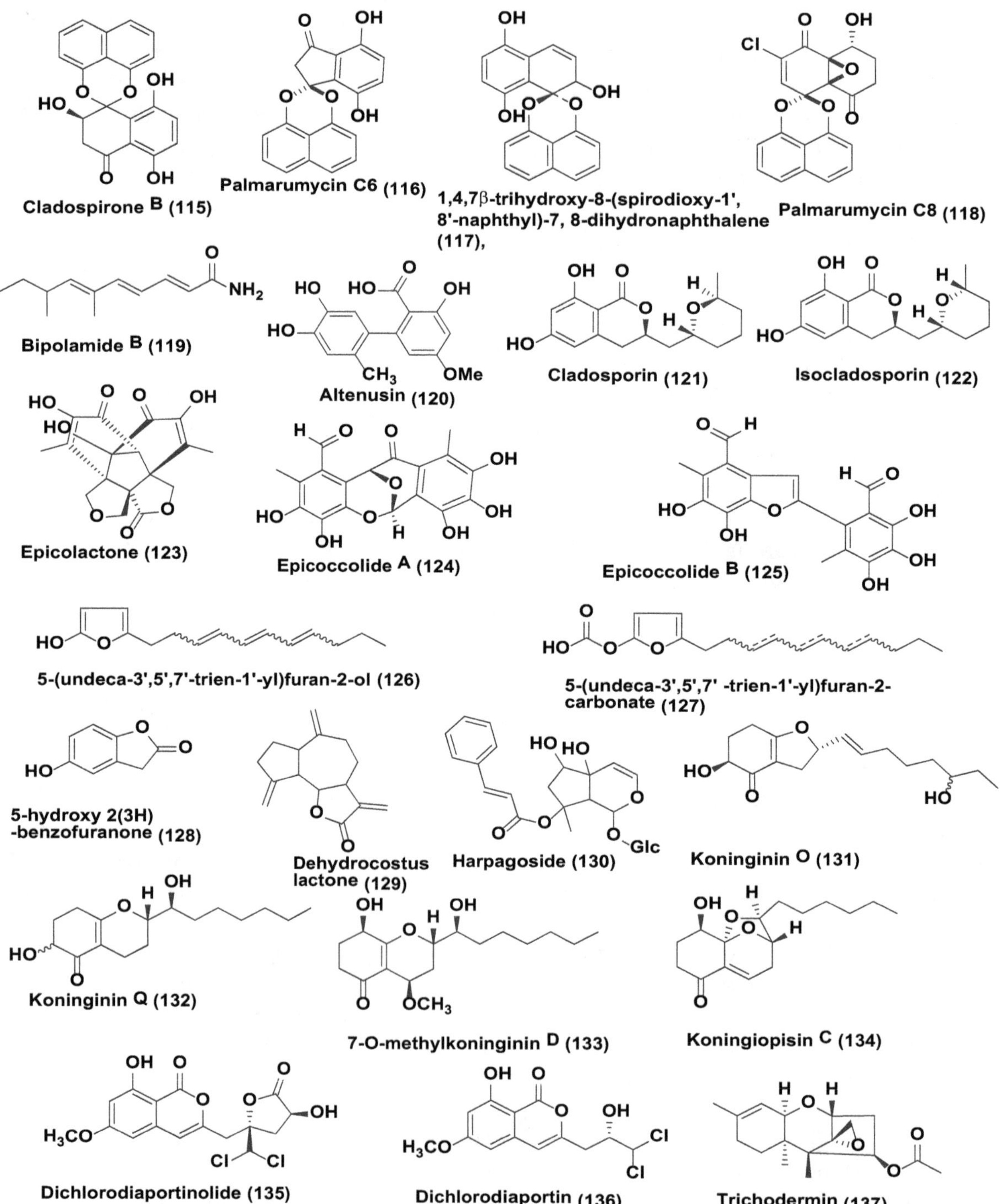

Figure 6. Structures of metabolites isolated from Ascomycetes (**115–127**) and Hyphomycetes (**128–137**).

Epicolactone (**123**) and epicoccolide A (**124**) and B (**125**) (Figure 6) polyoxygenated polyketides were obtained from an endophytic fungus, *Epicoccum* sp. CAFTBO, associated with stem bark and leaves of *Theobroma cacao* of Mount Kala, Republic of Cameroon. Compounds (**123–125**) showed good inhibitory effects on the mycelial growth of *Pythium ultimum* and *Aphanomyces cochlioides* and *Rhizoctonia solani* (MIC in the range of 20–80 μg per paper disc) [67].

5-methylmellein (**100**) (Figure 5) was isolated from endophytic fungus *Biscogniauxia mediterranea* Ohu 19B obtained from *Opuntia humifusa* (Cactaceae) of United States. Antifungal activity of compound **100** was evaluated using an in vitro microdilution broth assay against seven plant pathogens i.e., *Colletotrichum acutatum*, *C. fragariae*, *C. gloeosporioides*, *Fusarium oxysporum*, *B. cinerea*, *Phomopsis obscurans*, and *P. viticola*. *Phomopsis obscurans* was found to be being most susceptible (63.5% growth inhibition) at 150 μM at 120 h. The best growth inhibition (20.1%) to *F. oxysporum* was at 300 μM at 48 h. Lower doses (75 and 150 μM) of this compound caused stimulation of *B. cinera* and *C. fragariae*, while all doses caused stimulation of *C. acutatum* and *C. gloeosporioides* [68].

5-(undeca-3′,5′,7′-trien-1′-yl)furan-2-ol (**126**) and 5-(undeca-3′,5′,7′-trien-1′-yl)furan-2-carbonate (**127**) (Figure 6), two new alkylated furan derivatives, were recovered from the endophytic fungus *Emericella* sp. XL029 associated with the leaves of *Panax notoginseng* collected from Shijiazhuang, Hebei Province, China. Compound **126** displayed good antifungal activity against *Rhizoctorzia solani*, *Verticillium dahliae*, *Helminthosporium maydis*, *Fusarium oxysporum*, *Fusarium tricinctum*, *Botryosphaeria dothidea*, and *Alternaria fragriae* with MIC values ranging from 25 to 3.1 μg/mL, while compound **127** was found active against *V. dahliae*, *H. maydis*, *F.tricinctum*, *B. dothidea*, and *A. fragriae* with MIC values ranging from 50 to 12.5 μg/mL [69].

2.3. Compounds Produced by Hyphomycetes

5-hydroxy 2(3H)-benzofuranone (**128**), dehydrocostus lactone (**129**) and harpagoside (**130**) (Figure 6) were isolated from *Fusarium fujikuroi*, *Penicilium chrysogenum* and *Penicillium expensum* endophytes of *Eleusine coracana* grown under semi-hydroponic conditions Arkell Field Station, Arkell, ON, Canada. Compounds **128–130** exhibited antifungal activity against *F. graminearum* with the MIC of 31.25, 250.00 and 31.25 μg/mL, respectively. An in vitro interaction between each compound and *Fusarium* was investigated using light microscopy and vitality staining where the results proposed a mixed fungicidal/fungistatic mode of action [70].

Endophytic fungus *Trichoderma koningiopsis* YIM PH30002 harbored in *Panax notoginseng* collected from Wenshan, Yunnan Province, China was the source of koninginin O (**131**), koninginin Q (**132**) and 7-*O*-methylkoninginin D (**133**) (Figure 6). The antifungal activities of these compounds were tested against phytopathogenic fungi, *Fusarium oxysporum*, *F. solani*, *F. flocciferum*, *Plectosphaerella cucumerina* and *Alternaria panax* which are causes of pathogens of root rot diseases of *P. notoginseng*. Koninginin O (**131**) and koninginin Q (**132**) exhibited poor activity against *F. oxysporum* and *P. cucumerina* (MIC of 128 μg/mL). 7-*O*-methylkoninginin D (**133**) also showed poor activity against *P. cucumerina* (MIC 128 μg/mL). Nystatin positive control showed antifungal activity with MICs at 32 μg/mL [71]. Koningiopisin C (**134**) (Figure 6) was also isolated from the same fungus. Koningiopisin C showed antimicrobial activities against *F. oxysporum*, *A. panax*, *F. solani* and *P. cucumerina* with MICs at 32, 64, 32, and 16 μg/mL, respectively [72].

Dichlorodiaportinolide (**135**) and dichlorodiaportin (**136**) (Figure 6) were isolated from endophytic fungus *Trichoderma* sp. 09 obtained from the root of *Myoporum bontioides* A. Dichlorodiaportinolide (**135**) and dichlorodiaportin (**136**) showed weak to high antifungal activities with MIC values ranging from 6.25 to 150 μg/mL against *Colletotrichum musae* and *Rhizoctonia solani* and were inactive to *Penicillium italic* and *Fusarium graminearum* (MIC values > 200 μg/mL) [73].

Trichodermin (**137**) (Figure 6) was isolated from endophytic fungus strain, *Trichoderma brevicompactum* 0248 obtained from *Allium sativum*. Trichodermin showed potent inhibitory activity against *Rhizoctonia solani*, with an EC_{50} of 0.25 μg/mL and against *B. cinerea*, with an EC_{50} of 2.02 μg/mL but relatively poorly active against *Colletotrichum lindemuthianum* (EC_{50} = 25.60 μg/mL). Compound **137** exhibited good antifungal activity against the tested phytopathogens compared with the positive control Carbendazim [74].

Trichoderma koningiopsis YIM PH30002 collected at Wenshan, Yunnan Province of China was the source of two new metabolites koninginins R and S (**138–139**) (Figure 7). These isolated compounds showed certain antifungal activities against phytopathogens, *Fusarium flocciferum* and *Fusarium oxysporum*. Compound **138** possess the weak activity against *F. oxysporum* and *F. flocciferum* with the MICs at 128 μg/mL, while compound **139** displayed the poor activity against *F. oxysporum* with the MIC at 128 μg/mL [75].

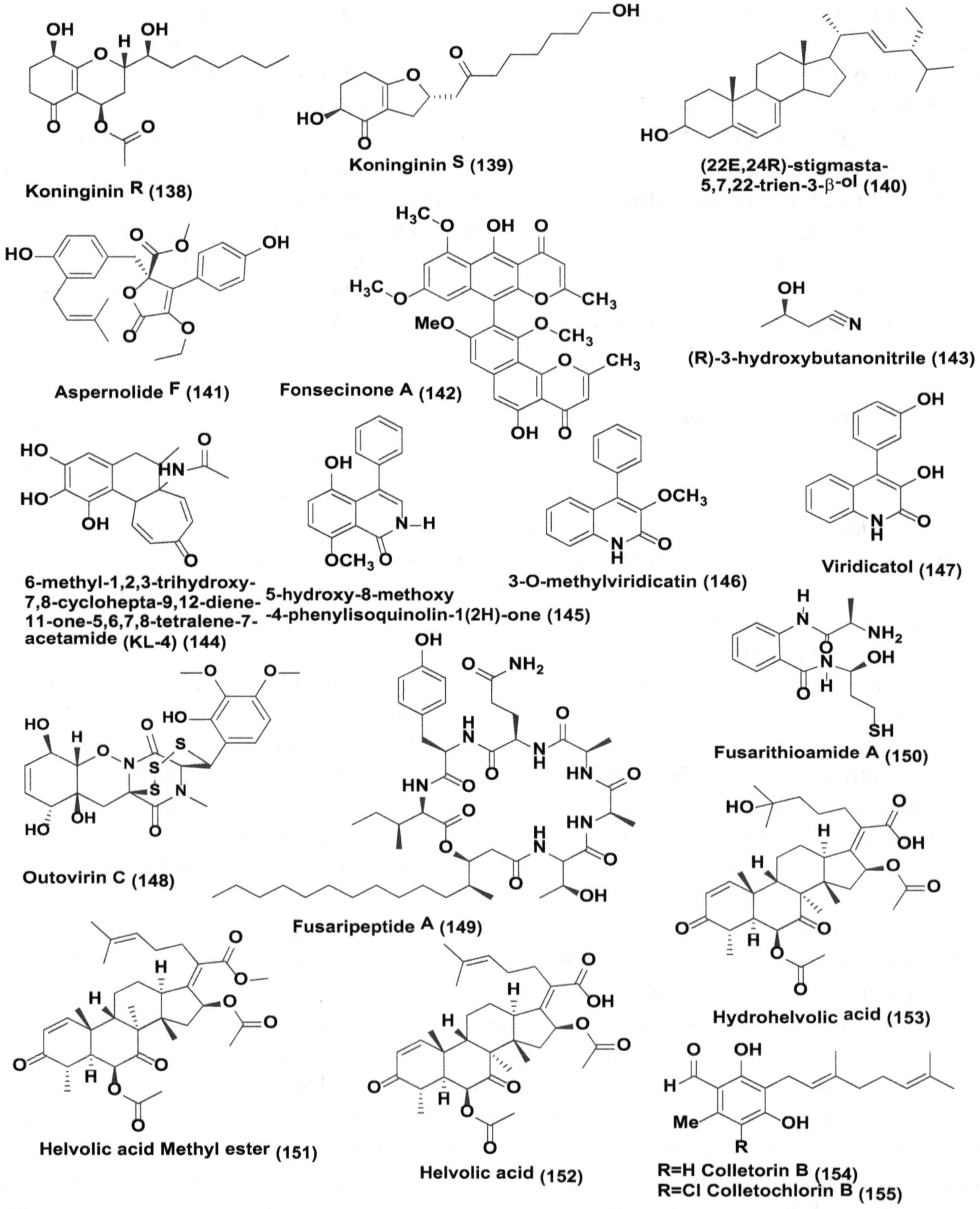

Koninginin R (138)
Koninginin S (139)
(22E,24R)-stigmasta-5,7,22-trien-3-β-ol (140)
Aspernolide F (141)
Fonsecinone A (142)
(R)-3-hydroxybutanonitrile (143)
6-methyl-1,2,3-trihydroxy-7,8-cyclohepta-9,12-diene-11-one-5,6,7,8-tetralene-7-acetamide (KL-4) (144)
5-hydroxy-8-methoxy-4-phenylisoquinolin-1(2H)-one (145)
3-O-methylviridicatin (146)
Viridicatol (147)
Outovirin C (148)
Fusaripeptide A (149)
Fusarithioamide A (150)
Helvolic acid Methyl ester (151)
Helvolic acid (152)
Hydrohelvolic acid (153)
R=H Colletorin B (154)
R=Cl Colletochlorin B (155)

Figure 7. Structures of metabolites isolated from Hyphomycetes (**138–155**).

Stigmasterol derivative (22E,24R)-stigmasta-5,7,22-trien-3-β-ol (**140**) and a new butyrolactones, aspernolide F (**141**) (Figure 7) were obtained from the endophytic fungus *Aspergillus terreus* associated with the roots of *Carthamus lanatus* collected at Assiut, Egypt. Compounds **140–141** exhibited good activity against *C. neoformans* with IC_{50} values of 4.38 and 5.19 μg/mL respectively, compared to amphotericin B (IC_{50} 0.34 μg/mL) [76].

Fonsecinone A (**142**), and (R)-3-hydroxybutanonitrile (**143**) (Figure 7), were obtained from *Aspergillus* sp. KJ-9 an endophytic fungus associated with *Melia azedarach* which was collected at Yangling, Shaanxi Province, China, and identified by spectroscopic methods. Compounds **142** and **143** were active against *Gibberella saubinetti, Magnaporthe grisea, B. cinerea, C. gloeosporioides* and *A. solani* (MIC range of 6.25–50 μM) [77].

6-methyl-1,2,3-trihydroxy-7,8-cyclohepta-9,12-diene-11-one-5,6,7,8-tetralene-7-acetamide (KL-4) (**144**) (Figure 7) was isolated from *Aspergillus* sp. obtained from the seeds of *Gloriosa superba* which were collected from Tirupati, India. KL-4 (**144**) exhibited good antifungal activity against *Saccharomyces cerevisiae, C. albicans* and *Cryptococcus gastricus* with MIC 25, 12.5, and 50 μg/mL respectively [78].

Endophytic fungus *Penicillium* sp. R22 associated with *Nerium indicum* collected from Qinling Mountain, Shaanxi Province, China was the source of 5-hydroxy-8-methoxy-4-phenylisoquinolin-1(2H)-one (**145**) a new isoquinolone alkaloid along with 3-*O*-methylviridicatin (**146**) and viridicatol (**147**) (Figure 7) two known quinolinone alkaloids. Compound **145** exhibited good antifungal activity against *Alternaria brassicae, A. alternata* and *Valsa mali* with MIC value of 31.2 μg/mL, compound **146** against *A. brassicae, B. cinerea and Valsa male* with MIC value of 31.2 μg/mL, compound **147** against *A. brassicae, A. alternata* and *B. cinerea* with MIC value of 31.2 μg/mL [79].

Trisulfide gliovirin-like compound Outovirin C (**148**) (Figure 7), an epithiodiketopiperazine natural product, was identified from *Penicillium raciborskii*, an endophytic fungus associated with *Rhododendron tomentosum* were collected at the test site of University of Oulu, Finland. Outovirin C (**148**) showed antifungal activity when assayed by micro-spectrophotometry using a dose response growth inhibition assay. Outovirin C inhibited the growth of *Fusarium oxysporum, B. cinerea,* and *Verticillium dahlia* at a low concentration of 0.38 mM (207 μg/mL) but a more significant growth inhibition was observed at the higher concentration of 0.76 mM (413 μg/mL). Compound **148** was most active against *B. cinerea* (57% inhibition) and slightly less effective against *V. dahliae* (45% inhibition) [80].

Fusaripeptide A (**149**) (Figure 7), a new cyclodepsipeptide, was isolated from the culture of the endophytic fungus *Fusarium* sp. associated with roots of *Mentha longifolia* growing in Saudi Arabia. Its structure was elucidated based on 1D and 2D NMR and HRESI and GC-MS experiments. The absolute configuration of the amino acid residues of **149** was assigned by chiral GC-MS and Marfey's analysis after acid hydrolysis. Compound **149** exhibited potent antifungal activity toward *C. albicans, C. glabrata, C. krusei,* and *A. fumigates* with IC_{50} values of 0.11, 0.24, 0.19, and 0.14 μM, respectively. Under similar condition control amphotericin B exhibited antifungal activity toward *C. albicans, C. glabrata, C. krusei,* and *A. fumigates* with IC_{50} values of 0.3, 0.6, 0.5, 0.7 μM, respectively [81].

Fusarithioamide A, a new benzamide derivative (**150**) (Figure 7) was isolated from *Fusarium chlamydosporium* associated with the leaves of *Anvillea garcinii* collected from Al-Azhar University, Saudi Arabia. Compound **150** exhibited good antifungal activity against *C. albicans* with inhibition zone diameters (IZD 16.2 mm and MIC 2.6 μg/mL which is comparable to the positive control substance clotrimazole (IZD 18.5 mm and MIC 3.7 μg/mL) [82].

A new helvolic acid derivative named helvolic acid methyl ester (**151**), together with two known helvolic acid compounds, helvolic acid (**152**) and hydrohelvolic acid (**153**) (Figure 7), were extracted from endophytic fungus *Fusarium* sp. associated with *Ficus carica* leaves collected from Qinling Mountain, Shaanxi Province, China. Compounds **151–153** exhibited good antifungal activity against *B. cinerea, C. gloeosporioides, F. oxysporum* f. sp. *niveum, Fusarium graminearum* and *Phytophthora capsici* with MIC value in the range of 12.5–25 μg/mL while Carbendazim the standard showed MIC value in the range of 32.2–62.5 μg/mL against the same fungi [83].

Colletorin B (**154**), colletochlorin B (**155**) (Figure 7), LL-Z1272β (llicicolin B) (**156**), and 4,5-dihydrodechloroascochlorin (**157**) (Figure 8) were extracted from endophytic fungus *Fusarium* sp. Colletorin B (**154**) and colletochlorin B (**155**) showed moderate antifungal activity towards *Ustilago violacea* and *F. oxysporum*. Compound **156** showed moderate antifungal activity towards *U. violacea* and *F. oxysporum*. Furthermore, 4,5-dihydrodechloroascochlorin (**157**) showed a very strong antifungal activity towards *Eurotium repens* [84].

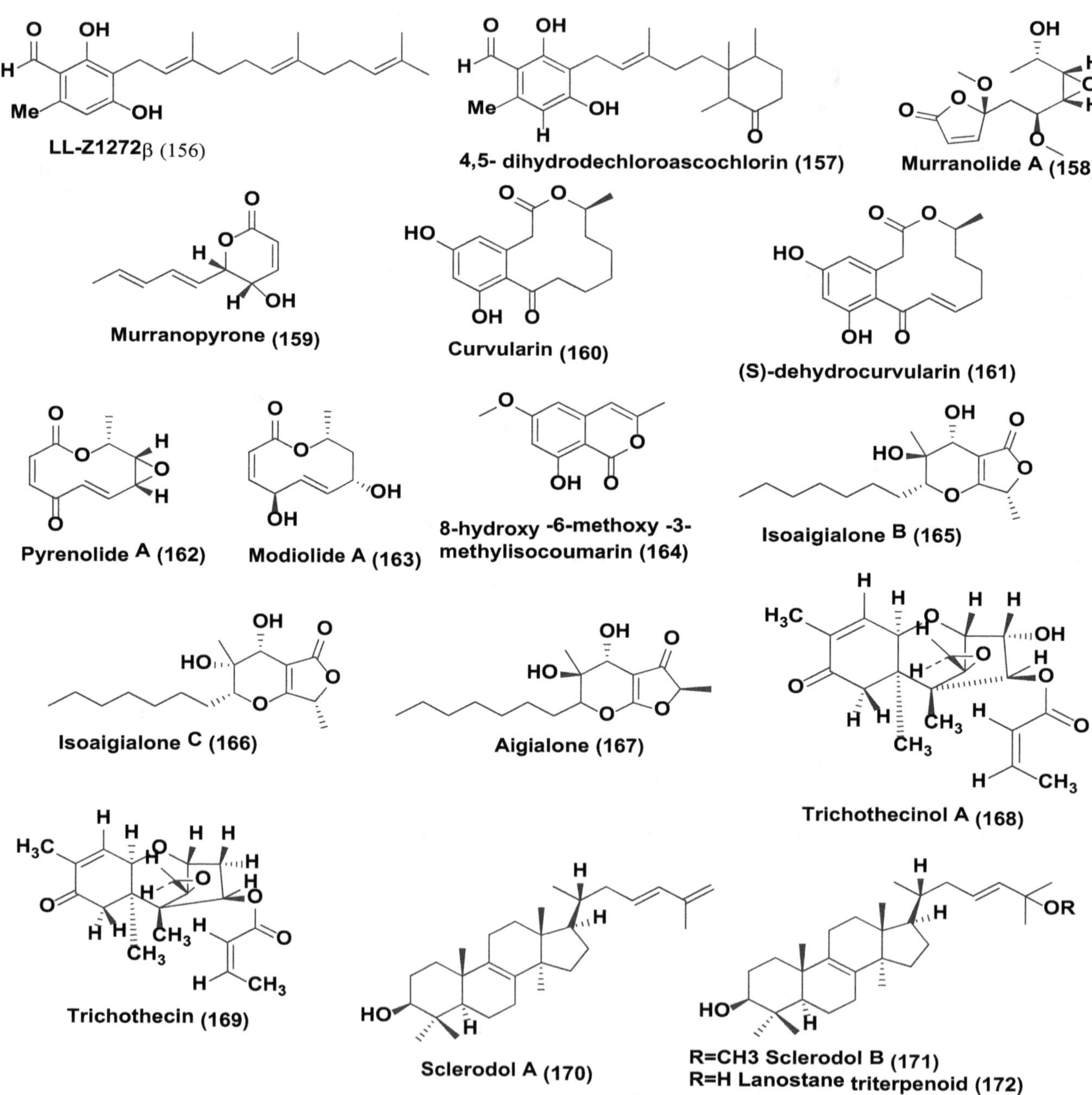

Figure 8. Structures of metabolites isolated from Hyphomycetes (**156–169**) and Basidiomycetes (**170–172**).

Murranolide A (**158**), murranopyrone (**159**), curvularin (**160**), (S)-dehydrocurvularin (**161**), pyrenolide A (**162**), modiolide A (**163**), and 8-hydroxy-6-methoxy-3-methylisocoumarin (**164**) (Figure 8) were extracted from the endophytic fungus *Curvularia* sp., strain M12, associated with the leaf of *Murraya koenigii* were collected from Rajshahi University, Bangladesh. Pyrenolide A (**162**) was observed to impair the mobility of *Phytophthora capsici* zoospores in a short time (30 min) at a low concentration (100% at 0.5 μg/mL). Murranolide A (**158**), murranopyrone (**159**), curvularin (**160**), (S)-dehydrocurvularin (**161**), modiolide A (**163**), and 8-hydroxy-6-methoxy-3-methylisocoumarin (**164**) exhibited zoospore motility impairment activity at higher concentrations (IC_{50}: 50–100 μg/mL) [85].

Two new isoaigialones, B (**165**) and C (**166**) (Figure 8), along with aigialone (**167**) (Figure 8), were obtained from *Phaeoacremonium* sp., an endophytic fungus associated with the leaves of *Senna spectabilis* was collected in the Araraquara Cerrado area, in June 2001, Araraquara, Sao Paulo state, Brazil. These compounds were evaluated against *Cladosporium cladosporioides* and *C. sphaerospermum* using direct bioautography. Compounds **165** and **167** exhibited antifungal activity, with a detection limit of 5 μg, for both fungi, while compound **166** displayed weak activity (detection limit > 5 μg), with a detection limit of 25 μg. Nystatin was used as a positive control, showing a detection limit of 1 μg [86].

Trichothecinol A (**168**) (Figure 8) was obtained from *Trichothecium* sp. an endophytic fungus isolated from *Phyllanthus amarus* collected from Pune India. Compound **101** showed activity against *Cryptococcus albidus* (NCIM 3372) up to 20 μg/mL [87]. Trichothecin **(169)** (Figure 8), a sesquiterpene, was isolated from endophytic fungus *Trichothecium* sp. residing inside the leaves of *Phyllanthus* sp. collected from Pune India. Compound **169** exhibited anti-fungal activity against *Saccharomyces cerevisiae, Cryptococcus albidus* var *diffluens* NCIM 3371, *Cryptococcus albidus* var *diffluens* NCIM 3372, *Fusarium oxysporum, Penicillium expansum, Trichoderma viride, Paecilomyces varioti* and *Aspergillus niger* with MIC of 6, 20, 12, 10, 30, 40, 20 and 12 μg/mL, respectively [88].

2.4. Compounds Produced by Basidiomycetes

Two lanostane triterpenoids, sclerodols A (**170**) and B (**171**), and a known related lanostane triterpenoid (**172**) (Figure 8) were isolated from *Scleroderma* UFSM Sc1 (Persoon) Fries an endophyte associated with *Eucalyptus grandis*. Both compounds were evaluated for their anti-candidal potential against *Candida albicans, C. tropicalis, C. crusei, C. parapsiosis* for activities. Compound **171** showed good anticandidal activity against *C. albicans, C. tropicalis, C. crusei, C. parapsiosis* with the MIC of 25.0, 25.0, 6.25 and 12.5 and MFC of 25.0, 25.0, 12.5 and 25.0 μg/mL respectively. Compounds **170** and **172** were less active against tested strain than compound **171** with the MIC in the range of 12.5–100 and MLC (minimal lethal concentratin) of >100.0 μg/mL. Control nystatin exhibited showed anti-candidal activities against tested strains with the MIC in the range of 0.77–1.52 μg/mL and MLC in the range of 3.12–6.25 μg/mL [89].

Table 1. Antifungal compounds reported from endophytic fungi.

Sr. No.	Fungus	Plantsource	Compounds Isolated	Biologicalactivity *	Refs.
			Comounds Produced by Coelomycetes		
1	*Pestalotiopsis foedan*	*Bruguiera sexangula* Hainan, China	(3R,4R,6R,7S)-7-hydroxyl-3,7-dimethyl-oxabicyclo[3.3.1]nonan- 2-one (**1**), (3R,4R)-3-(7-methylcyclohexenyl)-propanoic acid (**2**)	Compound **1** *B. cinerea* and *P. nicotianae* (MIC 3.1 and 6.3 μg/mL), ketoconazole (MIC 3.1 μg/mL each) Compound **2** *C. albicans* MIC 50 μg/mL) ketoconazole (MIC 6.3 μg/mL)	[15]
2	*Pestalotiopsis* sp. DO14	*Dendrobium officinale*, Yandang Mountain, Zhejiang Province, China.	(4S,6S)-6-[(1S,2R)-1,2-dihydroxybutyl]-4-hydroxy-4-methoxytetrahydro-2H-pyran-2-one (**3**) and (6S,2E)-6-hydroxy-3-methoxy-5-oxodec-2-enoic acid (**4**), LL-P880γ (**5**), LL-P880α (**6**)	Compounds **3–6** active against *C. albicans, C. neoformans, T. rubrum, and A. fumigates* (MIC ≤ 50 μg/mL) Compounds **3–4** active against *C. albicans, C. neoformans, T. rubrum*, and *A. fumigatus* (MIC, ≤ 25 μg/mL)	[16]
3	*Pestalotiopsis fici*	*Camellia sinensis* Hangzhou, China.	Ficipyrone A (**7**)	Compound **7** active against *G. zeae* (IC_{50} 15.9 μM), ketoconazole (IC_{50} 6.02 μM)	[17]
4	*Pestalotiopsis mangiferae*	*Mangifera indica* Maduravoyal, Tamil Nadu Province, India.	4-(2,4,7-trioxa-bicyclo[4.1.0]heptan-3-yl) phenol (**8**)	Compound **8** active against *C. albicans* (MIC, 0.039 μg/mL), Nystatin (MIC 10.0 μg/mL)	[18]
5	*Phomopsis* sp.	*Senna spectabilis* São Paulo, Brazil	Cytochalasin H (**9**)	Compound **9** active against *C. cladosporioides* and *C. sphaerosphermum* (MIC 10.0 and 25.0 μg, respectively), nystatin (MIC = 1.0 μg)	[19]
6	*Phomopsis* sp.	*Aconitum carmichaeli*, Huize County, Yunnan Province, China.	(14β,22E)-9,14-dihydroxyergosta-4,7,22-triene-3,6-dione (**10**), (5α,6β,15β,22E)-6-ethoxy-5,15-dihydroxyergosta-7,22-dien- 3-one (**11**), calvasterols A (**12**), and ganodermaside D (**13**)	Compound **10** active against *C. albicans, H. compactum*, and *A. niger*, (MIC, 64, 64, and 128 μg/mL, respectively). Compound **11** active against *C. albicans* and *F. avenaceum* (MIC = 128 μg/mL). Compounds **12** and **13** active against *F. avenaceum*. (MIC, 64 μg/mL), Compound **12** activie against *P. oryzae* and *T. gypseum* (MIC 128 and 256 μg/mL)	[20]
7	*Diaporthe maritima*	*Picea* sp., Acadian forest of Eastern Canada.	Phomopsolide A (**14**), B (**15**), and C (**16**), and a stable alpha-pyrone (**17**)	Compound **14** active against *M. violaceum* and *S. cerevisiae* at 25 μM, Compounds **15–17** demonstrated growth inhibition at 250 μM	[21]
8	*Phoma* sp.	*Fucus serratus,*	Phomalacton (**18**), (3R)-5-hydroxymellein (**19**) and emodin (**20**)	Compounds **18–20** active against *M. violaceum* with 5, 6 and 5mm zone of inhibition.	[23]
9	*Phoma* sp. WF4	*Eleusine coracana* Arkell Field Station, Arkell, ON, Canada	Viridicatol (**21**), tenuazonic acid (**22**), alternariol (**23**), and alternariol monomethyl ether (**24**)	Compounds **21–24** caused dramatic breakage of *F. graminearum* hyphae in vitro	[24]
10	*Rhizopycnis vagum* Nitaf 22	*Nicotiana tabacum*, China Agricultural University. Beijing 100193, China.	Rhizopycnin D (**25**) and TMC-264 (**26**)	Compounds **25–26** inhibited the spore germination of *M. oryzae* with IC_{50} values of 9.9 and 12.0 μg/mL, respectively	[25]
11	*Microsphaeropsis* sp. *Seimatosporium* sp.	*Salsola oppositifolia*, Playa del Ingles, Gomera, Spain	Microsphaerol (**27**) Seimatorone (**28**)	Compounds **27** and **28** active against *M. violaceum* with 9 and 5 mm zone of inhibition. In addition, there was some growth with in zone of inhibition	[26]
12	*Colletotrichum gloeosporioides*	*Michelia champaca* São Paulo State University, Araraquara, São Paulo, Brazil.	2-phenylethyl 1H-indol-3-yl-acetate (**29**)	Compound **29** active against *C. cladosporioides* and *C. sphaerospermum* comparable to that of the positive control nystatin	[27]
13	*Colletotrichum* sp.	Gomera (Spain).	Colletonoic acid (**30**)	Compound **30** active against *M. violaceum* with 7 mm zone of inhibition	[28]
14	*Coniothyrium* sp.,	*Salsola oppostifolia* Gomera in the Canary Islands.	1,7-dihydroxy3-methyl-9,10-anthraquinone (**31**), 1,6-dihydroxy-3 -methyl-9,10-anthraquinone (phomarin) (**32**), and 1-hydroxy-3-hydroxymethyl-9,10-anthraquinone (**33**) coniothyrinones A-D (**34–37**)	Compounds **31–37** active against *M. violaceum* with 7, 10, 8, 7.5, 6, 8 and 7.5 mm zone of inhibition. Compounds **32–34** active against *M. violaceum* (10 and 9 mm zone of inhibition) and *B. cinerea* (7.5 and 12.5 mm zone of inhibition) when tested under similar conditions	[29,30]

Table 1. *Cont.*

Sr. No.	Fungus	Plantsource	Compounds Isolated	Biologicalactivity *	Refs.
			Comounds Produced by Acsomycetes		
15	*Xylaria* sp. YM 311647	*Azadirachta indica*,Yuanjiang County, Yunnan Province, China,	(1S,4S,5R,7R,10R,11R)-Guaiane-5,10,11,12-tetraol (**38**) (1S,4S,5S,7R,10R,11S) -Guaiane-1,10,11,12-tetraol (**39**) (1S,4S,5R,7R,10R,11S)-Guaiane-5,10,11,12-tetraol (**40**) (1S,4S,5S,7R,10R,11R)-Guaiane-1,10,11,12-tetraol (**41**) (1R,3S,4R,5S,7R,10R,11S) -Guaiane-3,10,11,12-tetraol (**42**) (1R,3R,4R,5S,7R,10R,11R)-Guaiane-3, 10,11,12-tetraol (**43**) (1R,4S,5S,7S,9R,10S,11R)-Guaiane-9,10,11,12-tetraol (**44**) (1R,4S,5S,7R,10R,11S) -Guaiane-10,11,12-triol (**45**) (1R,4S,5S,7R,10R,11R)-Guaiane-10,11,12-triol (**46**), 14a,16-Epoxy-18-norisopimar-7-en-4a-ol (**47**), 16-*O*-Sulfo-18-norisopimar-7-en-4a, 16-diol (**48**), and 9-Deoxy-hymatoxin A (**49**)	Compounds **38–46** active against *C. albicans* and *H. compactum* (MIC in the range of 32 to 256 µg/mL), compounds **47–49** active against *C. albicans, A. niger, P. oryzae, F. avenaceum*, and *H. compactum* (MIC in the range of 16 to 256 µg/mL). Compound **49** exhibited the potent inhibitory activity against *C. albicans* and *P. oryzae* with MIC values of 16 µg/mL	[32]
16	*Xylaria* sp. YM 31164	*Azadirachta indica*, Yuanjiang County, Yunnan Province, China	(1S,2S,4S,5S,7R,10R)-Guaiane-2,10,11,12-tetraol (**50**), (1S,2S,4R,5R,7R,10R)-Guaiane-2,4,10,11,12-pentaol (**51**), (1S,4R,5S,7R,10R)-Guaiane-4,5,10,11,12-pentaol (**52**), (1R,4S,5R,7R,10R)-Guaiane-1,5,10,11,12-pentaol (**53**), (1R,4R,5R,7R,10R)-11-Methoxyguaiane-4,10,12-triol (**54**),	Compounds **50–54** active against *P. oryzae* and *H. compactum* (MIC in the range of 32–256 µg/mL). Compound **53** active against *P. oryzae* (MIC 32 µg/mL). Compounds **52** and **53** active against *H. compactum* with (MIC, 64 µg/mL), Compound **53** and **54** active against *C. albicans* (MIC 32 µg/mL). Compound **52** active against *C. albicans, A. niger*, and *H. compactum* (MIC, 64 µg/mL).	[33]
17	*X. feejeensis*	*Croton lechleri*.	Xyolide (**55**),	Compound **55** active against *P. ultimum* (MIC 425 µM)	[34]
18	*Xylaria* sp. XC-16	*Toona sinensis* Yangling, Shaanxi Province, China	Cytochalasin Z28 (**56**)	Compound **56** active against *G. saubinetti* (MIC of 12.5 µM), Hymexazol (MIC = of 25 µM)	Zhang et al. [35]
19	*Xylaria* sp. strain F0010, *Xylaria* sp. PSU-G12X. *cubensis*, 13 strains of *Xylaria* sp.	*Abies holophylla; Garcinia hombroniana ; Asimina triloba; Pinus strobus; Vaccinium angustifolium*, New Brunswick and Nova Scotia, Canada	Griseofulvin (**57**)	Griseofulvin (**57**) Inhibits *A. mali, B. cinerea, Colletotrichum gloeosporioides, Corticium sasaki, F. oxysporum* and *M. grisea* in in vitro (IC_{50} values of 18.0, 5.0, 1.7, 11.0, 30.0, and 1.7 µg/mL, respectively, Compound **57** active against *M. grisea, C. sasaki, B. cinerea, P. recondite* and *B. graminis* f. sp. *hordei* in in vivo, with % of fungal control of 95, 100, 60, 90 and 90, respectively, at 150 µg/mL.	[36–38]
20	*Xylaria* sp.		Dechlorogriseofulvin (**58**)	Compound **58** showed weak antifungal activity, with an IC_{50} value, 200 µg/mL against *M. grisea, C. sasaki, B. cinerea, P. recondite* and *B. graminis* f. sp. *hordei* in in vivo,	[36,38]
21	*Chaetomium globosum* CDW7	*Ginkgo biloba* China	Chaetoglobosin A (**59**) and D (**60**)	Compounds **59–60** active against *S. sclerotiorum* with IC_{50} values of 0.35 and 0.62 µg/mL, respectively, carbendazim (0.17 µg/mL)	[41]
22	*Chaetomium globosum*	Seeds of *Panax notoginseng* collected at the Wenshan, Yunnan, China	Chaetoglobosin A (**59**), Chaetomugilin A (**61**), Chaetomugilin D (**62**), Chaetoglobosin B (**63**), Chaetoglobosin E (**64**), Chaetoglobosin F (**65**) **and** Penochalasin G (**66**)	Compounds **59** and **61–66** active against *P. herbarum* (MIC in the range of 16–128 µg/mL) and, *E. nigrum* (MIC in the range of <1–16 µg/mL).	[42]
23	*Chaetomium cupreum* ZJWCF079	*Macleaya cordata*.	Ergosta-5, 7, 22-trien-3-beta-ol (**67**)	Compound **67** against *S. sclerotiorum* and *B. cinerea* with EC_{50} values of 125 µg/mL and 190 µg/mL respectively,	[43]
24	*Chaetomium globosum* No.04	Barks of *Ginkgo biloba*, Linyi, Shandong Province, China.	Chaetoglobosin A (**59**), D (**60**), E (**64**), C (**68**), G (**69**), R (**70**)	Compounds **59–60, 64**, and **68–70** active against *R. stolonifer* and *C. diplodiella* at a concentration of 20 µg/disk	[44]
25	*Botryosphaeria* sp. P483	*Huperzia serrata*, Xichou County, Yunnan Province, China	Botryosphaerin H (**71**) 13,14,15,16-tetranorlabd-7-en-19,6β:12,17-diolide (**72**)	At 100 µg/disk, compound **71** showed zone of inhibition of 9, 7, 7, 8, and 8 mm, against *G. graminis, F. solani, P. oryzae, F. moniliforme*, and *F. oxysporum*; compound **72** showed zone of inhibition of 12, 10, 10, 11, and 13 mm against *G. graminis, F. solani, P. oryzae, F. moniliforme*, and *F. oxysporum;* carbendazim (50 µg/disk) showed the zone of inhibition of 14, 18, 15, 17, 15 mm against *G. graminis F. solani P. oryzae F. moniliforme F. oxysporum*, respectively	[45]

Table 1. *Cont.*

Sr. No.	Fungus	Plantsource	Compounds Isolated	Biologicalactivity *	Refs.
26	*Botryosphaeriadothidea* KJ-1,	*Melia azedarach* Yangling, Shaanxi Province, China.	Pycnophorin (73), stemphyperylenol (**74**), chaetoglobosin C (**68**), djalonensone (**75**), alternariol (**76**), β-sitosterol glucoside (**77**), 5-hydroxymethylfurfural (**78**)	Compound **74** active against *A. solani* (MICs of 1.57 μM) Compounds **68**, **73**, and **75–78** active against *A. solani* (MICs of 6.25−25 μM)	[46]
27	*Mycosphaerella* sp.	*Eugenia bimarginata* DC. Brazil (savannah).	2-amino-3,4-dihydroxy-2-25-(hydroxymethyl)-14-oxo-6,12- eicosenoic acid (**79**), myriocin (**80**)	Compounds **79** active against several isolates of *C. neoformans* and *C. gattii*, with MIC values ranging from 1.3 to 2.50 μg/mL and 0.5 μg/mL, for compound **80**	[47]
				Compounds 79 active against several isolates of *C. neoformans* and *C. gattii*, with MIC values ranging from 0.49 to 7.82 μM and 0.48–1.95 μM for compound **80**. Compounds **79** and **80** cause deformities in cell shape, depressions on the surface, and withered cells.	[48]
28	*Guignardia* sp.,	*Euphorbia sieboldiana* collected from the campus of China Pharmaceutical University, Nanjing, Jiangsu, China	Guignardone N (**81**), guignardic acid (**82**)	At 6.3 μg/mL combined with 0.031 μg/mL of fluconazole, compounds **81** and **82** were found to have prominent inhibition on the growth of *C. albicans* with FIC index values of 0.23 and 0.19, respectively. Combined with fluconazole, both of them (40 μg/mL for (**81**) and 20 μg/mL for (**82**) could also inhibit *C. albicans* biofilms and reverse the tolerance of *C. albicans* biofilms to fluconazole	[49]
29	*Hyalodendriella* sp. Ponipodef 12	"Neva" hybrid of *Populus deltoides* Marsh × *P. nigra* L., Longhua in Hebei Province of China.	hyalodendriol C (**83**), rhizopycnin D (**84**), palmariol B (**85**), TMC-264 (**86**), penicilliumolide B (**87**) and alternariol 9-methyl ether (**88**)	Compound **88** exhibited spore germination of *M. oryzae* with IC_{50} value of 11.6 μg/mL, positive control, carbendazim (IC_{50} 6.9 μg/mL) Compounds **84–88** displayed antifungal effects against the spore germination of *M. oryzae*	[50–52]
30	*Pezicula* sp.	*Forsythia viridissima*, collected from Zhejiang Province, Southeast China	Mellein (**89**)	Compound **89** active against *B. cinerea*, *P. ultimum*, *F. oxysporium f.* sp. *cucumerinum*, *C. orbiculare*, *V. dahliae*, *P. oryzae*, *P. diospyri*, *S. sclerotiorum* and *F. fulva*, especially *B. cinerea* and *F. fulva* with EC_{50} values below 50 μg/mL	[53]
31	*Nodulisporium* sp. A21	Leaves of *Ginkgo biloba*. Nanjing in Jiangsu Province, China	Sporothriolide, (**90**)	The EC_{50} of compound **90** against *R.solani* was 3.04 μg/mL (11.6 μM), while the EC_{50} of carbendazim was 1.84 μg/mL (9.6 μM).	[54]
32	*Lopherdermium nitens* DAOM 250027	*Pinus strobus* Sussex, NB, Canada	Six phenolic bisabolane-type sesquiterpenoids (**91–96**), pyrenophorin (**97**)	Compound **97** significantly reduced the growth of *M. violaceum* and *S. cerevisiae* at 5 μM whereas sesquiterpenoids **91–96** active at 50 μM to both species tested	[55]
33	*Exserohilum* sp.	*Acer truncatum* Beijing, China.	Exserolide C (**98**), (12R)-12-hydroxymonocerin (**99**)	Compounds **98** and **99** active against *F. oxysporum*, both showing a MIC value of 20 μg/mL, Amphotericin B (MIC, 0.63 μg/mL)	[56]
34	*Biscogniauxia mediterranea* EPU38CA	*Echinacea purpurea* Missouri, USA,	(−)-5-methylmellein (**100**) and (−)-(3R)-8-hydroxy-6-methoxy-3,5-dimethyl-3, 4-dihydroisocoumarin (**101**)	Compound **100** active against *P. obscurans*, *P. viticola*, and *F.oxysporum*, and caused growth stimulation of *C. fragariae*, *C. acutatum*, *C. gloeosporioides*, and *B. cinerea*. Compound **101** was found to be slightly more active in the microtiter environment than 5-methylmellein	[57]
35	*Hypoxylon submonticulosum*	*Rubus idaeus* collected from Jordan Station, ON, Canada.	Trienylfuranol A (**102**) Complete hydrogenation of (**102**) yielded THF 7 (**103**)	THF 7 (**103**) inhibited the growth of *S. cerevisiae* (74 ± 4% inhibition) at a concentration of 250 μg/mL as compared with complete inhibition by nystatin at 10 μg/mL	[58]
36	*Phialophoramustea*	*Crocus sativus.*	Phialomustin C-D (**104**) (**105**)	Compounds **104–105** active against *C. albicans* ($IC_{50,}$ 14.3 and 73.6 μM)	[59]
37	unidentified Ascomycete,	*Melilotus dentatus.*	*cis*-4-acetoxyoxymellein (**106**) and 8-deoxy-6-hydroxy-*cis*-acetoxyoxymellein (**107**)	Compounds **106** and **107** displayed activities toward *M. violaceum*, *B. cinerea*, with 8 mm zone of inhibition for both fungi.	[60]
38	*Plectophomella* sp.		(−)-Mycorrhizin A (**108**)	Compound **108** active against *U. violacea* and *E. repens.*	[61]
39	*Physalospora* sp.		Cytochalasin E (**109**) and K (**110**)	Compound **109–110** active against *E. repens* and *M. microspora*	[61]

Table 1. *Cont.*

Sr. No.	Fungus	Plantsource	Compounds Isolated	Biologicalactivity *	Refs.
40	*Crataegus monogyna.*		Radicinin (**111**)	Radicinin (**111**) active against *E. repens* and *M. microspora*	[61]
41	*Berkleasmium* sp.,	*Dioscorea zingiberensis.* Hubei Province, China.	Diepoxin ζ (**112**), palmarumycin C11 (**113**), palmarumycin C12 (**114**), cladospirone B (**115**), palmarumycin C6 (**116**), 1,4,7β-trihydroxy-8-(spirodioxy-1′,8′-naphthyl)-7,8-dihydronaphthalene (**117**) and palmarumycin C8 (**118**)	Compounds **112–118** inhibited spore germination of *M. oryzae* (IC_{50} values in the range 9.1 − 124.5 μg/mL). Compound **118** showed the best inhibitory activity (IC_{50}, 9.1 μg/mL) among the compounds tested. Carbendazim (IC_{50} 6.3 μg/mL)	[62]
42	*Bipolaris* sp. MU34	*Gynura hispida* Bangkok, Thailand.	Bipolamide B (**119**)	Bipolamide B (**119**) active against *C. cladosporioides*, *C. cucumerinum*, *S. cerevisiae*, *A. niger* and *R. oryzae*, with MIC values of 16, 32, 32, 64 and 64 μg/mL, respectively	[63]
43	*Alternaria alternata* Tche-153	*Terminalia chebula* Rezt. Suanluang Rama IX Public Park, Bangkok, Thailand.	Altenusin (**120**)	Altenusin (**120**) in combination with each of three azole drugs, ketoconazole, fluconazole or itraconazole at their low sub-inhibitory concentrations exhibited potent synergistic activity against *C. albicans* with the FIC index range of 0.078 to 0.188	[64]
44	*Alternaria* sp. UFMGCB 55,	Leaves of *Trixis vauthieri* DC (Asteraceae).	Altenusin (**120**)	The altenusin (**120**) exhibited strong activity against 11 strains *P. brasiliensis* with MIC values ranging between 1.9 and 31.2 μg/mL MIC values found for amphotericin B were between 0.031 and 0.12 μg/mL. Additionally, *S. pombe* cells treated with altenusin were more rounded in shape than untreated cellssuggeststhat altenusin could act through the inhibition of cell wall synthesis or assembly in *P. brasiliensis* and *S. pombe*	[65]
45	*Cladosporium cladosporioides*		Cladosporin (**121**), Isocladosporin (**122**)	At 30 μM compound **121** exhibited 92.7, 90.1, 95.4, and 79.9% growth inhibition against *C. acutatum*, *C. fragariae*, *C. gloeosporioides* and *P. viticola* respectively. Compound **122** showed 50.4, 60.2, and 83.0% growth inhibition at 30 μM against *C. fragariae*, *C. gloeosporioides*, and *P. viticola*, respectively	[66]
46	*Epicoccum* sp. CAFTBO,	*Theobroma cacao* (Sterculiaceae) Mount Kala, near Yaoundé, Centre Province, Republic of Cameroon	Epicolactone (**123**), Epicoccolide A (**124**) and B (**125**)	Compounds **123–125** showed inhibitory effects on the mycelial growth of *P. ultimum* and *A. cochlioides* and *R. solani* (MIC in the range of 20–80 μg per paper disc)	[67]
47	*Biscogniauxiamediterranea* Ohu 19B	*Opuntia humifusa* (*Cactaceae*) from the United States	5-methylmellein (**100**)	Compound **100** 5-methylmellein was evaluated for antifungal activity against seven plant pathogens (*C. acutatum*, *C. fragariae*, *C. gloeosporioides*, *F. oxysporum*, *B. cinerea*, *P. obscurans*, and *P. viticola*) using an in vitro microdilution broth assay.	[68]
48	*Emericella* sp. XL029	Leaves of *Panax notoginseng* Shijiazhuang, Hebei Province, China.	5-(undeca-3′,5′,7′-trien-1′-yl)furan-2-ol (**126**) and 5-(undeca-3′,5′,7′-trien-1′-yl)furan-2-carbonate (**127**)	Compound **126** active against *R. solani*, *V. dahliae*, *H. maydis*, *F. oxysporum*, *F. tricinctum*, *B. dothidea*, *and A. fragriae* (MIC values from 25 to 3.1 μg/mL), while compound **127** displayed activity against *V. dahliae*, *H. maydis*, *F. tricinctum*, *B. dothidea*, *and A. fragriae* (MIC values from 50 to 12.5 μg/mL)	[69]
			Comounds Produced by Hyphomycetes		
49	*Fusarium fujikuroi* (WF5), *Penicilium chrysogenum* WF6, and *P. expensum* WF7	Finger millet Plants Arkell Field Station, Arkell, ON, Canada.	5-hydroxy 2(3H)-benzofuranone (**128**), dehydrocostus lactone (**129**) and harpagoside (**130**)	Compounds **128–130** active against *F. graminearum* with MIC of 31.25, 250.00 and 31.25 μg/mL, respectively.	[70]
50	*Trichoderma koningiopsis* YIM PH30002	*Panax notoginseng*. Wenshan, Yunnan Province, China.	Koninginin O (**131**), koninginin Q (**132**), 7-*O*-methylkoninginin D (**133**)	Compounds **131–132** active against *F. oxysporum and P. cucumerina*, with an MIC of 128 μg/mL. Compound **133** showed activity against *P. cucumerina* with an MIC of 128 μg/mL. Nystatin was active with MICs at 32 μg/mL	[71]
51	*Trichoderma koningiopsis* YIM PH30002	*Panax notoginseng*. Wenshan, Yunnan Province, China.	Koningiopisin C (**134**)	Compound **134** exhibited in vitro antifungal activity against *F. oxysporum*, *A. panax*, *F. solani* and *P. cucumerina* with MICs at 32, 64, 32, and 16 μg/mL, respectively	[72]

Table 1. *Cont.*

Sr. No.	Fungus	Plantsource	Compounds Isolated	Biologicalactivity *	Refs.
52	*Trichoderma* sp. 09	*Myoporum bontioides*	Dichlorodiaportinolide (**135**), dichlorodiaportin (**136**)	Compounds **135–136** active against *C. musae* and *Rhizoctoniasolani* (MIC values from 6.25 to 150 μg/mL)	[73]
53	*Trichoderma brevicompactum* 0248	*Allium sativum*	Trichodermin (**137**)	Compound **137** active against *R. solani*, *B. cinereal*, *C. lindemuthianum* with an EC_{50} of 0.25, 2.02 and 25.60 μg/mL respectively. Carbendazim showed, antifungal activity against *R. solani*, *B. cinereal*, with an EC_{50} of 0.36 and 10.35 μg/mL respectively	[74]
54	*Trichoderma koningiopsis* YIM PH30002	Wenshan, Yunnan Province of China.	Koninginin R (**138**) and S (**139**)	Compound **138** active against *F. oxysporum* and *F. flocciferum* with MICs at 128 μg/mL, while compound **139** displayed activity against *F. oxysporum* with MIC at 128 μg/mL	[75]
55	*Aspergillus terreus*	*Carthamus lanatus* Al-Azhar University campus, Assiut Branch, Assiut, Egypt.	(22E,24R)-stigmasta-5,7,22-trien-3-β-ol (**140**), aspernolides F (**141**)	Compound **140** active against *C. neoformans* with IC_{50} values of 4.38 μg/mL, amphotericin B (IC_{50} 0.34 μg/mL). Compound **141** showed good activity against *C. neoformans* (IC_{50} 5.19 μg/mL).	[76]
56	*Aspergillus* sp. KJ-9,	*Melia azedarach* which was collected at Yangling, Shaanxi Province, China	Fonsecinone A (**142**), (R)-3-hydroxybutanonitrile (**143**)	Compounds **142** and **143** were active against *G. saubinetti*, *M. grisea*, *B. cinerea*, *C. gloeosporioides* and *A. solani* with MIC range of 6.25–50 μM	[77]
57	*Aspergillus* sp.	*Gloriosa superba* Tirupati, India.	6-methyl-1,2,3-trihydroxy-7,8-cyclohepta-9,12-diene-11-one-5,6,7,8-tetralene-7-acetamide (KL-4) (**144**)	KL-4 (**144**) active against *S. cerevisiae*, *C. albicans* and *C. gastricus* with MIC 25, 12.5, and 50 μg/mL respectively	[78]
58	*Penicillium* sp. R22	*Nerium indicum* collected from Qinling Mountain, Shaanxi Province, China.	5-hydroxy-8-methoxy-4-phenylisoquinolin-1(2H)-one (**145**), 3-*O*-methylviridicatin (**146**) and viridicatol (**147**)	Compound **145** active against *A. brassicae*, *A. alternata and V. mali* with MIC value of 31.2 μg/mL, compound **146** against *A. brassicae*, *B. cinerea* and *V. male* with MIC value of 31.2 μg/mL, compound **147** against *A. brassicae*, *A. alternata* and *B. cinerea* with MIC value of 31.2 μg/mL	[79]
59	*Penicillium raciborskii*,	*Rhododendron tomentosum* were collected at the test site of University of Oulu, Finland.	Outovirin C (**148**)	Outovirin C (**148**) inhibited growth of *F. oxysporum*, *B. cinerea*, and *V. dahlia* at the concentration of 0.38 μM. Compound **148** active against *B. cinerea* (57% inhibition) and slightly less effective against *V. dahliae* (45% inhibition)	[80]
60	*Fusarium* sp.	*Mentha longifolia* Saudi Arabia.	Fusaripeptide A (**149**)	Compound **149** active against *C. albicans*, *C. glabrata*, *C. krusei*, and *A. fumigates* with IC_{50} values of 0.11, 0.24, 0.19, and 0.14 μM, respectively. Amphotericin B exhibited antifungal activity toward *C. albicans*, *C. glabrata*, *C. krusei*, and *A. fumigates* with IC_{50} values of 0.3, 0.6, 0.5, 0.7 μM, respectively	[81]
61	*Fusarium chlamydosporium*	*Anvillea garcinii* Al Madinah Al Munawwarah, Saudi Arabia.	Fusarithioamide A (**150**)	Compound **150** active with inhibition zone diameters 16.2 mm and MIC 2.6 μg/mL towards *C. albicans*. Clotrimazole (inhibition zone diameters 18.5 mm and MIC 3.7 μg/mL)	[82]
62	*Fusarium* sp.	*Ficus carica* Qinling Mountain, Shaanxi Province, China	Helvolic acid Methyl ester (**151**), helvolic acid (**152**) and hydrohelvolic acid (**153**)	Compounds **151–153** active against *B. cinerea*, *C. gloeosporioides*, *F. oxysporum* f. sp. *niveum*, *F. graminearum* and *P. capsici* (MIC in the range of 12.5–25 μg/mL), Carbendazim (MIC in the range of 32.2–62.5 μg/mL)	[83]
63	*Fusarium* sp.		Colletorin B (**154**), colletochlorin B (**155**), LL-Z1272β (ilicicolin B) (**156**) and 4,5-dihydrodechloroascochlorin (**157**)	Compounds **154–156** showed antifungal against *U. violacea* and *F. oxysporum*. Compound **157** showed antifungal activity towards *E. repens*,	Hussain et al. [84]
64	*Curvularia* sp., strain M12,	*Murraya koenigii* Rajshahi University, Bangladesh	Murranolide A (**158**), murranopyrone (**159**), Curvularin (**160**), (S)-dehydrocurvularin (**161**), pyrenolide A (**162**), modiolide A (**163**), and 8-hydroxy-6-methoxy-3-methylisocoumarin (**164**)	Pyrenolide A (**162**) showed a strong motility impairing activity against *Phytophthora capsici* zoospores at a low concentration (100% at 0.5 μg/mL) in a short time (30 min). Compounds **158–161** and **163–164** exhibited zoospore motility impairment activity at higher concentrations (IC_{50}: 50–100 μg/mL)	[85]

Table 1. *Cont.*

Sr. No.	Fungus	Plantsource	Compounds Isolated	Biologicalactivity *	Refs.
65	*Phaeoacremonium* sp.,	*Senna spectabilis* AraraquaraCerrado area, Araraquara, Sao Paulo state, Brazil.	Isoaigialone B (**165**), and C (**166**), aigialone (**167**)	Compounds **165** and **167** exhibited antifungal activity, with a detection limit of 5 µg, for *C. cladosporioides* and *C. sphaerospermum*, compound **166** exhibited weak activity (detection limit > 5 µg), with a detection limit of 25 µg. Nystatin, positive control, showing a detection limit of 1 µg	[86]
66	*Trichothecium* sp.	*Phyllanthus amarus* Pune India.	Trichothecinol A (**168**)	Compound **168** active against *C. albidus* up to 20 µg/mL	[87]
67	*Trichothecium* sp.	*Phyllanthus* sp. Pune India.	Trichothecin (**169**)	Trichothecin (**169**) active against *S. cerevisiae*, *C. albidus var diffluens* (NCIM 3371), *C. albidus* var *diffluens* (NCIM 3372), *F. oxysporum*, *P. expansum*, *T. viride*, *P. varioti* and *A. niger* with MIC of 6.0, 20.0, 12.0, 10.0, 30.0, 40.0, 20.0 and 12.0 µg/mL respectively	[88]
			Comounds Produced by Basidiomycetes		
68	*Scleroderma* UFSM Sc1(Persoon) Fries	*Eucalyptus grandis*.	Sclerodol A (**170**) and B (**171**) and related lanostane triterpenoid (**172**)	Compound **170** active against *C. albicans*, *C. tropicalis*, *C. crusei*, *C. parapsiosis* (MIC of 25.0, 25.0, 6.25 and 12.5 MFC 25.0 25.0, 12.5, 25.0 µg/mL) Compound **170** and **172** were active against tested strain (MIC in range of 12.5–100 µg/mL). Nystatin active against test strains (MIC in the range of 0.77–1.52 µg/mL).	[89]

3. Antifungal Potential of Volatile Organic Compounds (VOCs) from Endophytic Fungi

Volatile organic compounds (VOCs) are generally carbon compounds which exist in the gaseous phase at normal/ambient temperatures and pressures. Over 250 different VOCs produced by fungi comprising different chemical classes such as aldehydes, ketones, alcohols, phenols, thioesters, and so forth, have been identified in the context of the deterioration of fruits, vegetables, indoor environments (sick building syndrome); as chemotaxonomic markers; and in the morphogenesis and development of fungi.

However, bioprospecting fungal endophytes for the production of volatile antimicrobials came into the limelight with the discovery of *Muscodor albus* from the plant *Cinnamomum zeylanicum*, from Honduras. *M. albus* was found to produce an admixture of VOCs which could effectively kill a variety of pathogenic bacteria and fungi associated with plants and animals. This research garnered much attention and drove people to explore the volatile antibiotic properties of endophytic fungi for varied applications [90,91].

The genus *Muscodor* comprises of an endophytic fungi which is predominantly sterile, does not possess true reproductive structures like other fungi, and emanates a characteristic smell which is largely attribute to its VOC composition [92]. Since the report of *M. albus* in the late 1990s, to date, 20 species have been added to this genus, which have largely been identified based on their volatile signatures, molecular phylogeny, and morphological characteristics (Table 2). The characteristic VOC profile, therefore, is helpful in delineating the species, as well as playing a significant role in its anti-fungal and anti-bacterial properties. In this section, we only be highlight the anti-fungal potential of VOCs produced by these endophytic fungi.

The majority of the VOCs produced by the endophytic fungi comprises of a mixture of volatile components which generally has either a synergistic effect or an additive effect that enhances their bioactivity against pathogenic microbes. However, in a couple of studies, the major components of the volatile mixture were independently evaluated to understand their true antimicrobial/anti-fungal potential. These are generally synthetically generated and converted into a volatile form and subsequently evaluated for their bioactivity against the test microorganisms. For instance, *Sclerotina sclerotiorum* was completely inhibited by 2-methyl-1-butanol and 3-methyl-1-butanol with an EC_{50} value of 0.8 μL/mL. 2-methyl-1-butanol also inhibited *Penicillium digitatum* with an EC_{50} value of 0.48 μL/mL and *B. cinerea* with a value of 1.38 μL/mL. However, the volatile admixture of the *M. albus* VOC exhibited an IC_{50} range between 0.08 and 1.13 μL/mL, which clearly confirms the hypothesis of the synergistic/additive effects of the volatile components [93].

Recently, ethyl acetate has been reported to be the main VOC of yeasts *Wickerhamomyces anomalus*, *Metschnikowia pulcherrima*, and *Saccharomyces cerevisiae*, which inhibit the decay causing mold, as well as *B. cineria*. All three yeasts exhibit excellent biological control properties and were used for checking the mold and pathogenic attack in sweet cherries and strawberries. *W. anomalus* induced the highest killing activity amongst the three which was attributed to the higher production of Ethyl acetate. The role of the ethyl acetate was re-affirmed by using synthetic ethyl acetate from strawberry fruits to affirm the anti-fungal action [94].

Similarly, *Phaeosphaeria nodorum*, which existed as an endophyte in plum leaves (*Prunus domestica*) was found to inhibit the pathogen *Monilinia fruticola*. The major component of the VOC produced by

Phaeosphoran odorum comprised of 3-methyl-1-butanol, acetic acid, 2-propyn-1-ol, and 2-propenenitril [95]. Similarly, six VOCs from the endophytic fungus *Hypoxylon anthochroum* (that is, phenylethyl alcohol), 2-methyl-butanol and 3-methyl-1-butanol, eucalyptol, ocimene, and terpenoline were tested against *Fusarium oxysporum*. The results indicated that these compounds exhibited concentration-dependent anti-fungal activity individually but have better action and control synergistically. Thus, the mixture of six VOCs may be used for the control of *Fusarium oxysporum* in tomatoes [96].

The genus *Muscodor* is one of the best studied endophytic fungus which produces a synergistic mixture of VOCs having lethal effects against a wide variety of plant and human pathogenic fungi, nematodes, and bacteria as well as certain insects [97–100]. The volatility of the *Muscodor* species has been used to replace methyl bromide (MeBr)—a traditional soil fumigant—which has been globally banned as it causes the depletion of ozone layer. Different species of *Muscodor*, their major VOCs, and their anti-fungal spectrum are given in Table 2. Geographically, each *Muscodor* species has a characteristic signature volatility. For instance, the Indian *Muscodor* species invariably has 4-Octadecylmorpholine as a marker compound while 2-methyl propanoic acid is generally found in Muscodor isolated from North and South America.

The majority of the VOCs from the fungal endophytic fungi are used as biological control agents to prevent the fungal deterioration of crops, fruits, and vegetable, under both pre- and post-harvest conditions. However, the exploitation of these fungally volatile organic compounds (FVOCs) from endophytic fungi are not being actively applied to humans for the prevention of fungal infections.

There exists a huge scope in evaluating these FVOCs from endophytic fungi since they could be helpful in curing superficial skin infections, the sanitization of public toilets, and in night soil. They can also find applications in personal care products such as for the aroma/ fragrance in deodorants and sprays. They could presumably be helpful in the development of sprays for inhalation to treat fungal diseases like Aspergillosis in lungs.

Table 2. Spectrum of VOCs emitted by *Muscodor* species and their anti-fungal activity predominantly against plant pathogenic fungi.

No.	Name of the Endophytic Fungi	Geographic Area of Isolation	Major VOCs Produced	Anti-Fungal Activity	Refs.
1	*Muscodor albus*	Central America (Honduras)	2-methylpropanoic acid; 3-methyl-1-butanol; ethanol; acetic acid (methyl ester)	*Rhizoctonia solani; Phytophthora cinnamomi; Sclerotinia sclerotiorum; Fusarium solani, Verticillium dahliae*	[91]
2	*M. vitigenus*	South America (Peru)	naphthalene; caryophyllene; azulene	*R. solani; Phoma* sp.; *C. coefficola*	[98]
3	*M. roseus*	Australia	2-butenoic acid (ethyl ester); 1,2,4-tri-methyl-benzene; 2-nonadiene	Antifungal spectrum not reported	[101]
4	*M. yucatensis*	South America (Mexico)	caryophyllene; aromadendrene	*Botrytis cineria; R. solani; C. coefficola; Phoma sp.*	[102]
5	*M. fengyangensis*	China	2-methylpropionic acid; β-phellendrene	*B. cineria; Aspergillus clavatus; Colletotrichum fragiae; Sclerotium rolfsii*	[90]
6	*M. crispans*	South America (Bolivia)	2-methylpropanoic acid; ethanol; ethyl acetate	*B. cineria; Curvularia lunata; P. cinnamomi; S. sclerotiorum*	[103]
7	*M. sutura*	USA (Columbia)	butylated hydroxytoleuene; octacecanoic acid; thujopsene; 2-methylpropanoic acid; naphthalene	*Aspergillus fumigatus; Colletotrichum lagenarium; B. cineria; Cercospora beticola; Phytophthora palmivora; Fusarium solani*	[104]
8	*M. musae*	Thailand	3-methylbutanol acetate 2-methylpropanoic acid	*Alternaria porri; Alternaria solani; Colletotrichum gloereosporioides; Nigrospora oryzae*	[105]
9	*M. oryzae*	Thailand	3-methylbutan-1-ol; 2-methylpropanoic acid	*A. porri; A. solani; Aspergillus flavus; B. cineria; C. gloereosporioides; N. oryzae*	[105]
10	*M. suthepensis*	Thailand	3-methylpropanoic acid 3-methylbutan-1-ol	*A. porri; Alternaria alternata; Aspergillus flavus; B. cineria; C. gloereosporioides; Fusarium oxysporum; Fusarium solani; N. oryzae*	[105]
11	*M. equiseti*	Thailand	3-methylbutan-1-ol; 3-methylbutanoyl acetate; 2-methylpropanoic acid	*A. porri; A. solani; B. cineria; C. gloereosporioides; F. oxysporum; F. solani; N. oryzae*	[105]
12	*M. cinnamomi*	Thailand	2-methylpropanoic acid;2-methyl butanoic acid; azulene	*Rhizoctonia solani*	[105]
13	*M. kashayum*	India	1-methyl-4(1-methylethhylidene)-cyclohexane; 2(4-morpholinyl)ethylamine; 9-octadecanoic acid (methyl ester); 4-octadecylmorpholine	*Bionectria ochroleuca; Cercospora beticola; Chaetomium heterosporum; C. gloereosporioides; F. oxysporum; Fusarium equiseti; Curvularia lunata*	[106]
14	*M. darjeelingensis*	India	4-octadecylmorpholine; 2,6-*bis*(1,1-dimethylethyl)-4-(1-oxopropyl)phenol; beta-aminoethyl-morpholine	*Lasiodiplodia theobromae; A. alternata; Rhizoctonia solani; Cercospora beticola*	[107]
15	*M. strobelii*	India	4-octadecylmorpholine; tetraoxapropellan; aspidofractanine-3-methanol;viridiflorol	*Rhizoctonia solani; Colletotrichum gloereosporioides; Fusarium oxysporum; Lasiodiplodia theobromae*	[108]

Table 2. *Cont.*

No.	Name of the Endophytic Fungi	Geographic Area of Isolation	Major VOCs Produced	Anti-Fungal Activity	Refs.
15	*M. strobelii*	India	4-octadecylmorpholine; tetraoxapropellan; aspidofractanine-3-methanol;viridiflorol	*Rhizoctonia solani;* *Colletotrichum gloereosporioides; Fusarium oxysporum; Lasiodiplodia theobromae*	[108]
16	*M. tigrii*	India	4-octadecylmorpholine; 1-tetradecamine n, n-dimethyl 1,2-benzidicarboxylic acid mono(2-ethylhexyl)ester	*Alternaria alternate* *Cercospora beticola*	[109]
17	*M. heavae*	Thailand	2-phenylethanol; azulene	*Aspergillus niger; Phellinus noxius; Rigidoporus microporus*	[110]
18	*M. ghoomensis*	India	n,n-dimethyl-1-nonadecamine; 4-octadecylmorpholine	*Cercospora beticola*	[111]
19	*M. indica*	India	n, n-dimethyl-1-pentadecamine; 4-morpholinethanamine	*Cercospora beticola* *Penicillium marnaeffi*	[111]
20	*M. camphora*	India	tetracontane; 4-octadecylmorpholine; n, n-dimethyl-1-pentadecamine	*Colletotrichum gloereosporioides; Lasiodiplodia theobromae*	[112]

4. Methods Used for Activation of Silent Biosynthetic Genes

Several research studies confirm that most of the biosynthetic gene clusters are observed to be silent or expressed at a low (minimal) level upon employing conventional culturing conditions for growth/propagation of microorganisms [113]. To activate such silent biosynthetic genes, numerous strategies have been employed, such as the one strain many compounds (OSMAC) approach (activation mediated through modification in composition of medium, aeration, temperature or shape of culturing flask), co-culturing method (facilitating activation through interspecies crosstalk) and genomics based approaches (expression of orphan biosynthesis genes in a heterologous host). In recent times, the use of chemicals as modifiers to alter the epigenetic makeup/constitution of a microorganism to improve its biosynthetic potential has become a beneficial tool. The method uses a chemical that acts as DNA methyltransferase inhibitors (DNMTi) or histone deacetylase inhibitors (HDACi), thereby stimulating the transcription previously silent gene clusters and fostering the production of a spectrum of natural products. A comprehensive description of some of these methods are given below/highlighted in the subsequent section.

4.1. Epigenetic Modification

Endophytes have proven to be the prolific source of bioactive metabolites and offer a substitute and untapped reserve for the discovery of novel metabolites. Studies have led to findings that tell biosynthetic gene clusters of microorganisms are mostly silent or expressed at very low levels under standard culture conditions and are least expressed, but under stress condition may it be biological, chemical or physical their expression takes place. Epigenetic modulators lead to the expression of these silent or cryptic genes. Epigenetic gene regulation is mediated by covalent histone modification, DNA methylation chromatin modeling basically induced by DNA methyl transferase inhibitors such as 5-aza-2-deoxycytidine, 5-azacytidine, hydralazine, procaine and histone deacetylase [114].

Chromatic modification in fungi to enhance gene transcription has led to secondary metabolite production of anthraquinones, cladochromes, lunalides, mycotoxins, and nygerones [115]. Structural genes that control transcriptional factor regulates the synthesis of secondary metabolites in fungi, these genes mediate factors occupied in environmental signals like pH, nitrogen and carbon sources, temperature, light, etc. [116]. In lab condition, these gene clusters are mostly silent. Under which natural conditions these clusters become activated is still unexplained. As per genetic sequencing studies carried so far, it is estimated that the clusters of genes responsible for secondary metabolites have not yet been deciphered completely [117].

From endophytic fungi *Aspergillus fumigatus* (GA-L7) obtained from *Grewia asiatica* led to identification of seven metabolites namely pseurotin A (**173**), pseurotin D (**174**), pseurotin F2 (**175**), fumagillin (**176**), tryprostatin C (**177**), gliotoxin (**178**) and *bis*(methylthio)gliotoxin (**179**) (Figure 9). On addition of the valproic acid, increase in the production of fumiquinazoline C (**180**) up to 10 times was noticed along with a shift in the pattern of metabolite production. It was also observed that all the genes, i.e., Afua_6g 12040, Afua_6g 12050, Afua_6g 12060, Afua_6g 12070 and Afua_6g 12080, tangled in the biosynthesis of fumiquinazoline C (**180**), were upregulated significantly by 7.5, 8.8, 3.4, 5.6 and 2.1 folds, respectively [118].

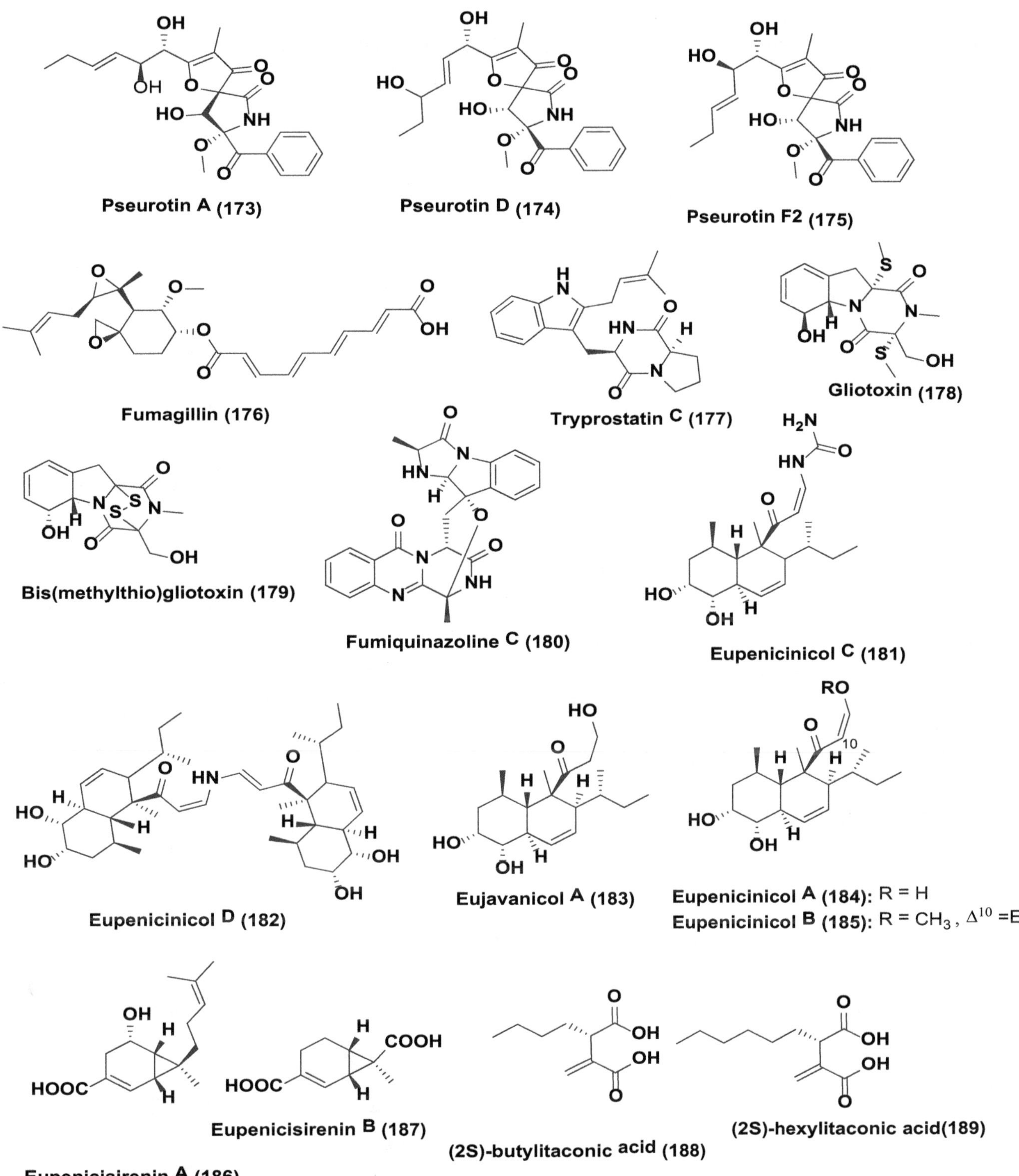

Figure 9. Structures of metabolites isolated from Epigenetic modification in endophytic fungi (**173–189**).

The NAD+-dependent HDAC inhibitor, nicotinamide, enhanced the yield of eupenicinicol C (**181**), and D (**182**) (Figure 9), decalin containing metabolites together with eujavanicol A (**183**), and eupenicinicol A (**184**) (Figure 9), biosynthetically related compounds by endophytic *Eupenicillium* sp. LG41, identified from the *Xanthium sibiricum*, a Chinese medicinal plant [119]. Under a similar condition without HDAC inhibitor, nicotinamide *Eupenicillium* sp. LG41 produce different decalin- containing compounds:

eupenicinicols A (**184**), and B (**185**); two new sirenin derivatives, eupenicisirenins A (**186**) and B (**187**); and other four known compounds, (2S)-butylitaconic acid (**188**), (2S)-hexylitaconic acid (**189**) (Figure 9), xanthomegnin (**190**), and viridicatumtoxin (**191**) (Figure 10) [120].

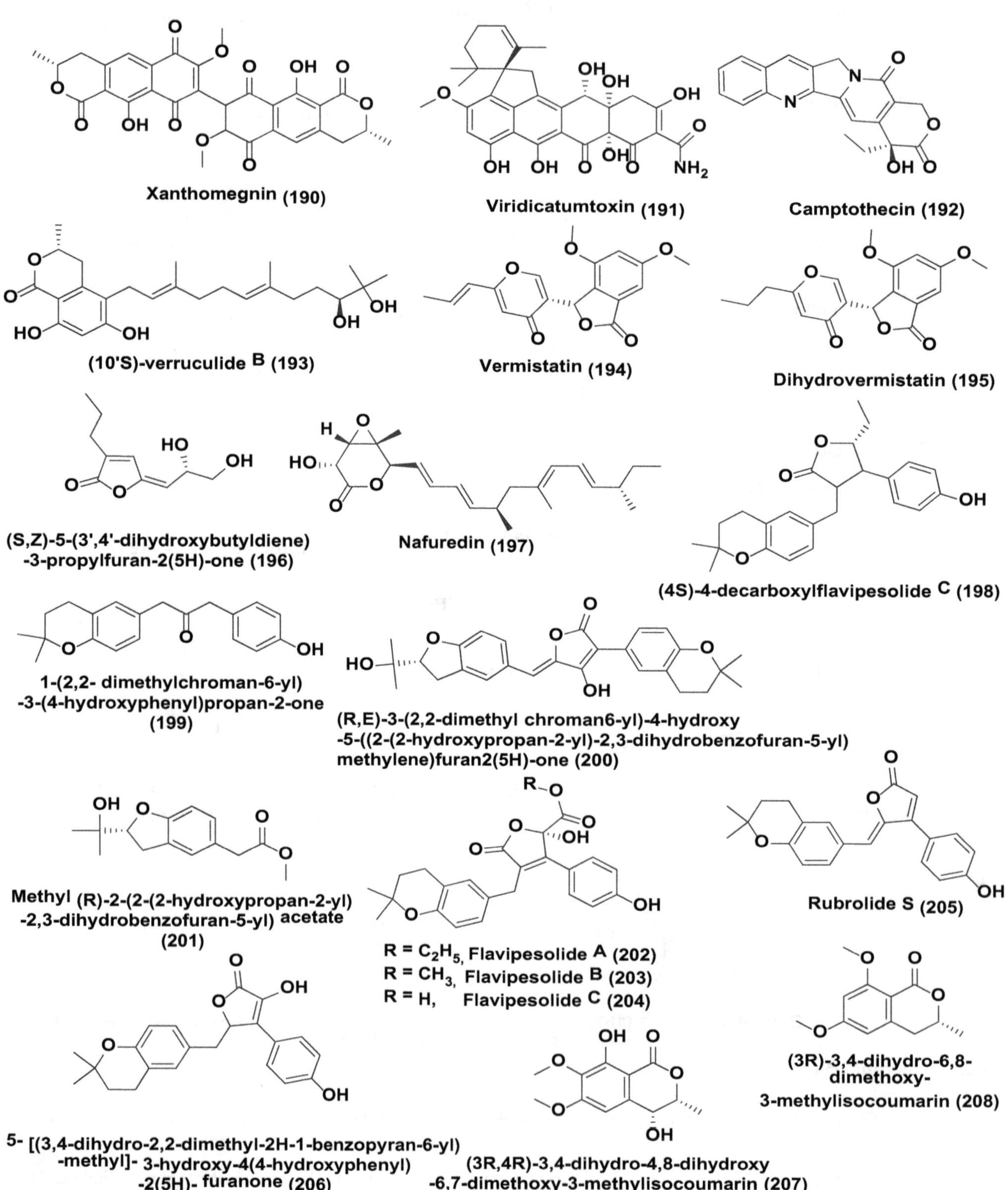

Figure 10. Structures of metabolites obtained from Epigenetic modification in endophytic fungi (**190**–**208**).

Vasanthakumari et al. [121] reported the attenuation of camptothecin (**192**) (Figure 10) yield in endophytic fungi obtained from camptothecin fabricating plants, *Nothapodytes nimmoniana* and *Miquelia dentata* incorporated with 5-azacytidine, a DNA methyltransferase inhibitor. It was indicated that the mode behind decrease of camptothecin yield in endophytic fungi could in principle be inverted by stimulating some signals from the tissue of the plant, probably the methylation or silencing of the genes liable for camptothecin production.

Metabolites such as (10′S)-verruculide B (**193**), vermistatin (**194**), and dihydrovermistatin (**195**) (Figure 10) were identified owing to the supplementation of HDAC inhibitor, suberoylanilide hydroxamic acid (SAHA) to endophytic fungi culture *Phoma* sp. nov. LG0217 isolated from *Parkinsonia microphylla*. However, in absence of SAHA, a novel metabolite (S,Z)-5-(3′,4′-dihydroxybutyldiene)-3-propylfuran-2(5H)-one (**196**), along with nafuredin (**197**) (Figure 10), was produced [122].

Four new meroterpenoids identified as (4S)-4-decarboxylflavipesolide C (**198**), 1-(2,2-dimethylchroman-6-yl)-3-(4-hydroxyphenyl)propan-2-one **(199)**, (R,E)-3-(2,2-dimethyl chroman6-yl)-4-hydroxy-5-((2-(2-hydroxypropan-2-yl)-2,3-dihydrobenzofuran-5-yl)methylene)furan2(5H)-one (**200**), methyl (R)-2-(2-(2-hydroxypropan-2-yl)-2,3-dihydrobenzofuran-5-yl) acetate (**201**), along with nine known compounds flavipesolides A−C (**202–204**), rubrolide S (**205**), 5-[(3,4-dihydro-2,2-dimethyl-2H-1-benzopyran-6-yl)-methyl]-3-hydroxy-4(4-hydroxyphenyl)-2(5H)-furanone (**206**), (3R,4R)-3,4-dihydro-4,8-dihydroxy-6,7-dimethoxy-3-methylisocoumarin (**207**), (3R)-3,4-dihydro-6,8-dimethoxy-3-methylisocoumarin (**208**) (Figure 10), terretonin C (**209**), and ergosterol (**210**) (Figure 11) were obtained using chemically modified epigenetic culture of *Aspergillus terreus* OUCMDZ-2739 with 10 μM trichostatin A (TSA). Under the similar parameters without TSA, *A. terreus* OUCMDZ-2739 yielded many compounds, i.e., aspernolide B (**211**), butyrolactone II (**212**), butyrolactone IV (**213**), butyrolactone I (**214**), aspernolide A (**215**), asterrelenin (**216**) and (+)-terrein (**217**) (Figure 11), supporting that fungal metabolite enrichment and chemodiversity using epigenetic modifiers can be done to obtain new products [123]. Induction of isosulochrin **(218)** (Figure 11) was also witnessed when *Chaetomium* sp. was supplemented with 5-azacytidine or SAHA on solid rice medium [124].

Asai et al. [125] reported six new benzophenones, cephalanones A–F (**219–224**), and 2-(2,6-dihydroxy-4-methylbenzoyl)-6-hydroxybenzoic acid (**225**) (Figure 11) from culture of *Graphiopsis chlorocephala*, f r o m *Paeonia lactiflora* in presence of HDAC inhibitors nicotinamide (10 μM) which resulted in significant increase in secondary metabolite production.

To express silent biosynthetic pathways, molecules such as HDAC and DNMT are used to enhance the fungal metabolites production. Different studies suggest an increase in chemical diversity of metabolites by induction with these epigenetic modifiers. For growth and acclimatization with the environment fungus are known to produce diverse secondary metabolites. Cross talk between microbes and plant lead to the expression of these pathways which stays silent in in vitro conditions. Metabolic profiles shift led by SMs induced modifier is due to expression of cryptic genes [126].

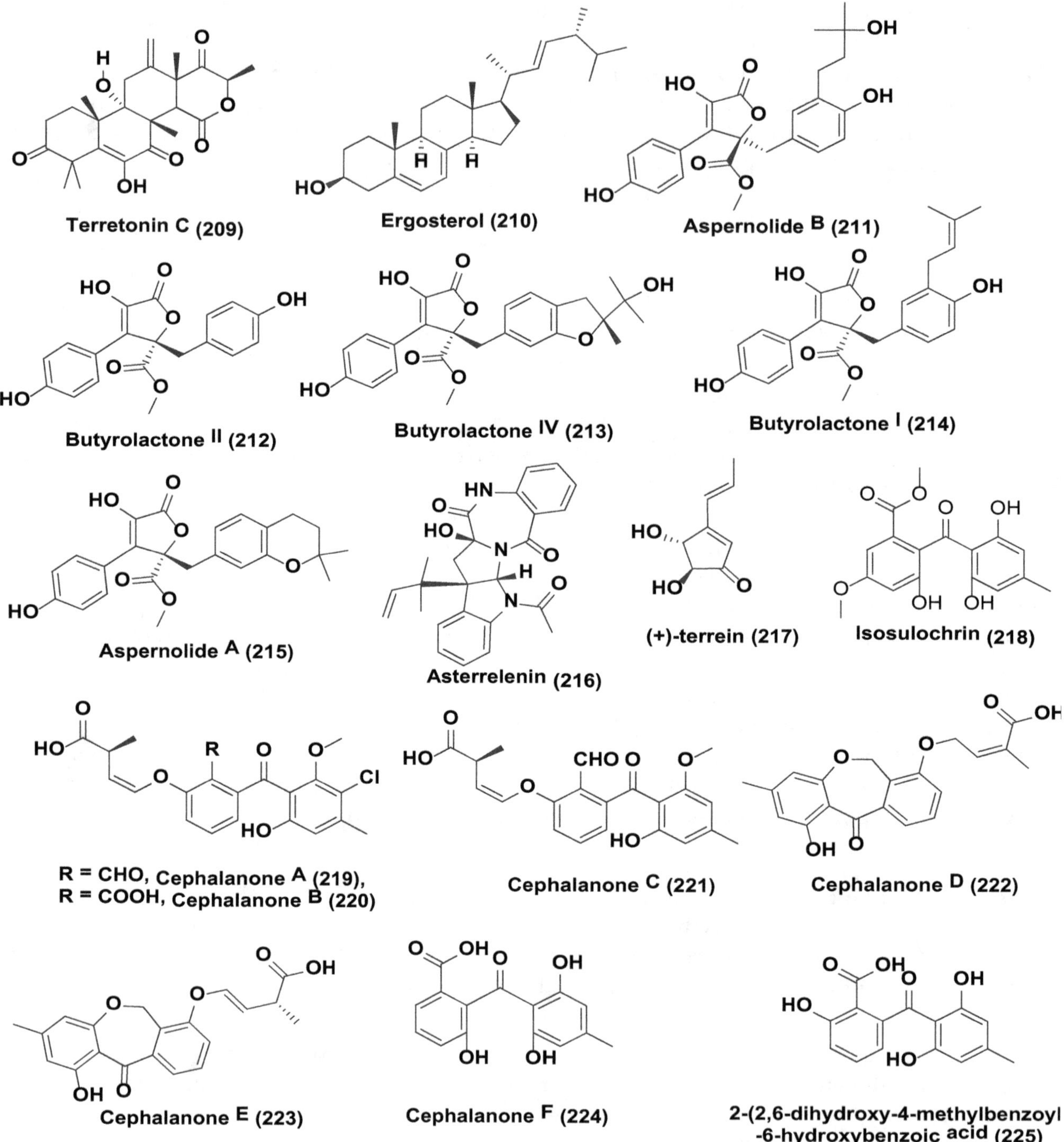

Figure 11. Structures of metabolites obtained from Epigenetic modification in endophytic fungi (**209–225**).

4.2. The Co-Culture Strategy

Interspecific interaction among different species leads to evolution and biodiversity, organism combines their genetic information for better adaptability. The cohabitation of different microorganisms that share similar niches competes with growth, morphology, adaptation, and development patterns [127,128]. The increased productions of metabolites in co-culture which are not produced in axenic culture are the result of competition or antagonism faced by the microorganism that leads to activation of cryptic genes. [129]. Co-cultivation is a way to provide natural habitat to fungi so that gene clusters become activated. In *Aspergillus nidulans,* the cryptic gene has been successfully activated leading to isolation of novel compounds [130].

In a study carried by Ola et al. [131], accumulation of secondary metabolites, i.e., lateropyrone (**226**), cyclic depsipeptides of the enniatin type (**227–229**), and the lipopeptide fusaristatin A (**230**) (Figure 12), was found to be enhanced by 78 folds by co-culturing *B. subtilis* 168 trpC2 with *Fusarium tricinctum* isolated from the *Aristolochia paucinervis*. This led to the identification of three new compounds macrocarpon C (**231**), 2-(carboxymethylamino)benzoic acid (**232**) and (−)-citreoisocoumarinol) (**233**), and a known compound, (−)-citreoisocoumarin (**234**), which was absent in axenic culture of bacterial or fungal control. On coculturing *Alternaria* sp. and *Phomopsis* sp. there was the enhancement of taxane (**235**) production by eight fold [132].

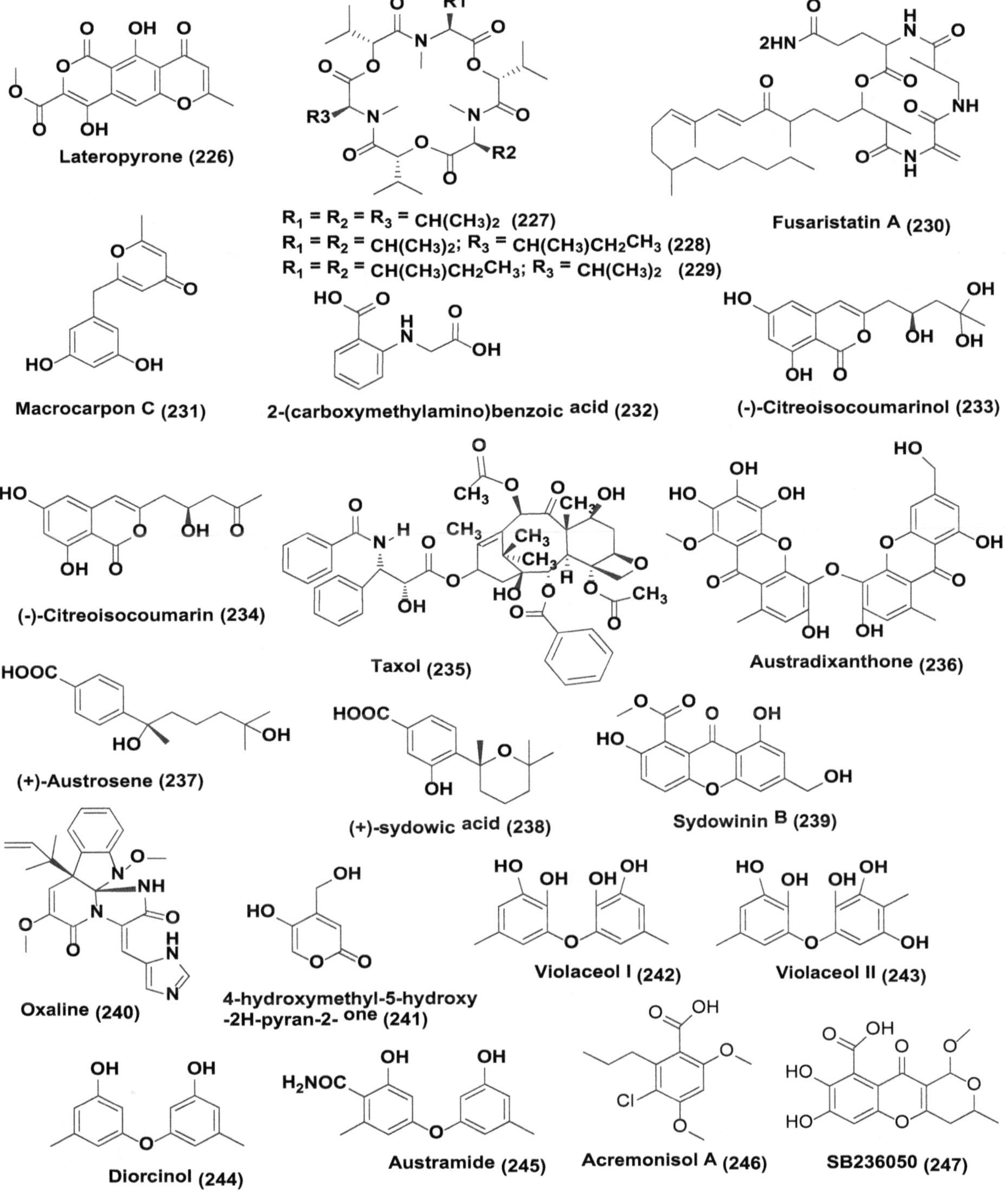

Figure 12. Structures of metabolites obtained from co-culture in endophytic fungi (**226–247**).

Aspergillus austroafricanus endophyte residing inside the leaves of *Eichhornia crassipes* was the source of two new metabolites, namely, xanthone dimer austradixanthone (**236**) and sesquiterpene (+)-austrosene (**237**), and five known compounds, (+)-sydowic acid (**238**), sydowinin B (**239**), oxaline (**240**), 4-hydroxymethyl-5-hydroxy-2H-pyran-2- one (**241**) (Figure 12), ergosterol (**210**) (Figure 13). However, the same endophyte grown in mixed cultures with *Bacillus subtilis* or *Streptomyces lividans* led to the identification of many diphenyl ethers i.e., violaceol I (**242**), violaceol II (**243**), and diorcinol (**244**) (Figure 12) along with new austramide (**245**) (Figure 12), increased up to 29 times [133].

Chaetomium sp. was isolated from *Sapium ellipticum* the Cameroonian medicinal plant. When *Chaetomium* sp. was cultured axenically on solid rice medium, average yields per culture flask were 2.8, 13.9, 132.7 and 14.6 mg of acremonisol A (**246**), SB236050 (**247**) (Figure 12), and SB238569 (**248**), respectively, and 1:1 mixture of 3- and 4-hydroxybenzoic acid methyl esters (**249–250**), respectively, (Figure 13) was observed. When Co-cultivation of *Chaetomium* sp. was undertaken with viable or autoclaved cultures of *Bacillus subtilis* there was a strong accumulation of the 1:1 mixture of (**249**), and (**250**), was observed, accounting for an 8.3 and 7.4-fold increase, respectively, compared to axenic fungal controls in both cases. SB236050 (**247**) and SB238569 (**248**), two major polyketides of *Chaetomium* sp., were not detected in co-cultures. Five new compounds, Shikimeran A (**251**), Bipherin A (**252**), Chorismeron (**253**), Quinomeran (**254**), and Serkydayn (**255**), and two known compounds, isosulochrin (**218**) and protocatechuic acid methyl ester (**256**) (Figure 12), were only detected in co-cultures of *Chaetomium* sp. with viable or autoclaved *B. subtilis* cultures, but were lacking in both fungal or bacterial controls when cultured axenically [124].

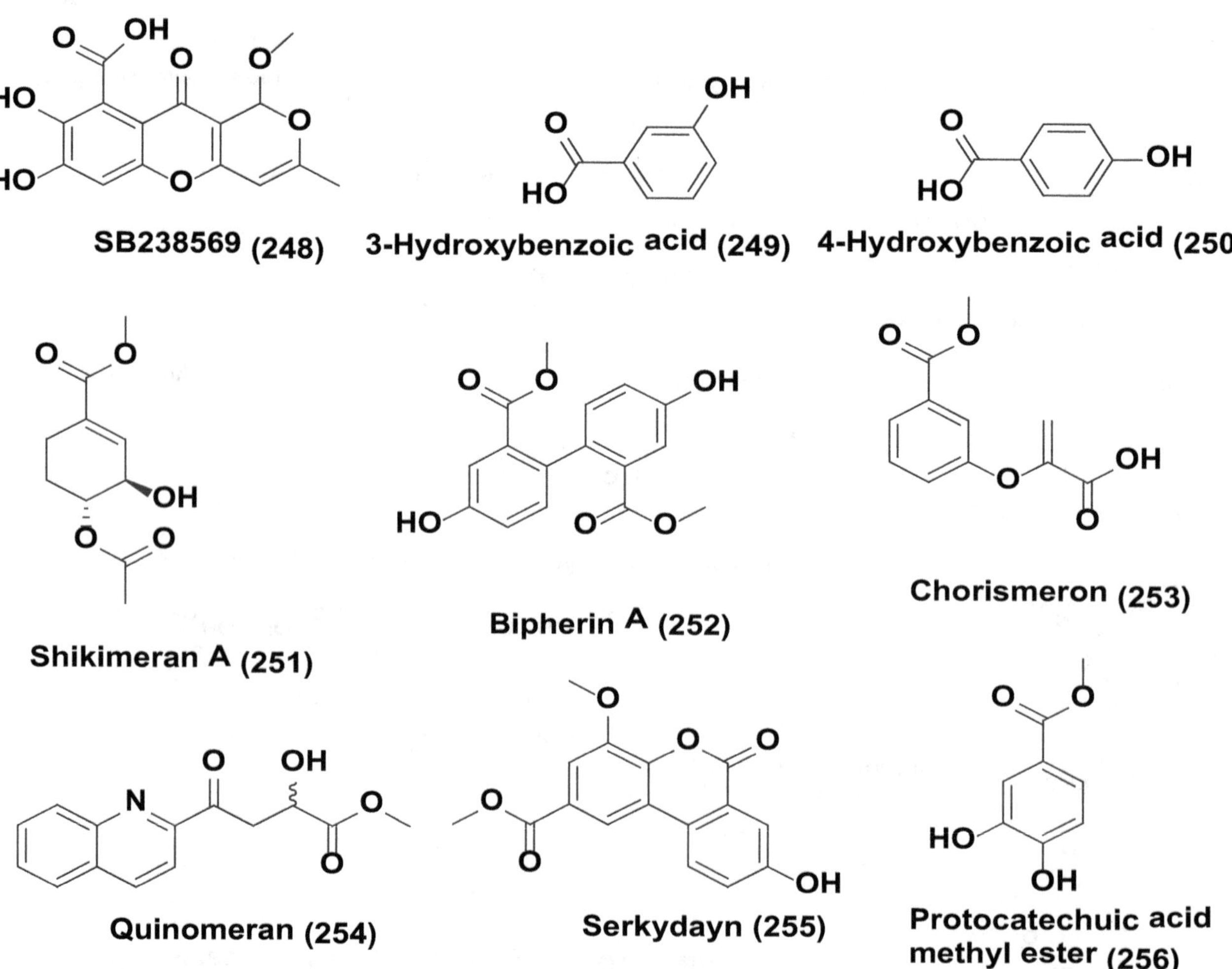

Figure 13. Structures of metabolites obtained from co-culture in endophytic fungi (**248–256**).

These studies indicate that co-culture generates a complex and promising environment to obtain new secondary metabolites as a response to the interaction between endophytic fungi. The above also indicates that the production of new natural products depends on stimuli.

5. Conclusions

Endophytic fungi are the ubiquitous source of novel chemical compounds having the potential to display antifungal activities. Interestingly, the active metabolites from endophytic fungi possess excellent antifungal activity not only against human fungal pathogens but also on plant fungal pathogens. In addition, the volatile organic compounds (VOCs) from genus *Muscodor* displayed significant antifungal as well as antibacterial properties and, therefore, they are used to prevent fungal deterioration of crops, fruits and vegetables. However, their application to control human fungal infection has not been explored. Fungal VOCs can be investigated for the development of sprays for inhalation to treat fungal diseases such as Aspergillosis in lungs, curing superficial skin infections and sanitization. Endophytic fungi are being studied to produce natural compounds which are originally produced from their host plants and, thus, emerging as an alternative and sustainable source of valuable natural products. It is important to investigate the interactions between endophytic fungi with the host plant and other endophytes which are very sensitive to the culture conditions and hence, provide an opportunity to tune the in vitro culture conditions to produce the desired range of secondary metabolites. It is possible to produce a compound of interest by varying the culture conditions such as media composition, aeration rate and temperature. In addition, cultivation of endophytic fungi in presence of bacteria or other fungi (co-cultivation) yield novel compounds which otherwise do not appear when fungi or bacteria are cultivated alone. Therefore, considerable research on endophytic fungi is required for the development of suitable co-culture system for the sustained production of the desired secondary metabolite.

Author Contributions: S.K.D., V.P., S.S. and M.K.G. reviewed the contents critically. V.P. and M.K.G. drew chemical structures and assisted in the preparation of Table 1. S.S. wrote the Antifungal potential of Volatile organic compounds of review. The manuscript has been read and approved by all named authors.

Acknowledgments: The authors are thankful to Alok Adholeya, Senior Director, Sustainable Agriculture Division, The Energy and Resources Institute (India) for continuous support.

References

1. Hawksworth, D.L. The fungal dimension of biodiversity: Magnitude, significance, and conservation. *Mycol. Res.* **1991**, *95*, 641–655. [CrossRef]
2. Schulz, B.; Boyle, C.; Draeger, S.; Römmert, A.K.; Krohn, K. Endophytic fungi: A source of novel biologically active secondary metabolites. *Mycol. Res.* **2002**, *106*, 996–1004. [CrossRef]
3. Nicoletti, R.; Fiorentino, A. Plant bioactive metabolites and drugs produced by endophytic fungi of Spermatophyta. *Agriculture* **2015**, *5*, 918–970. [CrossRef]
4. Strobel, G.; Daisy, B. Bioprospecting for microbial endophytes and their natural products. *Microbiol. Mol. Biol. Rev.* **2003**, *67*, 491–502. [CrossRef] [PubMed]
5. Gunatilaka, A.A.L. Natural products from plant-associated microorganisms: Distribution, structural diversity, bioactivity, and implications of their occurrence. *J. Nat. Prod.* **2006**, *69*, 509–526. [CrossRef] [PubMed]
6. Kharwar, R.N.; Mishra, A.; Gond, S.K.; Stierle, A.; Stierle, D. Anticancer compounds derived from fungal endophytes: Their importance and future challenges. *Nat. Prod. Rep.* **2011**, *28*, 1208–1228. [CrossRef] [PubMed]
7. Deshmukh, S.K.; Verekar, S.A. Fungal endophytes: A potential source of antifungal compounds. *Front. Biosci.* **2012**, *E4*, 2045–2070. [CrossRef]
8. Deshmukh, S.K.; Verekar, S.A.; Bhave, S.V. Endophytic fungi: A reservoir of antibacterials. *Front. Microbiol.* **2015**, *5*, 715. [CrossRef] [PubMed]

9. Gao, H.; Li, G.; Lou, H.X. Structural Diversity and Biological Activities of Novel Secondary metabolites from Endophytes. *Molecules* **2018**, *23*, 646. [CrossRef] [PubMed]
10. Carroll, G. Fungal endophytes in stems and leaves: From latent pathogens to mutualistic symbionts. *Ecology* **1988**, *69*, 2–9. [CrossRef]
11. Hallmann, J.; Sikora, R. Toxicity of fungal endophytic secondary metabolites to plant parasitic nematodes and soil borne plant pathogenic fungi. *Eur. J. Plant Pathol.* **1996**, *102*, 155–162. [CrossRef]
12. Sturz, A.V.; Nowak, J. An endophytic community of rhizobacteria and the strategies requires to create yield enhancing associations with crops. *Appl. Soil Ecol.* **2000**, *15*, 183–190. [CrossRef]
13. Azevedo, J.L.; Araujo, W.L. Diversity and applications of endophytic fungi isolated from tropical plants. In *Fungi Multifaceted Microbes*; Ganguli, B.N., Deshmukh, S.K., Eds.; Anamaya: New Delhi, India, 2007; pp. 189–207.
14. Deshmukh, S.K.; Prakash, V.; Ranjan, N. Recent advances in the discovery of bioactive metabolites from *Pestalotiopsis*. *Phytochem. Rev.* **2017**, *16*, 883–920. [CrossRef]
15. Xu, D.; Zhang, B.Y.; Yang, X.L. Antifungal monoterpene derivatives from the plant endophytic fungus *Pestalotiopsis foedan*. *Chem. Biodivers.* **2016**, *13*, 1422–1425. [CrossRef] [PubMed]
16. Wu, L.S.; Jia, M.; Chen, L.; Zhu, B.; Dong, H.X.; Si, J.P.; Peng, W.; Han, T. Cytotoxic and antifungal constituents isolated from the metabolites of endophytic fungus DO14 from *Dendrobium officinale*. *Molecules* **2016**, *21*, 14. [CrossRef] [PubMed]
17. Liu, S.; Liu, X.; Guo, L.; Che, Y.; Liu, L. 2H-Pyran-2-one and2H-Furan-2-one derivatives from the plant endophytic fungus *Pestalotiopsis fici*. *Chem. Biodivers.* **2013**, *10*, 2007–2013. [CrossRef] [PubMed]
18. Subban, K.; Subramani, R.; Muthumary, J. A novel antibacterial and antifungal phenolic compound from the endophytic fungus *Pestalotiopsis mangiferae*. *Nat. Prod. Res.* **2013**, *27*, 1445–1449. [CrossRef] [PubMed]
19. Chapla, V.M.; Zeraik, M.L.; Ximenes, V.F.; Zanardi, L.M.; Lopes, M.N.; Cavalheiro, A.J.; Silva, D.H.S.; Young, M.C.M.; da Fonseca, L.M.; Bolzani, V.S.; et al. Bioactive secondary metabolites from *Phomopsis* sp., an endophytic fungus from Senna spectabilis. *Molecules* **2014**, *19*, 6597–6608.
20. Wu, S.H.; Huang, R.; Miao, C.P.; Chen, Y.W. Two new steroids from an endophytic fungus *Phomopsis* sp. *Chem. Biodivers.* **2013**, *10*, 1276–1283. [CrossRef] [PubMed]
21. Tanney, J.B.; McMullin, D.R.; Green, B.D.; Miller, J.D.; Seifert, K.A. Production of antifungal and anti-insectan metabolites by the Picea endophyte *Diaporthe maritima* sp. nov. *Fungal Biol.* **2016**, *120*, 1448–1457. [CrossRef] [PubMed]
22. Rai, M.; Deshmukh, P.; Gade, A.; Ingle, A.; Kövics, G.J.; Irinyi, L. *Phoma* Saccardo: Distribution, secondary metabolite production and biotechnological applications. *Crit. Rev. Microbiol.* **2009**, *35*, 182–196. [CrossRef] [PubMed]
23. Hussain, H.; Kock, I.; Al-Harras, A.; Al-Rawahi, A.; Abbas, G.; Green, I.R.; Shah, A.; Badshah, A.; Saleem, M.; Draeger, S.; et al. Antimicrobial chemical constituents from endophytic fungus *Phoma* sp. *Asian. Pac. J. Trop. Med.* **2014**, *7*, 699–702. [CrossRef]
24. Mousa, W.K.; Schwan, A.; Davidson, J.; Auzanneau, F.I.; Strange, P.; Liu, H.; Zhou, T.; Raizada, M.N. An endophytic fungus isolated from finger millet (*Eleusine coracana*) produces anti-fungal natural products. *Front. Microbiol.* **2015**, *6*, 1157. [CrossRef] [PubMed]
25. Lai, D.; Wang, A.; Cao, Y.; Zhou, K.; Mao, Z.; Dong, X.; Tian, J.; Xu, D.; Dai, J.; Peng, Y.; et al. Bioactivedibenzo-α-pyrone derivatives from the endophytic fungus *Rhizopycnis vagum* Nitaf22. *J. Nat. Prod.* **2016**, *79*, 2022–2031. [CrossRef] [PubMed]
26. Hussain, H.; Root, N.; Jabeen, F.; Al-Harrasi, A.; Ahmad, M.; Mabood, F.; Hassan, Z.; Shah, A.; Green, I.R.; Schulz, B.; et al. Microsphaerol and seimatorone: Two new compounds isolated from the endophytic fungi, *Microsphaeropsis* sp. and *Seimatosporium* sp. *Chem. Biodivers.* **2015**, *12*, 289–294. [CrossRef] [PubMed]
27. Chapla, V.M.; Zeraik, M.L.; Leptokarydis, I.H.; Silva, G.H.; Bolzani, V.S.; Young, M.C.M.; Pfenning, L.H.; Araujo, A.R. Antifungal compounds produced by *Colletotrichum gloeosporioides*, an endophytic fungus from *Michelia champaca*. *Molecules* **2014**, *19*, 19243–19252. [CrossRef] [PubMed]
28. Hussain, H.; Root, N.; Jabeen, F.; Al-Harras, A.; Al-Rawahi, A.; Ahmad, M.; Hassan, Z.; Abba, G.; Mabood, F.; Shah, A.; et al. Seimatoric acid and colletonoic acid: Two new compounds from the endophytic fungi, *Seimatosporium* sp. and *Colletotrichum* sp. *Chin. Chem. Lett.* **2014**, *25*, 1577–1579. [CrossRef]

29. Sun, P.; Huo, J.; Kurtan, T.; Mandi, A.; Antus, S.; Tang, H.; Draeger, S.; Schulz, B.; Hussain, H.; Krohn, K.; et al. Structural and stereochemical studies of hydroxyanthraquinone derivatives from the endophytic fungus *Coniothyrium* sp. *Chirality* **2013**, *25*, 141–148. [CrossRef] [PubMed]
30. Macias-Rubalcava, M.L.; Sanchez-Fernandez, R.E. Secondary metabolites of endophytic *Xylaria* species with potential applications in medicine and agriculture. *World J. Microbiol. Biotechnol.* **2017**, *33*, 1–22. [CrossRef] [PubMed]
31. Song, F.; Wu, S.H.; Zhai, Y.Z.; Xuan, Q.C.; Wang, T. Secondary metabolites from the genus *Xylaria* and their bioactivities. *Chem. Biodivers.* **2014**, *11*, 673–694. [CrossRef] [PubMed]
32. Wu, S.H.; He, J.; Li, X.N.; Huang, R.; Song, F.; Chen, Y.W.; Miao, C.P. Guaianesesquiterpenes and isopimaranediterpenes from an endophytic fungus *Xylaria* sp. *Phytochemistry* **2014**, *105*, 197–204. [CrossRef] [PubMed]
33. Huang, R.; Xie, X.S.; Fang, X.W.; Ma, K.X.; Wu, S.H. Five new guaianesesquiterpenes from the endophytic fungus *Xylaria* sp. YM 311647 of *Azadirachta indica*. *Chem. Biodivers.* **2015**, *12*, 1281–1286. [CrossRef] [PubMed]
34. Baraban, E.G.; Morin, J.B.; Phillips, G.M.; Phillips, A.J.; Strobel, S.A.; Handelsman, J. Xyolide, a bioactive nonenolide from an Amazonian endophytic fungus, *Xylaria feejeensis*. *Tetrahedron Lett.* **2013**, *54*, 4058–4060. [CrossRef] [PubMed]
35. Zhang, Q.; Xiao, J.; Sun, Q.Q.; Qin, J.C.; Pescitelli, G.; Gao, J.M. Characterization of cytochalasins from the endophytic *Xylaria* sp. and their biological functions. *J. Agric. Food. Chem.* **2014**, *62*, 10962–10969. [CrossRef] [PubMed]
36. Rukachaisirikul, V.; Buadam, S.; Sukpondma, Y.; Phongpaichit, S.; Sakayaroj, J.; Hutadilok-Towatana, N. Indanone and mellein derivatives from the Garcinia-derived fungus *Xylaria* sp. PSU-G12. *Phytochem. Lett.* **2013**, *6*, 135–138. [CrossRef]
37. Sica, V.P.; Rees, E.R.; Tchegnon, E.; Bardsley, R.H.; Raja, H.A.; Oberlies, N.H. Spatial and temporal profiling of griseofulvin production in *Xylaria cubensis* using mass spectrometry mapping. *Front. Microbiol.* **2016**, *7*, 544. [CrossRef] [PubMed]
38. Richardson, S.N.; Walker, A.K.; Nsiama, T.K.; McFarlane, J.; Sumarah, M.W.; Ibrahim, A.; Miller, J.D. Griseofulvin producing *Xylaria* endophytes of *Pinus strobus* and *Vaccinium angustifolium*: Evidence for a conifer-understory species endophyte ecology. *Fungal Ecol.* **2014**, *11*, 107–113. [CrossRef]
39. Zhang, Q.; Li, H.Q.; Zong, S.C.; Gao, J.M.; Zhang, A.L. Chemical and bioactive diversities of the genus *Chaetomium* secondary metabolites. *Mini-Rev. Med. Chem.* **2012**, *12*, 127–148. [CrossRef] [PubMed]
40. Wang, X.Y.; Yan, X.; Fang, M.J.; Wu, Z.; Wang, D.; Qiu, Y.K. Two new cytochalasan derivatives from *Chaetomium globosum* SNSHI-5, a fungus derived from extreme environment. *Nat. Prod. Res.* **2017**, *31*, 1669–1675. [CrossRef] [PubMed]
41. Zhao, S.S.; Zhang, Y.Y.; Yan, W.; Cao, L.L.; Xiao, Y.; Ye, Y.H.; Zhao, S.S.; Zhang, Y.Y.; Yan, W.; Cao, L.L.; et al. *Chaetomium globosum* CDW7, a potential biological control strain and its antifungal metabolites. *FEMS Microbiol. Lett.* **2017**, *364*. [CrossRef] [PubMed]
42. Li, W.; Yang, X.; Yang, Y.; Duang, R.; Chen, G.; Li, X.; Li, Q.; Qin, S.; Li, S.; Zhao, L.; et al. Anti-phytopathogen, multi-target acetylcholinesterase inhibitory and antioxidant activities of metabolites from endophytic *Chaetomium globosum*. *Nat. Prod. Res.* **2016**, *30*, 2616–2619. [CrossRef] [PubMed]
43. Wang, J.; Zhang, Y.Y.; Ding, D.D.; Yu, S.P.; Wang, L.W. A study on the secondary metabolites of endophytic fungus *Chaetomium cupreum* ZJWCF079 in *Macleaya cordata*. *Health Res.* **2013**, *33*, 94–96.
44. Zhang, G.; Zhang, Y.; Qin, J.; Qu, X.; Liu, J.; Li, X.; Pan, H. Antifungal metabolites produced by *Chaetomiumglobosum* No.04, an endophytic fungus isolated from *Ginkgo biloba*. *Indian J. Microbiol.* **2013**, *53*, 175–180. [PubMed]
45. Chen, Y.M.; Yang, Y.H.; Li, X.N.; Zou, C.; Zhao, P.J. Diterpenoids from the endophytic fungus *Botryosphaeria* sp. P483 of the Chinese herbal medicine *Huperzia serrata*. *Molecules* **2015**, *20*, 16924–16932. [CrossRef] [PubMed]
46. Xiao, J.; Zhang, Q.; Gao, Y.Q.; Tang, J.J.; Zhang, A.L.; Gao, J.M. Secondary metabolites from the endophytic *Botryosphaeria dothidea* of *Melia azedarach* and their antifungal, antibacterial, antioxidant, and cytotoxic activities. *J. Agric. Food Chem.* **2014**, *62*, 3584–3590. [CrossRef] [PubMed]

47. Pereira, C.B.; de Oliveira, D.M.; Hughes, A.F.S.; Kohlhoff, M.; Vieira, M.L.A.; Martins Vaz, A.B.; Ferreira, M.C.; Carvalho, C.R.; Rosa, L.H.; Rosa, C.A.; et al. Endophytic fungal compounds active against *Cryptococcus neoformans* and *C. gattii*. *J. Antibiot.* **2015**, *68*, 436–444. [CrossRef] [PubMed]
48. Pereira, C.; de Sá, N.P.; Borelli, B.M.; Rosa, C.A.; Barbeira, P.J.S.; Cota, B.B.; Johann, S. Antifungal activity of eicosanoic acids isolated from the endophytic fungus *Mycosphaerella* sp. against *Cryptococcus neoformans* and *C. gattii*. *Microb. Pathog.* **2016**, *100*, 205–212. [CrossRef] [PubMed]
49. Li, T.X.; Yang, M.H.; Wang, X.B.; Wang, Y.; Kong, L.Y. Synergistic antifungal meroterpenes and dioxolanone derivatives from the endophytic fungus *Guignardia* sp. *J. Nat. Prod.* **2015**, *78*, 2511–2520. [CrossRef] [PubMed]
50. Mao, Z.; Lai, D.; Liu, X.; Fu, X.; Meng, J.; Wang, A.; Wang, X.; Sun, W.; Liu, Z.L.; Zhou, L.; et al. Dibenzo-α-pyrones: A new class of larvicidal metabolites against *Aedes aegypti* from the endophytic fungus *Hyalodendriella* sp. Ponipodef12. *Pest Manag. Sci.* **2017**, *73*, 1478–1485. [CrossRef] [PubMed]
51. Meng, X.; Mao, Z.; Lou, J.; Xu, L.; Zhong, L.; Peng, Y.; Zhou, L.; Wang, M. Benzopyranones from the endophytic fungus *Hyalodendriella* sp. Ponipodef12 and their bioactivities. *Molecules* **2012**, *17*, 11303–11314. [CrossRef] [PubMed]
52. Rosseto, P.; Costa, A.T.; Polonio, J.C.; da Silva, A.A.; Pamphile, J.A.; Azevedo, J.L. Investigation of mycoviruses in endophytic and phytopathogenic strains of *Colletotrichum* from different hosts. *Genet. Mol. Res.* **2016**, *15*, 1–11. [CrossRef] [PubMed]
53. Wang, J.; Wang, G.; Zhang, Y.; Zheng, B.; Zhang, C.; Wang, L. Isolation and identification of an endophytic fungus *Pezicula* sp. in *Forsythia viridissima* and its secondary metabolites. *World. J. Microbiol. Biotechnol.* **2014**, *30*, 2639–2644. [CrossRef] [PubMed]
54. Cao, L.L.; Zhang, Y.Y.; Liu, Y.J.; Yang, T.T.; Zhang, J.L.; Zhang, Z.G.; Shen, L.; Liu, J.Y.; Ye, Y.H. Anti-phytopathogenic activity of sporothriolide, a metabolite from endophyte *Nodulisporium* sp. A21 in *Ginkgo biloba*. *Pestic. Biochem. Physiol.* **2016**, *129*, 7–13. [CrossRef] [PubMed]
55. McMullin, D.R.; Green, B.D.; Miller, J.D. Antifungal sesquiterpenoids and macrolides from an endophytic *Lophodermium* species of *Pinus strobus*. *Phytochem. Lett.* **2015**, *14*, 148–152. [CrossRef]
56. Li, R.; Chen, S.; Niu, S.; Guo, L.; Yin, J.; Che, Y. Exserolides A-F, new isocoumarin derivatives from the plant endophytic fungus *Exserohilum* sp. *Fitoterapia* **2014**, *96*, 88–94. [CrossRef] [PubMed]
57. Carvalho, C.R.; Wedge, D.E.; Cantrell, C.L.; Silva-Hughes, A.F.; Pan, Z.; Moraes, R.M.; Madoxx, V.L.; Rosa, L.H. Molecular phylogeny, diversity, and bioprospecting of endophytic fungi associated with wild ethnomedicinal North American plant *Echinacea purpurea* (Asteraceae). *Chem. Biodivers.* **2016**, *13*, 918–930. [CrossRef] [PubMed]
58. Burgess, K.M.N.; Ibrahim, A.; Soerensen, D.; Sumarah, M.W. Trienylfuranol A and trienylfuranone A.B: Metabolites isolated from an endophytic fungus, *Hypoxylon submonticulosum*, in the raspberry *Rubus idaeus*. *J. Antibiot.* **2017**, *70*, 721–725. [CrossRef] [PubMed]
59. Nalli, Y.; Mirza, D.N.; Wani, Z.A.; Wadhwa, B.; Mallik, F.A.; Raina, C.; Chaubey, A.; Riyaz-Ul-Hassan, S.; Ali, A. Phialomustin A-D, new antimicrobial and cytotoxic metabolites from an endophytic fungus, *Phialophora mustea*. *RSC Adv.* **2015**, *5*, 95307–95312. [CrossRef]
60. Hussain, H.; Jabeen, F.; Krohn, K.; Al-Harras, A.; Ahmad, M.; Mabood, F.; Shah, A.; Badshah, A.; Rehman, N.U.; Green, I.R.; et al. Antimicrobial activity of two mellein derivatives isolated from an endophytic fungus. *Med. Chem. Res.* **2015**, *24*, 2111–2114. [CrossRef]
61. Hussain, H.; Kliche-Spory, C.; Al-Harrasi, A.; Al-Rawahi, A.; Abbas, G.; Green, I.R.; Schulz, B.; Krohn, K.; Shah, A. Antimicrobial constituents from three endophyticfungi. *Asian Pac. J. Trop. Med.* **2014**, *7*, S224–S227. [CrossRef]
62. Shan, T.; Tian, J.; Wang, X.; Mou, Y.; Mao, Z.; Lai, D.; Dai, J.; Peng, Y.; Zhou, L.; Wang, M. Bioactive spirobisnaphthalenes from the endophytic fungus *Berkleasmium* sp. *J. Nat. Prod.* **2014**, *77*, 2151–2160. [CrossRef] [PubMed]
63. Siriwach, R.; Kinoshita, H.; Kitani, S.; Igarashi, Y.; Pansuksan, K.; Panbangred, W.; Nihira, T. Bipolamides A and B, triene amides isolated from the endophytic fungus *Bipolaris* sp. MU34. *J. Antibiot.* **2014**, *67*, 167–170. [CrossRef] [PubMed]
64. Phaopongthai, J.; Wiyakrutta, S.; Meksuriyen, D.; Sriubolmas, N.; Suwanborirux, K. Azole-synergistic anti-candidal activity of altenusin, a biphenyl metabolite of the endophytic fungus *Alternaria alternata* isolated from *Terminalia chebula* Retz. *J. Microbiol.* **2013**, *51*, 821–828. [CrossRef] [PubMed]

65. Johann, S.; Rosa, L.H.; Rosa, C.A.; Perez, P.; Cisalpino, P.S.; Zani, C.L.; Cota, B.B. Antifungal activity of altenusin isolated from the endophytic fungus *Alternaria* sp. against the pathogenic fungus *Paracoccidioides brasiliensis*. *Rev. Iberoam. Micol.* **2012**, *29*, 205–209. [CrossRef] [PubMed]
66. Wang, X.; Radwan, M.M.; Tarawneh, A.H.; Gao, J.; Wedge, D.E.; Rosa, L.H.; Cutler, H.G.; Cutler, S.J. Antifungal activity against plant pathogens of metabolites from the endophytic fungus *Cladosporium cladosporioides*. *J. Agric. Food Chem.* **2013**, *61*, 4551–4555. [CrossRef] [PubMed]
67. Talontsi, F.M.; Dittrich, B.; Schueffler, A.; Sun, H.; Laatsch, H. Epicoccolides: Antimicrobial and antifungal polyketides from an endophytic fungus *Epicoccum* sp. associated with *Theobroma cacao*. *Eur. J. Org. Chem.* **2013**, *2013*, 3174–3180. [CrossRef]
68. Silva-Hughes, A.F.; Carvalho, C.R.; Wedge, D.E.; Cantrell, C.L.; Pan, Z.; Moraes, R.M.; Madoxx, V.L.; Rosa, L.H. Diversity and antifungal activity of the endophytic fungi associated with the native medicinal cactus *Opuntia humifusa* (Cactaceae) from the United States. *Microbiol. Res.* **2015**, *175*, 67–77. [CrossRef] [PubMed]
69. Wu, X.; Pang, X.J.; Xu, L.L.; Zhao, T.; Long, X.Y.; Zhang, Q.Y.; Qin, H.L.; Yang, D.F.; Yang, X.L. Two new alkylated furan derivatives with antifungal and antibacterial activities from the plant endophytic fungus *Emericella* sp. XL029. *Nat. Prod. Res.* **2017**, 1–7. [CrossRef] [PubMed]
70. Mousa, W.K.; Schwan, A.L.; Raizada, M.N. Characterization of antifungal natural products isolated from endophytic fungi of finger millet (*Eleusine coracana*). *Molecules* **2016**, *21*, 1171. [CrossRef] [PubMed]
71. Liu, K.; Yang, Y.B.; Chen, J.L.; Miao, C.P.; Wang, Q.; Zhou, H.; Chen, Y.W.; Li, Y.Q.; Ding, Z.T.; Zhao, L.X. Koninginins N-Q, polyketides from the endophytic fungus *Trichoderma koningiopsis* Harbored in *Panax notoginseng*. *Nat. Prod. Bioprospect.* **2016**, *6*, 49–55. [CrossRef] [PubMed]
72. Liu, K.; Yang, Y.; Miao, C.P.; Zheng, Y.K.; Chen, J.L.; Chen, Y.W.; Xu, L.H.; Guang, H.L.; Ding, Z.T.; Zhao, L.X. Koningiopisins A-H, polyketides with synergistic antifungal activities from the endophytic fungus *Trichoderma koningiopsis*. *Planta Medica* **2016**, *82*, 371–376. [CrossRef] [PubMed]
73. Li, W.; Xu, J.; Li, F.; Xu, L.; Li, C. A New antifungal isocoumarin from the endophytic fungus *Trichoderma* sp. 09 of *Myoporum bontioides* A. Gray. *Pharmacogn. Mag.* **2016**, *12*, 259–261. [PubMed]
74. Shentu, X.; Zhan, X.; Ma, Z.; Yu, X.; Zhang, C. Antifungal activity of metabolites of the endophytic fungus *Trichoderma brevicompactum* from garlic. *Braz. J. Microbiol.* **2014**, *45*, 248–254. [CrossRef] [PubMed]
75. Hu, M.; Li, Q.L.; Yang, Y.B.; Liu, K.; Miao, C.P.; Zhao, L.X.; Ding, Z.T. Koninginins R-S from the endophytic fungus *Trichoderma koningiopsis*. *Nat. Prod. Res.* **2017**, *31*, 835–839. [CrossRef] [PubMed]
76. Ibrahim, S.R.M.; Elkhayat, E.S.; Mohamed, G.A.; Khedr, A.I.M.; Fouad, M.A.; Kotb, M.H.R.; Ross, S.A. Aspernolides F and G, new butyrolactones from the endophytic fungus *Aspergillus terreus*. *Phytochem. Lett.* **2015**, *14*, 84–90. [CrossRef]
77. Xiao, J.; Zhang, Q.; Gao, Y.Q.; Shi, X.W.; Gao, J.M. Antifungal and antibacterial metabolites from an endophytic *Aspergillus* sp. associated with *Melia azedarach*. *Nat. Prod. Res.* **2014**, *28*, 1388–1392. [PubMed]
78. Budhiraja, A.; Nepali, K.; Sapra, S.; Gupta, S.; Kumar, S.; Dhar, K.L. Bioactive metabolites from an endophytic fungus of *Aspergillus* species isolated from seeds of *Gloriosa superba* Linn. *Med. Chem. Res.* **2013**, *22*, 323–329. [CrossRef]
79. Ma, Y.M.; Qiao, K.; Kong, Y.; Li, M.Y.; Guo, L.X.; Miao, Z.; Fan, C. A new isoquinolone alkaloid from an endophytic fungus R22 of *Nerium indicum*. *Nat. Prod. Res.* **2017**, *31*, 951–958. [CrossRef] [PubMed]
80. Kajula, M.; Ward, J.M.; Turpeinen, A.; Tejesvi, M.V.; Hokkanen, J.; Tolonen, A.; Hakkanen, H.; Picart, P.; Ihalainen, J.; Sahl, H.G.; et al. Bridged epipolythiodiketopiperazines from *Penicillium raciborskii*, an endophytic fungus of *Rhododendron tomentosum* Harmaja. *J. Nat. Prod.* **2016**, *79*, 685–690. [CrossRef] [PubMed]
81. Ibrahim, S.R.M.; Abdallah, H.M.; Elkhayat, E.S.; Al Musayeib, N.M.; Asfour, H.Z.; Zayed, M.F.; Mohamed, G.A. Fusaripeptide A: New antifungal and anti-malarial cyclodepsipeptide from the endophytic fungus *Fusarium* sp. *J. Asian Nat. Prod. Res.* **2018**, *20*, 75–85. [CrossRef] [PubMed]
82. Ibrahim, S.R.M.; Elkhayat, E.S.; Mohamed, G.A.A.; Fat'hi, S.M.; Ross, S.A. Fusarithioamide A, a new antimicrobial and cytotoxic benzamide derivative from the endophytic fungus *Fusarium chlamydosporium*. *Biochem. Biophys. Res. Commun.* **2016**, *479*, 211–216. [CrossRef] [PubMed]
83. Liang, X.A.; Ma, Y.M.; Zhang, H.C.; Liu, R.A. new helvolic acid derivative from an endophytic *Fusarium* sp. of *Ficus carica*. *Nat. Prod. Res.* **2016**, *30*, 2407–2412. [CrossRef] [PubMed]

84. Hussain, H.; Drogies, K.H.; Al-Harrasi, A.; Hassan, Z.; Shah, A.; Ran, U.A.; Green, I.R.; Draeger, S.; Schulz, B.; Krohn, K. Antimicrobial constituents from endophytic fungus *Fusarium* sp. *Asian Pac. J. Trop. Dis.* **2015**, *5*, 186–189. [CrossRef]
85. Mondol, M.A.M.; Farthouse, J.; Islam, M.T.; Schueffler, A.; Laatsch, H. Metabolites from the endophytic fungus *Curvularia* sp. M12 act as motility inhibitors against *Phytophthora capsici* Zoospores. *J. Nat. Prod.* **2017**, *80*, 347–355. [CrossRef] [PubMed]
86. Silva, G.H.; Zeraik, M.L.; de Oliveira, C.M.; Teles, H.L.; Trevisan, H.C.; Pfenning, L.H.; Nicolli, C.P.; Young, M.C.M.; Mascarenhas, Y.P.; Abreu, L.M.; et al. Lactone derivatives produced by a *Phaeoacremonium* sp., an endophytic fungus from *Senna spectabilis*. *J. Nat. Prod.* **2017**, *80*, 1674–1678. [CrossRef] [PubMed]
87. Taware, R.; Abnave, P.; Patil, D.; Rajamohananan, P.R.; Raja, R.; Soundararajan, G.; Kundu, G.C.; Ahmad, A. Isolation, purification and characterization of Trichothecinol-A produced by endophytic fungus *Trichothecium* sp. and its antifungal, anticancer and antimetastatic activities. *Sustain. Chem. Process.* **2014**, *2*, 8. [CrossRef]
88. Taware, R.; Abnave, P.; Patil, D.; Rajamohananan, P.R.; Raja, R.; Soundararajan, G.; Kundu, G.C.; Kharat, M.K.D.; Pai, K.; Ahmad, A. Trichothecin from endophytic fungus *Trichothecium* sp. and its anticancer effect on murine melanoma and breast cancer cell lines. *Curr. Biochem. Eng.* **2015**, *2*, 73–80. [CrossRef]
89. Morandini, L.M.B.; Neto, A.T.; Pedroso, M.; Antoniolli, Z.I.; Burrow, R.A.; Bortoluzzi, A.J.; Mostardeiro, M.A.; da Silva, U.F.; Dalcol, I.I.; Morel, A.F. Lanostane-type triterpenes from the fungal endophyte Scleroderma UFSMSc1 (Persoon) Fries. *Bioorg. Med. Chem. Lett.* **2016**, *26*, 1173–1176. [CrossRef] [PubMed]
90. Zhang, C.L.; Wang, G.P.; Mao, L.J.; Komon-Zelazowska, M.; Yuan, Z.L.; Lin, F.C.; Druzhinina, I.S.; Kubicek, C.P. *Muscodor fengyangensis* sp. nov. from southeast China: Morphology, physiology and production of volatile compounds. *Fungal Biol.* **2010**, *114*, 797–808. [CrossRef] [PubMed]
91. Woropong, J.; Strobel, G.A.; Ford, E.J.; Li, J.Y.; Baird, G.; Hess, W.M. *Muscodor albusanam.* nov., an endophyte from *Cinnamomum zeylanicum*. *Mycotaxon* **2001**, *79*, 67–79.
92. Strobel, G.A. *Muscodor* species-endophytes with biological promise. *Phytochem. Rev.* **2011**, *10*, 165–172. [CrossRef]
93. DiFranscesco, A.; Ugolini, L.; Lazzeri, L.; Mari, M. Production of volatile organic compounds by *Aureobasidium pullulans* as a potential mechanism against post-harvest fruit pathogens. *Biol. Control* **2015**, *81*, 8–14. [CrossRef]
94. Oro, L.; Feliziani, E.; Ciani, M.; Romanazzi, G.; Comitini, F. Volatile organic compounds of *Wickerhamomyces anomalus*, *Metschinikowia pulcherrima* and *Saccharomyces cerevisiae* inhibit growth of decay causing fungi and post-harvest disease of strawberries. *J. Food Microbiol.* **2018**, *265*, 18–22. [CrossRef] [PubMed]
95. Pimenta, R.S.; da Silva, J.F.M.; Buyer, J.S.; Janiesiewicz, W.J. Endophytic fungi from Plums (*Prunis domestica*) and their antifungal activity against *Monilinia fruticola*. *J. Food Prot.* **2012**, *75*, 1883–1889. [CrossRef] [PubMed]
96. Medina-Romaro, Y.M.; Roque-Flores, G.; Macias-Rubalcava, M.L. Volatile organic compounds from endophytic fungi as innovative post-harvest control of *Fusarium oxysporum* in cherry tomato fruits. *Appl. Microbiol. Biotechnol.* **2017**, *101*, 8209–8222. [CrossRef] [PubMed]
97. Strobel, G.A.; Dirkse, E.; Sears, J.; Markworth, C. Volatile antimicrobials from *Muscodor albus*, a novel endophytic fungus. *Microbiology* **2001**, *147*, 2943–2950. [CrossRef] [PubMed]
98. Daisy, B.H.; Strobel, G.A.; Castillo, U.; Ezra, D.; Sears, J.; Weaver, D.K.; Runyon, J.B. Naphthalene, an insect repellent, is produced by *Muscodor vitigenus*, a novel endophytic fungus. *Microbiology* **2002**, *148*, 3737–3741. [CrossRef] [PubMed]
99. Grimme, E.; Zidack, N.K.; Sikora, R.A.; Strobel, G.A.; Jacobsen, B.J. Comparison of *Muscodor albus* volatiles with abiorational mixture for control of seedling diseases of sugar beet and root knot nematode on tomato. *Plant Dis.* **2007**, *91*, 220–225. [CrossRef]
100. Strobel, G.A.; Knighton, B.; Kluck, K.; Ren, Y.; Livinghouse, T.; Griffin, M.; Spakowicz, D.; Sears, J. The production of myco-diesel hydrocarbons and their derivatives by the endophytic fungus *Gliocladiumroseum* (NRRL 50072). *Microbiology* **2008**, *154*, 3319–3328. [CrossRef] [PubMed]
101. Worapong, J.; Strobel, G.; Daisy, B.; Castillo, U.F.; Baird, G.; Hess, M.W. *Muscodor roseousanam.* Sp. Nov. and endophyte from *Grevillea pteridifolia*. *Mycotaxon* **2002**, *81*, 463–475.
102. González, M.C.; Anaya, A.L.; Glenn, A.E.; Macías-Rubalcava, M.L.; Hernández-Bautista, B.E.; Hanlin, R.T. *Muscodor yucatanensis*, a new endophytic ascomycete from Mexican chakah, *Bursera simaruba*. *Mycotaxon* **2009**, *110*, 363–372. [CrossRef]

103. Mitchell, A.M.; Strobel, G.A.; Moore, E.; Robison, R.; Sears, J. Volatile antimicrobials from *Muscodor crispans*, a novel endophytic fungus. *Microbiology* **2010**, *156*, 270–277. [CrossRef] [PubMed]
104. Kudalkar, P.; Strobel, G.; Riyaz-Ul-Hassan, S.; Geary, B.; Sears, J. *Muscodor sutura*, a novel endophytic fungus with volatile antibiotic activities. *Mycoscience* **2012**, *53*, 319–325. [CrossRef]
105. Suwannarach, N.; Kumla, J.; Bussaban, B.; Hyde, K.D.; Matsui, K.; Lumyong, S. Molecular and morphological evidence support four new species in the genus *Muscodor* from northern Thailand. *Ann. Microbiol.* **2013**, *63*, 1341–1351. [CrossRef]
106. Meshram, V.; Kapoor, N.; Saxena, S. *Muscodor kashayum* sp. nov.—A new volatile anti-microbial producing endophytic fungus. *Mycology* **2013**, *4*, 196–204. [CrossRef] [PubMed]
107. Saxena, S.; Meshram, V.; Kapoor, N. *Muscodor darjeelingensis*, a new endophytic fungus of *Cinnamomum camphora* collected from north eastern Himalayas. *Sydowia* **2014**, *66*, 55–67.
108. Meshram, V.; Saxena, S.; Kapoor, N. *Muscodor strobelii*, a new endophytic species from South India. *Mycotaxon* **2014**, *128*, 93–104. [CrossRef]
109. Saxena, S.; Meshram, V.; Kapoor, N. *Muscodor tigerii* sp. nov. Volatile antibiotic producing endophytic fungus from the North eastern Himalayas. *Ann. Microbiol.* **2015**, *65*, 47–57. [CrossRef]
110. Siri-udom, S.; Suwannarach, N.; Lumyong, S. Existence of *Muscodor vitigenus*, *M. equiseti* and *M. heveae* sp. nov.in leaves of the rubber tree (*Hevea brasiliensis* Müll. Arg.), and their biocontrol potential. *Ann. Microbiol.* **2015**, *66*, 437–448. [CrossRef]
111. Meshram, V.; Gupta, M.; Saxena, S. *Muscodor ghoomensis* and *Muscodor indica*: New endophytic species based on morphological features, molecular and volatile organic analysis from northeast India. *Sydowia* **2016**, *67*, 133–146.
112. Meshram, V.; Kapoor, N.; Chopra, G.; Saxena, S. *Muscodor camphora*, a new endophytic species from *Cinnamomum camphora*. *Mycosphere* **2017**, *8*, 568–582. [CrossRef]
113. Scherlach, K.; Hertweck, C. Triggering cryptic natural product biosynthesis in microorganisms. *Org. Biomol. Chem.* **2009**, *7*, 1753–1760. [CrossRef] [PubMed]
114. Williams, R.B.; Henrikson, J.C.; Hoover, A.R.; Lee, A.E.; Cichewicz, R.H. Epigenetic remodeling of the fungal secondary metabolome. *Org. Biomol. Chem.* **2008**, *6*, 1895–1897. [CrossRef] [PubMed]
115. Cichewicz, R.H. Epigenome manipulation as a pathway to new natural product scaffolds and their congeners. *Nat. Prod. Rep.* **2010**, *27*, 11–22. [CrossRef] [PubMed]
116. Fox, E.M.; Howlett, B.J. Secondary metabolism, regulation and role in fungal biology. *Curr. Opin. Microbiol.* **2008**, *11*, 481–487. [CrossRef] [PubMed]
117. Fisch, K.M.; Gillaspy, A.F.; Gipson, M.; Henrikson, J.C.; Hoover, A.R.; Jackson, L.; Najar, F.Z.; Wägele, H.; Cichewicz, R.H. Chemical induction of silent pathway transcription in *Aspergillus niger*. *J. Ind. Microbiol. Biotechnol.* **2009**, *36*, 1199–1213. [CrossRef] [PubMed]
118. Magotra, A.; Kumar, M.; Kushwaha, M.; Awasthi, P.; Raina, C.; Gupta, A.P.; Shah, B.A.; Gandhi, S.G.; Chaubey, A. Epigenetic modifier induced enhancement of fumiquinazoline C production in *Aspergillus fumigatus* (GA-L7): An endophytic fungus from *Grewia asiatica* L. *AMB Express* **2017**, *7*, 43. [CrossRef] [PubMed]
119. Li, G.; Kusari, S.; Golz, C.; Laatsch, H.; Strohmann, C.; Spiteller, M. Epigenetic modulation of endophytic *Eupenicillium* sp. LG41 by a histone deacetylase inhibitor for production of decalin-containing compounds. *J. Nat. Prod.* **2017**, *80*, 983–988. [CrossRef] [PubMed]
120. Li, G.; Kusari, S.; Lamshöft, M.; Schüffler, A.; Laatsch, H.; Spiteller, M. Antibacterial secondary metabolites from an endophytic fungus, *Eupenicillium* sp. LG41. *J. Nat. Prod.* **2014**, *77*, 2335–2341. [CrossRef] [PubMed]
121. Vasanthakumari, M.M.; Jadhav, S.S.; Sachin, N.; Vinod, G.; Singh, S.; Manjunatha, B.L.; Kumara, P.M.; Ravikanth, G.; Nataraja, K.N.; Uma Shaanker, R. Restoration of camptothecine production in attenuated endophytic fungus on re-inoculation into host plant and treatment with DNA methyltransferase inhibitor. *World. J. Microbiol. Biotechnol.* **2015**, *31*, 1629–1639. [CrossRef] [PubMed]
122. Gubiani, J.R.; Wijeratne, E.M.K.; Shi, T.; Araujo, A.R.; Arnold, A.E.; Chapman, E.; Gunatilaka, A.A.L. An epigenetic modifier induces production of (10′ S)-verruculide B, an inhibitor of protein tyrosine phosphatases by *Phoma* sp. nov. LG0217, a fungal endophyte of *Parkinsonia microphylla*. *Bioorg. Med. Chem.* **2017**, *25*, 1860–1866. [CrossRef] [PubMed]
123. Sun, K.; Zhu, G.; Hao, J.; Wang, Y.; Zhu, W. Chemical-epigenetic Method to Enhance the Chemodiversity of the Marine Algicolous Fungus, *Aspergillus terreus* OUCMDZ-2739. *Tetrahedron* **2018**, *74*, 83–87. [CrossRef]

124. Akone, S.H.; Mandi, A.; Kurtan, T.; Hartmann, R.; Lin, W.; Daletos, G.; Proksch, P. Inducing secondary metabolite production by the endophytic fungus *Chaetomium* sp. through fungal-bacterial co-culture and epigenetic modification. *Tetrahedron* **2016**, *72*, 6340–6347. [CrossRef]
125. Asai, T.; Otsuki, S.; Sakurai, H.; Yamashita, K.; Ozeki, T.; Oshima, Y. Benzophenones from an endophytic fungus, *Graphiopsis chlorocephala*, from Cultivated in the Presence of an NAD^+-Dependent HDAC Inhibitor. *Org. Lett.* **2013**, *15*, 2058–2061. [CrossRef] [PubMed]
126. Gonzalez-Menendez, V.; Perez-Bonilla, M.; Perez-Victoria, I.; Martin, J.; Munoz, F.; Reyes, F.; Tormo, J.R.; Genilloud, O. Multicomponent analysis of the differential induction of secondary metabolite profiles in fungal endophytes. *Molecules* **2016**, *21*, 234. [CrossRef] [PubMed]
127. Rico-Gray, V. Interspecific interaction. *Encycl. Life Sci.* **2001**. [CrossRef]
128. Sandland, G.J.; Rodgers, J.K.; Minchella, D.J. Interspecific antagonism and virulence in hostsex-posed to two parasite species. *J. Invertebr. Pathol.* **2007**, *96*, 43–47. [CrossRef] [PubMed]
129. Marmann, A.; Aly, A.H.; Lin, W.; Wang, B.; Proksch, P. Co-Cultivation—A powerful emerging tool for enhancing the chemical diversity of microorganisms. *Mar. Drugs* **2014**, *12*, 1043–1065. [CrossRef] [PubMed]
130. Brakhage, A.A.; Schroeckh, V. Fungal secondary metabolites–strategies to activate silent gene clusters. *Fungal Genet. Biol.* **2011**, *48*, 15–22. [CrossRef] [PubMed]
131. Ola, A.R.; Thomy, D.; Lai, D.; Brötz-Oesterhelt, H.; Proksch, P. Inducing secondary metabolite production by the endophytic fungus *Fusarium tricinctum* through co-culture with Bacillus subtilis. *J. Nat. Prod.* **2013**, *76*, 2094–2099. [CrossRef] [PubMed]
132. Kusari, S.; Singh, S.; Jayabaskaran, C. Rethinking production of Taxol® (paclitaxel) using endophyte biotechnology. *Trends Biotechnol.* **2014**, *32*, 304–311. [CrossRef] [PubMed]
133. Ebrahim, W.; El-Neketi, M.; Lewald, L.I.; Orfali, R.S.; Lin, W.; Rehberg, N.; Kalscheuer, R.; Daletos, G.; Proksch, P. Metabolites from the Fungal Endophyte *Aspergillus austroafricanus* in Axenic Culture and in Fungal-Bacterial Mixed Cultures. *J. Nat. Prod.* **2016**, *79*, 914–922. [CrossRef] [PubMed]

8

The Emergence of Endophytic Microbes and their Biological Promise

Gary Strobel

Department of Plant Sciences, Montana State University, Bozeman, MT 59717, USA; uplgs@montana.edu

Abstract: As is true with animal species, plants also have an associated microflora including endophytes as well as microbes associated with the phyllosphere and rhizosphere (plant surfaces) and this is considered the plant microbiome. However, those organisms within virtually all tissues and organs of the plant are known as endophytes. Most often fungi are the most frequently recovered endophytes from plant tissues, but bacterial forms generally occur in greater numbers, but not in species varieties. The exact biological/biochemical role of the endophyte in the plant and how it interacts with the plant and other endophytes and plant associated organisms has not been intensely and carefully examined. However, this has not stopped investigators in exploring the direct utility of endophytes in boosting plant production, and discovering that endophytes can directly influence the plant to resist temperature extremes, drought, as well as the presence of disease causing organisms. Also, because of the relationships that endophytes seem to have with their host plants, they make a myriad of biologically active compounds some of which are classified as antibiotics, antioxidants, anticancer agents, volatile antimicrobial agents, immunosuppressive compounds, plant growth promoting agents, and insecticides. These endophytic compounds represent a wide range of organic molecules including terpenoids, peptides, carbohydrates, aromatics, hydrocarbons and others and it seems that these compounds may have a role in the host microbe relationship. Most recently and quite surprisingly, some endophytes have been discovered that make hydrocarbons of the types found in diesel and gasoline fuels. In addition, recently discovered are epigenetic factors relating to the biology and biochemistry of endophytes. Interestingly, only about 1–2% of the entire spectrum of 300,000 known plants have been studied for their endophyte composition. Additionally, only a few plants have ever been completely studied including all tissues for the microbes within them. Likewise, the vast majority of plants, including those in oceans and lower plant forms, have never been examined for their endophytes. Furthermore, endophytes representing the "microbiome" of world's major food plants as they exist in their native "centers of origin" are largely unknown. This non-classical review is intended to provide background information on aspects of developments in endophyte biology and more importantly the identification of new questions in this field that need to be addressed. The review is primarily based on the author's long held experience in this field.

Keywords: fungi; endophytes; natural products; plant microbiome; mycodiesel

---

1. Introduction

Endophyte biology is an emerging field. With many new developments in natural product characterization, along with their increasing use in biological control and the application of these organisms to sustain and assist crop production, the field has recently seen a huge jump in scientific attention and interest which has sprung from approaches and methods used in classical plant pathology [1]. These organisms are obtained from all types of plant tissues that seemed to bear no external evidence for the presence of any life forms within them. The advent of important and novel strides being made in the make-up and subsequently the importance of the human microbiome has sparked a huge interest in the plant microbiome (endophytes) and how these organisms may influence the development and, the ability of a plant to resist disease, drought, heat, cold and other insults [2]. In addition, there has been renewed interest in endophytes since many of them have been shown to produce important compounds of pharmaceutical and commercial interest [3]. This list includes novel anticancer agents, antibiotics, novel immunosuppressive compounds, volatile antibiotic mixtures, antioxidants and most recently—fuel related hydrocarbons [3,4]. It appears that the field is ripe for scientific discovery and invention in a number of important scientific areas.

Since the early times, the definition of an endophyte has more or less remained the same—"a microorganism associated with living plant tissues that produces no apparent indication of its presence in the plant and seems not to cause harm to the host [5]." Endophytes have been isolated from virtually all plant organs (roots, stems, leaves, flowers, fruits, and seeds) and the most commonly observed endophytes are the fungi (usually Ascomycteous fungi associated with Fungi Imperfecti), as well as bacteria and this includes the filamentous bacteria which are the Actinomycetes. Periodically, a Zygomycete, and a Basidiomycete and are also isolated. In the future, it may be the case that other life forms may be found that can be considered endophytes such as the mycoplasmas. There is some certainty that all plant forms, including those in the world's oceans, are hosts or are potential hosts for one or more endophytes [6]. Additionally, many of the lower plant forms such as mosses and liverworts are also hosts to endophytes [7]. This review will primarily focus on fungal endophytes of higher plant forms.

As with most biological sciences, the initial work on endophytes began as an observational science with investigators who were keenly interested in isolating, identifying every possible endophyte from a given plant and then moving on to the next plant species [5,8]. Much of this early work was being done by the Petrini's and their group in Switzerland [9]. However, in spite of all of the reports on endophytes from Europe, Canada and the USA, it appears that hundreds, if not thousands of novel endophytic species still remain to be discovered especially in those remote, unique and untouched forests, fields and even oceans of the world. Thus, recent efforts in India, Thailand, Brazil and China have resulted in a plethora of new information on fungal endophytes of tropical plant species. In addition, investigators, at the ecological level, have begun to study how endophytes may influence the biology of the plant enabling it to resist biological as well as environment stresses [2,10–12]. Collectively, these efforts have crystallized thinking about these organisms and how they represent the "plant microbiome." This switch in thinking from endophytes as strictly an academic subject to that of practical biology now has the potential for crop associated endophytes to be useful biological tools in promoting plant growth and assisting the plant to better cope with environmental and biological stresses.

On another note, studies on the distribution of endophytes quickly led to some attention be paid to the relationship of endophytes to their host plants. It has been learned that endophytes are mutualists and some, on occasion may become pathogenic when environmental and physiological conditions of the plant may allow [5,8]. However, the exact nature of the interaction with the plant and the products and processes involved in the interactions has gone missing from the literature. What has appeared however is conjecture on what might be happening in the microbe-plant interaction based on tangential observations and speculation.

The likelihood of finding taxonomic novel endophytic microbes is the greatest in the world's tropical and temperate rainforests and also with plants growing in peculiar and unexpected places such as harsh deserts, high alpine areas, or even in the world's oceans [3]. It is to be noted that novel endophytes probably do not commonly exist in botanical gardens in population dense areas because of the pollutants in the air that can and do affect microbial growth and activity. The same can be said of unusual plants artificially placed on university campuses where the same nasty pollution elements undoubtedly also exist and researchers can find easy access to plant material but take no thought of the microbes. Ultimately, with taxonomic novelty, it seems, the chances of finding chemical novelty greatly increases [3]. In addition, with the discovery of novel organisms possessing unique biochemical properties there are opportunities for the development of uses of such organisms.

This necessarily implies the extreme likelihood for patenting possibilities for new organism and new product development.

This review is certainly not intended to cover all aspects and all discoveries related to fungal endophytes, but it will cover the main recent focal points in this field from the perspective of work done in the author's own laboratory. Then, as part of the content of each focal point area the author has put forth his ideas and suggestions on what he sees as the potential for future studies in the given area and provides suggestions and guidance on how work on endophytes should be encouraged in some areas and discouraged in others. Specific examples that are given are used to illustrate important considerations when working with endophytes, their discovery, their biology, and their utility. Also discussed and referenced are certain critical techniques that apply to the science of endophyte biology and how they may be useful and applicable to those working in this field.

2. Endophytes, Isolation, Storage, Fundamental Biology

Many of the commonly isolated fungal genera associated with plants as pathogens can also be isolated as fungal endophytes and upon inoculation of the plant with the endophyte one does not witness the development of disease symptoms. Some of these fungal organisms may represent new endophytic species or biotypes. Some of the commonly fungal genera that we have commonly isolated include *Fusarium* sp., *Colletotrichum* sp., *Phoma* sp., *Pestalotiopsis* sp., *Xylaria* sp., *Cladosporium* sp., *Curvularia* sp., and a host of others. The most common represented group of fungi isolated as endophytes are the Fungi Imperfecti whose perfect stage is commonly the Ascomycetes. Less frequently isolated members of the groups Basidiomycetes and Zygomycetes.

The methods of isolating these organisms are well known and usually involve the use of a surface treatment solution to rid the selected plant material of unwanted contaminants. This usually involves a thorough rinse in a 70% ethanolic solution followed by flaming the tissue or treatment with a Clorox solution. One must be careful as to not over treat the tissue to kill or harm the endophytes associated with epidermal tissues. Plant tissues are placed on water agar or other agars that may represent selective media containing certain antibiotics to discourage bacterial growth. Over the years we have realized that plant associated mites can cause major damage to endophyte based experimentation in the laboratory and can easily be handled by adding a small crystal or two taken from a common moth ball and placed into a plastic box containing Petri plates from which the endophytes will eventually be selected (Strobel–unpublished, 2018). Mite contamination of the laboratory can literally halt all scientific progress on endophytes until it is cleaned up. Once obtained in pure culture, we have learned that most tropical endophytes may be best preserved for future work by growing them on doubly autoclaved barley seed (hydrated), placed directly in small freezer vials and stored at −70 °C [13].

In order to do taxonomic work on novel fungi it is important to have one or more spore stages of the organism available for description. Frequently endophytes show no sporulation on potato dextrose agar. While a choice of other agars may provide some help in encouraging sporulation, the best approach is to use gamma irradiated carnation leaves placed on water agar as a medium of choice [14]. In addition, as an alternative one may use pieces of the plant host material that has been autoclaved on a water medium that may encourage fructification of the organism. One could also even make a host based concoction agar medium from stem, leaf or root materials and eventually use more purified plant compounds that are critical for fructification [15]. Ideally, one can use the fungus growing and fructifying on these media as a source of biological material for scanning electron and light microscopy as aids in identifying the organism. Limited molecular techniques such sequencing targeting the ITS region are an important aid in identification and should be used to assist, but not used exclusively in trying to describe an organism. When no spore forms are evident under a multitude of growth conditions, one must rely on a host of other methods to assist in identification and description at the taxonomic level. This includes a complement of molecular gene sequence information starting at the ITS region and moving to specifically targeted gene sequence information involving certain enzymes, other analytical chemical data entries including an analysis of volatile chemicals, and scanning electron microscopy of any unique hyphal/spore/fruiting body characteristics. Ultimately, any data gathered must be placed in context with data that exists in the mycological literature for other already known and previously described fungal species. It is to be noted that one must be prepared to face the taxonomic literature and scrutiny of the mycological community when working in the field of endophyte biology especially if one is intent on finding and thence describing novel microbes.

Microorganisms requiring plant tissues to complete their life cycle are classified as "obligate." Well-documented examples of obligate endophytes are found among mycorrhizal fungi and members of the fungal genera Balansia, Epichloë, and Neotyphodium, from the family Clavicipitaceae (Ascomycota). Interestingly, there is little work demonstrating if there are other obligate endophytes, that are not specifically mycorrhizal fungi or members of the Clavicipitaceae. It is generally suspected that they do exist based on what is known and understood from obligate plant pathogens such as the rust and powdery mildew causing fungi and assuming that the number of host species is constant, the total number of host specific endophytic species can be extrapolated from the number of plant species [16]. Metagenomic analyses for the total population of endophytes in plants is a challenge since there is the technical limitation of separating fungal from host DNA. Basically, the plant DNA is much more abundant than fungal DNA which confounds the analysis. Additionally, we have observed that frequently slow growing unidentified endophytes do arise from many plant species. Upon the second transfer to a potato dextrose medium they again begin to grow very slowly and then gradually die as if some critically important nutrient was missing from the medium. Obviously, further work on such organisms is confounded by absence of the organism. We have seen hints of this phenomenon over the years in my laboratory but lack of available host plants (since many were obtained in the world's jungles) to substantiate the initial observations made further work on the problem virtually impossible to study.

Also needed are studies showing the complete endophyte characterization of a given plant species. This should include mention of all organisms associated with all tissues and organs of the plant including flowers, fruits and roots which are sometimes difficult specimens to find and study. It is known that the make-up of the population of microbes in a plant will vary as a function of season, the organs being studied, and the host itself. It appears, however, that it is exceedingly interesting to find, identify, and describe totally novel fungal genera and species that are acting as endophytes. The biological potential of such studies is great given the fact, as previously mentioned, that novel taxonomy portends the presence of novel chemistry.

Specifically, I do not completely encourage studies describing the endophytes of a previously unstudied plant that show no novelty in the endophytic composition of the plant, any novel organism or other unique scientific contribution resulting from the study since so little new is to be learned. On the other hand, if a novel fungal genus is isolated from the plant then it is worthwhile to do the work that properly describes the new organism. In addition, if new uses and concepts are found with already known organisms this too can be valuable information. Furthermore, it must be recognized that the use of limited molecular genetic techniques to assist in the identification of the endophytic flora of a plant can be helpful but misleading if no attempts are made at the actual isolation and examination of the host endophytes. It is well understood that identification of fungi at the species level is really not possible with ITS sequence information alone. The use of classical mycological methodologies come in handy in helping with fungal species identification and this is not to be ignored.

3. Novel Endophytes

The opportunity to find novel endophytic fungi at the genus or species level can be done if one's attention is properly directed. Many of the frequently appearing endophytic fungi have their counterpart relatives existing as plant pathogens. This is especially true of the genus Fusarium which is often isolated from numerous plant species as pathogens but it seems that they are even more common as endophytes. Most interesting however, are reports of truly novel fungi as endophytes that represent novel fungal genera. Such reports require all of the work necessary to list a valid new genus/species. This necessitates getting a pure culture, and then doing the comparative taxonomic work with other related fungi at all levels including structural, developmental, and molecular as previously mentioned. Depositing the organism in a national, state or well-maintained university culture collection is also a requirement and one that is critically important. Novel endophytes can best be found in endemic plant species which are numerous in certain locations especially those hosting great biodiversity. Usually these places exist in temperate or tropical rainforests and maybe remote.

To illustrate the point of how to approach the discovery of novel endophytes I have shown below several examples of how we have solved the issue of biological discovery by visiting far reaching places in the world to find novel fungal genera. For instance, the fungus—*Seimatoantlerium tepuiense* was discovered as an endophyte of *Maguireothamnus speciosus* on top of the Roraima tepui on the border junction of Guyana, Brazil and Venezuela. The area is very remote and the host plant is endemic to this location and only two other small locations in tropical America. To get to this spot required the use of a rented helicopter and some knowledge of the local flora. The conidia of this endophyte resemble those of the commonly isolated endophyte-*Pestalotiopsis macrospora* however, the condiospore develops an appendage that is much more complex than the helicopter-like appendage in *P. macrospora* (Figure 1). In this case it forms an elk antler-like appendage and this structure easily dehisces from the spore (Figure 1) [17]. Presumably this allows for easy distribution by birds visiting the N.E. South American tepui plateaus since the spores can easily become attached to the fine structures of their feathers and eventually fall as the appendage becomes detached from the spore [17]. The fungus also makes small amounts of the anticancer drug-taxol. Thus, in this case not only was novel taxonomy discovered but so too was a product of extreme medical importance also found quite to our surprise.

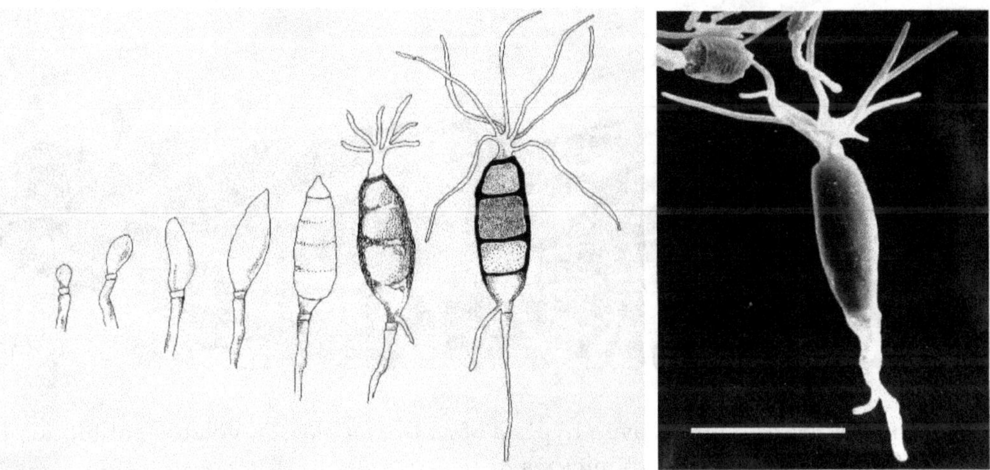

Figure 1. Development of the conidiospore of *Seimatoanlerium tepuiense* a novel endophyte discovered on a tepui in the Brazil, Venezuela, Guyana area of South America. The actual spore is shown to the right as a scanning electron micrograph. The bar is equivalent to 20 microns.

One of the most important novel endophytes that we have ever isolated is *Muscodor albus*. This whitish, non-spore producing endophyte was originally obtained from a cinnamon tree (*Cinnamon zeylanicum*) in Hondouras [18]. This endophyte is relatively slow growing on potato dextrose agar and it makes a plethora of volatile organic compounds that possess strong antimicrobial activities [19]. In fact, the volatiles in split plate agar assays are not only inhibitory to other microbes but in most cases are lethal [19]. This makes for a likely prospect that the fungus itself might be useful as a biological control measure. In fact, in November of 2016, the US—Environmental Protection Agency approved the use of this organism for applications in US agriculture. Field testing by Marrone Bioinnovations Co. in Davis, California, has shown efficacy of *M. albus* as a pre-plant soil additive to effectively eliminate root rot causing pathogens in fields of strawberry and celery. Numerous other applications of this fungus to control disease problems in various crops has been demonstrated by Mercier and his group [20,21]. The fungus has the potential to replace the use of methyl bromide as a soil sterilant. It is the first broadly active antimicrobial producing endophyte to officially be on the market in 2017.

Since its discovery almost twenty years ago many other species of this genus have also been described and now number over twenty. All members of this group have several things in common—(1) they have close homology at the ITS r DNA level [22], (2) they each produce bioactive volatile substances and (3) they are all producing a sterile (non-spore) producing mycelium that is commonly convoluted, interwoven, and sometimes possess unique hyphal projections (Figure 2) [23]. One interesting example of a novel species of this fungus is *M. vitigenus,* which was obtained from *Paullinia paullinoides* in the Peruvian Amazon, and it makes only one volatile which is naphthalene, an insecticide [24]. Cultures of *M. vitigenus* were extremely active in repelling insects in Y-tube testing experiments [25].

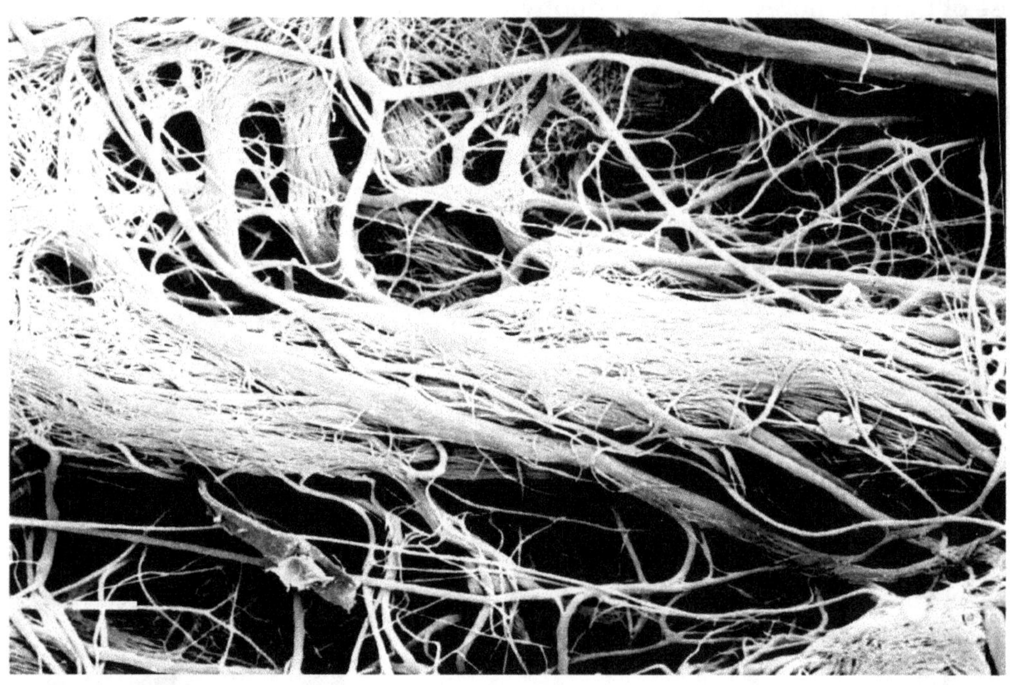

Figure 2. The convoluted and interwoven hyphae of *Muscodor albus*, a volatile antibiotic producing fungus. The white bar is equivalent to 5 microns.

Another unique species is *Muscodor crispans* whose volatile composition is extremely antimicrobial, and the majority of its active compounds are on the FDA-GRAS list, that is Generally Recognized As Safe [26,27]. Thus, the bioactive volatiles can be either synthesized or made individually by fermentation processes using other microbes to make large quantities of the product in a liquid state (N. Gandhi unpublished, 2018). It has been learned that all of these volatile compounds in the fungal gas mixture are not needed for biological activity, but certain ones act in a synergistic manner to cause microbial inhibition and death of target microbes. Those that have such activity have been termed synergistans [28]. The products of *M. crispans* (Jeneil Biotech name of Flavorzon) at the 1% level has antimicrobial activity that parallels that of Clorox or quaternary ammonium [29]. This makes the solution usable on biological or industrial surfaces that may be contaminated or have the potential to be infected by or contaminated with organisms such as *Salmonella* sp. [29]. Since it is a safe product it has been used to decontaminate rooms where food-grade work is performed. Presently, another version of the Flavorzon formula is being tested for the prevention of decay in post-harvest fruits/vegetables. Still another version of the fungus gas formula is now being used to enhance the shelf life of certain products such as soy milk and is in the market place.

It turns out that each of these *Muscodor* spp. that were unique species, was patented by us at the time that they were discovered. Likewise, claims were made for the volatile chemistry that was discovered in each organism. In each case the US and many International patent offices approved the applications. It seems as if this intellectual property approach represents the best mechanism to make the discovery process useful to humankind. It is something that needs to be considered as one does all of the work to make important, valuable, and useful discoveries. Without intellectual property, a discovery such as the ones related to Muscodor are interesting but are rendered useless unless protection is sought and product development is pursued. Ultimately, the general public benefits from all of the efforts made in work of this type.

As a side light and most pleasing to me is the efforts of a group of Indian scientists who have recently described *M. strobelii* from *Cinnamon zeylanicum*, the same species from which the original isolate of muscodor was obtained, only in this case the host tree was growing in southern India, which is most likely the same source area for the Honduran tree mentioned above. The organism was unique from other species in this genus by virtue of its production of very unique volatiles, its antimicrobial activity and its unique hyphal structure [30].

4. Bioactive Products of Endophytes and Their Utility

Most work on the secondary products of endophytes arose in the 1980s with a focus on the toxic compounds associated with diseases in livestock. These symptoms were associated with fescue grass that was infected with *Neotyphodium* sp. that produces toxic metabolites causing abortion and even death in animals grazing especially on this grass species [31]. Endophytes, during that era, received notoriety and the greatest interest in them was for the damage that they caused in livestock. The broader picture of their benefit to plants and mankind was not evident until later. Then with the advent of the discovery of *Taxomyces andreanae* in Montana in a yew tree, making the anticancer drug taxol, the concept of endophytes as a source of novel and useful drugs was introduced (Figure 3) [32,33]. At the outset, it seems only logical that such possibilities exist in nature if one assumes that the endophyte and the host plant are living in some sort of symbiotic relationship and horizontal gene exchange may occur between the symbionts. Thus, any contribution to the survival of a plant has the potential to be translated into products that also might be beneficial to mankind once isolation of the endophyte is achieved along with its biologically active products. As it turns out, taxol is a potent anti-oomyceteous compound and its mode of action on fungal pathogenic oomycetes is identical to its activity against rapidly dividing cancer cells [34]. It seems likely that taxol and its hundreds of related compounds are produced by *Taxus* sp. to protect the plant from attack by "water molds."

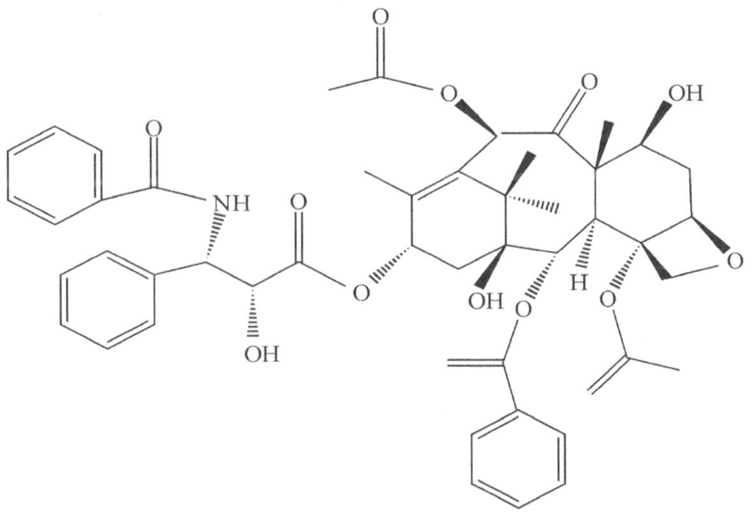

Figure 3. The structure of taxol.

In my experience, any search for a natural product (with a given bioactivity) is driven by what system of bioassays might be in place or available at the time of fermentation of the fungal culture and the purification procedures to which it may eventually be applied—bioassay guided fractionation. A good and effective bioassay can greatly assist in the eventual isolation and purification of a bioactive product by allowing an assessment of where on a TLC plate or a column eluent the product of interest may be located. To this end, over the past 20 years, a plethora of review papers have appeared on the secondary natural products of endophytes and their associated biological activities [35–39]. These reports nicely show that endophytes are a wonderfully unique and growing source of biologically active molecules and one must be prepared to work on virtually any class of biologically active molecules as one follows the bioactivity through the purification processes.

Thus, once an endophyte is isolated in pure culture some form of fermentation methodology is employed in order to find and characterize biologically active secondary products made by the organism. Normally, a more defined medium such as the M-I-D medium is used rather than more complex media [3]. The use of a more defined medium necessarily eliminates many medium products from the purification processes. Generally, we carry out fermentation in still culture in large flasks for a period of 2–3 weeks. Initial extraction is done with methylene chloride, or ethyl acetate or n—butanol. This step is followed by flash chromatography, then preparative thin layer chromatography and HPLC. After each step the various fractions are subjected to the biological assay that has been selected. The final and most desirable goal is to get a product that is crystalline so that the product may be subjected to X-ray crystallographic analysis. Ultimately, all other spectroscopic analytical data should be in agreement with the X-ray data.

A few of the products of endophytes on which we have worked are discussed and illustrated below. The first three products were all obtained from various endophytic isolates of *Pestalotiopsis microspora* that were found in various plants in various locations around the world. This fungus is one of the most frequently isolated endophytic fungi from tropical and temperate rainforests plants. It has been seen many times and assigned many different species names, usually solely based on the host plant from which it was isolated. Until recently little work has been done on the biological chemistry of this organism. It turns out to have a treasure trove of interesting novel and bio-active molecules. Each of the products, shown below, has a unique biological activity and the chemistry of the products is also unique. This nicely illustrates the point that one must be prepared to deal with a variety of techniques in separation science and chemical characterization when dealing with the bioactive products of endophytes.

4.1. Pesatcin

This novel benzofuran was obtained from a culture of *Pestalotiopsis microspora*, an endophytic fungus obtained from *Taxus wallichiana* sampled in the Himalayan foothills at which time we were engaged in looking for other endophytes making taxol and were surprised in finding an endophte making both taxol as well as two novel antioxidants. This fungal endophyte produces a new 1,3-dihydro-isobenzofuran which exhibits antioxidant activity eleven times greater than the vitamin E derivative troxol. The compound also has moderate antifungal activities [40] (Figure 4). Isolation of pestacin was achieved by extraction of culture fluid with methylene chloride followed by silica gel chromatography. Its structure was established by X-ray diffraction and ^{13}C and ^{1}H NMR spectroscopies. The X-ray data demonstrated that pestacin occurs naturally as a racemic mixture. Mechanisms for antioxidant activity and post-biosynthetic racemization have been proposed. Isopestacin is also produced by this endophyte and it also possesses similar bioactivities as pestacin [41].

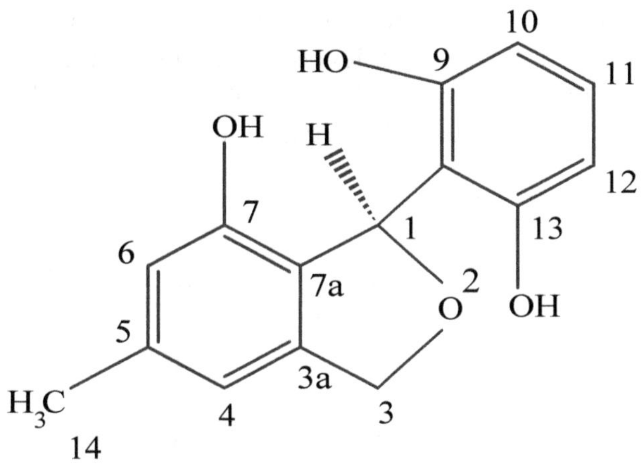

Figure 4. The structure of pestacin.

4.2. Ambuic Acid

We were greatly surprised to find a highly functionalized cyclohexenone being produced by an isolate of *Pestalotiopsis microspora* as an endophyte of *Fagraea bodenii* found in the highlands of Papua New Guinea [42] (Figure 5). The compound possesses weak antifungal properties. It was first natural product to have its absolute structure established by solid state NMR methods allowing a spatial assignment to the –OH group on carbon 7 [43,44] (Figure 5). Quite surprisingly, after the initial work on the isolation and structural determination of ambuic acid, it was later learned that it is one of the best compounds known for its anti-quorum sensing activity in Gram-positive bacteria [45]. Ambuic acid inhibits the biosynthesis of the cyclic peptide quormones of *Staphylococcus aureus* and *Listeria innocua*. Ambuic acid is a lead compound in the search for anti-pathogenic drugs that target quorum sensing—mediated virulence expression of Gram positive bacteria. Once again, what initially was supposed as a weak antifungal agent turned out to have a totally unsuspected biological activity as an inhibitor of quorum sensing in bacteria and this point was not established by us but by other investigators in Japan. An assay for anti-quorum sensing activity was not originally in our repertoire of bioassays.

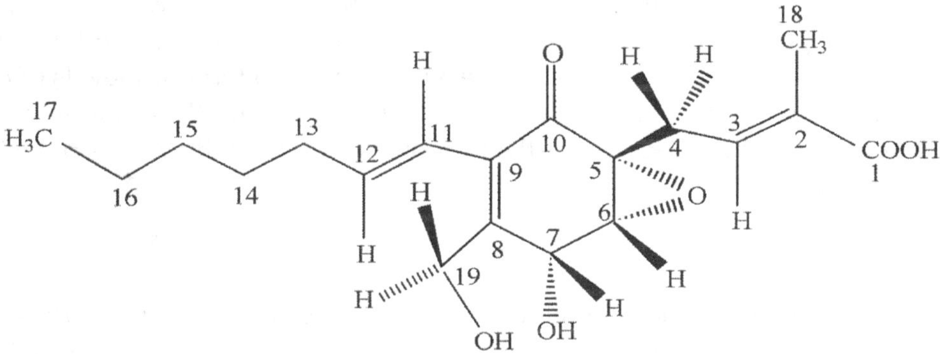

Figure 5. The structure of ambuic acid.

4.3. Torreyanic Acid

The Florida Torreya is a rare and threatened plant and for this reason it was sampled for the presence of unusual endophytes. Torreyanic acid was isolated and characterized as a dimeric quinone obtained from the endophyte, *Pestalotiopsis microspora* originally isolated from *Torreya taxifolia* in Northern Florida [46] (Figure 6). The compound was cytotoxic against 25 different human cancer cell lines with an average IC_{50} value of 9.4 μg/mL, ranging from 3.5 (NEC) to 45 (A549) μg/mL. Torreyanic acid is 5–10 times more potent in cell lines sensitive to protein kinase C (PKC) agonists, 12-*O*-tetradecanoyl phorbol-13-acetate (TPA), and was shown to cause cell death via apoptosis. Torreyanic acid also promoted G1 arrest of G0 synchronized cells at 1–5 μg/mL levels, depending on the cell line. It has been proposed that the eukaryotic translation initiation factor EIF-4a is a potential biochemical target for the natural compound. Additionally, it has been prepared by organic synthetic techniques.

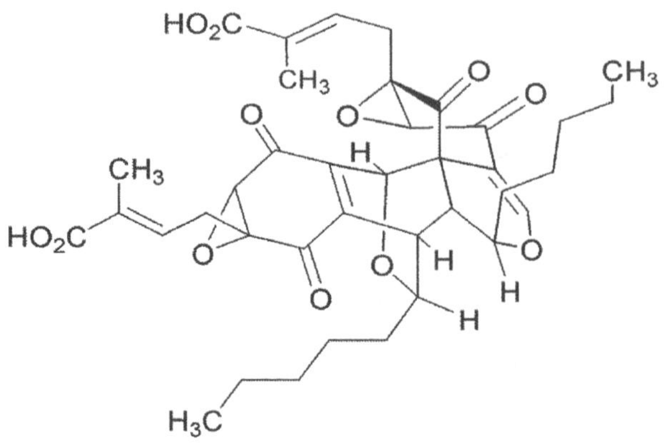

Figure 6. The structure of torreyanic acid.

4.4. Colutellin A

Colletotrichum dematium is an endophytic fungus recovered from a *Pteromischum* sp. growing in a tropical forest in Costa Rica [4]. Strangely enough this was the only endophyte isolated from samples of this plant and this is unusual circumstance especially in dealing with tropical plant species. This fungus makes a novel peptide antimycotic, Colutellin A, with minimum inhibitory concentrations of 3.6 μg/mL (48 h) against *Botrytis cinerea* and *Sclerotinia sclerotiorum*, respectively. This peptide has a mass of 1127.70 and contains residues of Ile, Val, Ser, N-methyl-Val, and β-amino-isobutryic acid in nominal molar ratios of 3:2:1:1:1, respectively. Independent lines of evidence suggest that the peptide is cyclic and sequences of val-ileu-ser-isoleu as well as ileu-pro-val have been deduced by MS/MS as well as Edman degradation methods. Colutellin A inhibited CD4—T cell activation of IL-2 production with an IC_{50} of 167.3, whereas cyclosporine A, in the same test yielded a value of 61.8 nM. Since IL-2 production is inhibited by Colutellin A, at such a low concentration, this is an effective measure of the potential immunosuppressive activity of this compound. On the other hand, in repeated experiments, cyclosporin A at or above 8 μg/mL exhibited high levels of cytotoxicity on human peripheral blood mononuclear cells whereas, Colutellin A or DMSO alone, after 24 and 48 h of culture, exhibited no toxicity. Because of these properties Colutellin A has potential as a novel immunosuppressive drug [47].

4.5. Cryptocin

Tripterygium wilfordii is an Asiatic plant with strong immunosuppressive properties. It was targeted for study to learn if any of its endophytes might also make the same or related biologically active molecules. This rationale for this work follows that of the fungal taxol story. Instead, one of the endophytic fungi that was isolated was a unique species of *Cryptosporiopsis*. It was unique in that its large conidiospore was segmented quite unlike other members of this group [48]. This organism possessed unusual antifungal activities that were related to two novel compounds—cryptocandin (a novel lipopeptide) and cryptocin, an unusual tetramic acid. The acid was isolated and characterized by X-ray crystallography and other spectroscopic data (Figure 7) [48]. The compound was active against many plant pathogenic fungi (<1 μg per mL) and much less impressive activity against human pathogenic fungi (>50 μg per mL). However, to be noted is that the most sensitive fungus was *Pyricularia oryzae*, the causal agent of rice blast one of the most important pathogens in the world as it relates to food production.

The search for these valuable products has just begun. Usually, in my experience, there are several initial clues that make an organism a potential candidate for a producer of novel products. If it had been isolated from a unique environment or an endemic species, if it is a slow grower on common lab media and if it is coming from a totally unique fungal genus, the chances for finding something new are greatly improved. It is not, however, uncommon to work on an organism only to find that the bioactive product already been discovered in the same or related organism by someone else. Thus, novelty is a major goal! Repeated again, once the work is completed and a novel bioactive product has been discovered, one should make an effort to get patent coverage and find an interested commercial partner in order for some aspect of the general public to directly benefit from the discovery.

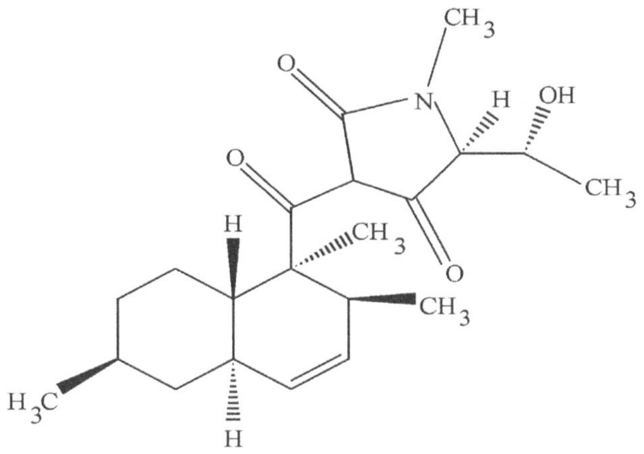

Figure 7. The structure of cryptocin.

5. Fungal Hydrocarbons and Fuels

We have relatively recently described endophytic fungi that make one or more hydrocarbons that have potential as fuels. Such products have been dubbed –mycodiesel [49]. This dramatically differs from standard yeast fermentation processes that utilizes sugars or starch to produce ethanol which is a proven and useful source of fuel, but by no means is it ideal. A number of endophytic fungi have been isolated and described that make compounds such as mono-terpenoids, alkanes, cyclohexanes, cyclopentanes, and alkyl alcohols/ketones, benzenes and polyaromatic hydrocarbons [4]. Many of these compounds are either identical to or are closely related to those specific classes of molecules that are found in diesel/gasoline [4,50]. Most importantly, these organisms make hydrocarbons while utilizing cellulose, hemicellulose and other polymers found in all plant-based agricultural wastes [51–53]. Endophytes are a prime source of hydrocarbon producers because they are the first microbes to begin the processes of tissue degradation, when a plant dies, to yield products with fuel–potential [4,53]. Examples of fungi that have been discovered that make fuel related hydrocarbons include—*Ascocoryne sarcoides*, *Gliocladium* sp., *Hypoxylon* sp. (*Nodulisporium* sp.), *Annulohypoxylon* sp., *Phoma* sp., *Phomopsis* sp., and *Daldinia* sp. [13,50–55]. The volatiles produced by these organisms are usually possessing antibiotic activities which conceivably contribute to their role as symbionts. Finally, it seems possible that endophytic fungi may have an additional attribute of having contributed to the formation of crude oil in the first place and experiments have been done to demonstrate this phenomenon [53].

Muscodor albus, as described above, was used to discover other fungi making bioactive volatiles that turned out to be hydrocarbon-like molecules. Plant tissues of carefully selected plants were placed on half plates containing 10 day old cultures of *Muscodor albus* whose gases killed most fungi other than those also making bioactive volatiles. As an example of the use of this selection technique was the sole appearance of *Gliocladium roseum* (now classified as *Ascocoryne sarcoides*) [49,54] in Petri plates (PDA) supporting the growth of *Muscodor albus* showed that this fungus was able to survive and grow in the presence of the inhibitory and lethal volatiles (VOCs) of *M. albus* [49]. Further testing of the recovered *A. sarcoides* culture revealed that its VOCs were active against other test fungi and that some of its volatile products had fuel potential [49]. Most interesting was the appearance of a series acetic acid esters of straight chained compounds including those of hexyl, heptyl, octyl, sec-octyl and decyl alcohols [49]. In addition, many other hydrocarbons were also noted in the GC/MS analyses of the VOCs of this fungus. The straight chained hydrocarbons (alcohols in the reduced form) are the backbone compounds found in all diesel fuels that we have investigated from widely differing parts of the world.

Since those observations were made it has become increasingly evident that many other endophytic fungi making volatile hydrocarbons are also resistant to the *M. albus* VOCs and their VOCs are biologically active [4]. Thus, this one selection technique has the potential of eliminating 80–90% of endophytic microbes that probably would not be of interest relative to VOC production since they would succumb to the VOCs of *M. albus* while growing out of the plant tissues on the Petri plates. This technique has merit if one does not wish to deal with a plethora of microbes that may not be of interest and waste time doing GC/MS analyses on organisms that would not be of interest. Nevertheless, it is to be noted that some important hydrocarbon producing fungi are sensitive to the VOCs of *M. albus* and would have been missed if this technique had been exclusively used for selection [55].They can be usually selected on the basis of the biological activity of their VOCs.

An endophytic fungal strain of *Hypoxylon* sp. was isolated from *Persea indica* an evergreen tree native to the Canary Islands where it grows not in abundance but is found on several islands including Tenerife in the Laurisilva [13]. This organism was isolated in its imperfect stage as *Nodulisporium* sp. from a small stem in the crown of the tree and it readily grows in the presence of the VOCs of *M. albus* which should facilitate its isolation from other plant sources. When grown on PDA—Petri plates, the VOCs produced by this fungus were primarily 1,8-cineole; 1-methyl-1,4-cyclohexadiene, and (+)-.alpha.-methylene-.alpha.-fenchocamphorone (Figure 8). Not only these but many of the compounds made by this organism are of interest because of their high energy densities and thus the potential they might have as Mycodiesel fuels [13].

Six-day-old cultures of *Hypoxylon* sp. (imperfect stage—*Nodulisporium* sp.) displayed maximal VOC-antimicrobial activity against *Botrytis cinerea*, *Phytophthora cinnamomi*, *Cercospora beticola*, and *Sclerotinia sclerotiorum* suggesting that the VOCs may play some role in the biology of the fungus and its survival in its host plant. In fact, this discovery has implications in developing methodology for strain improvement via mutation/selection techniques [13]. Media containing starch-or sugar related substrates best supported VOC production by the fungus. Continuous direct on-line quantification of VOCs was measured by proton transfer mass spectrometry (PTR-MS) covering a 12 day range with optimum VOC production occurring at 6 days at 145 ppmv with a rate of production of 7.65 ppmv/h [13]. The production of 1,8-cineole from a fungal source is of significant interest given the fact that this compound does not have any previously known biological sources aside from plant tissue, and has thus far limited the compound's availability for fuel purposes (Figure 8).

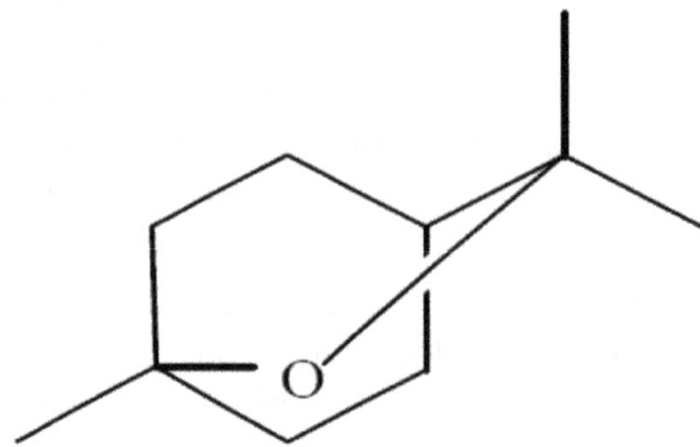

Figure 8. The structure of 1,8-cineole.

In a related *Hypoxylon* sp. a specific cineole synthase has been identified along with 11 new terpene synthase genes [56]. The discovery is an important first step in identifying a complete fungal pathway for the synthesis of 1,8-cineole and related monoterpenes. Currently, the yields of these terpenes and other hydrocarbons from fungal cultures does not allow for the production of these compounds at an economical level. These problems may eventually be overcome with conventional mutagenseis techniques and or genetic manulipation combined with complete sequencing and annotation of important fungal genomes such as *Ascocoryne sarcoides,* which is the best annotated endophyte thus far studied [57]. In combination with these approaches it will be necessary to develop non-conventional fermentation methods which may employ novel solid state methods allowing for a constant yield of volatile products via an airsteam and appropiate gas traps (Strobel unpublished, 2018).

Furthermore, as a practical matter cineole has been examined in multiple ways for its utility as a fuel. Studies on 1,8-cineole have shown prevention of phase separation when used as an additive in ethanol-gasoline fuel blends [58]. Furthermore, when fuels comprised of a gasoline/eucalyptus oil mixture (with 1,8-cineole as the major fuel component up to 80% or more) there was an improved octane number and reduced carbon monoxide exhaust [59]. Thus, 1,8-cineole is a worthy target molecule for study and scale up and it has amazing potential for replacing fossil-based hydrocarbons as a fuel additive.

Finally, it is to be noted that to do the best work on fungal volatiles does require the use of some unique methods and techniques in qualitative and quantitative analysis of these gaseous compounds. Standard practice is the use of the SPME fiber trapping technique combined with GC/MS to acquire an idea of the compounds being produced. The advent of PTR/MS has permitted the evaluation of on line-real time analysis and quantification of fungal gas production and now PTR/TOF-MS allows for this plus qualitative analysis of the gas stream [4]. Gas trapping in stainless steel columns as well as PTR /MS in line with a platinum catalysis will permit a total organic volatile compound production analysis—again in real time [60].

6. Control of Secondary Product Formation in Fungi

In many cases, once a product has been isolated and identified, the fungus may become attenuated in the production of the wanted compound (s). This is an event that has happened at least 50% of the time in my experience and can cause concern and self-doubt in the mind of the investigator. This is best exemplified by the attenuation of helminthosporoside (a plant host specific toxin) produced by the plant pathogen-*Helminthosporium sacchari.* When the organism was continuously cultured and transferred on a semisynthetic medium the production of this fungal toxin totally ceased [61]. However, when amounts of leaf exudate—concentrate were placed into the cultural medium, toxin production fully resumed. It turned out that several secondary products of sugarcane metabolism were responsible for the regulation of toxin production by this pathogenic fungus. One of these compounds was isolated and characterized as serinol (primary alcohol of serine) and the other was a sugarcane lipid-like compound [62]. It seems that the best place to look for answers to problems surrounding attenuation of secondary product formation in endophytic or plant pathogenic fungi is the host itself.

Similar events happened during the course of the discovery of taxol in an endophytic *Periconia* sp. which was isolated from *Torreya grandifolia* (a relative of yew that does not synthesize taxol) in China. This fungus, not previously known as a tree endophyte, was isolated from the inner bark of a small lower limb. When freshly isolated from the tree and placed in a semi-synthetic medium, the fungus produced readily detectable quantities of the anticancer drug taxol [63]. The production of taxol by *Periconia* sp. was demonstrated unequivocally via spectroscopic and immunological methods [63]. However, successive transfers of the fungus on a semi-synthetic medium resulted in the gradual attenuation of taxol production until minimal amounts of it were produced even though fungal growth was relatively unaffected. Several compounds, known previously as activators of microbial metabolism, including serinol, *p*-hydroxybenzoic acid, and a mixture of phenolic acids, were capable of fully or partially restoring taxol production to otherwise taxol-attenuated cultures. The compound with the most impressive ability to activate taxol production was benzoic acid at 0.01 mM. To our knowledge, benzoic acid is not a direct taxol precursor [63].

The fungal fuel producing fungus—*Nodulisporium* sp. also falls into the category of an organism that will attenuate secondary product formation. It appears as if there are factors controlling 1,8-cineole and other hydrocarbon molecule production in this fungus based on the observation that serial transfer of the fungus on PDA resulted in a dramatic reduction of VOC production including 1,8-cineole [64]. Subsequently, when the attenuated organism was placed on certain plant parts and plant extracts the VOC production resumed to normal as measured by PTR-MS [64]. *Nodulisporium* sp. (Ti-13) in question had been isolated as an endophyte from *Cassia fistula* in Thailand. The fungus produces a spectrum of volatile organic compounds (VOCs) that includes ethanol, acetaldehyde, and 1,8-cineole as major components [64]. Initial observations of the fungal isolate suggested that reversible attenuation of the organism via removal from the host and at least 5 successive transfers in pure culture resulted in a 50% decrease in cineole production unrelated to an overall alteration in fungal growth. A compound (CPM_1) was obtained from *Betula pendula* (silver birch) that increases the production of 1,8-cineole by an attenuated Ti-13 strain to its original level as measured by a novel bioassay method employing a 1,8-cineole sensitive fungus-*Sclerotinia sclerotiorum* [64]. The host plant *Cassia fistula* also produces similar compounds possessing this activity. Bioactivity assays with structurally similar compounds such as ferulic acid, gallic acid and others suggested that the CPM_1 does not to act as a simple precursor to the biosynthesis of 1,8-cineole. Nuclear magnetic resonance spectroscopy and high-performance liquid chromatography electrospray ionization mass spectrometry indicated that the CPM_1 is a para-substituted benzene with alkyl and carboxyl substituents [64]. The VOCs of Ti-13, especially 1,8-cineole, have potential applications in the industrial, fuel, and medical fields.

These observations are suggestive of one of more regulatory mechanisms involved in hydrocarbon production that seems to be influenced by the host plant. Obviously, these considerations on the regulation of secondary product formation will be vital in making hydrocarbon—based fuel production from fungi a reality in the future. The work also has relevance to literally any plant microbe that is under the regulation of secondary product formation by host-related substances.

7. Endophytes and Epigenetics

We have recently learned that the regulation of the production of fuel –like compounds in fungi is under epigenetic control. *Hypoxylon* sp. (*Nodulisporium* sp.) is of interest because of its ability to make hydrocarbon-like compounds that may serve as fuels. Thus, experimental work was undertaken to affect the gene expression of this organism through compounds known to act as epigenetic modifiers and explore the possibility of the appearance of new products and/or the increased or decreased production of other VOCs already known from this organism [65]. For instance, it is realized that the complete analyses of fungal genomes in recent times has indicated that many putative biosynthetic gene clusters are located in the distal regions of the chromosomes and exist in a heterochromatin state with the constitutive genes often transcriptionally controlled by epigenetic regulation such as histone deacetylation and DNA methylation [66,67]. Several studies have demonstrated that the inhibition of histone deacetylase activity, through gene disruption or use of epigenetic modulators, leads to the transcriptional activation of gene clusters resulting in enhanced production of secondary metabolites [66,67]. Fungi treated with DNA methyl transferase- and histone deacetylase inhibitors exhibited natural product profiles with enhanced chemical diversity demonstrating that the small-molecule epigenetic modifiers are effective tools for rationally controlling the native expression of fungal biosynthetic pathways and generating biomolecules not previously associated with the organism [66,67]. Thus, the fungus was exposed to the epigenetic modulators- histone deacetylase- and a DNA methyltransferase inhibitor -SAHA (suberoylanilide hydroxamic acid) as well as 5-azacytidine (AZA), respectively. Subsequently, the organism displayed striking cultural changes including variations in pigmentation, growth rates and odor, in addition to significant differences in the bioactivities of its VOCs [65]. Analyses (GC/MS) of the VOCs produced by the modulated fungus showed considerable variation with the emergence of several compounds not previously observed in this fungus, particularly an array of tentatively identified terpenes such as ∞-thujene, sabinene, δ-4-carene, γ-terpinene, ∞-terpinolene and β-selinene, in addition to several primary and secondary alkanes, alkenes, organic acids and derivatives of benzene. The spectrum of identifiable compounds, in this study, was greatly enhanced by virtue of the organism being altered with the epigenetic modulators [65]. It is likely that the same phenomenon would occur with other fungi but the result would be an entirely different set of secondary products.

8. Biology of Endophytes and Their Ultimate Utility

This review has considered only a limited range of topics concerning endophytes. Their vastness and diversity literally covers the entire planet since they are associated with all macroscopic plant life forms. Only a small portion of their range and diversity has been sampled with entire vast reaches of the world never having been sampled at all. This would include areas of central Africa, the lower reaches of the Amazon, many area of the rainforests of South Eastern Asia, as well as the expansive forests of Russia and central Asia. In addition to these sources are the macro algae of the world's oceans along with an abundance of other plant forms such as the ferns, mosses, lichens that also need attention. Only about 1% of the world's botanical sources of endophytes have ever sampled and incompletely at that since only a limb or leaf samples were the subject of study. It is well known that less accessible roots, fruits and the inner tissues of the stem each contains its own endophytic assemblage. The patterns of abundance and diversity of these organisms then becomes an issue for study as per Arnold et al. [68]. The biology of endophytes is also influenced by leaf or plant age, canopy cover, seasonal factors and even the presence of other endophytes [69].

Most interesting are observations that suggest that endophytes may play some role in the protection of the host plant from plant pathogens or other agents of destruction including insects and herbivores [11]. This brings into focus the potential roles of endophytes in enabling the host plant to be best fit for its survival and the potential that exists for harnessing these organisms for the benefit of agriculture [11,70]. Thus, managing the plant microbiome is a goal of several companies that wish to take advantage of the role that endophytes may play in the health, fitness, growth and development of the plant in its environment. Some of these are Indigo (Cambridge, MA, USA), Bayer/Ginko Bioworks (West Sacramento, CA, USA) and MBI (Davis, CA, USA). Many in the world of agricultural business would agree that managing a living organism as a product can be fraught with some problems. These are usually lack of reliability and consistency of outcomes when placing the live product on the market.

Some guidance for the development, and testing endophytes for eventual commercialization has recently been presented [71]. This work was done because of the problems mentioned above. Murphy et al. [71] began their work on barley by finding candidate source endophytes in a wild relative of *Hordeum mirnum* followed by testing in a carefully controlled environment and finally doing multi-year field testing. In their hands this approach was successful. It is to be strongly stressed in that what is generally missing in most studies of endophytes of crop plants is exactly what was avoided by Murphy et al. [71]. The organisms of interest to them were recovered by an extensive search and description of the endophytic flora of wild host plants growing in or near the center of origin of that species. It is entirely likely that that most plants serving mankind as food sources have long since lost their complement of normal flora as a result of movement from their centers of origin, intensive genetic selection techniques dealing only with the seeds or fruits of the plant and exposure of the crop to a myriad of pesticides. Evidence for this concept has been demonstrated in wild wheat related species vs. *Triticum aestivum* [72]. The wild types relatives of wheat contain a plethora of taxonomically diverse fungal endophytes that are not found in modern wheat. It turns out that most wild or native food source plants have yet to have their microflora studied. It behooves us to find and carefully preserve all natural sites that harbor wild relatives of mankind's major and minor food plants not only for the genetic diversity that they may hold but also for the unique microflora of these plants that may exist in such places.

9. Endophyte Biology as a Platform for Teaching Undergraduate Students

Finally, it should be mentioned that the concepts and techniques involved in isolating and characterizing endophytes and studying their biology represents an excellent opportunity for independent undergraduate participation. Several studies have demonstrated that an independent research experience is the best way to get undergraduates excited about a career in science [73]. In fact, the vast majority (around 80%) of the science faculties in US universities had an undergraduate research experience during their college careers. In this case a student can do her/his own collecting, plant identification, finding GPS coordinates and bringing the samples to the lab for further work. Endophyte isolation, microbial identification using classical methods and modern molecular methods offer a student a real opportunity to acquire independent project ownership which is a great way to encourage progress on a project. Ultimately, the student will have the possibility to develop novel bioassays for endophyte products or learn a myriad of other techniques involving chemistry, molecular biology, statistical methods and spectroscopic methods among others as applied to the microbes that were isolated. The student may have an opportunity to publish her/his work and possibly be involved in the patenting process. In fact, in this review I have cited many papers from my lab as the reference list to this paper. The names of Mitchell, Daisy, Tomsheck, Schaible, Mends and Nigg had all worked as undergraduate students in my laboratory and the majority of these students were women. Each of these students has had one or more senior authored papers from one or more individual research projects and their names appear in the Reference section of this paper [13,24–28,55]. The concept of giving undergraduate students and opportunity to work on endophyte projects has been as rewarding for them as it has for me.

Acknowledgments: This research received no specific grant.

References

1. Walker, J.C. *Plant Pathology*; McGraw-Hill: New York, NY, USA, 1957.
2. ASM. *How Microbes Can Help Feed the World*; American Academy of Microbiology Colloquium Report; ASM: Washington, DC, USA, 2013.
3. Strobel, G.A.; Daisy, B. Bioprospecting for microbial endophytes and their natural products. *Microbiol. Mol. Biol. Rev.* **2003**, *67*, 491–502. [CrossRef] [PubMed]
4. Strobel, G.A. Methods of discovery and techniques to study endophytic fungi producing fuel-related hydrocarbons. *Nat. Prod. Rep.* **2014**, *39*, 259–272. [CrossRef] [PubMed]
5. Bacon, C.W.; White, J.F. *Microbial Endophytes*; Marcel Dekker Inc.: New York, NY, USA, 2000.
6. Jones, E.B.G.; Stanley, S.J.; Pinruan, U. Marine endophytes sources of new chemical natural products: A review. *Bot. Mar.* **2008**, *51*, 163–170. [CrossRef]
7. Davis, C.; Franklin, J.B.; Shaw, A.J. Biogeographic and phylogenetic patterns in diversity of liverwort-associated endophytes. *Am. J. Bot.* **2003**, *90*, 1661–1667. [CrossRef] [PubMed]
8. Redlin, S.C.; Carris, L.M. *Endophytic Fungi in Grasses and Woody Plants*; APS Press: St Paul, MN, USA, 1996.
9. Petrini, A.E.; Petrini, O. Xylarious fungi as endophytes. *Sydowia* **1985**, *38*, 216–234.
10. Faeth, S.H.; Fagan, W.F. Fungal endophytes: Common host plant symbionts but uncommon mutalists. *Integr. Comp. Biol.* **2002**, *42*, 360–368. [CrossRef] [PubMed]
11. Arnold, A.E.; Mejia, L.C.; Kyllo, D.; Rojas, E.I.; Maynard, Z.; Robbins, N.; Herre, E.A. Fungal endophytes limit pathogen damage in a tropical tree. *Proc. Natl. Acad. Sci. USA* **2003**, *100*, 15649–15654. [CrossRef] [PubMed]
12. Redman, R.S.; Sheehan, K.B.; Stout, R.; Rodriguez, R.J.; Henson, J. Thermotolerance generated by plant/fungal symbiosis. *Science* **2002**, *298*, 1581. [CrossRef] [PubMed]
13. Tomsheck, A.; Strobel, G.A.; Booth, E.; Geary, B.; Spakowicz, D.; Knighton, B.; Floerchinger, C.; Sears, J.; Liarzi, O.; Ezra, D. *Hypoxylon* sp. an endophyte of *Persea indica*, producing 1,8-cineole and other bioactive volatiles with fuel potential. *Microb. Ecol.* **2010**, *60*, 903–914. [CrossRef] [PubMed]

14. Strobel, G.A.; Ford, E.; Worapong, J.; Hess, W.M. *Pestalotiopsis jesterii* sp. nov. an endophyte from *Fragraea bodeni* from the southern highlands of Papua New Guinea. *Mycotaxon* **2000**, *76*, 257–266.
15. Metz, A.; Haddad, A.; Worapong, J.; Long, D.; Ford, E.; Hess, W.M.; Strobel, G.A. Induction of the sexual stage of *Pestalotiopsis microspora*, a taxol producing fungus. *Microbiology* **2000**, *146*, 2079–2089. [CrossRef] [PubMed]
16. Porras-Alfaro, A.; Bayman, P. Hidden fungi, emergent properties: Endophytes and microbiomes. *Ann. Rev. Phytopathol.* **2011**, *49*, 291–315. [CrossRef] [PubMed]
17. Strobel, G.A.; Ford, E.; Li, J.Y.; Sears, J.; Sidhu, R.S.; Hess, W.M. *Seimatoantlerium tepuiense* gen. nov., a unique epiphytic fungus producing taxol from the Venezuelan Guayana. *Syst. Appl. Microbiol.* **1999**, *22*, 426–433. [CrossRef]
18. Worapong, J.; Strobel, G.A.; Ford, E.J.; Li, J.Y.; Baird, G.; Hess, W.M. *Muscodor albus* gen. et sp. nov., an endophyte from *Cinnamomum zeylanicum*. *Mycotaxon* **2001**, *79*, 67–79.
19. Strobel, G.A.; Dirksie, E.; Sears, J.; Markworth, C. Volatile antimicrobials from a Novel Endophytic Fungus. *Microbiology* **2001**, *147*, 2943–2950. [CrossRef] [PubMed]
20. Mercier, J.; Jiménez, J.I. Control of fungal decay of apples and peaches by the biofumigant fungus *Muscodor albus*. *Postharvest Biol. Technol.* **2004**, *31*, 1–8. [CrossRef]
21. Mercier, J.; Smilanick, J.L. Control of green mold and sour rot of stored lemon by biofumigation with *Muscodor albus*. *Biol. Control* **2005**, *32*, 401–407. [CrossRef]
22. Meshram, V.; Kapoor, N.; Saxena, S. *Muscodor kashayum* sp. nov.—A new volatile anti-microbial producing endophytic fungus. *Mycology* **2013**, *4*, 196–204. [CrossRef] [PubMed]
23. Strobel, G.A. *Muscodor albus* and its biological promise. *Phytochem. Rev.* **2011**, *10*, 165–172. [CrossRef]
24. Daisy, B.; Strobel, G.; Ezra, D.; Castillo, U.; Baird, G.; Hess, W.M. *Muscodor vitigenus* sp. nov. an endophyte from *Paullinia paullinoides*. *Mycotaxon* **2002**, *84*, 39–50.
25. Daisy, B.H.; Strobel, G.A.; Castillo, U.; Sears, J.; Weaver, D.K.; Runyon, J.B. Naphthalene production by *Muscodor vitigenus*, a novel endophytic fungus. *Microbiology* **2002**, *148*, 3737–3741. [CrossRef] [PubMed]
26. Mitchell, A.M.; Strobel, G.A.; Hess, W.M.; Vargas, P.N.; Ezra, D. *Muscodor crispans*, a novel endophyte from *Ananas ananassoides* in the Bolivian Amazon. *Fungal Divers.* **2008**, *31*, 37–43.
27. Mitchell, A.M.; Strobel, G.A.; Moore, E.; Robison, R.; Sears, J. Volatile antimicrobials from *Muscodor crispans*. *Microbiology* **2010**, *156*, 270–277. [CrossRef] [PubMed]
28. Schaible, G.A.; Strobel, G.A.; Mends, M.T.; Geary, B.; Sears, J. Characterization of an endophytic *Gloeosporium* sp. and its novel bioactivity with "synergistans". *Microb. Ecol.* **2015**, *70*, 41–50. [CrossRef] [PubMed]
29. Leveille, L. Efficacy of a Food Grade Mixture of Volatile Compounds to Reduce Salmonella Levels on Food Contact Surfaces. Ph.D. Thesis, University of Georgia Graduate School, Athens, GA, USA, 2016. Available online: http://purl.galileo.usg.edu/uga_etd/leveille_laurie_201605_ms (accessed on 1 April 2018).
30. Meshram, V.; Saxena, S.; Kapoor, N. Muscodor strobelii, a new endophytic species from South India. *Mycotaxon* **2014**, *128*, 93–104. [CrossRef]
31. Lyons, P.C.; Plattner, R.D.; Bacon, C.W. Occurrence of peptide and clavine ergot alkaloids in tall fescue grass. *Science* **1986**, *232*, 487–489. [CrossRef] [PubMed]
32. Strobel, G.A.; Stierle, A.; Stierle, D.; Hess, W.M. *Taxomyces andreanae*, a proposed new taxon for a bulbilliferous hyphomycete associated with Pacific yew. *Mycotaxon* **1993**, *47*, 71–78.
33. Stierle, A.; Strobel, G.; Stierle, D. Taxol and taxane production by *Taxomyces andreanae*, an endophytic fungus of Pacific yew. *Science* **1993**, *260*, 154–155. [CrossRef]
34. Young, D.H.; Michelotti, E.J.; Sivendell, C.S.; Krauss, N.E. Antifungal properties of taxol and various analogues. *Experientia* **1992**, *48*, 882–885. [CrossRef] [PubMed]
35. Strobel, G.A. Rainforest endophytes and bioactive products. *Crit. Rev. Biotechnol.* **2002**, *22*, 315–333. [CrossRef] [PubMed]
36. Strobel, G.A.; Daisy, B.; Castillo, U.; Harper, J. Natural products from endophytic fungi. *J. Nat. Prod.* **2004**, *67*, 257–268. [CrossRef] [PubMed]
37. Guo, B.; Wang, Y.; Sun, X.; Tang, K. Bioactive natural products from endophytes: A review. *Appl. Biochem. Microbiol.* **2008**, *44*, 136–144. [CrossRef]
38. Verma, V.C.; Kharwar, R.N.; Strobel, G.A. Chemical and functional diversity of natural products from plant associated endophytic fungi. *Nat. Prod. Commun.* **2009**, *4*, 1511–1532. [PubMed]

39. Gouda, S.; Das, G.; Sen, S.; Shin, H.S.; Patra, J.K. Endophytes: A treasure house of bioactive compounds of medicinal importance. *Front. Microbiol.* **2016**, *7*, 1538–1548. [CrossRef] [PubMed]
40. Harper, J.K.; Ford, E.J.; Strobel, G.A.; Arif, A.; Grant, D.M.; Porco, J.; Tomer, D.P.; Oneill, K. Pestacin: A 1,3-dihydro isobenzofuran from *Pestalotiopsis microspora* possessing antioxidant and antimycotic activities. *Tetrahedron* **2003**, *59*, 2471–2476. [CrossRef]
41. Strobel, G.A.; Ford, E.J.; Worapong, J.; Harper, J.K.; Arif, A.M.; Grant, D.M.; Chau, R.M.W. Isopestacin, a unique isobenzofuranone from *Pestalotiopsis microspora* possessing antifungal and antioxidant properties. *Phytochemistry* **2002**, *60*, 179–183. [CrossRef]
42. Li, J.Y.; Harper, J.K.; Grant, D.M.; Tombe, B.O.; Bashyal, B.; Hess, W.M.; Strobel, G.A. Ambuic acid, a highly functionalized cyclohexenone with bioactivity from *Pestalotiopsis* spp. and *Monochaetia* sp. *Phytochemistry* **2001**, *56*, 463–468. [CrossRef]
43. Harper, J.K.; Li, J.Y.; Grant, D.M.; Strobel, G.A. Characterization of stereochemistry and molecular conformation using solid-state NMR tensors. *J. Am. Chem. Soc.* **2001**, *123*, 9837–9842. [CrossRef] [PubMed]
44. Harper, J.K.; Barich, D.H.; Hu, J.Z.; Strobel, G.A.; Grant, D.M. Stereochemical Analysis by Solid-State NMR: Structural Predictions in Ambuic Acid. *J. Org. Chem.* **2003**, *68*, 4609–4614. [CrossRef] [PubMed]
45. Nakayama, J.; Uemura, Y.; Nishiguchi, K.; Yoshimura, N.; Igarashi, Y.; Sonomoto, K. Ambuic Acid Inhibits the Biosynthesis of Cyclic Peptide Quormones in Gram-Positive Bacteria. *Antimicrob. Agents Chemother.* **2009**, *53*, 580–586. [CrossRef] [PubMed]
46. Lee, J.C.; Strobel, G.A.; Lobkovsky, E.; Clardy, J.C. Torreyanic acid: A selectively cytotoxic quinone dimer from the endophytic fungus *Pestalotiopsis microspora*. *J. Org. Chem.* **1996**, *61*, 3232–3233. [CrossRef]
47. Ren, Y.; Strobel, G.A.; Graff, J.C.; Jutila, M.; Park, S.G.; Gosh, S.; Moore, E. Colutellin A, an immunosuppressive peptide from *Colletotrichum dematium*. *Microbiology* **2008**, *154*, 1973–1979. [CrossRef] [PubMed]
48. Li, J.; Strobel, G.A.; Harper, J.; Lobkovsky, E.; Clardy, J. Cryptocin, a potent tetramic acid antimycotic from the endophytic fungus *Cryptosporiopsis quercina*. *Org. Lett.* **2000**, *2*, 767–770. [CrossRef] [PubMed]
49. Strobel, G.A.; Knighton, B.; Kluck, K.; Ren, Y.; Livinghouse, T.; Griffen, M.; Spakowicz, D.; Sears, J. The production of myco-diesel hydrocarbons and their derivatives by the endophytic fungus *Gliocladium roseum* (NRRL 50072). *Microbiology* **2008**, *154*, 3319–3328. [CrossRef] [PubMed]
50. Strobel, G.A. The story of mycodiesel. *Curr. Opin. Microbiol.* **2014**, *19*, 52–58. [CrossRef] [PubMed]
51. Mallette, N.D.; Pankrantz, E.M.; Busse, S.; Strobel, G.A.; Carlson, R.P.; Peyton, B. Evaluation of cellulose as a substrate for hydrocarbon fuel production by *Ascocoryne sarcoides* (NRRL 50072). *J. Sustain. Bioenergy Syst.* **2014**, *4*, 33–49. [CrossRef]
52. Yu, E.T.; Tran-Gyamfi, M.; Strobel, G.A.; Taatjes, C.; Hadi, M.Z. *VOC Profile of Endophytic Fungi Is Altered by Nature of Lignocellulosic Biomass Feedstock*; NASA Report; USA-National Aeronautics and Space Administration: Washington, DC, USA, 2014.
53. Strobel, G.A.; Booth, E.; Schaible, G.; Mends, M.T.; Sears, J.; Geary, B. The Paleobiosphere: A novel device for the in vivo testing of hydrocarbon production utilizing microorganisms. *Biotechnol. Lett.* **2013**, *35*, 539–552. [CrossRef] [PubMed]
54. Strobel, G.A.; Tomsheck, A.; Geary, B.; Spakowicz, D.; Strobel, S.; Mattner, S.; Mann, R. Endophytic strain NRRL 50072 producing volatile organics is a species of Ascocoryne. *Mycology* **2010**, *1*, 187–194. [CrossRef]
55. Mends, M.T.; Yu, E.; Strobel, G.A.; Hassan, S.R.U.; Booth, E.; Geary, B.; Sears, J.; Taatjes, C.A.; Hadi, M. An endophytic *Nodulisporium* sp. producing volatile organic compounds having bioactivity and fuel potential. *J. Pet. Environ. Biotechnol.* **2012**. [CrossRef]
56. Shaw, J.J.; Berbasova, T.; Sasaki, T.; Jefferson-George, K.; Spakowicz, D.J.; Dunican, B.F.; Strobel, S.A. Identification of a fungal 1,8-cineole synthase from *Hypoxylon* sp. with specificity determinants in common with the plant synthases. *J. Biol. Chem.* **2015**, *290*, 8511–8526. [CrossRef] [PubMed]
57. Gianolis, T.A.; Griffin, M.A.; Spakowicz, D.J.; Dunican, B.F.; Sboner, A.; Sismour, A.M.; Strobel, S.A. Genomic analysis of the hydrocarbon-producing celluloytic fungus *Ascocoryne sarcoides*. *PLoS Genet.* **2012**, *8*, e1002558. [CrossRef]
58. Barton, A.; Tjandra, J. Eucalyptus oil as a cosolvent in water-ethanol-gasoline mixtures. *Fuel* **1989**, *68*, 11–17. [CrossRef]
59. Sugito, K.; Takeda, S. Fuel Composition. Patent No. 4,297,109, 27 October 1981.

60. Schoen, H.R.; Peyton, B.M.; Knighton, W.B. Rapid total volatile organic carbon quantification from microbial fermentation using a platinum catalyst and proton transfer reaction-mass spectrometry. *AMB Express* **2016**, *6*, 90. [CrossRef] [PubMed]
61. Pinkerton, F.; Strobel, G.A. Serinol as an activator of toxin production in attenuate cultures of *H. sacchari*. *Proc. Natl. Acad. Sci. USA* **1976**, *73*, 4007–4011. [CrossRef] [PubMed]
62. Matern, U.; Beier, R.C.; Strobel, G.A. A novel plant glycolipid serves as an activator of toxin production in *H. sacchari*. *Biochem. Int.* **1982**, *4*, 655–661.
63. Li, J.Y.; Sidhu, R.S.; Ford, E.; Hess, W.M.; Strobel, G.A. The induction of taxol production in the endophytic fungus—*Periconia* sp. from *Torreya grandifolia*. *J. Ind. Microbiol.* **1998**, *20*, 259–264. [CrossRef]
64. Nigg, J.; Strobel, G.A.; Knighton, B.; Hilmer, J.; Geary, B.; Hassan, S.R.; Harper, J.; Valenti, D.; Wang, Y. Functionalize para-substituted. Benzene as 1,8-cineole production modulators (CPMs) in an endophytic *Nodulisporium* sp. *Microbiology* **2014**, *160*, 1772–1782. [CrossRef] [PubMed]
65. Hassan, R.; Strobel, G.A.; Booth, E.; Knighton, B.; Floerchinger, C.; Sears, J. Modulation of Volatile Organic Compound Formation in the Mycodiesel producing endophyte *Hypoxylon* sp. CI-4. *Microbiology* **2012**, *158*, 465–473. [CrossRef] [PubMed]
66. Cichewicz, R.H. Epigenome manipulation as a pathway to new natural product scaffolds and their congeners. *Nat. Prod. Rep.* **2010**, *27*, 11–22. [CrossRef] [PubMed]
67. Williams, R.B.; Henrikson, J.C.; Hoover, A.R.; Lee, A.E.; Cichewicz, R.H. Epigenetic remodeling of the fungal secondary metablome. *Org. Biomol. Chem.* **2008**, *6*, 1895–1897. [CrossRef] [PubMed]
68. Arnold, A.E.; Maynard, Z.; Gilbert, G. Fungal endophytes in dicotyledonous neotropical trees: Patterns of abundance and diversity. *Mycol. Res.* **2001**, *105*, 1502–1507. [CrossRef]
69. Arnold, A.E.; Herre, E.A. Canopy cover and leaf age affect colonization by tropical fungal endophytes: Ecological pattern and process in *Theobroma cacao* (Malvaceae). *Mycologia* **2003**, *93*, 388–398. [CrossRef]
70. Rodriguez, R.; White, J.; Arnold, A.E.; Redman, R. Fungal endophytes: Diversity and ecological roles. *New Phytol.* **2009**, *182*, 314–330. [CrossRef] [PubMed]
71. Murphy, B.R.; Doohan, F.M.; Hodkinson, T.R. From concept to commerce: Developing a successful fungal endophyte inoculant for agricultural crops. *J. Fungi* **2018**, *4*, 24. [CrossRef] [PubMed]
72. Oflek-Lalzar, M.; Gur, Y.; Ben-Moshe, S.; Sharon, O.; Kosman, E.; Mochli, E.; Sharon, A. Diversity of fungal endophytes in recent and ancient wheat ancestors *Triticum dicoccoides* and *Aegilops sharonensis*. *FEMS Microbiol. Ecol.* **2016**, *92*. [CrossRef]
73. Strobel, S.A.; Strobel, G.A. Plant endophytes as a platform for discovery-based undergraduate science education. *Nat. Chem. Biol.* **2007**, *3*, 356–359. [CrossRef] [PubMed]

Antiplasmodial Properties and Cytotoxicity of Endophytic Fungi from *Symphonia globulifera* (Clusiaceae)

Joël E. T. Ateba [1], Rufin M. K. Toghueo [2], Angelbert F. Awantu [3], Brice M. Mba'ning [1,*], Sebastian Gohlke [4], Dinkar Sahal [5], Edson Rodrigues-Filho [6], Etienne Tsamo [1], Fabrice F. Boyom [2], Norbert Sewald [4] and Bruno N. Lenta [7,*]

[1] Department of Organic Chemistry, Faculty of Science, University of Yaoundé 1, P.O. Box 812, 237 Yaoundé, Cameroon; atebaterence@yahoo.fr (J.E.T.A.); etsamo@yahoo.fr (E.T.)

[2] Department of Biochemistry, Faculty of Science, University of Yaoundé I, Cameroon; P.O. Box 812, 237 Yaoundé, Cameroon; toghueo.rufin@yahoo.fr (R.M.K.T.); fabrice.boyom@fulbrightmail.org (F.F.B.)

[3] Department of Chemistry, Faculty of Science, University of Bamenda, P.O. Box 39, 237 Bambili, Cameroon; aawantu@gmail.com

[4] Department of Chemistry, Organic and Bioorganic Chemistry, Bielefeld University, P.O. Box 100131, D-33501 Bielefeld, Germany; sebastian.gohlke@uni-bielefeld.de (S.G.); norbert.sewald@uni-bielefeld.de (N.S.)

[5] International Centre for Genetic Engineering and Biotechnology, Aruna Asaf Ali Marg, New Delhi 110067, India; dsahal@gmail.com

[6] Departamento de Química-Universidade Federal de São Carlos-CP 676, São Carlos-SP 13565-905, Brazil; edinho@pq.cnpq.br

[7] Department of Chemistry, Higher Teacher Training College, University of Yaoundé 1, P.O. Box 47, 237 Yaoundé, Cameroon

* Correspondence: brice_mbaning@yahoo.fr (B.M.M.); lentabruno@yahoo.fr (B.N.L.)

Abstract: There is continuing need for new and improved drugs to tackle malaria, which remains a major public health problem, especially in tropical and subtropical regions of the world. Natural products represent credible sources of new antiplasmodial agents for antimalarial drug development. Endophytes that widely colonize healthy tissues of plants have been shown to synthesize a great variety of secondary metabolites that might possess antiplasmodial benefits. The present study was carried out to evaluate the antiplasmodial potential of extracts from endophytic fungi isolated from *Symphonia globulifera* against a chloroquine-resistant strain of *Plasmodium falciparum (PfINDO)*. Sixty-one fungal isolates with infection frequency of 67.77% were obtained from the bark of *S. globulifera*. Twelve selected isolates were classified into six different genera including *Fusarium*, *Paecilomyces*, *Penicillium*, *Aspergillus*, *Mucor*, and *Bipolaris*. Extracts from the 12 isolates were tested against *PfINDO*, and nine showed good activity ($IC_{50} < 10\ \mu g \cdot mL^{-1}$) with three fungi including *Paecilomyces lilacinus* ($IC_{50} = 0.44\ \mu g \cdot mL^{-1}$), *Penicillium janthinellum* ($IC_{50} = 0.2\ \mu g \cdot mL^{-1}$), and *Paecilomyces* sp. ($IC_{50} = 0.55\ \mu g \cdot mL^{-1}$) showing the highest promise. These three isolates were found to be less cytotoxic against the HEK293T cell line with selectivity indices ranging from 24.52 to 70.56. Results from this study indicate that endophytic fungi from *Symphonia globulifera* are promising sources of hit compounds that might be further investigated as novel drugs against malaria. The chemical investigation of active extracts is ongoing.

Keywords: malaria; chloroquine-resistant *Plasmodium falciparum*; *Symphonia globulifera*; endophytic fungi; antiplasmodial IC_{50}; selectivity index

1. Introduction

Malaria remains a major cause of morbidity and mortality with more than 3.3 billion people living worldwide in countries with ongoing transmission at risk [1]. In 2016, 91 countries reported a total of 216 million cases of malaria and 445,000 deaths. The most affected populations were located in tropical and subtropical regions of the world, particularly in sub-Saharan Africa and Southeast Asia, where almost 80% of malaria cases are caused by *Plasmodium falciparum* [1]. The widespread appearance of drug-resistant malaria parasites, even to newly developed second and third generation therapeutics such as artemisinin and its derivatives, makes the development of novel antimalarial drug treatments all the more urgent [2].

Natural products remain a consistent source of antimalarial drug leads, with the best examples being quinine and artemisinin that have acted as gifts of medicinal plants for victims of malaria. Therefore, investigating natural resources for antiplasmodial drug discovery continues to be one of the best scientific approaches that can lead to the identification of novel lead compounds against malaria [3]. Moreover, in comparison to other natural sources like plants, microorganisms are highly diverse but narrowly explored. Indeed, microbes often coexist with plants and animals making it difficult to assess if the real hosts of the promising metabolites identified are from higher organisms themselves or the tiny microbes residing in them. However, studies based on estimation of microbial populations have revealed that only about 1% of bacteria and 5% of fungi have been characterized and the rest remain unexplored for their potential to offer novel drugs against diverse diseases [4]. One of these groups being explored in recent times for their drug discovery potential is endophytic fungi from medicinal plants.

In fact, during the last 20 years it has been observed that much of the wealth of microbial biodiversity with novel biochemistry and secondary metabolite production resides in endophytic association with plant tissues [5]. Interest in such microorganisms, termed endophytes, increased immensely with the discovery of an endophytic fungus, from *Taxus brevifolia*, producing the billion dollar anti-cancer drug taxol [6]. Numerous bioactive molecules have been isolated from endophytic fungi since this groundbreaking discovery [7,8]. Endophytes are metabolically more active than their free counterparts due to their specific functions in nature and activation of various metabolic pathways needed to survive in the host tissues [4,9,10]. Therefore, investigation of endophytic fungi from medicinal plants used to treat malaria can lead to new antimalarial drug discovery.

Symphonia globulifera L. f., an evergreen tree of the Clusiaceae family, is widely distributed in Central and South America, and in tropical Africa, from Sierra Leone to Uganda, Zambia, and Angola. In traditional medicine, a decoction of leaves from this plant is used to manage malaria and several other diseases, including diabetes and skin diseases [11]. In addition, decoctions and extracts of bark are used as diuretics and antiparasitics, and to treat river blindness, chest complaints, cough in children, gonorrhea, scabies, intestinal worms, and prehepatic jaundice [12,13]. Therefore, the present study was designed to investigate for the first time antiplasmodial activity of endophytic fungi isolated from the bark of *S. globulifera*.

2. Materials and Methods

2.1. Collection of Plant Material

Healthy and mature stem bark of *Symphonia globulifera* L. f. was collected at Nomayos in the Centre Region of Cameroon in January 2015, and was identified at the National Herbarium of Cameroon, where a voucher specimen was deposited under the number 32192/HNC. Plant material was directly brought to the laboratory in sterile bags and processed within a few hours after sampling.

2.2. Isolation of Endophytic Fungi

Stem bark was rinsed with tap water and cut into small pieces, approximately 5 mm length. Sterilization was assessed by means of a 5 min rinse with ethanol, followed by treatment with a 1%active chlorine solution for 15 min, 2 min in ethanol, and a final rinse in sterile water [14].

After surface sterilization, six plates of potato dextrose agar (PDA) supplemented with chloramphenicol (200 mg·L^{-1}), each containing 15 pieces of individual plant material, were prepared and kept in the dark at room temperature (22–26 °C). Fragments of mycelium emerging from plant pieces were transferred to new PDA plates without chloramphenicol to obtain pure cultures for identification. To assess whether disinfection methods were effective in eliminating surface fungi, imprints of treated fragments were made by pressing them against the surface of fresh PDA plates that were incubated without plant parts. These plates were checked for fungi emerging from the prints [14].

2.3. Identification of Endophytic Fungi

Fungal cultures were maintained at room temperature (i.e., 22–26 °C) under a natural photoperiod for 10–21 days and then examined visually for macroscopic (i.e., morphology, size, and coloration of the mycelium and agar media) and microscopic (i.e., presence of spores or other reproductive structures) characteristics. Colonies were analyzed with respect to their average diameter, sporulation, and the size and coloration of the conidia. Isolates having similar culture characteristics were grouped into morphotypes. From morphotype groups, only one isolate was processed for further identification.

Identification of endophytic isolates representative of each morphotype was based on the nucleotide sequence of the ITS1-5.8S rRNA-ITS2 region. DNA was extracted from samples of mycelium from the fungal cultures using a commercial kit (RedExtract-N-Amp Plant PCR, Sigma Aldrich, USA). The ITS1-5.8S rRNA-ITS2 region was amplified in a polymerase chain reaction (PCR) using primers ITS4 and ITS5 and the protocol described by White et al. [15]. Amplicons were purified by filtration (MSB Spin PCRapace, Invitek, Germany) and sequenced.

The FASTA algorithm was used to find sequences similar to those obtained from fungal isolates. The criteria for identification of isolates were based on the similarity of their sequences to those of reliable reference isolates included in public nucleotide databases. To visualize the diverse fungal taxa identified sequences, a dendrogram was made with the ITS1-5.8S rRNA-ITS2 nucleotide sequences of the isolates and those of reference strains deposited in CBS (Centraalbureau voor Schimmelcultures), ATCC (American Type Culture Collection), or other fungal collections, as well as some used in published works of fungal taxonomy. Sequences were aligned using the Clustal X 2.1, and the dendrogram was made with MEGA 6.06 software using the neighbor-joining method with Kimura 2-parameter distances. Groups of sequences at close proximity within the same branch of the dendrogram were individually aligned with Clustal X 2.1 to determine their percentage of similarity. Sequences with a similarity greater than 99% were considered to belong to the same species [14].

2.4. Fermentation and Extraction

Each fungus was cultivated on 2 kg of rice by placing agar blocks of actively growing pure culture (3 mm in diameter) in a 1000 mL Erlenmeyer flask. Each flask was incubated at 25 ± 2 °C for 30 days. After incubation, moldy rice was macerated with ethyl acetate and filtered through three layers of muslin cloth. The organic phase was collected and the solvent was then removed by evaporation under reduced pressure at 40 °C using a rotary vacuum evaporator. The dry solid residues were examined for their antiplasmodial activity and cytotoxicity.

2.5. Biological Assays

2.5.1. In Vitro Cultivation of *Plasmodium falciparum*

A chloroquine-resistant *PfINDO* strain of *P. falciparum* was maintained in a continuous culture following the method of Trager & Jensen, 1976 [16] with minor modifications. Cultures were maintained in fresh O positive human erythrocytes suspended at 4% (*v/v*) haematocrit in complete medium (16.2 g·L^{-1} RPMI 1640 (Sigma) containing 25 mM HEPES, 11.11 mM glucose, 0.2% sodium bicarbonate (Sigma), 0.5% Albumax I (Gibco), 45 mg·L^{-1} hypoxanthine (Sigma), and 50 mg·L^{-1} gentamicin (Gibco)) and incubated at 37 °C in a gas mixture consisting of 5% O_2, 5% CO_2, and 90% N_2. The spent medium was replaced with fresh complete medium every day to propagate the culture. Giemsa-stained blood smears were examined microscopically to monitor cell cycle transitions and parasitemia.

2.5.2. Preparation of Stock Solution of Chloroquine and Fungal Extracts

Stock solutions of fungal extracts were prepared in Dimethyl sulfoxide (DMSO) at 25 mg·mL^{-1} and that of chloroquine phosphate (CQ) (Sigma) was prepared at 1 mM in Milli-Q grade water. The required drug concentrations were achieved by diluting the stocks with incomplete RPMI (Roswell Park Memorial Institute) 1640 medium. The solutions of drugs and extracts were placed in 96-well flat-bottom tissue culture grade plates (Corning). For each extract, the concentrations (μg·mL^{-1}) tested were 0.195, 0.39, 0.78, 1.562, 3.125, 6.25, 12.5, 25, 50, and 100.

2.5.3. In Vitro Antiplasmodial Assay

Before each experiment, synchronized ring stage parasites were obtained by 5% (*w/v*) sorbitol treatment [17]. It is important to note that use of synchronized cultures over mixed-stage cultures can enable the test molecules to interact with all three stages (i.e., ring, trophozoite, and schizont) of the 48 h long life cycle of *P. falciparum* in culture. Moreover, starting the experiment with synchronized ring stage culture provides the distinct advantage of observing growth inhibitory effects without a rise in parasitemia during the ring-trophozoite-schizont transitions.

For drug screening, the SYBR green I based fluorescence assay was used [18]. The ability of SYBR green to give strong fluorescence only in the presence of DNA forms the basis for its use to assess cell proliferation. The absence of a nucleus in human red blood cells where the malaria parasite proliferates allows the use of SYBR green for the specific monitoring of the growth of malarial parasite.

Sorbitol-synchronized ring stage parasites (haematocrit: 2%, parasitemia: 1%, 96 μL) under normal culture conditions were incubated in the presence or absence of increasing concentrations of the extracts. Four μL CQ (1 mM) was used as a positive control and 0.4% DMSO (*v/v*), which was found to be non-toxic to the parasite, was used as vehicle control. After 48 h of incubation, 100 μL of SYBR Green I buffer {0.2 μL of 10,000 × SYBR Green I (Invitrogen) per mL of lysis buffer (Tris (20 mM; pH 7.5), EDTA (5 mM), saponin (0.008%; *w/v*), and Triton X-100 (0.08%; *v/v*)} was added to each well, mixed twice gently with multi-channel pipette and incubated in the dark at 37 °C for 1 h. Fluorescence was measured using a Victor fluorescence multi-well plate reader (Perkin Elmer) with excitation and emission at 485 and 530 nm, respectively. Fluorescence counts for CQ representing zero growth were deducted from counts in each well. A dose–response curve was constructed by plotting fluorescence counts against the drug concentration and IC_{50} (dose of a drug required to retard the growth of a cell population by 50%) was determined using IC Estimator-version 1.2 (http://www.antimalarial-icestimator.net/MethodIntro.htm). In this experiment, no drug (control) corresponds to 100% growth while 40 μM chloroquine (sufficient to cause total arrest of growth) corresponds to 0% growth. Giemsa-stained smears of extract-treated parasite cultures were visualized microscopically to validate the results from fluorescence-based assay. Experiments were done in triplicate and means with standard deviation were calculated. Extracts were classified according to criteria of the antiplasmodial activity based upon good ($IC_{50} < 10$ μg·mL^{-1}), moderate ($IC_{50} > 10$ to < 25 μg·mL^{-1}), and inactive ($IC_{50} > 25$ μg·mL^{-1}) activity as proposed by Bagavan et al. [19].

2.5.4. Cytotoxicity Assay

The cytotoxic effects of the most potent crude extracts were determined by a functional assay [20] using the human embryonic kidney HEK239T cells cultured in complete medium containing 16.2 g·L^{-1} DMEM, 10% fetal bovine serum, 0.2% sodium bicarbonate (*w*/*v*) (Sigma), and 50 μg·mL^{-1} gentamycin. Cells (5 × 10^3 cells/200 μL/well) were seeded into 96-well flat-bottom tissue culture plates in complete medium. After 24 h of cells seeding, test extracts were added and cells incubated for 48 h in a humidified atmosphere at 37 °C and 5% CO_2. DMSO (as positive inhibitor) was added at 10% *v*/*v*. Twenty microliters of a stock solution of MTT [3-(4,5-dimethylthiazol-2-yl)-2,5-diphenyltetrazolium bromide] (5 mg·mL^{-1} in 1× phosphate buffered saline) was added to each well, gently mixed, and incubated for another 4 h. The supernatant was thereafter removed and 100 μL DMSO (quench agent) was added to the cell pellet. Formazan formation was measured using a microtiter plate reader (VersaMax tunable multiwell plate reader) at 570 nm. The 50% cytotoxic concentration (CC_{50}) of drugs was determined by analysis of dose–response curves. Selectivity index (CC_{50}/IC_{50}) was calculated for each extract (tested in triplicate).

3. Results

3.1. Isolation and Identification of Fungi

From a total of 90 fragments of the stem bark of *Symphonia globulifera*, 61 isolates were obtained with an infection frequency of 67.77%. They were subsequently grouped into 14 morphotypes and one isolate from each morphotype was identified by sequence analysis of the Internal Transcribed Spacer (ITS) region. Twelve analyzed isolates were classified into six different genera including, *Fusarium*, *Paecilomyces*, *Penicillium*, *Aspergillus*, *Mucor*, and *Bipolaris* (Table 1; Figure 1). The most representative genera were *Fusarium* and *Penicillium* (Table 1).

Table 1. Identification of endophyte isolates and percentage homology of sequences.

Endophyte Isolates	% Sequence Homology	Organism with the Highest Sequence Identity, GenBank Acc. No.
Fusarium decemcellulaire	98	*Fusarium decemcellulaire* KM231809
Paecilomyces sp.	100	*Paecilomyces* sp. FJ196762
Fusarium oxysporum	100	*Fusarium oxysporum* KT794176
Paecilomyces lilacinus	100	*Paecilomyces lilacinus* GU980023
Penicillium janthinellum	99	*Penicillium janthinellum* AY373921
Aspergillus tamarri	100	*Aspergillus tamarri* HQ340111
Penicilium sp. (1)	99	*Penicillium* sp. JN021538
Bipolaris sorokiniana	94	*Bipolaris sorokiniana* KU194490
Fusarium lateritium	94	*Fusarium lateritium* KC453998
Penicillium sp. (2)	98	*Penicillium* sp. JN021538
Aspergillus aculeatus	100	*Aspergillus aculeatus* AY585556
Mucor falcatus	89	*Mucor falcatus* NR103647

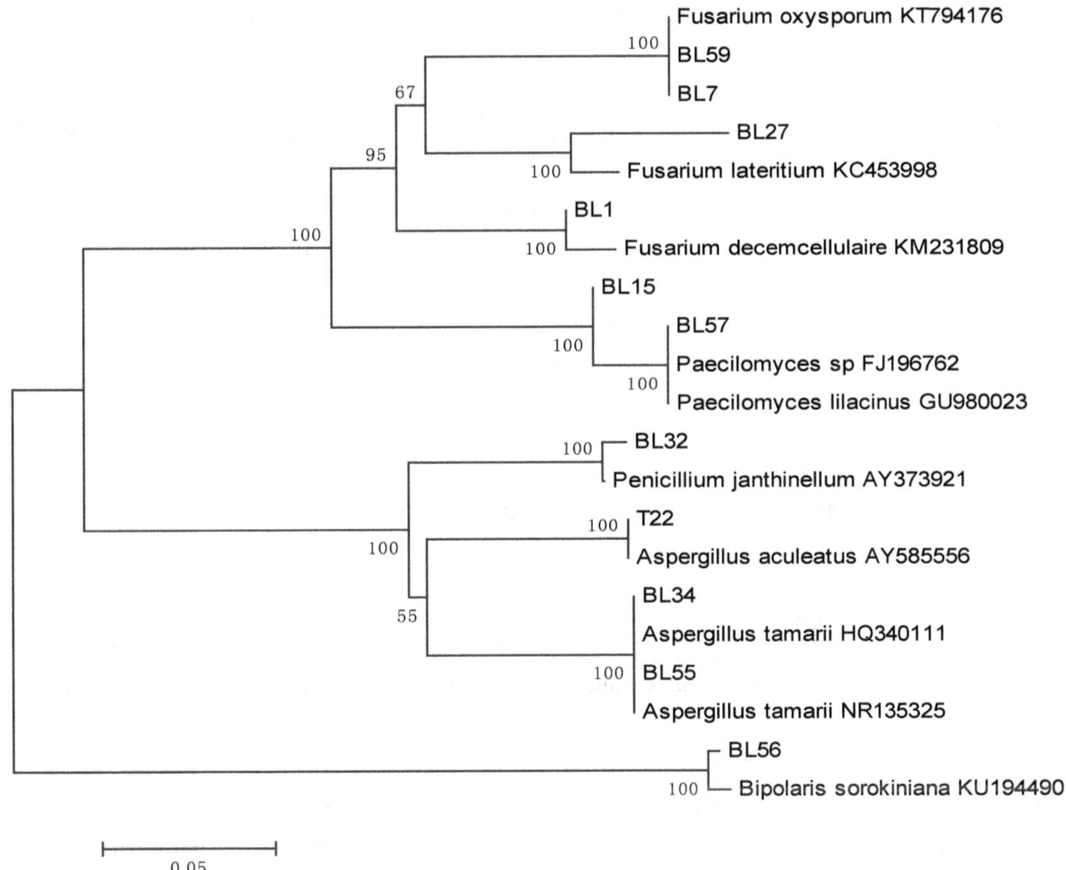

Figure 1. Classification of fungal endophytes isolated from *Symphonia globulifera* based on a neighbor-joining phylogenetic tree made with ITS1-5.8S rDNA-ITS2 nucleotide sequences. Taxa with accession numbers correspond to reference strains. Numbers at branch nodes are bootstrap values based on 500 replications.

3.2. *In Vitro Antiplasmodial Activity of Endophytic Extracts*

Twelve endophytic fungi were cultured in a rice medium at 25 ± 2 °C for 30 days and subsequently extracted using ethyl acetate as a solvent. The extraction yields ranged from 6.30 g to 21.70 g with *A. tamarri* producing the higher yield.

The IC_{50} of the tested extracts against the chloroquine-resistant *Plasmodium falciparum* INDO (*PfINDO*) strain ranged from 0.2 to >100 $\mu g \cdot mL^{-1}$ (Table 2). Out of the 12 tested extracts, nine showed good activity ($IC_{50} < 10$ $\mu g \cdot mL^{-1}$). Extracts from *Paecilomyces lilacinus, Penicillium janthinellum*, and *Paecilomyces* sp. exerted highly potent activities with $IC_{50} < 1$ $\mu g/mL$. Extracts from *Mucor falcatus* and *Aspergillus aculeatus* showed moderate potency with IC_{50} between 10 and 25 $\mu g \cdot mL^{-1}$, while the extract from *Aspergillus tamarri* with $IC_{50} > 100$ $\mu g \cdot mL^{-1}$ was considered as inactive.

3.3. *Cytotoxicity of Extracts*

The cytotoxicity of the extracts was tested against HEK239T cells. As shown in Table 2, CC_{50} values ($\mu g\ mL^{-1}$) ranged from 3.06 to 70.91 leading selectivity indices to lie between 0.5 and 70.56. *Penicillium* sp. (2). (SI = 0.69), *Bipolaris sorokiniana* (SI = 0.504), and *Penicilliun* sp. (1) (SI = 0.842) were found to be highly cytotoxic against HEK239T cells. Conversely, the most potent fungal extracts from *Paecilomyces lilacinus* (CC_{50} = 10.79 $\mu g\ mL^{-1}$; SI = 24.52), *Penicillium janthinellum* (CC_{50} = 9.14 $\mu g\ mL^{-1}$; SI = 45.7), and *Paecilomyces* sp. (CC_{50} = 38.88 $\mu g\ mL^{-1}$; SI = 70.56) were less cytotoxic.

Table 2. Antiplasmodial and cytotoxic activities of extracts of endophytic fungi from the bark of *Symphonia globulifera*.

Sample	Dry Mass of Extract (g)	IC_{50} against *PfINDO* Strain ($\mu g \cdot mL^{-1} \pm SD$)	CC_{50} against *HEK239T* Cells ($\mu g \cdot mL^{-1} \pm SD$)	SI (CC_{50}/IC_{50})
Fusarium decemcellulaire	10.81	2.19 ± 0.07	70.91 ± 0.18	32.37
Paecilomyces sp.	14.66	0.55 ± 0.01	38.81 ± 0.19	70.56
Fusarium oxysporum	8.23	1.70 ± 0.22	18.24 ± 0.11	10.729
Paecilomyces lilacinus	6.20	0.44 ± 0.03	10.79 ± 0.2	24.52
Penicillium janthinellum	6.30	0.20 ± 0.01	9.14 ± 0.12	45.7
Aspergillus tamarri	21.70	>100	***	NA
Penicilium sp.	14.57	3.63 ± 0.02	3.06 ± 0.36	0.842
Bipolaris sorokiniana	9.89	6.10 ± 0.37	3.08 ± 0.19	0.504
Fusarium lateritium	11.67	6.61± 0.01	2.82± 0.23	0.426
Penicillium sp.(2)	16.32	9.08 ± 0.13	6.28 ± 1.23	0.69
Aspergillus aculeatus	8.74	22.08 ± 0.43	***	NA
Mucor falcatus	9.22	17.51 ± 0.19	***	NA
Chloroquine	NA	*400	***	NA

Data are presented as mean values ± standard deviation of triplicate experiments; SD: standard deviation; ***: not tested; NA: not applicable. SI: Selectivity index (ratio of CC_{50} to IC_{50}); * IC_{50} of chloroquine given in nM.

4. Discussion

Endophytic fungi have been claimed to be responsible for the medicinal properties of several medicinal plants [21]. Therefore, in this study we have explored the antiplasmodial potential of endophytic fungi associated with *Symphonia globulifera* used in several African countries for treatment of malaria. More specifically, towards discovery of new drugs against malaria, this study was done to perform antiplasmodial screening of endophytic fungi obtained from the bark of this plant.

A high prevalence of endophyte infection of 67.77% was found in bark fragments of *Symphonia globulifera*. This high level of endophyte colonization is corroborated by previous findings which indicated that endophyte prevalence in some tissues of Cameroonian medicinal plants is quite high [14]. Moreover, Arnold and Lutzoni [22] reported that the prevalence of endophyte species in plant tissues of tropical regions is greater than in plant species of temperate and boreal forests. The identification based on sequence analysis indicated that all of the analyzed endophytic isolates obtained from bark of *S. globulifera* belong to six different genera including *Fusarium, Paecilomyces, Penicillium, Aspergillus, Mucor,* and *Bipolaris*. All these genera have also been reported in previous endophyte surveys in different plant species [14,23–26].

Although only 12 endophyte isolates were screened in the present study, as many as nine of them produced antiplasmodial compounds [27,28]. The finding that as high as 75% gave good activity (IC_{50} < 10 $\mu g \cdot mL^{-1}$) with 33.3% (3) of this portion showing highly potent antiplasmodial activity (IC_{50} < 1 $\mu g \cdot mL^{-1}$) suggests that deeper exploration of endophytic fungi may reveal a rich repertoire of antiplasmodial molecules. Previous studies reported that endophyte fungi of *Penicillium* and *Paecilomyces* species are important sources of antiplasmodial secondary metabolites [28]. In fact, β-Resorcyclic acid lactones, cyclodepsipeptides, and tropane derivatives with potent antiplasmodial activities have earlier been isolated from *Paecilomyces* [29–31] and *Penicillium* [27] species.

Our findings have indicated that endophytic fungi belonging to the genera *Fusarium, Paecilomyces, Penicillium,* and *Bipolaris* isolated from this plant species should be studied intensely for novel antiplasmodial compounds. Therefore, further studies on the antiplasmodial activity guided isolation and characterization of active metabolites produced by promising fungi necessary for the discovery of new drugs against malaria are ongoing.

5. Conclusions

This study is the first report on the antiplasmodial potential of endophytic fungi from the bark of *Symphonia globulifera*. In this investigation, 9 out of 12 fungi showed promising antiplasmodial potency with *Paecilomyces lilacinus*, *Penicillium janthinellum*, and *Paecilomyces* sp. being the most promising. These results obtained in this study highlight endophytic fungi from *S. globulifera* as a promising source of novel lead antimalarial compounds.

Author Contributions: B.N.L., E.T., and F.F.B. conceived and designed the experiments; J.E.T.A., R.M.K.T., B.M.M., A.F.A., S.G., and B.N.L. performed the experiments; E.R.-F. analyzed the data; N.S., D.S., and B.N.L. contributed reagents/materials/analysis tools; J.E.T.A. wrote the paper.

Acknowledgments: The authors wish to acknowledge the Alexander von Humboldt Foundation for providing a fellowship to Bruno N. Lenta. The authors are also grateful to the TWAS-CNPq (The World Academy of Science-Conselho Nacional de Desenvolvimento Científico e Tecnológico) for the fellowships awarded to Brice M. Mba'ning (Award No. 190644/2015-0).

References

1. *World Malaria Report 2017*; Licence: CC BY-NC-SA 3.0 IGO; World Health Organization: Geneva, Switzerland, 2017.
2. D'Alessandro, U. Existing antimalarial agents and malaria-treatment strategies. *Expert Opin. Pharmacother.* **2009**, *10*, 1291–1306. [CrossRef] [PubMed]
3. Dhar, R.; Zhang, K.; Talwar, G.P.; Garg, S.; Kumar, N. Inhibition of the growth and development of asexual and sexual stages of drug-sensitive and resistant strains of the human malaria parasite *Plasmodium falciparum* by Neem (*Azadirachta indica*) fractions. *J. Ethnopharmacol.* **1998**, *61*, 31–39. [CrossRef]
4. Strobel, G.A.; Daisy, B. Bioprospecting for microbial endophytes and their natural products. *Microbiol. Mol. Biol. Rev.* **2003**, *67*, 491–502. [CrossRef] [PubMed]
5. Porras-Alfaro, A.; Bayman, P. Hidden fungi, emergent properties: Endophytes and microbiomes. *Ann. Rev. Phytopathol.* **2011**, *49*, 291–315. [CrossRef] [PubMed]
6. Stierle, A.; Strobel, G.A.; Stierle, D.B. Taxol and taxane production by *Taxomyces andreanae*, an endophytic fungus of Pacific yew. *Science* **1993**, *260*, 214–216. [CrossRef] [PubMed]
7. Wang, L.W.; Zhang, Y.L.; Lin, F.C.; Hu, Y.Z.; Zhang, C.L. Natural products with antitumor activity from endophytic fungi. *Mini Rev. Med. Chem.* **2011**, *11*, 1056–1074. [CrossRef] [PubMed]
8. Zhang, Y.; Han, T.; Ming, Q.; Wu, L.; Rahman, K.; Qin, L. Alkaloids produced by endophytic fungi: A review. *Nat. Prod. Commun.* **2012**, *7*, 963–968. [PubMed]
9. Strobel, G. Harnessing endophytes for industrial microbiology. *Curr. Opin. Microbiol.* **2006**, *9*, 240–244. [CrossRef] [PubMed]
10. Riyaz-Ul-Hassan, S.; Strobel, G.A.; Booth, E.; Knighton, B.; Floerchinger, C.; Sears, J. Modulation of volatile organic compound formation in the Mycodiesel producing endophyte- *Hypoxylon* sp. C1-4. *Microbiology* **2012**, *158*, 464–473. [CrossRef]
11. Ajibesin, K.; Ekpo, B.; Bala, D.; Essien, E.; Adesanya, S. Ethnobotanical survey of AkwaIbom State of Nigeria. *J. Ethnopharmacol.* **2008**, *115*, 387–408. [CrossRef] [PubMed]
12. Ssegawa, P.; Kasenene, J.M. Medicinal plant diversity and uses in the Sango bay area, Southern Uganda. *J. Ethnopharmacol.* **2007**, *113*, 521–540. [CrossRef] [PubMed]
13. Fromentin, Y.; Cottet, K.; Kritsanida, M.; Miche, S.; Gaboriaud-Kolar, N.; Lallemand, M.-C. *Symphoniaglobulifera*, a widespread Source of Complex Metabolites with Potent Biological Activities. *Planta Med.* **2015**, *81*, 95–107. [CrossRef] [PubMed]
14. Toghueo, R.M.K.; Zabalgogeazcoa, I.; Vázquez de Aldana, B.R.; Boyom, F.F. Enzymatic activity of endophytic fungi from the medicinal plants *Terminalia catappa*, *Terminalia mantaly* and *Cananga odorata*. *S. Afr. J. Bot.* **2017**, *109*, 146–153. [CrossRef]
15. White, T.J.; Bruns, T.; Lee, S.; Taylor, J. Amplification and Direct Sequencing of Fungalribosomal RNA Genes for Phylogenetics. In *PCR Protocols. A Guide to Methods and Applications*; Innis, M.A., Gelfland, D.H., Sninsky, J.J., White, T.J., Eds.; Academic Press: Cambridge, MA, USA, 1990; pp. 315–322.

16. Trager, W.; Jensen, J.B. Human malaria parasites in continuous culture. *Science* **1976**, *193*, 673–675. [CrossRef] [PubMed]
17. Lambros, C.; Vanderberg, J.P. Synchronization of *Plasmodium falciparum* erythrocytic stages in culture. *J. Parasitol.* **1979**, *65*, 418–420. [CrossRef] [PubMed]
18. Smilkstein, M.; Sriwilaijaroen, N.; Kelly, J.X.; Wilairat, P.; Riscoe, M. Simple and inexpensive fluorescence-based techniquefor high-throughput antimalarial drug screening. *Antimicrob. Agents Chemother.* **2004**, *48*, 1803–1806. [CrossRef] [PubMed]
19. Bagavan, A.; Rahuman, A.A.; Kumaraj, C.; Kaushik, N.K.; Mohanahrishnan, D.; Sahal, D. Antiplasmodial activity of botanical extracts against *Plasmodium falciparum*. *Parasitol. Res.* **2011**, *108*, 1099–1109. [CrossRef] [PubMed]
20. Mosmann, T. Rapid colorimetric assay for cellular growth and survival: Application to proliferation and cytotoxicity assays. *J. Immunol. Methods* **1983**, *65*, 55–63. [CrossRef]
21. Zou, W.X.; Meng, J.C.; Lu, H.; Chen, G.X.; Shi, G.X.; Zhang, T.Y.; Tan, R.X. Metabolites of *Colletotrichum gloeosporioides*, an endophytic fungus in *Artemisia mongolica*. *J. Nat. Prod.* **2000**, *63*, 1529–1530. [CrossRef] [PubMed]
22. Arnold, A.E.; Lutzoni, F. Diversity and host range of foliar fungal endophytes: Are tropical leaves biodiversity hotspots? *Ecology* **2007**, *88*, 541–549. [CrossRef] [PubMed]
23. Sánchez, M.S.; Bills, G.F.; Zabalgogeazcoa, I. The endophytic mycobiota of the grass *Dactylis glomerata*. *Fungal Divers.* **2007**, *27*, 171–195.
24. Sun, X.; Ding, Q.; Hyde, H.D.; Guo, L.D. Community structure and preference of endophytic fungi of three woody plants in a mixed forest. *Fungal Ecol.* **2012**, *5*, 624–632. [CrossRef]
25. Felber, C.A.; Orlandelli, C.R.; Rhoden, S.A.; Garcia, A.; Costa, A.T.; Azevedo, J.L.; Pamphile, J.A. Bioprospecting foliar endophytic fungi of *Vitisla brusca* L., cv. Bordô and Concord. *Ann. Microbiol.* **2015**, *26*, 765–775. [CrossRef]
26. Arnold, A.E.; Maynard, Z.; Gilbert, G.S.; Coley, P.D.; Kursar, T.A. Are tropical fungal endophytes hyperdiverse? *Ecol. Lett.* **2000**, *3*, 267–274. [CrossRef]
27. Iwatsuki, M.; Takada, S.; Mori, M.; Ishiyama, A.; Namatame, M.; Nishihara-Tsukashima, A.; Nonaka, K.; Masuma, R.; Otoguro, K.; Shiomi, K.; et al. In vitro and in vivo antimalarial activity of puberulic acid and its new analogs, viticolins A-C, produced by *Penicillium* sp. FKI-4410. *J. Antibiot.* **2011**, *64*, 183–188. [CrossRef] [PubMed]
28. Kaushik, N.K.; Murali, T.S.; Sahal, D.; Suryanarayanan, T.S. A search for antiplasmodial metabolites among fungal endophytes of terrestrial and marine plants of southern India. *Acta Parasitol.* **2014**, *59*, 745–757. [CrossRef] [PubMed]
29. Xu, L.; He, Z.; Xue, J.; Chen, X.; Wei, X. B-Resorcyclic acid lactones from a *Paecilomyces* fungus. *J. Nat. Prod.* **2010**, *73*, 885–889. [CrossRef] [PubMed]
30. Isaka, M.; Palasarn, S.; Lapanun, S.; Sriklung, K. Paecilodepsipeptide A, an antimalarial and antitumor cyclohexadepsipeptide from the insect pathogenic fungus *Paecilomyces cinnamomeus* BCC 9616. *J. Nat. Prod.* **2007**, *70*, 675–678. [CrossRef] [PubMed]
31. Nilanonta, C.; Isaka, M.; Kittakoop, P.; Palittapongarnpim, P.; Kamchonwongpaisan, S.; Pittayakhajonwut, D.; Tanticharoen, M.; Thebtaranonth, Y. Antimycobacterial and antiplasmodial cyclodepsipetides fron the insect pathogenic fungus *Paecilomyces tenuipes* BCC 1614. *Planta Med.* **2000**, *66*, 756–758. [CrossRef] [PubMed]

10

Endophytic Mycoflora and their Bioactive Compounds from *Azadirachta Indica*

Eyob Chukalo Chutulo [1,2] and Raju Krishna Chalannavar [3,*]

1 Department of Biology, Wolaita Sodo University, Wolaita Sodo 138, Ethiopia; eyob.chukalo@wsu.edu.et
2 Department of Biosciences, Mangalore University, Mangalagangothri 574199, India
3 Department of Applied Botany, Mangalore University, Mangalagangothri 574199, India
* Correspondence: rajukrishna@mangaloreuniversity.ac.in

Abstract: Plants are all inhabited by endophytic fungi in the interior of their tissues. The neem tree *Azadirachta* is an Indian lilac used for various therapeutic purposes in different forms of preparations. This plant hosts different types of endophytic fungi. In some cases, different tissues of a given plant are inhabited by different endophytic fungi which are discussed in this paper. Recently, there have been new reports on endophytic fungi and their bioactive compounds from *Azadirachta indica*. The biological function of bioactive compounds was discussed in view of their future industrial prospects. There are a number of different research investigations that examine the endophytes isolated and screened for their potential bioactive secondary metabolites from neem, but there is no comprehensive review on neem endophytes and their secondary metabolites to bring all trends from different researchers together. Therefore, in this review, we have discussed the endophytic fungi from the different tissues of neem, in view of the latest understandings of antimicrobial, antioxidant, and pathogenicity target compounds. Importantly, tracing the previous findings would pave the way to forecast the missing link for future work by researchers.

Keywords: endophytic fungi; plant tissues; bioactive compounds; biological activities

---

1. Introduction

The term "endophyte" refers to all microorganisms that colonize internal plant tissues for all or part of their lifetime [1]. They cause unapparent and asymptomatic infection and live entirely within plant tissues. They cause no symptoms of disease [2]. Endophytes are the chemical synthesizers inside plants and plants have been extensively investigated for their endophytic microbial complement [3].

Endophytes are an under-investigated group of microorganisms that represent a plentiful and renewable source of bioactive and chemically new compounds with potential for exploitation in a wide variety of medical, agricultural, and industrial realms [4]. They are a taxonomically and ecologically heterogeneous group of organisms; mainly belonging to Ascomycota, coelomycetes, and hyphomycetes [5–8].

All plants are inhabited internally by diverse microbial communities comprising bacteria, archaea, fungi, and protista [1,9]. The association between fungal endophytes and their host plant is due to the result of unique adaptations which enable the endophytes to harmonize their growth with that of their host [10]. These endophytic communities are accountable to either partial or complete biosynthesis of host plant secondary metabolites [8,11].

Endophytic biotechnology can be used for the efficient production of agriculturally, industrially and economically important plants and plant products. The rational application of endophytes to manipulate the microbiota, intimately associated with plants, can help in enhancement of production of the agricultural product, increased production of key metabolites in medicinal and aromatic plants, as well as adaption to new bio-geographic regions through tolerance to various biotic and abiotic conditions [12].

Endophytes have recently generated significant interest in the microbial chemistry community due to their great potential to contribute to the discovery of new bioactive compounds. It has been suggested that the close biological association between endophytes and their plant host results in the production of a great number and diversity of biologically active molecules compared to epiphytes or soil-related microbes [13]. Moreover, the symbiotic nature of this relationship indicates that endophytic bioactive compounds are less toxic to the cell, as these chemicals do not kill the eukaryotic host system. This is particularly important to the medical community as potential drugs may not adversely affect human cells [14].

Azadirachta indica is widely used for the investigation of endophytes and their secondary metabolites, but it remains crucial to further extend the study as it is one of the major traditional medicinal plants used by about 80% of developing nations. This is because the discovery of new bioactive compounds as well as new endophytic fungi from *Azadirachta indica* has been reported recently. Hence, this review discusses the latest research advances on endophytic fungi from the neem tree, including types of endophytic fungi identified, described, bioactive compounds, and biological activities of endophytes to trace the current scope of knowledge of endophytes. It gives collective recent knowledge on antimicrobial activities of endophytic fungi of neem and on recent advances concerning their efficacy as potential antimicrobials to fight the current health problems due to emerging multidrug-resistant pathogens. It also gives focus to the well-studied endophytic sources of potential antimicrobials with emphasis on their future prospects of endophytes on human pathogens based on the recent research outputs.

The chance of getting new bioactive compounds from endophytes depends on several factors, for example, culture condition, tissue type, and age of host, the degree of interaction of endophytes with their photobionts in nature, the genetic basis of mycobionts and photobionts. Leaving aside most of these factors, by modifying the culture condition different investigators are reporting new endophytic isolates [15] and new bioactive compounds in recent studies. This signifies the continuous nature of exploiting the endophytes even from largely exploited hosts as well. Importantly, to our knowledge, there is no review on the endophytic fungi and their bioactive compounds from neem plants. Therefore, the need to review the endophytic fungi and their bioactive compound from *Azadirachta indica* is thus apparent. Ultimately, we believe that this review provides relevant research trends on the subject.

1.1. *Plant-Endophyte Metabolism*

Plant and endophyte are under continuous interaction in nature in their symbiotic existence. Their metabolism can interact on many levels: (a) the endophyte induces host metabolism, (b) the host induces endophyte metabolism, (c) host and endophyte share parts of a specific pathway and contribute partially, (d) the host can metabolize products from the endophyte and vice versa, (e) the endophyte can metabolize secondary compounds from the host. The two latter possibilities can be accomplished by only one, several or all enzymatic steps for biochemical transformation [11].

Endophytes can influence their host plant's metabolism, but one can speculate that the established host range could also alter or influence the pattern of secondary metabolites in endophytic fungi. Moreover, the host plants can influence the metabolite pattern in pathogenic fungi [11].

1.2. *Effects of Endophytes on Neem Plants*

Endophytes form a symbiotic relationship with their plant host. It is believed that in many cases the microbes function as the biological defense for the plant against foreign phytopathogens due to the fact that the majority of secondary metabolites occurring in endophytic microorganisms have antimicrobial activity and these have been implicated in protecting the host plant against [16]. The protection mechanism of the endophytes are exerted directly, by releasing metabolites to attack any antagonists or lyse affected cells, and indirectly, by either inducing host defense mechanisms or promoting growth [14]. Antibiotics or hydrolytic enzymes can be released by endophytes to prevent colonization of microbial plant pathogens [13,17].

2. Endophytic Fungi from *Azadirachta indica*

Different tissue samples of a given plant in several locations yield the greatest species diversity of endophytes. *Periconia*, *Stenella*, and *Drechslera* were endophytes reported from neem plant [6]. Moreover, the authors of [6] reported *Phomopsis oblonga*, *Cladosporium cladosporioides*, *Pestalotiopsis* sp., *Trichoderma* sp., and *Aspergillus* sp. as the dominant endophytic fungal isolates. The huge and substantial collection of endophytic fungi from neem may represent a peculiar source of the interesting and useful bioactive compounds associated with *Azadirachta indica* such as the azadirachtins and related tetranortriterpenoids [6].

The diversity of endophytic fungi is likely determined by the age of the plant parts [18]. Some endophytes are isolated only from certain specific locations irrespective of the difference in tissue specificity [19]. Therefore, in this section, endophytic fungi from different tissues of *Azadirachta indica* are reviewed and presented in Table 1.

2.1. Endophytic Fungi of Leaf

The leaf is an ideal tissue for exploring the endophytes and their secondary metabolites. The leaf has shown the highest report on species richness and colonization of endophytes [6]. The authors of [6] reported the occurrence of five fungal endophytes from asymptomatic green and senescent leaves of neem plant. From these, four were sterile forms and one was *Fusarium avenaceum*. The frequency of occurrence of endophytes was significantly higher in the basal leaflets than in the apical or middle leaflets and in the main vein of the leaflet than in the lamina tissue. The colonization frequency of green leaves by endophytes increased during the rainy season although no new endophyte species could be recovered by their study. The antifungal metabolites present in the leaves could be accounted for by the restricted number of endophytic fungal genera and the absence of common endophytic fungi in the neem leaves. The occurrence of foliar endophytes in tropical trees is influenced by environment, types and chemistry of the host tissue [20], and the nature of interaction between the mycobiont and phycobiont. They have isolated and recorded *Fusarium avenaceum* with some other sterile mycelia. Another study conducted by [21] isolated four endophytic fungal species (*Trichoderma* sp., *Colletotrichum* sp., *Curvularia* sp. and *Chaetomium* sp.) and *Alternaria alternata* [22] from the leaves of *Azadirachta indica* A. Juss.

In most studies of endophyte fungi, the sterile mycelia fungi are reported as mycelia-sterile. Unlike those, the three mycelia-sterile isolated were identified as *Fusarium solani* and *Chaetomium globosum*. Additionally, leaf isolated mycelia-sterile was identified as *F. solani* (soil fungus) by Verma [7], and it shows the vertical traveling nature of this fungus from root to upper tissues of the host by the same study.

Tenguria and Khan [23] have isolated 85 endophytic fungi of 10 genera from 200 segments of fresh *Azadirachta indica* leaves. These are *Chaetomium globosum*, *Pestalotiopsis* spp., *Phoma* sp., *Aspergillus flavus*, *Aspergillus niger*, *Alternaria alternata* (Fr.) Keissl., *Fusarium* spp., *Penicillium* spp., *Trichoderma* spp., and Sterile mycelia. Their isolates are almost similar to fungi reported by Taware & Rajurkar [24]. The endophyte *Geotrichum* sp. AL4 was cultivated from the leaf of *Azadirachta indica* and produced novel active components with nematicidal activities [25]. Abubakar and Ndana [26] have isolated five species of endophytic fungi from leaf tissue of *Azadirachta indica* with *Cladosporium* spp. dominated the leaves with the frequency of colonization of 11.3%.

2.2. Endophytic Fungi of Stem

The endophytic fungus *Xylaria* sp. YM 311647 was isolated from stems of healthy *Azadirachta indica* collected in China [27]. Another endophytic fungus *Phomopsis* sp. YM 311483 was isolated from surface-sterilized fresh stems of an apparently healthy *Azadirachta indica* specimen collected in China.

This fungus produced two new ten-membered lactones which have antifungal potential against plant pathogenic fungi [28].

Wu et al. [29] have isolated and reported the fungal genus *Xylaria* sp. YC-10 from stems of *Azadirachta indica* A. Juss. Abubakar and Ndana [26] have isolated seven species of endophytic fungi from stems (twig) tissue of *A. indica* with *Cryptococcus* spp. reported being the most dominant isolate with high percentage frequency of colonization 13.9%.

2.3. Endophytes of Fruit

The endophytic fungi from *A. indica* are mostly known from leaves, bark, and stems but it is uncommon from fruits of this host. Verma et al. [7] have isolated endophytic fungi from unripe fruit and roots. They have isolated 105 endophytic fungi from these unripe fruits, from 29 taxa, at a rate of 68.0%. Their isolates were composed of 11.06% mycelia sterilia, 7.25% coelomycetes, and 81.69% hyphomycetes, while *Humicola, Drechslera,* and *Colletotrichum* sp. were obtained exclusively from fruit samples.

2.4. Endophytic Fungi of Bark

Mahesh et al. [5] have investigated the endophytic fungi from the inner bark of *A. indica*. They have isolated 77 endophytic fungal isolates belonging to 15 genera composed of hyphomycetes, coelomycetes, ascomycetes and sterile mycelia (Table 1). The endophytes *Curvularia, Cochlonema, Gliomastix* and *Verticillium* spp. were reported in their study. Most importantly, *Trichoderma, Penicillium* and *Pestalotiopsis* spp. were the most dominant endophytes recorded. Verma et al. [6] have isolated and characterized 18 different taxa of endophytic fungi from leaf, stem, and bark of *Azadirachta indica* A. Juss. Their isolates are composed of 62.2% hyphomycetes, coelomycetes (27.4%) and mycelia sterilia (7.7%). Importantly, the endophytic fungi occur more often in leaf segments than stem and bark tissues [6].

Tejesvi et al. [19] have isolated the endophytic fungi *Gliomastix, Curvularia, Phoma eupyrena,* and *Phyllosticta* from the inner bark of *A. indica*. In another investigation, they have reported three isolates of *Pestalotiopsis microspora* and two isolates of *Bartalinia robillardoides* from the inner bark of *A. indica* [30].

2.5. Endophytic Fungi of Root

The endophytic fungus *Chloridium* sp. was isolated from the roots of *A. indica* A. Juss. in Varanasi, India. The fungus produces a highly functionalized javanicin, with promising antibacterial activity [31] that might prevent its host from invasion by pathogenic microorganisms.

Verma et al. [7] have isolated 167 endophytic fungi from roots, at a rate of 68.0%. Their isolates are identified into 29 taxa; those composed of mycelia sterilia accounted for 11.06%, coelomycetes 7.25%, while hyphomycetes showed the maximum number of representative isolates (81.69%). They have recorded species such as *Chaetomium globosum, Chloridium, Scytalidium, Nigrospora* and *Verticillium* exclusively from the root. Kusari et al. [32] have reported the isolation and characterization of a novel endophytic fungus *Eupenicillium parvum* in their study.

2.6. Endophytic Fungi of Seed and Twigs

Although endophytic fungi isolated from different tissue of plants in several investigations, the reports of endophytic fungi from twigs of neem plant, unlike from other plants, are very few. Ashkezari and Fotouhifar [15] have isolated nineteen new endophytic fungi from a common yew plant. The pathogenic endophyte *Phomopsis azadirachtae*, the cause of devastating die-back disease, were reported from seeds and twigs of neem plants [33]. It is important to note that the twigs of *Azadirachta indica* are a poorly exploited surface for investigation of endophytes. Therefore, it is imperative to conduct further study on the endophytic fungi and their bioactive prospects in the search for potential applications.

Table 1. Endophytic fungi from different parts of neem plant.

Division (Subdivision)	Class	Subclass	Order	Family	Genus (Former Name)	Identified by	Plant Tissue	References
Ascomycota	Sordariomycetes	Hypocreomycetidae	Hypocreales	Nectriaceae	*Fusarium*	ITS-5.8S rDNA analysis	R,L,S,B,F	[6,7,19,34–36]
(Pezizomycotina)				Hypocreaceae	*Trichoderma*	LSU, SSU, TEF and RPB2 sequence data	R,L,S,B,F	[7,36]
					Gliomastix	Phylogeny	L,B	[5,6,19,36]
					Acremonium	Phylogeny and DNA-based identification	R,B,F	[7,19,37]
			Glomerellales	Plectosphaerellaceae	*Verticillium*	LSU, SSU, TEF and RPB2 sequence data	R	[7,36]
				Glomerellaceae	*Colletotrichum*	ITS-5.8S rDNA analysis	L,F	[7,21]
		Sordariomycetidae	Sordariales	Chaetomiaceae	*Chaetomium*	ITS-5.8S rDNA analysis	R,L,B	[7,19,23]
					Humicola	ITS-5.8S rDNA analysis	F	[7,36]
			Chaetospheriales	Chaetosphaeriaceae	*Chloridium*	ITS	R	[7]
		Xylariomycetidae	Xylariales	Xylariaceae	*Xylaria*	Phylogenetic analysis; Molecular clock analysis	S	[29,34–36]
				Amphisphaeriaceae	*Pestalotiopsis*	Phylogenetic analysis; Molecular clock analysis	B,L,S	[5,6,23,30,36]
				Bartaliniaceae	*Bartalinia*	ITS rDNA analysis	B	[38,39]
		Diaporthomycetidae	Trichosphaeriales	Trichosphaeriaceae	*Nigrospora*	ITS-5.8S rDNA analysis	R,S,L,B	[6,7,36]
			Diaporthales	Diaporthaceae	*Diaporthe* (*Phomopsis*)	six-gene phylogeny	S,L,B	[6,34,36,40]
	Leotiomycetes				*Scytalidium*	ITS-5.8S rDNA analysis	R	[7]
	Eurotiomycetes				*Aspergillus*	six-gene phylogeny	R,S,L,B,F	[34]
					Penicillium	ITS-5.8S rDNA analysis	R,F	[7]
					Eupenicillium	five-gene phylogeny	R,S,L,B	[41,42]
	Dothideomycetes		Pleosporales		*Curvularia*	ITS-5.8S rDNA analysis	R,B,F	[6,7,19]
					Alternaria	ITS-5.8S rDNA analysis	R,L,F	[6,7]
					Drechslera	ITS-5.8S rDNA analysis	B,F	[6,7]
					Phoma	Phylogenetic analysis	L	[36]
					Periconia	Phylogenetic analysis; Molecular clock analysis	B	[6,36]
			Capnodiales	Davidiellaceae	*Cladosporium*	ITS-5.8S rDNA analysis	R,L,B,F	[7,43,44]
					Stenella		B	[6]
			Botryospheriales	Botryosphaeriaceae	*Phyllosticta*	ITS-5.8S rDNA analysis	R,F	[7,39,45]
Basidiomycota (Agaricomycotina)	Tremellomycetes	Tremellomycetidae	Filobasidiales		*Cryptococcus*	six-gene phylogeny	S	[34,46]
Zygomycota (Zoopagomycotina)			Zoopagales	Cochlonemataceae	*Cochlonema*	Phylogeny	B	[5,19,46]
				Mucoraceae	*Cercinella*	ITS-5.8S rDNA analysis	R	[7]

Note: "R": root; "L": Leaf; "B": Bark; "S": Stem; "F": Fruit. The endophytes are isolated from the tissue types.

3. Current Trends on Bioactive Metabolites from Neem Endophytic Fungi

The production of antimicrobial bioactive compounds by endophytes is currently receiving urgent concern due to the emergence of multidrug-resistant pathogens in chemotherapy. Unlike the chemical synthetics, due to health and environmental issues, there is a strong need of endophytic sources of antimicrobials for their biological and chemical safety effects. There exist many appreciable possibilities for the exploitation of endophytic fungi for the generation of abundant novel biologically active secondary metabolites. Secondary metabolites from microorganisms have proven to be tremendous and enduring sources [47]. The endophytes isolated from the medicinal plants have the potential as an alternative source of these bioactive secondary metabolites [48]. Some of the bioactive compounds produced by endophytic fungi of the neem tree and their biological activities are discussed hereunder.

3.1. Melanin Pigment

Diaporthe (former name: *phomopsis*) is a phellophytic fungus that lives in the outer layer of the bark [49]. Found in *Azadirachta indica*, it synthesized and deposited DOPA (3,4- dihydroxyphenylalanine) type of melanin on their hyphae. Melanin produced by the endophyte, *Phomopsis*, provides adaptations, which include mechanisms for overcoming host barriers, successful competition with other phylloplane fungi and surviving harsh environmental conditions [50] and in the widespread occurrence of the endophyte [51].

Melanin pigment production is catalyzed by the enzyme tyrosinase produced by endophytic fungi *Fusarium* spp. from *Azadirachta indica*. The fungi produced 2.8 U/mol (L-DOPA) extracellular tyrosinase activity. This enzyme is essential for pigmentation, an important factor in wound healing and primary immune response. It has a great application in environmental, pharmaceuticals, cosmetics and food industries [52].

Even though endophytes are known for their antibacterial, antitumor, immunomodulatory, anti-inflammatory and antiviral activities, there are few reports on endophytes for tyrosinase production from neem plants [52]. However, extracellular production of fungal tyrosinase was reported, firstly, from a filamentous fungus, *Trichoderma reesei* [53]. Therefore, future research should focus on investigating tyrosinase from neem endophytes due to its higher industrial applications.

3.2. Antioxidant Metabolites

Kumaresan et al. [21] have screened endophytic fungi, from the leaf, for their antioxidant bioactive compounds. They have investigated the constituents of antioxidants like phenols, tannin, flavonoid, ascorbic acid, and β-carotene. All isolated endophytic fungi studied by their team possessed in vitro antioxidant activities. Similarly, evaluation of secondary metabolites from *Penicillium species* revealed the presence of saponins in addition to those mentioned above [54]. This is one of the very few reports from neem endophytes for their antioxidant activities. Another study by Zhao et al. [55] reported the natural antioxidant cajaninstilbene acid (100.5 ± 9.4 μg/g dry weight of mycelium) from endophytic fungi *F. solani* isolated from pigeon pea. This Fusarium was also found in neem tree as mentioned in Table 1.

3.3. Antimicrobial Secondary Metabolites

Wu et al. [56] reported two solanapyrone analogues (solanapyrones N and O) compounds isolated from the fermentation culture of *Nigrospora* sp. YB-141, an endophytic fungus isolated from *Azadirachta indica* A. Juss growing in the tropical region of Southwest China.

Verma et al. [57] have screened and evaluated the efficacy of six fungal endophytic strains from *Azadirachta indica* A. Juss against dermatophytic fungi *Trichophyton* and *Microsporum*. They have reported production of active compounds by their isolates but identification of the compound is not yet reported. Similarly, evaluation of secondary metabolites from *Penicillium species* isolated from the leaf of neem plant showed that they possessed some degree of antibacterial and antifungal activities [54].

Five 10-membered lactones were isolated from the endophytic fungus, *Phomopsis* sp. YM 311483, obtained from the stem of *Azadirachta indica*. Of these, two compounds (8R-acetoxy-5R-hydroxy-7-oxodecan-9-olide and 7R-acetoxymultiplolide A) were newly recorded by Wu et al. [28]. Two compounds identified as 7R,8R-dihydroxy-3,5-decadien-10-olide and 8R-acetoxymultiplolide A are identical to compounds reported by Tan et al. [58]. All the compounds were evaluated for their antifungal activity against seven plant pathogens, *Aspergillus niger*, *Botrytis cinerea*, *Fusarium avenaceum*, *Fusarium moniliforme*, *Helminthosporium maydis*, *Penicillium islandicum*, and *Ophiostoma minus*. Compound 8R-acetoxymultiplolide A showed the most potent antifungal activities with MIC values in the range of 31.25–500 μg/mL. Interestingly, 8R-acetoxymultiplolide A was more potent [58] than 7R-acetoxymultiplolide A even though their structures differed only in the position of the acetoxy substituent [40]. On the contrary, Tan et al. [58] have reported that compounds 7α,8α-Dihydroxy-3,5-decadien-10-olide and 8α-Acetoxymultiplolide A have not shown any antifungal activities against *Candida albicans*. Therefore, further antimicrobial activity bioassay is deemed necessary to reconcile the discrepancy of the report and to utilize the potential of the isolates and their compounds.

Five guaiane sesquiterpenes isolated from the culture broth of endophytic fungus *Xylaria* sp. YM 311647 showed moderate or weak antifungal activities in a broth microdilution assay against five pathogenic fungi, but no obvious inhibitory activities against *Fusarium avenaceum* [27]. Nine oxygenated guaiane-type sesquiterpenes and three isopimarane diterpenes were reported from this fungus, by Tejesvi et al. [30]. Suggested by Wu et al. [59], 18-norisopimarane diterpene and the diterpene sulfate are effective against human pathogenic fungi.

The extracts of *Pestalotiopsis microspora* isolated by Tejesvi et al. [30] demonstrated the potential of antibacterial activities as therapeutic agents against various pathogens tested. The same species from different isolates by the same work demonstrated antioxidant and antihypertensive properties. From this, it is possible to deduce that this isolate needs to be thoroughly investigated for its commercial use as an antioxidant or antihypertensive agent for the treatment of the respective diseases.

An investigation of the secondary metabolites from *Chloridium* sp. root isolates by Kharwar et al. [31] resulted in the identification of javanicin. This compound was identical in all respects to a previously described naphthaquinone, javanicin. However, its crystallized structure was confirmed by X-ray crystallography. Javanicin was tested for its antimicrobial potential against human and plant pathogens. Tests demonstrated that javanicin had antimicrobial potential against human and plant pathogens. It was either slightly active or not active against fungi such as *Pythium ultimum*, *Phytophthora infestans*, *Botrytis cinerea*, and *Ceratocystis ulmi*, whereas it was active against *C. albicans*, *Escherichia coli*, *Bacillus* sp., and *Fusarium oxysporum* at higher MIC values ranging from 20 and 40 μg/mL. Its activity was recorded against *Rhizoctonia solani* and *Verticillium dahliae* at 10 μg/mL while it was active at 5 μg/mL against *Cercospora arachidicola*. The bacteria that were the most sensitive to the javanicin (2 μg/mL) were *P. aeruginosa* and *P. fluorescens*. On account of this, it may be ideal for compounds, particularly for valuable selective antibiotic formulations for human and plant pathogens [31].

3.4. Natural Insecticides

The production of natural insecticides, azadirachtin A and B, from endophytic fungus *Eupenicillium parvum* was reported by Kusari et al. [42]. These compounds are exclusive to the neem tree, *Azadirachta indica* A. Juss, from where they are currently originated. The production of this compound by the fungus is not only in the mycelium but also released into the culture media. The discovery of this azadirachtin-producing endophytic fungus has vast implications for further research from the ecological and biochemical point of view. Another study conducted by Kaur et al. [22] reported antifeedant and toxic activity against tobacco caterpillar *Spodoptera litura* by *Alternaria alternata*. The larvae fed on the diet supplemented with fungal extract significantly reduced the survival and influence the development and reproduction of *S. litura* due to potent toxicity of *A. alternata*. Additionally, the fungus was known for its immunomodulatory effects on *Spodoptera litura* [60].

The genus *Xylaria* sp. YC-10 isolated from stems of *Azadirachta indica* A. Juss. produced eleven compounds from which eight compounds (5-Methylmellein, 5-Carboxylmellein, Hymatoxin C, Hymatoxin D, Halorosellinic acid, Cerebroside C, Cerevisterol and (2S,3S,4R,2′R)-2-(2′-Hydroxytetracosanoylamino)-octadecane-1,3,4-triol) exhibited weak insecticidal activity against *Plutella xylostella* [29].

3.5. Nematicides

Li et al. [25] isolated four compounds from the endophytic fungal strain *Geotrichum* sp. AL4, cultivated from the leaves of the neem tree. They reported two compounds, chlorinated oxazinane derivate (1-[(2R*,4S*,5S*)-2-chloro-4-methyl-1,3-oxazinan-5-yl] ethenone) and an epimer of the former (1-[(2R*,4S*,5R*)-2-chloro-4-methyl-1,3-oxazinan-5-yl] ethanone) and two other known compounds. The two compounds reported by their team were assessed for nematicidal activities against the nematodes *Bursaphelenchus xylophilus* and *Panagrellus redivivus*, and showed noticeable bioactivities. Of known compounds, compound 1-(2,4-dihydroxyphenyl)-ethanone is commonly isolated from plants. However, they have reported, for the first time, this compound from a microbial source showing nematicidal activity. Therefore, the *Geotrichum* sp. AL4 can be an ideal source for the formulation of antinematicidal agents.

3.6. Antiparasites

Verma et al. [61] have investigated several endophytic fungal strains from *Azadirachta indica* A. Juss and came up with the endophytic fungal strain *Pestalotiopsis* sp. with a significant anticestodal potential against hydatid cysts *Echinococcus granulosus*. The anticestodal activity observed with *Pestalotiopsis* sp. showed promising scolicidal activity of up to 97% mortality within 30 min of incubation. Since this was the first report that dealt with fungal endophytes for the anticestodal potential activities, further studies to reaffirm its anticestodal biopotential activities through optimized culture condition are thus needed.

4. Conclusions

Currently, the demand for health services is growing dramatically, particularly in developing countries, due to the emergence of drug resistance by pathogenic microorganisms. Moreover, the highly-increased recurrence of cancer and other infectious diseases makes the situation a greater tragedy. Hence, it is urgently necessary to investigate new bioactive compounds effective against drug-resistant pathogens for the remedy of the aforementioned diseases. Therefore, endophytic fungi, producer of a wide array of secondary bioactive metabolites with their peculiar potential compounds, namely, melanin, antimicrobials, antioxidant, anti-inflammatory, insecticides, nematicides, etc., are the ideal targets. These bioactive compounds have a range of potential to combat etiologic agents of plants and animal's disease, prevent cell damage due to reactive oxygen species, crop pests, animal pests, and to target pathogenicity traits of pathogenic microbes for necessary remediation.

To date, very few structures of bioactive compounds are characterized and identified from endophytic fungi of neem, but most of the investigations are restricted to the level of fungal identification and bioactivities assay. The different types of Ascomycota, Basidiomycota and Zygomycota endophytes found in the neem tree are phylogenetically synthesized using modern taxonomy from different systematic reviews (Table 1). Some of the endophytic isolates which failed to sporulate, sterile mycelia, are placed to their natural group based on modern taxonomy. Most importantly, the uncultured endophytic fungi from the neem tree were not reported in the literature surveyed so far. As a result, non-culturable endophytes have not been looked for with a metagenomics method. Metagenomics approaches, if used for neem tree, will allow identifying uncultured endophyte fungi that are ignored when isolation and cultivation of endophytes are done before identification. Similarly, too many authors do not make sufficient use of the modern taxonomy based on phylogenetic analysis for identification of their endophytic isolates. The compounds are identified by few investigators and this may be due to a lack of advanced equipment, but possible cooperation with the institutions with the tools are crucial for further discovery.

Even though there are several records on the antioxidant activities of endophytic fungi from different plant hosts, very few endophytes having these effects are known from *Azadirachta indica*. However, this host is widely used for its various biological activities, as antioxidant properties, in different forms of the formulation. Therefore, more work on the antioxidant activities of its endophytes can exert a profound effect on the search for novel antioxidant bioactive compounds.

Acknowledgments: The authors are thankful to Mangalore University for providing access to necessary resources. The departments of Biosciences and Applied Botany are acknowledged.

Author Contributions: Eyob Chukalo Chutulo—conceived of the presented idea. Wrote the manuscript with support from the idea of Raju Krishna Chalannavar. Eyob Chukalo Chutulo took the lead in writing the manuscript. Raju Krishna Chalannavar–Encouraged Eyob Chukalo Chutulo to investigate [a specific aspect] and supervised the findings of this work. Supervised development of work, helped in data interpretation and manuscript evaluation and acted as the corresponding author. Both authors discussed the results and contributed to the final manuscript.

References

1. Hardoim, P.R.; van Overbeek, L.S.; Berg, G.; Pirttilä, A.M.; Compant, S.; Campisano, A.; Döring, M. The Hidden World within Plants: Ecological and Evolutionary Considerations for Defining Functioning of Microbial Endophytes. *Microbiol. Mol. Biol. Rev.* **2015**, *79*, 293–320. [CrossRef] [PubMed]
2. Wilson, D. Endophyte: The Evolution of a Term, and Clarification of Its Use and Definition. *Oikos* **1995**, *73*, 274–276. [CrossRef]
3. Ramesha, A.; Srinivas, C. Antimicrobial activity and phytochemical analysis of crude extracts of endophytic fungi isolated from *Plumeria acuminata* L. and *Plumeria obtusifolia* L. *Eur. J. Exp. Biol.* **2014**, *4*, 35–43.
4. Strobel, G.; Daisy, B.; Castillo, U.; Harper, J. Natural Products from Endophytic Microorganisms. *J. Nat. Prod.* **2004**, *67*, 257–268. [CrossRef] [PubMed]
5. Mahesh, B.; Tejesvi, M.V.; Nalini, M.S.; Prakash, H.S.; Kini, K.R.; Subbiah, V.; Shetty, H.S. Endophytic mycoflora of inner bark of *Azadirachata indica*. *Curr. Sci.* **2005**, *88*, 218–220.
6. Verma, V.C.; Gond, S.K.; Kumar, A.; Kharwar, R.N.; Strobel, G. The endophytic mycoflora of bark, leaf, and stem tissues of *Azadirachta indica* A. Juss (Neem) from Varanasi (India). *Microb. Ecol.* **2007**, *54*, 119–125. [CrossRef] [PubMed]
7. Verma, V.C.; Gond, S.K.; Kumar, A.; Kharwar, R.N.; Boulanger, L.A.; Strobel, G.A. Endophytic Fungal Flora from Roots and Fruits of an Indian Neem Plant *Azadirachta indica* A. Juss., and Impact of Culture Media on their Isolation. *Indian J. Microbiol.* **2011**, *51*, 469–476. [CrossRef] [PubMed]
8. Rajagopal, K.; Maheswari, S.; Kathiravan, G. Diversity of endophytic fungi in some tropical medicinal plants—A report. *Afr. J. Microbiol. Res.* **2012**, *6*, 2822–2827.
9. Verma, V.C.; Gond, S.K.; Kumar, A.; Mishra, A.; Kharwar, R.N.; Gange, A.C. Endophytic Actinomycetes from *Azadirachta indica* A. Juss.: Isolation, Diversity, and Anti-microbial Activity. *Microb. Ecol.* **2008**, *57*, 749–756. [CrossRef] [PubMed]
10. Verma, V.C.; Singh, S.K.; Kharwar, R.N. Histological Investigation of Fungal Endophytes in Healthy Tissues of *Azadirachta indica* A. Juss. *Kasetsart J. Nat. Sci.* **2012**, *46*, 229–237.
11. Ludwig-Müller, J. Plants and endophytes: Equal partners in secondary metabolite production? *Biotechnol. Lett.* **2015**, *37*, 1325–1334. [CrossRef] [PubMed]
12. Wani, Z.A.; Ashraf, N.; Mohiuddin, T.; Riyaz-Ul-Hassan, S. Plant-endophyte symbiosis, an ecological perspective. *Appl. Microbiol. Biotechnol.* **2015**, *99*, 2955–2965. [CrossRef] [PubMed]
13. Strobel, G.A. Endophytes as sources of bioactive products. *Microbes Infect.* **2003**, *5*, 535–544. [CrossRef]
14. Alvin, A.; Miller, K.I.; Neilan, B.A. Exploring the potential of endophytes from medicinal plants as sources of antimycobacterial compounds. *Microbiol. Res.* **2014**, *169*, 483–495. [CrossRef] [PubMed]
15. Ashkezari, S.J.; Fotouhifar, K.-B. Diversity of endophytic fungi of common yew (*Taxus baccata* L.) in Iran. *Mycol. Prog.* **2017**, *16*, 247–256. [CrossRef]

16. Gunatilaka, A.A.L. Natural Products from Plant-Associated Microorganisms: Distribution, Structural Diversity, Bioactivity, and Implications of Their Occurrence. *J. Nat. Prod.* **2006**, *69*, 509–526. [CrossRef] [PubMed]
17. Berg, G.; Hallmann, J. *Control of Plant Pathogenic Fungi with Bacterial Endophytes*; Schulz, P.D.B.J.E., Boyle, D.C.J.C., Sieber, D.T.N., Eds.; Springer: Berlin/Heidelberg, Germany, 2006; pp. 53–69.
18. Qi, F.; Jing, T.; Zhan, Y. Characterization of Endophytic Fungi from Acer ginnala Maxim. in an Artificial Plantation: Media Effect and Tissue-Dependent Variation. *PLoS ONE* **2012**, *7*, e46785. [CrossRef] [PubMed]
19. Tejesvi, M.V.; Mahesh, B.; Nalini, M.S.; Prakash, H.S.; Kini, K.R.; Subbiah, V.; Shetty, H.S. Fungal endophyte assemblages from ethnopharmaceutically important medicinal trees. *Can. J. Microbiol.* **2006**, *52*, 427–435. [CrossRef] [PubMed]
20. Rajagopal, K.; Suryanarayanan, T.S. Isolation of endophytic fungi from leaves of neem (*Azadirachta indica* A. Juss.). *Curr. Sci.* **2000**, *78*, 1375–1378.
21. Kumaresan, S.; Karthi, V.; Senthilkumar, V.; Balakumar, B.S.; Stephen, A. Biochemical Constituents and Antioxidant Potential of Endophytic Fungi isolated from the Leaves of *Azadirachta indica* A. Juss (Neem) from Chennai. *J. Acad. Ind. Res.* **2015**, *3*, 355–361.
22. Kaur, H.P.; Singh, B.; Kaur, A.; Kaur, S. Antifeedent and toxic activity of endophytic *Alternaria alternata* against tobacco caterpillar *Spodoptera litura*. *J. Pest Sci.* **2013**, *86*, 543–550. [CrossRef]
23. Tenguria, R.K.; Khan, F.N. Distribution of Endophytic Fungi in Leaves of *Azadirachta indica* A. JUSS. (Neem) of Panchmarhi Biosphere Reserve. *Curr. Bot.* **2011**, *2*, 27–29.
24. Taware, A.S.; Rajurkar, S.K. Diversity assessment of endophytic fungi from *Azadirachta indica* A. Juss. from various regions of Aurangabad, Maharashtra (India). *Int. J. Innov. Sci. Eng. Technol.* **2015**, *2*, 96–107.
25. Li, G.H.; Yu, Z.F.; Li, X.; Wang, X.B.; Zheng, L.J.; Zhang, K.Q. Nematicidal metabolites produced by the endophytic fungus *Geotrichum* sp. AL4. *Chem. Biodivers.* **2007**, *4*, 1520–1524. [CrossRef] [PubMed]
26. Abubakar, S.; Ndana, R.W. Preliminary study of Endomycodiversity among three ethnomedicinal plants from family Meliaceae in Nigeria. *J. BioSci. Biotechnol.* **2016**, *5*, 195–201.
27. Huang, R.; Xie, X.S.; Fang, X.W.; Ma, K.X.; Wu, S.H. Five new guaiane sesquiterpenes from the endophytic fungus *Xylaria* sp. YM 311647 of *Azadirachta indica*. *Chem. Biodivers.* **2015**, *12*, 1281–1286. [CrossRef] [PubMed]
28. Wu, S.H.; Chen, Y.W.; Shao, S.C.; Wang, L.D.; Li, Z.Y.; Yang, L.Y.; Li, S.L.; Huang, R. Ten-membered lactones from *Phomopsis* sp., an endophytic fungus of *Azadirachta indica*. *J. Nat. Prod.* **2008**, *71*, 731–734. [CrossRef] [PubMed]
29. Wu, S.-H.; Chen, Y.-W.; Miao, C.-P. Secondary metabolites of endophytic fungus *Xylaria* sp. YC-10 of *Azadirachta indica*. *Chem. Nat. Compd.* **2011**, *47*, 858–861. [CrossRef]
30. Tejesvi, M.V.; Kini, K.R.; Prakash, H.S.; Subbiah, V.; Shetty, H.S. Antioxidant, antihypertensive, and antibacterial properties of endophytic *Pestalotiopsis* species from medicinal plants. *Can. J. Microbiol.* **2008**, *54*, 769–780. [CrossRef] [PubMed]
31. Kharwar, R.N.; Verma, V.C.; Kumar, A.; Gond, S.K.; Harper, J.K.; Hess, W.M.; Lobkovosky, E.; Ma, C.; Ren, Y.; Strobel, G.A. Javanicin, an antibacterial naphthaquinone from an endophytic fungus of neem, *chloridium* sp. *Curr. Microbiol.* **2009**, *58*, 233–238. [CrossRef] [PubMed]
32. Kusari, S.; Singh, S.; Jayabaskaran, C. Biotechnological potential of plant-associated endophytic fungi: Hope versus hype. *Trends Biotechnol.* **2014**, *32*, 297–303. [CrossRef] [PubMed]
33. Vedashree, S.; Sateesh, M.K.; Lakshmeesha, T.R.; Mohammed, S.S. Screening and assay of extracellular enzymes in *Phomopsis azadirachtae* causing die-back disease of neem. *J. Agric. Technol.* **2013**, *9*, 915–927.
34. James, T.Y.; Kauff, F.; Schoch, C.L.; Matheny, P.B.; Hofstetter, V.; Cox, C.J.; Celio, G.; Gueidan, C.; Fraker, E.; Miadlikowska, J.; et al. Reconstructing the early evolution of Fungi using a six-gene phylogeny. *Nature* **2006**, *443*, 818–822. [CrossRef] [PubMed]
35. Hongsanan, S.; Maharachchikumbura, S.S.N.; Hyde, K.D.; Samarakoon, M.C.; Jeewon, R.; Zhao, Q.; Al-Sadi, A.M.; Bahkali, A.H. An updated phylogeny of Sordariomycetes based on phylogenetic and molecular clock evidence. *Fungal Divers.* **2017**, *84*, 25–41. [CrossRef]
36. Maharachchikumbura, S.S.; Hyde, K.D.; Jones, E.G.; McKenzie, E.H.; Huang, S.K.; Abdel-Wahab, M.A.; Daranagama, D.A.; Dayarathne, M.; D'souza, M.J.; Goonasekara, I.D.; et al. Towards a natural classification and backbone tree for Sordariomycetes. *Fungal Divers.* **2015**, *72*, 199–301.

37. Summerbell, R.C.; Gueidan, C.; Schroers, H.J.; de Hoog, G.S.; Starink, M.; Rosete, A.Y.; Guarro, J.; Scott, J.A. Acremonium phylogenetic overview and revision of *Gliomastix*, *Sarocladium*, and *Trichothecium*. *Stud. Mycol.* **2011**, *68*, 139–162. [CrossRef] [PubMed]
38. Wijayawardene, N.N.; Hyde, K.D.; Wanasinghe, D.N.; Papizadeh, M.; Goonasekara, I.D.; Camporesi, E.; Bhat, D.J.; McKenzie, E.H.C.; Phillips, A.J.L.; Diederich, P.; et al. Taxonomy and phylogeny of dematiaceous coelomycetes. *Fungal Divers.* **2016**, *77*, 1–316.
39. Senanayake, I.C.; Maharachchikumbura, S.S.N.; Hyde, K.D.; Bhat, J.D.; Jones, E.B.G.; McKenzie, E.H.C.; Dai, D.Q.; Daranagama, D.A.; Dayarathne, M.C.; Goonasekara, I.D.; et al. Towards unraveling relationships in Xylariomycetidae (Sordariomycetes). *Fungal Divers.* **2015**, *73*, 73–144. [CrossRef]
40. Wu, S.-H.; Chen, Y.-W.; Shao, S.-C.; Wang, L.-D.; Li, Z.-Y.; Yang, L.-Y.; Li, S.-L.; Huang, R. Ten-Membered Lactones from *Phomopsis* sp., an Endophytic Fungus of *Azadirachta indica*. *J. Nat. Prod.* **2008**, *71*, 731–734. [CrossRef] [PubMed]
41. Spatafora, J.W.; Sung, G.-H.; Johnson, D.; Hesse, C.; O'Rourke, B.; Serdani, M.; Spotts, R.; Lutzoni, F.; Hofstetter, V.; Miadlikowska, J.; et al. A five-gene phylogeny of *Pezizomycotina*. *Mycologia* **2006**, *98*, 1018–1028. [CrossRef] [PubMed]
42. Kusari, S.; Verma, V.C.; Lamshoeft, M.; Spiteller, M. An endophytic fungus from *Azadirachta indica* A. Juss. that produces azadirachtin. *World J. Microbiol. Biotechnol.* **2012**, *28*, 1287–1294. [CrossRef] [PubMed]
43. Schubert, K.; Groenewald, J.Z.; Braun, U.; Dijksterhuis, J.; Starink, M.; Hill, C.F.; Zalar, P.; de Hoog, G.S.; Crous, P.W. Biodiversity in the *Cladosporium herbarum* complex (Davidiellaceae, Capnodiales), with standardisation of methods for *Cladosporium* taxonomy and diagnostics. *Stud. Mycol.* **2007**, *58*, 105–156. [CrossRef] [PubMed]
44. Crous, P.W.; Braun, U.; Schubert, K.; Groenewald, J.Z. Delimiting *Cladosporium* from morphologically similar genera. *Stud. Mycol.* **2007**, *58*, 33–56. [CrossRef] [PubMed]
45. Wikee, S.; Lombard, L.; Nakashima, C.; Motohashi, K.; Chukeatirote, E.; Cheewangkoon, R.; McKenzie, E.H.C.; Hyde, K.D.; Crous, P.W. A phylogenetic re-evaluation of *Phyllosticta* (Botryosphaeriales). *Stud. Mycol.* **2013**, *76*, 1–29. [CrossRef] [PubMed]
46. Hibbett, D.S.; Binder, M.; Bischoff, J.F.; Blackwell, M.; Cannon, P.F.; Eriksson, O.E.; Huhndorf, S.; James, T.; Kirk, P.M.; Lücking, R.; Lumbsch, H.T.; et al. A higher-level phylogenetic classification of the Fungi. *Mycol. Res.* **2007**, *111*, 509–547. [CrossRef] [PubMed]
47. Chowdhary, K.; Kaushik, N.; Coloma, A.G.; Raimundo, C.M. Endophytic fungi and their metabolites isolated from Indian medicinal plant. *Phytochem. Rev.* **2012**, *11*, 467–485. [CrossRef]
48. Prakash, H.S.; Ruma, K.; Shailasree, S.; Kumara, K.; Sampath, S.; Ramachandrappa, N.; Prakash, H.S. Diversity of Fungal Endophytes from Two Endemic Tree Species *Artocarpus hirsutus* Lam. And *Vateria indica* Linn. of Western Ghats, India. *World J. Agric. Sci.* **2011**, *7*, 577–582.
49. Manawasinghe, I.S.; Phillips, A.J.L.; Hyde, K.D.; Chethana, K.W.T.; Zhang, W.; Zhao, W.S.; Yan, J.Y.; Li, X. Mycosphere Essays 14: Assessing the aggressiveness of plant pathogenic Botryosphaeriaceae. *Mycosphere* **2016**, *7*, 883–892. [CrossRef]
50. Rajagopal, K.; Kathiravan, G.; Karthikeyan, S. Extraction and characterization of melanin from *Phomopsis*: A phellophytic fungi Isolated from *Azadirachta indica* A. Juss. *Afr. J. Microbiol. Res.* **2011**, *5*, 762–766.
51. Suryanarayanan, T.S.; Ravishankar, J.P.; Venkatesan, G.; Murali, T.S. Characterization of the melanin pigment of a cosmopolitan fungal endophyte. *Mycol. Res.* **2004**, *108*, 974–978. [CrossRef] [PubMed]
52. Zaidi, K.U.; Mani, A.; Ali, A.S.; Ali, S.A. Evaluation of Tyrosinase Producing Endophytic Fungi from *Calotropis gigantea*, *Azadirachta indica*, *Ocimum tenuiflorum* and *Lantana camara*. *Annu. Rev. Res. Biol.* **2013**, *3*, 389–396.
53. Selinheimo, E.; Saloheimo, M.; Ahola, E.; Westerholm-Parvinen, A.; Kalkkinen, N.; Buchert, J.; Kruus, K. Production and characterization of a secreted, C-terminally processed tyrosinase from the filamentous fungus *Trichoderma reesei*. *FEBS J.* **2006**, *273*, 4322–4335. [CrossRef] [PubMed]
54. Abubakar, S.; Ndana, R.W.; Afolabi, A.S. Bioprospective potentials of endophytic fungi *Penicillium spp.* isolated from leaves of *Azadirachta indica* (A. JUSS). *Int. J. Biol. Res.* **2017**, *5*, 15–21.
55. Zhao, J.; Fu, Y.; Luo, M.; Zu, Y.; Wang, W.; Zhao, C.; Gu, C. Endophytic fungi from pigeon pea [*Cajanus cajan* (L.) Millsp.] produce antioxidant cajaninstilbene acid. *J. Agric. Food Chem.* **2012**, *60*, 4314–4319. [CrossRef] [PubMed]

56. Wu, S.-H.; Chen, Y.-W.; Shao, S.-C.; Wang, L.-D.; Yu, Y.; Li, Z.-Y.; Yang, L.-Y.; Li, S.-L.; Huang, R. Two New Solanapyrone Analogues from the Endophytic Fungus *Nigrospora* sp. YB-141 of *Azadirachta indica*. *Chem. Biodivers.* **2009**, *6*, 79–85. [CrossRef] [PubMed]
57. Verma, V.; Gond, S.; Mishra, A.; Kumar, A.; Kharwar, R. Selection of Natural Strains of Fungal Endophytes from *Azadirachta indica* A. Juss, with Anti-Microbial Activity Against Dermatophytes. *Curr. Bioact. Compd.* **2008**, *4*, 36–40. [CrossRef]
58. Tan, Q.; Yan, X.; Lin, X.; Huang, Y.; Zheng, Z.; Song, S.; Lu, C.; Shen, Y. Chemical Constituents of the Endophytic Fungal Strain *Phomopsis* sp. NXZ-05 of Camptotheca acuminata. *Helv. Chim. Acta* **2007**, *90*, 1811–1817. [CrossRef]
59. Wu, S.-H.; He, J.; Li, X.-N.; Huang, R.; Song, F.; Chen, Y.-W.; Miao, C.-P. Guaiane sesquiterpenes and isopimarane diterpenes from an endophytic fungus *Xylaria* sp. *Phytochemistry* **2014**, *105*, 197–204. [CrossRef] [PubMed]
60. Kaur, H.P.; Singh, B.; Thakur, A.; Kaur, A.; Kaur, S. Studies on immunomodulatory effect of endophytic fungus *Alternaria alternata* on *Spodoptera litura*. *J. Asia. Pac. Entomol.* **2015**, *18*, 67–75. [CrossRef]
61. Verma, V.C.; Gangwar, M.; Yashpal, M.; Nath, G. Anticestodal activity of endophytic *Pestalotiopsis* sp. on protoscoleces of hydatid cyst *Echinococcus granulosus*. *Biomed Res. Int.* **2013**, *2013*, 308515. [CrossRef] [PubMed]

Antifungal Activities of Volatile Secondary Metabolites of Four *Diaporthe* Strains Isolated from *Catharanthus roseus*

Dong-Hui Yan [1,*], Xiaoyu Song [1], Hongchang Li [1], Tushou Luo [2], Guiming Dou [1] and Gary Strobel [3,*]

1 Research Institute of Forest Ecology, Environment and Protection, Chinese Academy of Forestry, Hai Dian District, Beijing 100091, China; songxiaoyucaf@gmail.com (X.S.); lhc1994224@163.com (H.L.); dgmgogogo@126.com (G.D.)
2 Research Institute of Tropical Forestry, Chinese Academy of Forestry, Tianhe District, Guangzhou 510520, China; luots@126.com
3 Department of Plant Sciences, Montana State University, Bozeman, MT 59717, USA
* Correspondence: yandh@caf.ac.cn (D.-H.Y.); uplgs@montana.edu (G.S.)

Abstract: Four endophytic fungi were isolated from the medicinal plant, *Catharanthus roseus*, and were identified as *Diaporthe* spp. with partial translation elongation factor 1-alpha (*TEF1*), beta-tubulin (*TUB*), histone H3 (*HIS*), calmodulin (*CAL*) genes, and rDNA internal transcribed spacer (ITS) region (*TEF1-TUB-HIS–CAL*-ITS) multigene phylogeny suggested for species delimitation in the *Diaporthe* genus. Each fungus produces a unique mixture of volatile organic compounds (VOCs) with an abundant mixture of terpenoids analyzed by headspace solid-phase microextraction (SPME) fiber-GC/MS. These tentatively-detected terpenes included α-muurolene, β-phellandrene, γ-terpinene, and α-thujene, as well as other minor terpenoids, including caryophyllene, patchoulene, cedrene, 2-carene, and thujone. The volatile metabolites of each isolate showed antifungal properties against a wide range of plant pathogenic test fungi and oomycetes, including *Alternaria alternata*, *Botrytis cinerea*, *Colletotrichum gloeosporioides*, *Fusarium graminearum*, and *Phytophthora cinnamomi*. The growth inhibition of the pathogens varied between 10% and 60% within 72 h of exposure. To our knowledge, the endophytic *Diaporthe*-like strains are first reported from *Catharanthus roseus*. VOCs produced by each strain of the endophytic *Diaporthe* fungi were unique components with dominant monoterpenes comparing to known *Diaporthe* fungal VOCs. A discussion is presented on the inhibitive bioactivities of secondary metabolites among endophytic *Diaporthe* fungi and this medicinal plant.

Keywords: endophytic fungi; *Diaporthe* spp.; *Catharanthus roseus*; volatile organic compounds (VOCs); antifungal bioactivity; inhibition; terpene; pathogens

---

1. Introduction

Many plants remain unexplored for their endophytic fungi and the potentially important products that they may produce [1]. *Catharanthus roseus* is known as a pharmaceutical plant containing rich anticancer alkaloids. The extracts of many organs of this plant also exhibit antimicrobial effects [2–6]. It turns out that *Catharanthus roseus* is host to a diverse group of endophytic fungi [7–10]. Some endopytic fungi were found to produce several metabolites biosynthesized by the host *Catharanthus roseus*. The endophytic fungi *Curvularia* sp. CATDLF5 and *Choanephora infundibulifera* CATDLF6 isolated from leaf issues were able to enhance leaf vindoline production content of *C. roseus* cv.

prabal by 2.29–4.03 times through root inoculation [8]. Endophytic *Fusarium* spp. from stem issues seemed to facilitate the host plant to produce secondary metabolites [9]. Additionally, some endophytic fungi from the plant produced antimicrobial compounds. For example, the compounds hydroxyemodin, citreoisocoumarin, citreoisocoumarinol, and cladosporin from endophytic fungi of leaves were effective in inhibiting fungal pathogens [10]. *Diaporthe* are commonly found as endophytes in a wide range of plants around the world [11–15]. These endophytes are prolific producers of antimicrobial metabolites [15,16]. *D. endophytica* and *D. terebinthifolii*, isolated from the medicinal plants *Maytenus ilicifolia* and *Schinus terebinthifolius*, had an inhibitory effect against *Pseudomonas citricarpa* in vitro and in detached fruits [12,13]. The crude extracts of *Diaporthe* sp. MFLUCC16-0682 and *Diaporthe* sp. MFLUCC16-0693 exhibited notable antibacterial and antioxidant activities [14]. An endophytic *Phomopsis* (asexual state of *Diaporthe*) fungus isolated from the stems of *Ficus pumila*, exhibited broad-spectrum antimicrobial activity against Gram-positive and Gram-negative human and phytopathogenic bacteria and fungi [15]. Thus, the genus *Diaporthe* is a potential source of metabolites that can be used in a variety of applications [14]. However, endophytic *Diaporthe* fungi have not been recorded from *Catharanthus roseus* to the present.

Volatile organic compounds (VOCs) have noted biofumigative effects especially from the endophytic fungus—*Muscodor albus* [17]. These observations opened a unique venue for the application of endophytic microorganisms to the ecological-friendly biocontrol of pests [17]. The inhibitive bioactive compounds were also found in a few isolates of endophytic *Diaporthe* [18]. An endophytic *Phomopsis* isolate of *Odontoglossum* sp. in Northern Ecuador was reported to produce a unique mixture of volatile organic compounds (VOCs) with sabinene, 1-butanol, 3-methyl; benzeneethanol; 1-propanol, 2-methyl, and 2-propanone. The VOCs showed antifungal bioactivities on a wide range of plant pathogenic fungi, such as *Sclerotinia*, *Rhizoctonia*, *Fusarium*, *Botrytis*, *Verticillium*, *Colletotrichum* and oomycetes *Pythium*, and *Phytophthora* [18]. The PR4 strain of an endophytic *Phomopsis* obtained from the medicinal plant *Picrorhiza kurroa* also produced a unique set of bioactive VOCs inhibitive to plant pathogenic fungi growth. The dominant compounds in VOCs of the PR4 strain were reported as menthol, phenylethyl alcohol, isomenthol, β-phellandrene, β-bisabolene, limonene, 3-pentanone and 1-pentanol [19]. In view of the antimicrobial properties of the extracts from the medicinal plant *Cantharatus roseus*, and limited knowledge on endophytic *Diaporthe* species in this host, we conducted an investigation on the antifungal bioactivity of VOCs from four endophytic *Diaporthe* strains isolated from wild *Catharanthus roseus* in China. The combined sequences of five loci, elongation factor 1-alpha (*TEF1*), beta-tubulin (*TUB*), histone H3 (*HIS*), calmodulin (*CAL*) genes, and the rDNA internal transcribed spacer (ITS) region were used for the strains' phylogenetic analyses within genus *Diaprothe*. Inhibitory bioactivity executed volatile organic compounds from the strains were observed on growths of tested plant pathogens in co-culture. Active components of VOCs were analyzed and inferred using headspace solid-phase microextraction (SPME) fiber-GC/MS and based on their reported properties.

2. Materials and Methods

2.1. Endophytic Fungal Isolation

The four endophytic fungi were isolated from wild plants, *Catharanthus roseus*, growing in the National Natural Conservation Area of TongGu Mountain, located in Wenchang city of Hainan Province. Several stem segments (5–10 cm in length) were collected for the eventual isolation of endophytes. Retrieving endophytic fungi followed a previously described procedure [20]. Briefly, the external tissues of segments were cleaned with tap water and scrubbed with 70% ethanol prior to excision of internal tissues. Then the segments were excised into smaller fragments about 0.2–0.5 cm in length. The fragments were thoroughly exposed to 75% ethanol for 60 s, 3% NaClO for 90 s, and sterile water for 60 s by agitation. The fragments at the last step were drained on sterile filter papers and put on water agar in Petri plates for growing endophytes. Further, pure isolates were obtained in potato dextrose agar media and stored on sterilized, inoculated barley seeds at 4 °C and −80 °C.

The four fungi of interest were assigned with our laboratory acquisition number-ID FPYF3053-3056 and deposited in China Forestry Culture Collection Center assigned IDs of CFCC 52704-52707.

2.2. DNA Extraction, PCR, and Sequencing

Fungal genomic DNA was extracted from colonies growing on PDA for one week with the CTAB procedure [20]. The extracted DNA was further purified through Mini Purification kit (Tiangen Biotech (Beijing) Co., Ltd., Beijing, China) following the manufacture's protocols. The DNA quality and concentration were determined with a NanoDrop 2000 (Thermo Fisher Scientific Inc., Waltham, MA, USA) after the DNA was checked with Genegreen nucleic acid dye (Tiangen Biotech (Beijing) Co., Ltd.) in an electrophoresis on 1% agarose gel stained under ultraviolet light. The extracted DNA was used as a template for the further PCR amplification ITS sequence and *TEF1*, *CAL*, *TUB*, and *HIS* genes regions. The primers were used to amplify the ITS targets, namely, the ITS1 and ITS4 [21], *TEF1* with EF1-688F/EF1-1251R [22], *CAL* with CL1F/CL2A or CAL563F/CL2A [23], *TUB* with T1/Bt-2b or Bt2a/Bt-2b [24,25], and *HIS* with HISdiaF/HISdiaR, sequences that were 5′-GGCTCCCCGYAAGCAGCTCGCCTCC-3 and 5′-ATYCCGACTGGATGGTCACACGCTTGG-3, respectively. All PCR reaction mixtures and conditions were followed as per the Taq PCR MasterMix kits (Tiangen Biotech (Beijing) Co., Ltd.) according to the manufacture's protocol. A PCR reaction system consisted of 0.5 μL of each primer (10 μM), 3 μL (15–80 ng) of DNA template, 12.5 μL of 2 × Taq PCR MasterMix (Tiangen Biotech (Beijing) Co., Ltd.), and 8.5 μL of double distilled water in total of 25 μL. The ITS thermal cycling program was as follows: 94 °C for 5 min, followed by 35 amplification cycles of 94 °C for 60 s, 55 °C for 30 s and 72 °C for 1 min, and a final extension step of 72 °C for 5 min. The annealing temperature at 55 °C for 45 s was changed in this program for *CAL*, *β-tubulin* and *TEF* amplification. For amplification of *HIS*, the program was changed with a cycling program of 32 cycles and an annealing temperature at 55 °C for 60 s. PCR products were visualized on 1.5% agarose gels mixed with Genegreen Nucleic Acid Dye and purified with a quick Midi Purification kit (Tiangen Biotech (Beijing) Co., Ltd.) according to the manufacturer's instructions. Sequencing PCR products were cycle-sequenced the BigDye® Terminator Cycle Sequencing Kit v. 3.1 (Applied Biosystems, Foster City, CA, USA) in an ABI Prism 3730 DNA Sequencer (Applied Biosystems, Foster City, CA, USA) at Biomed Company in Beijing. Then sequence data collected by ABI 3730 Data collection v. 3.0 (Applied Biosystems, Foster City, CA, USA) and ABI Peak Scanner Software v. 1.0 (Applied Biosystems, Foster City, CA, USA), were assembled with forward and reverse sequences by BioEdit. The gene sequences were submitted and awarded access numbers in GenBank of NCBI (Table 1).

Table 1. Access numbers for ITS, translation elongation factor 1-alpha (*TEF1*), beta-tubulin (*TUB*), histone H3 (*HIS*), calmodulin (*CAL*) genes region sequences of the four endophytic *Diaporthe* fungi in the GenBank of NCBI.

Isolate	ITS	*TEF1*	*Tublin*	*CAL*	*HIS*
FPYF3053	MH203054.1	MH220826.1	MH220836.1	MH220831.1	MH220839.1
FPYF3054	MH203055.1	MH220827.1	MH220833.1	MH220832.1	MH220840.1
FPYF3055	MH203056.1	MH220828.1	MH220834.1	MH220829.1	MH220837.1
FPYF3056	MH203057.1	MH220825.1	MH220835.1	MH220830.1	MH220838.1

2.3. Sequence Alignment and Phylogenetic Analysis

In order to determine the phylogenetic locations of the four isolates within the *Diaporthe* genus, 143 reference taxa [26] (Table 2) together with the four isolates were used for building a phylogenetic a tree with *Diaporthella corylina* as a root outgroup species [23]. The evolutionary relationships were taken on a five-gene concatenated alignment of ITS, *TEF1*, *CAL*, *HIS*, and *TUB* regions by maximum likelihood (ML) and maximum parsimony (MP) phylogenetic analyses. Sequences were aligned using the MAFFTv.7 online program with default parameters [27]. A partition homogeneity test implemented in PAUP* v.4.0 (Sinauer Associates, Sunderland, MA, USA) was applied to determine if the five sequence data could be combined. The best evolutionary model for the partitioning analysis was performed on the concatenated sequences by PartitionFinder 2.1.1 [28]. A concatenated alignment for the five gene regions was made from SequenceMatrix [29]. The inference methods of maximum likelihood and maximum parsimony in Mega 6.0 [30] were applied to estimate phylogeny for the concatenated sequences, with the evolutionary models GTR and AIC for ML and MP, respectively, with a bootstrap support of 1000 replicates. Evidence on the trees were visualized and edited by TreeGraph 2 [31].

Table 2. Reference sequences of *Diaporthe* strains with NCBI access numbers for phylogenetic analysis.

Source	ITS	*TEF1*	*TUB*	*CAL*	*HIS*
*Diaporthe acaciigena*_CBS 129521	KC343005.1	KC343731.1	KC343973.1	KC343247.1	KC343489.1
*Diaporthe acerina*_CBS 137.27	KC343006.1	KC343732.1	KC343974.1	KC343248.1	KC343490.1
*Diaporthe alleghaniensis*_CBS 495.72	KC343007.1	KC343733.1	KC343975.1	KC343249.1	KC343491.1
*Diaporthe alnea*_CBS 146.46	KC343008.1	KC343734.1	KC343976.1	KC343250.1	KC343492.1
*Diaporthe alnea*_CBS 159.47	KC343009.1	KC343735.1	KC343977.1	KC343251.1	KC343493.1
*Diaporthe ambigua*_CBS 114015	KC343010.1	KC343736.1	KC343978.1	KC343252.1	KC343494.1
*Diaporthe ambigua*_CBS 117167	KC343011.1	KC343737.1	KC343979.1	KC343253.1	KC343495.1
*Diaporthe amygdali*_CBS 126679	KC343022.1	KC343742.1	KC343984.1	KC343258.1	KC343506.1
*Diaporthe ampelina*_CBS 111888	KC343016.1	KC343748.1	KC343990.1	KC343264.1	KC343500.1
*Diaporthe amygdali*_CBS 111811	KC343019.1	KC343745.1	KC343987.1	KC343261.1	KC343503.1
*Diaporthe anacardii*_CBS 720.97	KC343024.1	KC343750.1	KC343992.1	KC343266.1	KC343508.1
*Diaporthe angelicae*_CBS 111592	KC343027.1	KC343753.1	KC343995.1	KC343269.1	KC343511.1
*Diaporthe angelicae*_CBS 123215	KC343028.1	KC343754.1	KC343996.1	KC343270.1	KC343512.1
*Diaporthe cucurbitae*_CBS 136.25	KC343031.1	KC343757.1	KC343999.1	KC343273.1	KC343515.1
*Diaporthe arecae*_CBS 161.64	KC343032.1	KC343758.1	KC344000.1	KC343274.1	KC343516.1
*Diaporthe arecae*_CBS 535.75	KC343033.1	KC343759.1	KC344001.1	KC343275.1	KC343517.1
*Diaporthe arengae*_CBS 114979	KC343034.1	KC343760.1	KC344002.1	KC343276.1	KC343518.1
*Diaporthe aspalathi*_CBS 117169	KC343036.1	KC343762.1	KC344004.1	KC343278.1	KC343520.1
*Diaporthe aspalathi*_CBS 117168	KC343035.1	KC343761.1	KC344003.1	KC343277.1	KC343519.1
*Diaporthe australafricana*_CBS 111886	KC343038.1	KC343764.1	KC344006.1	KC343280.1	KC343522.1
*Diaporthe australafricana*_CBS 113487	KC343039.1	KC343765.1	KC344007.1	KC343281.1	KC343523.1
*Diaporthe batatas*_CBS 122.21	KC343040.1	KC343766.1	KC344008.1	KC343282.1	KC343524.1
*Diaporthe beckhausii*_CBS 138.27	KC343041.1	KC343767.1	KC344009.1	KC343283.1	KC343525.1
*Diaporthe bicincta*_CBS 121004	KC343134.1	KC343860.1	KC344102.1	KC343376.1	KC343618.1
*Diaporthe brasiliensis*_CBS 133183	KC343042.1	KC343768.1	KC344010.1	KC343284.1	KC343526.1
*Diaporthe brasiliensis*_LGMF926	KC343043.1	KC343769.1	KC344011.1	KC343285.1	KC343527.1
*Diaporthe carpini*_CBS 114437	KC343044.1	KC343770.1	KC344012.1	KC343286.1	KC343528.1
*Diaporthe caulivora*_CBS 127268	KC343045.1	KC343771.1	KC344013.1	KC343287.1	KC343529.1
*Diaporthe caulivora*_CBS 178.55	KC343046.1	KC343772.1	KC344014.1	KC343288.1	KC343530.1
*Diaporthe celastrina*_CBS 139.27	KC343047.1	KC343773.1	KC344015.1	KC343289.1	KC343531.1
*Diaporthe chamaeropis*_CBS 454.81	KC343048.1	KC343774.1	KC344016.1	KC343290.1	KC343532.1
*Diaporthe chamaeropis*_CBS 753.70	KC343049.1	KC343775.1	KC344017.1	KC343291.1	KC343533.1
*Diaporthe cinerascens*_CBS 719.96	KC343050.1	KC343776.1	KC344018.1	KC343292.1	KC343534.1
*Diaporthe citri*_CBS 199.39	KC343051.1	KC343777.1	KC344019.1	KC343293.1	KC343535.1
*Diaporthe citri*_CBS 230.52	KC343052.1	KC343778.1	KC344020.1	KC343294.1	KC343536.1
*Diaporthe convolvuli*_CBS 124654	KC343054.1	KC343780.1	KC344022.1	KC343296.1	KC343538.1
*Diaporthe crataegi*_CBS 114435	KC343055.1	KC343781.1	KC344023.1	KC343297.1	KC343539.1
*Diaporthe crotalariae*_CBS 162.33	KC343056.1	KC343782.1	KC344024.1	KC343298.1	KC343540.1
*Diaporthe cuppatea*_CBS 117499	KC343057.1	KC343783.1	KC344025.1	KC343299.1	KC343541.1
*Diaporthe cynaroidis*_CBS 122676	KC343058.1	KC343784.1	KC344026.1	KC343300.1	KC343542.1
*Diaporthe decedens*_CBS 109772	KC343059.1	KC343785.1	KC344027.1	KC343301.1	KC343543.1
*Diaporthe decedens*_CBS 114281	KC343060.1	KC343786.1	KC344028.1	KC343302.1	KC343544.1
*Diaporthe detrusa*_CBS 109770	KC343061.1	KC343787.1	KC344029.1	KC343303.1	KC343545.1
*Diaporthe detrusa*_CBS 114652	KC343062.1	KC343788.1	KC344030.1	KC343304.1	KC343546.1
*Diaporthe elaeagni*_CBS 504.72	KC343064.1	KC343790.1	KC344032.1	KC343306.1	KC343548.1
*Diaporthe endophytica*_CBS 133811	KC343065.1	KC343791.1	KC344033.1	KC343307.1	KC343549.1
*Diaporthe endophytica*_LGMF928	KC343068.1	KC343794.1	KC344036.1	KC343310.1	KC343552.1
*Diaporthe eres*_CBS 439.82	KC343090.1	KC343816.1	KC344058.1	KC343332.1	KC343574.1
*Diaporthe eres*_CBS 101742	KC343073.1	KC343799.1	KC344041.1	KC343315.1	KC343557.1
*Diaporthe eres*_CBS 109767	KC343075.1	KC343801.1	KC344043.1	KC343317.1	KC343559.1
Diaporthe cf. *nobilis* RG-2013_CBS 113470	KC343146.1	KC343872.1	KC344114.1	KC343388.1	KC343630.1
Diaporthe cf. *nobilis* RG-2013_CBS 116953	KC343147.1	KC343873.1	KC344115.1	KC343389.1	KC343631.1

Table 2. *Cont.*

Source	ITS	*TEF1*	*TUB*	*CAL*	*HIS*
Diaporthe cf. *nobilis* RG-2013_CBS 200.39	KC343151.1	KC343877.1	KC344119.1	KC343393.1	KC343635.1
*Diaporthe eugeniae*_CBS 444.82	KC343098.1	KC343824.1	KC344066.1	KC343340.1	KC343582.1
*Diaporthe fibrosa*_CBS 109751	KC343099.1	KC343825.1	KC344067.1	KC343341.1	KC343583.1
*Diaporthe fibrosa*_CBS 113830	KC343100.1	KC343826.1	KC344068.1	KC343342.1	KC343584.1
*Diaporthe foeniculacea*_CBS 123208	KC343104.1	KC343830.1	KC344072.1	KC343346.1	KC343588.1
*Diaporthe foeniculacea*_CBS 111553	KC343101.1	KC343827.1	KC344069.1	KC343349.1	KC343585.1
*Diaporthe foeniculacea*_CBS 187.27	KC343107.1	KC343833.1	KC344075.1	KC343343.1	KC343591.1
*Diaporthe ganjae*_CBS 180.91	KC343112.1	KC343838.1	KC344080.1	KC343354.1	KC343596.1
*Diaporthe gardeniae*_CBS 288.56	KC343113.1	KC343839.1	KC344081.1	KC343355.1	KC343597.1
*Diaporthe helianthi*_CBS 592.81	KC343115.1	KC343841.1	KC344083.1	KC343357.1	KC343599.1
*Diaporthe helianthi*_CBS 344.94	KC343114.1	KC343840.1	KC344082.1	KC343356.1	KC343598.1
*Diaporthe hickoriae*_CBS 145.26	KC343118.1	KC343844.1	KC344086.1	KC343360.1	KC343602.1
*Diaporthe hongkongensis*_CBS 115448	KC343119.1	KC343845.1	KC344087.1	KC343361.1	KC343603.1
*Diaporthe hordei*_CBS 481.92	KC343120.1	KC343846.1	KC344088.1	KC343362.1	KC343604.1
*Diaporthe impulsa*_CBS 114434	KC343121.1	KC343847.1	KC344089.1	KC343363.1	KC343605.1
*Diaporthe impulsa*_CBS 141.27	KC343122.1	KC343848.1	KC344090.1	KC343364.1	KC343606.1
*Diaporthe inconspicua*_LGMF922	KC343124.1	KC343849.1	KC344091.1	KC343365.1	KC343607.1
*Diaporthe inconspicua*_CBS 133813	KC343123.1	KC343850.1	KC344092.1	KC343366.1	KC343608.1
*Diaporthe infecunda*_CBS 133812	KC343126.1	KC343852.1	KC344094.1	KC343368.1	KC343610.1
*Diaporthe infecunda*_LGMF933	KC343132.1	KC343858.1	KC344100.1	KC343374.1	KC343616.1
*Diaporthe longispora*_CBS 194.36	KC343135.1	KC343861.1	KC344103.1	KC343377.1	KC343619.1
*Diaporthe lusitanicae*_CBS 123212	KC343136.1	KC343862.1	KC344104.1	KC343378.1	KC343620.1
*Diaporthe lusitanicae*_CBS 123213	KC343137.1	KC343863.1	KC344105.1	KC343379.1	KC343621.1
*Diaporthe manihotia*_CBS 505.76	KC343138.1	KC343864.1	KC344106.1	KC343380.1	KC343622.1
*Diaporthe mayteni*_CBS 133185	KC343139.1	KC343865.1	KC344107.1	KC343381.1	KC343623.1
*Diaporthe megalospora*_CBS 143.27	KC343140.1	KC343866.1	KC344108.1	KC343383.1	KC343624.1
*Diaporthe melonis*_CBS 507.78	KC343142.1	KC343868.1	KC344110.1	KC343384.1	KC343626.1
*Diaporthe melonis*_CBS 435.87	KC343141.1	KC343867.1	KC344109.1	KC343382.1	KC343625.1
*Diaporthe musigena*_CBS 129519	KC343143.1	KC343869.1	KC344111.1	KC343385.1	KC343627.1
*Diaporthe neilliae*_CBS 144.27	KC343144.1	KC343870.1	KC344112.1	KC343386.1	KC343628.1
*Diaporthe neoarctii*_CBS 109490	KC343145.1	KC343871.1	KC344113.1	KC343387.1	KC343629.1
*Diaporthe nomurai*_CBS 157.29	KC343154.1	KC343880.1	KC344122.1	KC343396.1	KC343638.1
*Diaporthe novem*_CBS 127270	KC343156.1	KC343882.1	KC344124.1	KC343398.1	KC343640.1
*Diaporthe novem*_CBS 354.71	KC343158.1	KC343884.1	KC344126.1	KC343400.1	KC343642.1
*Diaporthe oncostoma*_CBS 109741	KC343161.1	KC343887.1	KC344129.1	KC343403.1	KC343645.1
*Diaporthe oncostoma*_CBS 100454	KC343160.1	KC343886.1	KC344128.1	KC343402.1	KC343644.1
*Diaporthe oxe*_CBS 133186	KC343164.1	KC343890.1	KC344132.1	KC343406.1	KC343648.1
*Diaporthe oxe*_CBS 133187	KC343165.1	KC343891.1	KC344133.1	KC343407.1	KC343649.1
Diaporthe padi var. *padi*_CBS 114200	KC343169.1	KC343895.1	KC344137.1	KC343411.1	KC343653.1
Diaporthe padi var. *padi*_CBS 114649	KC343170.1	KC343896.1	KC344138.1	KC343412.1	KC343654.1
*Diaporthe paranensis*_CBS 133184	KC343171.1	KC343897.1	KC344139.1	KC343413.1	KC343655.1
*Diaporthe perjuncta*_CBS 109745	KC343172.1	KC343898.1	KC344140.1	KC343414.1	KC343656.1
*Diaporthe perseae*_CBS 151.73	KC343173.1	KC343899.1	KC344141.1	KC343415.1	KC343657.1
*Diaporthe phaseolorum*_CBS 116019	KC343175.1	KC343901.1	KC344143.1	KC343417.1	KC343659.1
*Diaporthe phaseolorum*_CBS 116020	KC343176.1	KC343902.1	KC344144.1	KC343418.1	KC343660.1
*Diaporthe pseudomangiferae*_CBS 101339	KC343181.1	KC343907.1	KC344149.1	KC343423.1	KC343665.1
*Diaporthe pseudomangiferae*_CBS 388.89	KC343182.1	KC343908.1	KC344150.1	KC343424.1	KC343666.1
*Diaporthe pseudophoenicicola*_CBS 462.69	KC343184.1	KC343910.1	KC344152.1	KC343426.1	KC343668.1
*Diaporthe pseudophoenicicola*_CBS 176.77	KC343183.1	KC343909.1	KC344151.1	KC343425.1	KC343667.1
*Diaporthe pustulata*_CBS 109784	KC343187.1	KC343913.1	KC344155.1	KC343429.1	KC343671.1
*Diaporthe pustulata*_CBS 109742	KC343185.1	KC343911.1	KC344153.1	KC343427.1	KC343669.1
*Diaporthe raonikayaporum*_CBS 133182	KC343188.1	KC343914.1	KC344156.1	KC343430.1	KC343672.1
*Diaporthe rhoina*_CBS 146.27	KC343189.1	KC343915.1	KC344157.1	KC343431.1	KC343673.1
*Diaporthe saccarata*_CBS 116311	KC343190.1	KC343916.1	KC344158.1	KC343432.1	KC343674.1
*Diaporthe schini*_CBS 133181	KC343191.1	KC343917.1	KC344159.1	KC343433.1	KC343675.1
*Diaporthe schini*_LGMF910	KC343192.1	KC343918.1	KC344160.1	KC343434.1	KC343676.1
*Diaporthe sclerotioides*_CBS 296.67	KC343193.1	KC343919.1	KC344161.1	KC343435.1	KC343677.1
*Diaporthe sclerotioides*_CBS 710.76	KC343194.1	KC343920.1	KC344162.1	KC343436.1	KC343678.1
*Diaporthe scobina*_CBS 251.38	KC343195.1	KC343921.1	KC344163.1	KC343437.1	KC343679.1
*Diaporthe sojae*_CBS 100.87	KC343196.1	KC343922.1	KC344164.1	KC343438.1	KC343680.1
Diaporthe longicolla isolate PL4	HM347700.1	HM347685.1	KC344167.1	KC343441.1	KC343683.1
*Diaporthe sojae*_CBS 116017	KC343197.1	KC343923.1	KC344165.1	KC343439.1	KC343681.1
*Diaporthe sojae*_CBS 180.55	KC343200.1	KC343926.1	KC344168.1	KC343442.1	KC343684.1
*Diaporthe subordinaria*_CBS 101711	KC343213.1	KC343938.1	KC344180.1	KC343454.1	KC343696.1
*Diaporthe subordinaria*_CBS 464.90	KC343214.1	KC343939.1	KC344181.1	KC343455.1	KC343697.1

Table 2. *Cont.*

Source	ITS	TEF1	TUB	CAL	HIS
*Diaporthe tecomae*_CBS 100547	KC343215.1	KC343940.1	KC344182.1	KC343456.1	KC343698.1
*Diaporthe terebinthifolii*_CBS 133180	KC343216.1	KC343941.1	KC344184.1	KC343457.1	KC343699.1
*Diaporthe terebinthifolii*_LGMF907	KC343217.1	KC343942.1	KC344183.1	KC343458.1	KC343700.1
*Diaporthe toxica*_CBS 534.93	KC343220.1	KC343943.1	KC344185.1	KC343459.1	KC343701.1
*Diaporthe toxica*_CBS 535.93	KC343221.1	KC343946.1	KC344188.1	KC343462.1	KC343704.1
*Diaporthe vaccinii*_CBS 160.32	KC343228.1	KC343947.1	KC344189.1	KC343463.1	KC343705.1
*Diaporthe vaccinii*_CBS 122112	KC343224.1	KC343954.1	KC344196.1	KC343470.1	KC343712.1
*Diaporthe vexans*_CBS 127.14	KC343229.1	KC343950.1	KC344192.1	KC343466.1	KC343708.1
*Diaporthe rudis*_CBS 113201	KC343234.1	KC343955.1	KC344197.1	KC343471.1	KC343713.1
*Diaporthe rudis*_CBS 109768	KC343233.1	KC343960.1	KC344202.1	KC343476.1	KC343718.1
*Diaporthe woodii*_CBS 558.93	KC343244.1	KC343959.1	KC344201.1	KC343475.1	KC343717.1
*Diaporthe woolworthii*_CBS 148.27	KC343245.1	KC343970.1	KC344212.1	KC343486.1	KC343728.1
Diaporthe cf. *heveae* 1 RG-2013_CBS 852.97	KC343116.1	KC343971.1	KC344213.1	KC343487.1	KC343729.1
Diaporthe cf. *heveae* 2 RG-2013_CBS 681.84	KC343117.1	KC343842.1	KC344084.1	KC343358.1	KC343600.1
Diaporthe sp. 1 RG-2013_CBS 119639	KC343202.1	KC343843.1	KC344085.1	KC343359.1	KC343601.1
Diaporthe sp. 1 RG-2013_LGMF947	KC343203.1	KC343928.1	KC344170.1	KC343444.1	KC343686.1
Diaporthe sp. 2 RG-2013_LGMF932	KC343204.1	KC343929.1	KC344171.1	KC343445.1	KC343687.1
Diaporthe sp. 3 RG-2013_CBS 287.29	KC343205.1	KC343930.1	KC344172.1	KC343446.1	KC343688.1
Diaporthe sp. 4 RG-2013_LGMF944	KC343206.1	KC343931.1	KC344173.1	KC343448.1	KC343689.1
Diaporthe sp. 5 RG-2013_CBS 125575	KC343207.1	KC343932.1	KC344174.1	KC343447.1	KC343690.1
Diaporthe sp. 6 RG-2013_CBS 115584	KC343208.1	KC343933.1	KC344175.1	KC343449.1	KC343691.1
Diaporthe sp. 6 RG-2013_CBS 115595	KC343209.1	KC343934.1	KC344176.1	KC343450.1	KC343692.1
Diaporthe sp. 7 RG-2013_CBS 458.78	KC343210.1	KC343935.1	KC344177.1	KC343451.1	KC343693.1
Diaporthe sp. 8 RG-2013_LGMF925	KC343211.1	KC343936.1	KC344178.1	KC343452.1	KC343694.1
*Diaporthella corylina*_CBS 121124	KC343004.1	KC343937.1	KC344179.1	KC343453.1	KC343695.1
*Diaporthe stictica*_CBS 370.54	KC343212.1	KC343730.1	KC343972.1	KC343246.1	KC343488.1

2.4. Antifungal Activity Tests for Fungal VOCs

The antifungal activity of the VOCs was determined by the methods previously described [17,18,20]. The four endophytic fungal strains of *Diaporthe* and targeted plant pathogenic microorganisms were paired opposite to each other in Petri plates containing PDA with a diameter of 90 mm. The agar was divided into two halves by removing a 2 cm wide strip in the center. An endophytic test fungus was inoculated onto one half-moon of the agar and incubated at 25 °C for five days for optimum production of volatile compounds before the antagonism bioassay. A test pathogen was inoculated onto the opposite half-moon part of the agar at the fifth day. The plates were then wrapped with parafilm and incubated at 25 °C in dark for 72 h. Growth of filamentous pathogenic fungi were quantitatively assessed after 24 h, 48 h, and 72 h based on multiple measurements of growth relative to controls, as described previously [17,18]. The colony diameter was measured in an average of four diameters on hours 24, 48, and 72 h, disregarding the initial inoculum size. Percentage of growth inhibition was calculated as the formula: $|(a - b/b)| \times 100$, a = mycelial colony diameter in control plate; b = mycelial colony diameter in the antagonism treatment plate. Statistical significance ($p < 0.01$) was evaluated by analysis of variance (ANOVA) followed by the Tukey 5% test. Antifungal activity of VOCs was tested against the plant pathogenic fungi *Alternaria alternata*, *Botryosphaeria dothidea*, *Botrytis cinerea*, *Cercospora* sp., *Colletotrichum gloeosporioides*, *Fusarium graminearum*, *Sphaeropsis sapinea*, and *Valsa sordida*, in addition to the oomycete *Phytophthora cinnamomi*. All tests were made in quintuplicate. Control cultures were obtained by growing each plant pathogen alone, under the same conditions.

2.5. Qualitative Analyses on Volatiles of the Four Endophytic Cultures

VOCs in the air space above the endophytic fungal colonies grown for five days at 25 ± 2 °C on PDA were analyzed using the solid phase microextraction (SPME) fiber technique according to previously described protocols [17,18,20]. Control PDA Petri plates not inoculated with the strain was used to subtract compounds contributed by the medium. All treatments and checks were done in triplicate. A fiber syringe of 50/30 divinylbenzene/carboxen on polydimethylsiloxane (Supelco, Bellefonte, PA, USA) was conditioned for 40 min at 200 °C, exposed to the vapor phase inside Petri during 40 min through a small hole (0.5 mm in diameter) drilled on the sides of the Petri plate. The fiber was directly inserted into the TRACE DSQ inlet (Thermo Electron Corporation, Beverly, MA, USA), at 200 °C, splitless mode. The desorption time was 40 s and the desorbed compounds were separated on a 30.0 m × 0.25 mm × 0.25 μm, HP-5MS capillary column, using the following GC oven temperature program: 2 min at 35 °C up to 220 °C at 7 °C/min. Helium was used as the carrier gas at a flow rate of 1 mL/min. The electronic ionization energy was 70 eV and the mass range scanned was 41–560 uma. The scan rate was 5 spec/s. Transfer line and ionization chamber temperatures were 250 °C and 200 °C respectively. Tentative identification of the volatile compounds produced by the four endophytic *Diaporthe* fungi was made via library comparison using the NIST database and all chemical compounds were described in this report following the NIST database chemical identity. Tentative compound identity was based on at least a 70% quality match with the NIST database information for each compound. Data acquisition and data processing were performed with the Hewlett Packard ChemStation software system (Version 2.0, Scientific Instrument Services, Inc., Ringoes, NJ, USA). Relative amounts of individual components of the treatments were determined and expressed as percentages of the peak area within the total peak area and as an average of the three replicates.

3. Results

3.1. The Identification on the Four Endophytic Isolates within the Diaporthe Genus

Each of the four isolates falling within the genus *Diaporthe* were further defined using molecular analyses as they appeared different, morphologically (Figure 1). For instance, strain FPYF3053 had compact mycelia with crenate margins, these colonies developed a brownish yellow pigmentation in the center on the underside having a growth rate of 18.3 mm day^{-1} (Figure 1a). On the other hand, strain FPYF3054 had aerial mycelium forming concentric rings with grey and dark pigmentation at the center showing a growth rate of 30.97 mm day^{-1} (Figure 1b). Strain FPYF3055 had vigorously-growing aerial hyphae near the margin, but loose hyphae scattered inside with aging, with a growth rate of 23 mm day^{-1} (Figure 1c). Finally, strain FPYF3056 had a compact mycelium with a crenate margin, but no pigmentation with a growth rate of 21.7 mm day^{-1} (Figure 1d).

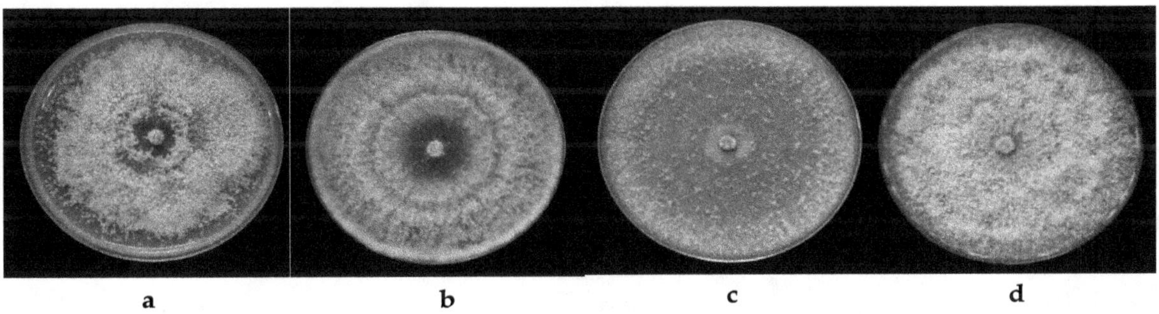

Figure 1. The colony cultures for the four endophytic *Diaporthe* fungi and their plant host. (**a**) FPYF3053; (**b**) FPYF3054; (**c**) FPYF3055; and (**d**) FPYF3056.

A combined alignment of five loci ITS, *TUB*, *TEF1*, *HIS*, and *CAL* was used for ML and MP phylogenic analyses. Based on the multi-locus phylogeny (Figure 2), the four endophytic *Diaporthe* strains could not be placed in one species only because they are distinct from each other and from all reference species listed (Table 2, Figure 2). Strains FPYF3055 and FPYF3056 were clustered by giving a high bootstrap support (BS = 82) from MP inference (Figure S1) while both separated from each other in ML inference (Figure S2). The reference sequences used to construct the phylogenetic tree were listed in Table 2 with their Genbank accession numbers. The alignment was uploaded in Treebase assigned with SI 22757.

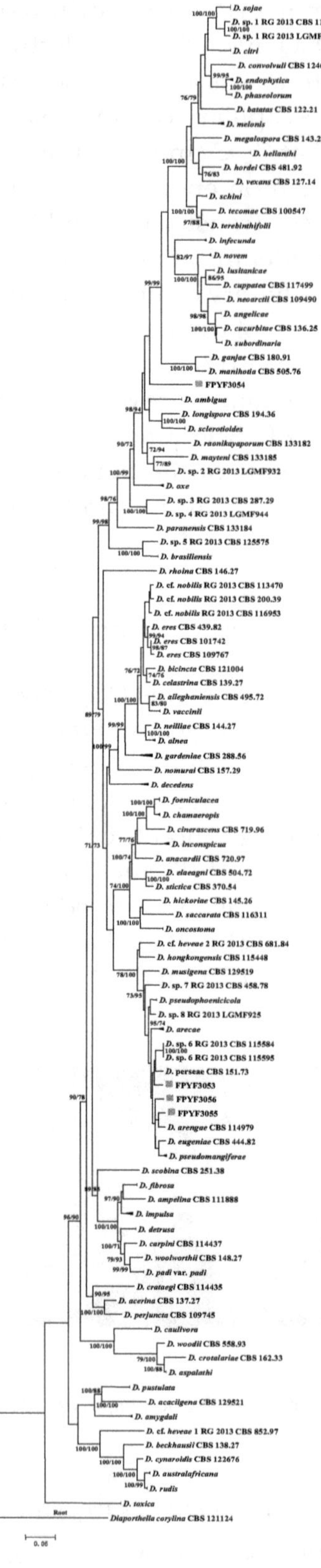

Figure 2. Phylogenetic tree based on combinedITS, *CAL*, *TEF1*, *HIS*, and *TUB* sequence alignment generated from a maximum parsimony and maximum likelihood analyses. Values near the branches represent parsimony/likelihood bootstrap support values (>70%), respectively. The tree is rooted with *Diaporthella corylina*. The four endophytic isolates were each named with strain ID marked green box. Compressed branches were used for saving space. The complete phylogenetic trees of MP and ML can be found in Figures S1 and S2, respectively.

3.2. The VOCs' Bioactivities of the Four Diaporthe Strains against Plant Fungal Pathogens

All of the four strains were observed to inhibit the growth of nine selected fungal pathogens by producing volatile compounds in the PDA medium (Table 3). The nine pathogens, *Alternaria alternata, Botryosphaeria dothidea, Botrytis cinerea, Cercospora asparagi, Colletotrichum gloeosporioides, Fusarium graminearum, Phytophthora cinnamomi, Sphaeropsis sapinea*, and *Valsa sordida*, are important causal agents to major trees, such as poplars and pines, or agricultural crops in China and elsewhere. All FPYF strains showed different inhibitory activities along the measurements, an exception was observed for the case of strain FPYF3053, which promoted the growth of *Phytophthora cinnamomi* (Table 3). Furthermore, all selected pathogens, except *V. sordid*, achieved obvious growth inhibition over around 10% during the testing period. After 24 h, *B. cinerea* was the most sensitive to VOCs emitted by all endophytic strains, reaching percent inhibitions of more than 55% when dual cultured with each strain. *B. dothidea* and *A. alternata* were highly sensitive to VOCs of all the endophytic strains, getting percent inhibitions of more than 30% with an exception to 28% of *A. alternata* in VOCs of the strain FPYF3053. *V. sordida* had the least sensitive or insensitive performance in VOCs from all the strains, showing percent inhibitions around 3% when dual cultured with FPYF3056. The inhibitive intensity of FPYF strains' VOCs on growth of pathogens decreased in times to most duel cultures. The maximum drop of the intensity was by 31% in percent inhibition on the pathogen *B. cinerea* duel culturing with strain FPYF3056. The obvious increase in intensity occurred in the pathogen *F. graminearum* duel culturing with FPYF3055 and FPYF3056, increasing by around 10% during 72 h. Some pathogens grew fast without percent inhibition records after 24 h (*V. sordida*) or 72 h (*B. dothidea* and *F. graminearum*).

Table 3. Growth inhibition percentage of plant pathogens by VOC bioassays of four *Diaporthe* strains. Percent of inhibition is shown as the means of four measurements of diameters with standard deviation ($n = 4$).

Pathogen	Day	FPYF3053		FPYF3054		FPYF3055		FPYF3056	
		Percentage Inhibition	*p*-Value	Percentage Inhibition	*p*-Value	Percentage Inhibition	*p*-Value	Percentage Inhibition	*p*-Value
Alternaria alternata	24 h	28.77 ± 2.26	0.0003	41.41 ± 1.65	0.0001	37.68 ± 5.6	0.0078	34.51 ± 2.03	0.0002
	48 h	22.42 ± 2.34	0.0003	30.19 ± 1.56	0.0009	30.25 ± 5.12	0.0113	26.50 ± 3.42	0.0006
	72 h	15.25 ± 2.59	0.0019	23.43 ± 2.27	0.0039	22 ± 4.03	0.0148	16.34 ± 2.36	0.0010
Botryosphaeria dothidea	24 h	46.3 ± 4.30	0.0030	50.17 ± 2.43	0.0006	43.74 ± 2.15	0.0000	37.88 ± 3.80	0.0002
	48 h	45.28 ± 2.63	0.0000	45.14 ± 2.35	0.0000	42.78 ± 0.43	0.0000	38.99 ± 0.98	0.0000
	72 h	NE *							
Botrytis cinerea	24 h	64.47 ± 1.05	0.0000	55.26 ± 4.82	0.0000	60.72 ± 1.91	0.0001	55.26 ± 4.71	0.0000
	48 h	50.42 ± 1.79	0.0000	35.02 ± 1.22	0.0000	39.44 ± 3.70	0.0000	32.41 ± 3.75	0.0003
	72 h	36.55 ± 2.81	0.0000	24.27 ± 3.08	0.0001	30.10 ± 3.54	0.0005	24.25 ± 4.62	0.0025
Cercospora asparagi	24 h	31.46 ± 4.11	0.0003	22.64 ± 2.86	0.0074	23.21 ± 4.54	0.0202	18.57 ± 5.01	0.0086
	48 h	33.81 ± 2.97	0.0000	24.32 ± 2.21	0.0008	22.02 ± 2.96	0.0000	16.34 ± 1.53	0.0000
	72 h	23.32 ± 2.17	0.0007	19.09 ± 2.73	0.0031	11.94 ± 3.54	0.0083	4.56 ± 0.85	0.0032
Colletotrichum gloeosporioides	24 h	26.83 ± 4.78	0.0153	9.75 ± 2.33	0.0009	10.68 ± 1.14	0.0001	10.74 ± 2.38	0.0229
	48 h	20.94 ± 3.33	0.0051	11.76 ± 2.35	0.0006	8.91 ± 1.24	0.0001	9.78 ± 2.36	0.0017
	72 h	20.68 ± 1.56	0.0024	9.39 ± 2.78	0.0023	6.29 ± 0.80	0.0001	6.82 ± 2.01	0.0388
Fusarium graminearum	24 h	25.68 ± 1.13	0.0031	14.68 ± 2.05	0.0147	20.45 ± 3.62	0.0046	12.96 ± 1.42	0.0103
	48 h	29.99 ± 5.29	0.0086	12.9 ± 4.50	0.0284	31.12 ± 3.57	0.0203	21.59 ± 4.57	0.0425
	72 h	NE							
Phytophthora cinnamomi	24 h	−5.01 ± 1.14 **	0.0029	19.21 ± 4.54	0.0036	31.02 ± 2.58	0.0001	8.38 ± 3.10	0.0154
	48 h	−15.65 ± 6.36	0.0186	12.19 ± 3.30	0.0500	25.21 ± 4.29	0.0050	11.32 ± 4.22	0.0302
	72 h	−19.70 ± 4.19	0.0153	11.91 ± 2.12	0.0209	21.03 ± 2.80	0.0031	8.94 ± 2.03	0.0013
Sphaeropsis sapinea	24 h	23.39 ± 4.25	0.0147	22.69 ± 5.23	0.0239	21.84 ± 7.61	0.0491	7.41 ± 2.68	0.0364
	48 h	20.93 ± 1.04	0.0009	23.85 ± 1.68	0.0023	18.33 ± 5.22	0.0367	9.53 ± 0.60	0.0024
	72 h	NE							
Valsa sordida	24 h	5.96 ± 1.61	0.0115	9.73 ± 2.79	0.0014	5.14 ± 1.02	0.0153	3.15 ± 1.00	0.0177
	48 h	NE							
	72 h	NE							

* No data. ** Negative values mean growth stimulation.

Table 4. Chemical composition of volatiles obtained from mycelial cultures of the four endophytic *Diaporthe* fungi using solid–phase microextraction (SPME).

Retention Time (min)	Molecular Weight	Compound	Quality (%) [a]				Abundance (Relative) [b]			
			FPYF3053	FPYF3054	FPYF3055	FPYF3056	FPYF3053	FPYF3054	FPYF3055	FPYF3056
6.17	106	Ethylbenzene		91.8	75.5	77.9		0.42	0.80	0.92
7.67	136	α-Thujene	91.9		89.3	89.8	30.57		37.10	36.19
9.88	136	1,3-Cyclohexadiene, 1-methyl-4-(1-methylethyl)-	84.7		80.2	81.2	6.26		5.92	1.87
9.89	136	2-Carene	86			84.9	1.80			6.20
10.21	136	α-Phellandrene		74.9	78			0.74	5.15	
10.22	136	β-Phellandrene	88.4	90.9	75.2	87.7	12.55	56.07	2.35	18.70
10.92	136	γ-Terpinene	89.4		85.6	88.7	21.15		16.82	19.21
11.63	136	Cyclohexene, 1-methyl-4-(1-methylethylidene)-			81.1				1.66	
11.89	154	2-Cyclohexen-1-ol, 1-methyl-4-(1-methylethyl)-	81				0.65			
12.29	152	Unknown	63.4		68.2		0.29		1.24	
12.3	152	Thujone				71.5				0.65
13.21	154	l-Menthone		90.4				27.91		
13.68	156	Cyclohexanol, 5-methyl-2-(1-methylethyl)-		87.6				10.29		
13.76	154	3-Cyclohexen-1-ol, 4-methyl-1-(1-methylethyl)-, (*R*)-	88.5		89.7	85.6	5.53		22.38	5.88
13.91	208	2,4,4-Trimethyl-3-(3-methylbutyl)cyclohex-2-enone		70.1				0.44		
14.12	352	Unknown	66.3				0.28			
14.13	240	Unknown				64.4				0.59
14.14	170	Unknown			64.9				1.35	
17.42	388	Unknown				60.9				0.31
18.41	188	Biphenylene, 1,2,3,6,7,8,8a,8b-octahydro-4,5-dimethyl-	79	73.9	68.5	72.2	7.05	1.62	2.06	6.81
18.92	204	Caryophyllene	73.6				0.46			
19.11	204	Unknown	69.7				0.31			
19.74	204	Patchoulene	76.8				0.45			
20.01	222	Unknown				63.1				0.41
20.13	204	Cedrene	78.2				1.11			
20.46	204	α-Muurolene	92.2	81.1	81	81.6	11.54	2.51	3.17	2.26

Notes: Data are averages of two cultures grown on the same medium with subtracting those from the control PDA plate. [a] The quality match is the % likelihood that the compound is identical to that which is listed on the table based on the NIST database. Compounds assigned as unknown with lower than 70% quality match. [b] The abundance figure presents the percentage amount of each compound in total area relative to all listed compounds detected for one strain.

3.3. The Qualification on VOCs of the Four Endophytic Diaporthe Strains

Each of the *Diaporthe* isolates showed a unique VOC profile as measured by SPME (Table 4). Nineteen VOC components from the four fungi were identified and seven compounds were unidentified according our set standard of a 70% quality match with the GC-MS. Generally, the terpenoids were the major components in the VOCs of each strain. The main terpenes included α-thujene, β-phellandrene, γ-terpinene, l-menthone, cyclohexanol, 5-methyl-2-(1-methylethyl)-, α-muurolene. The amounts of each component of these monoterpenes had a relative area over 10% of the total of its VOCs. There also existed other minor terpenoids at very low amounts, including carene, α-phellandrene, thujone, caryophyllene, patchoulene, etc. Two monoterpenes, β-phellandrene and α-muurolene, and a chemical biphenylene,1,2,3,6,7,8,8a,8b-octahydro-4,5-dimethyl, which were detected in VOCs of all four strains. Four chemicals were common to VOCs from FPYF3053, FPYF3055, and FPYF3056, including α-thujene, 1,3-cyclohexadiene, 1-methyl-4-(1-methylethyl)-, γ-terpinene and 3-cyclohexen-1-ol, and 4-methyl-1-(1-methylethyl)-,(*R*)-. However, each strain produced a unique mixture of volatile organic compounds. The strain FPYF3053 produced 15 volatile compounds with three prominent components, α-thujene, β-phellandrene, and α-muurolene. FPYF3054 was able to synthesize eight compounds with three prominent components of β-phellandrene, l-menthone, and cyclohexanol,5-methyl-2-(1-methylethyl)- in VOC mixtures. Strains FPYF3055 and 3056 generated relatively close chemical compositions in amount and quality of VOCs compared to FPYF3053 and FPYF3054. However, FPYF3055 had three prominent components, α-thujene, γ-terpinene, and 3-cyclohexen-1-ol,4-methyl-1-(1-methylethyl)-,(*R*)-, in 12 compounds of the VOCs, while FPYF3056 had three prominent components—namely α-thujene β-phellandrene, and γ-terpinene—of 13 compounds in its VOCs.

4. Discussion

4.1. Endophytic Diaporthe spp. from Catharanthus roseus

Four isolates of endophytes in the genus *Diaporthe* were obtained from the medicinal plant *Catharanthus roseus* growing in a conservation area of Southern China. In order to best distinguish these individual organisms they were subjected to a combined analysis of five-loci alignment of *TEF1-TUB-CAL-HIS*-ITS which gave a more robust isolate identification [23]. Adding our four endophytic isolates did not affect the congruency in each locus, partition homogeneity for the combination and the best evolutionary model for the five-locus concatenated alignment reported. *Diaporthe* fungi are one of the most common endophytic fungal communites found in plants [11]. However, the previous work on endophytic fungi from *C. roseus* [7,8,32–42] did not record strains of the *Diaporthe* genus. *Alternaria alternata* was determined as the dominant endophytic species in leaf tissue of *C. roseus* along with associated fungi from the following genera, *Aspergillus, Fusarium, Penicillium,* and *Helminthosporium* [33]. In addition the endophytes of root tissue appeared including *Colletotrichum* sp., *Macrophomina phaseolina, Nigrospora sphaerica,* and *Fusarium solani* [7]. Other isolated endophytic fungi from this plant included *Colletotrichum truncatum, Drechsclera* sp., *Cladosporium* sp., and *Myrothecium* sp. [43]. To our four *Diaporthe* strains, no reproductive structures were obtained in the employed conditions. They were designated *Diaporthe* sp. strains (FPYF3053-3056) without spore characterization strictly using phylogenetic analysis. The strains seemed not to share a close phylogenetic relationship to any other species based on the five-locus alignment study (Figure 2, [12,23,26]). The robust inference on the strains will take place when fruits bodies appear combined with full species phylogeny in the genus *Diaporthe.*

4.2. VOCs Antifungal Effects of Endophytic Diaporthe spp. from Catharanthus roseus

Compounds extracted from *Catharanthus roseus* [4,5] and extracts from some endophytes of this plant [10,44] have been shown to have antimicrobial bioactivities to some human microbial pathogens and plant fungal pathogens, including *Staphylococcus aureus, Pseudomonas aeruginosa, Bacillus subtilis,*

Escherichia coli, *Aspergillus fumigatus*, *Candida albicans*, etc. However, the VOCs or essential oils from *Catharanthus roseus* in the literature is scarce results on antimicrobial activities [45,46]. The previous work on the other endophytic fungi of this host plant did not consider that VOCs of the endophyte may have antimicrobial activities [7,8,32–42]. However, this work shows that VOCs produced by four endophytic *Diaporthe* fungi from the plant are able to functionally inhibit the growth of a number of specifically-targeted fungal pathogens (Table 3).

In the past there have been three endophytic *Diaporthe* strains recorded with their VOCs [18,19,47]. Two of them were reported to be inhibitory to plant pathogens [18,19]. One strain PR4 was isolated from a medicinal plant growing in Kashimir, Himalayas [19]; the other strain EC-4 was isolated from *Odontoglossum* sp. in Northern Ecuador [18]. With our four strains, the volatile compounds from endophytic *Diaporthe* fungi varied in degrees of inhibition against selected pathogenic fungi and test timings depending on the endophytic strain (Tables 3 and 5). However, the maximal inhibition of fungal growth of *Diaporthe* was from strain PR4, which reduced growth of *Rhizoctonia solani* by 100%. FPYF strains' and EC-4 VOCs also appeared effective in the inhibition of growth of *Botrytis cinerea* by more than 30% with a maximal of 50.42 ± 1.8%. During the test course of 72 h, to most cases, FPYF strains' VOCs showed strong bioactivities in the first day and then decreased inhibition on the pathogens in following two days (Table 3). PR4 VOCs were effective in reducing radial growth of *Pythium ultimatum* by 13.3%; EC-4 VOCs were effective in reducing radial growth of *Pythium ultimatum*, *Phytophthora cinnamomi*, and *Phytophthora palmivora* by 59.1 ± 0.9%, 42.0 ± 0.5%, and 5.6 ± 0.5%, respectively. FPYF3054-3056's VOCs were effective against *Phytophthora cinnamomi* in a range of 25.21 ± 4.3 ~11.32 ± 4.2%. The alcohol compounds such as 1-propanol,2-methyl- and 1-butanol,3-methyl- might made the oomycete *P. cinnamomi* more sensitive to EC-4's VOCs [18], which were lack in VOCs of all FPYF strains (Table 4). The two alcohol compounds had antimicrobial activities in VOCs of endophytic *Phomopsis* sp. strain EC-4 [18]. The sensitivity of the pathogen *F. graminearum* to VOCs from *Diaprothe* spp. might be analogous even though the VOCs components were not similar among *Diaprothe* strains. Two *Diaporthe* strains FPYF3053, 3055 (Table 2) and *Diaporthe* strain PR4 [19] had percent inhibition of *F. graminearum* growth of around 30% under their VOCs bioactivities. However, only beta-phellandrene was a common compound found in VOCs among them (Table 4, [19]). Contrast to cytochalasins as a predominantly common component in soluble secondary metabolites of *Diaporthe* strains [16], the genus-specific or predominant conserved components of fungal VOCs of genus *Diaporthe* should be proposed to illustrate further. The experimental data suggests that the VOCs of FPYF strains are both biologically active and biologically selective. Finally, isolate FPYF3053 were showed no effective inhibition of *Phytophthora cinnamomi* growth. In this study, we attempt to understand the VOCs inhibitory impacts from the endophytic *Diaporthe* strains without consideration of interaction between the strains and pathogenes. Future research is proposed to investigate the dual interaction in the VOCs' levels and other molecules between fungal interactions [48].

The headspace analyses of the four *Diaporthe* strains in potato dextrose medium revealed that three monoterpenes—β-phellandrene, biphenylene,1,2,3,6,7,8,8a,8b-octahydro-4,5-dimethyl and α-muurolene—seemed to be characteristic compounds of endophytic *Diaporthe* strains endophytic to *Catharathus roseus*. However, among all monoterpenes mentioned above, only 1-menthone can be found in volatile compounds of *Catharathus roseus* flowers, the essential oil of which is high in limonene and other monoterpenes [45,46]. Menthol and β-phellandrene were also found in VOCs of *Diaporthe* strain PR4 with very low relative amounts of less than 1.0% [19]. No chemicals were shared in VOCs between our FPYF strains and *Phomopsis* strain EC-4 (Table 4, [18]). Therefore, the antifungal VOCs from the four endophytic *Diaporthe* Chinese strains possesses unique VOC compositions compared with known *Diaporthe* VOCs. Although many fungi were reported to produce many terpene compounds in their VOCs [49], our *Diaporthe* fungi maybe of some interest as a source of some other monoterpenes, which often only have been thought to originate from specific plants. For instance, essential oils from many plants containing more or less such monoterpenes as α-thujene, β-phellandrene [50–52], γ-terpinene [53,54], l-menthone [55,56], cyclohexanol, α-muurolene, thujone, and caryophyllene have some antifungal activities. For example, γ-terpinene, singly or in mixtures with sabinene in oil from coastal redwood leaves, has strong antifungal activity on some endophytic fungi [53]. Therefore, it could be rational to infer the terpenes in FPYF strains synergistically played a main role in their inhibition pathogenic fungi growths. In addition, the high content of monoterpenes in the *Diaporthe* VOCs does have potential for the biofuel industry [18,20,57].

Table 5. Comparison VOCs' inhibitive effect among *Diaporthe* strains.

Pathogens	Percent Growth Inhibition		
	***Phomopsis* sp. EC-4 [17] ***	***Diaporthe* Strain PR4 [18]**	**FPYF3053-3056 ****
Aspergillus flavus	/ ***	34.6	/
Aspergillus fumigatus	57.0 ± 0.5	/	/
Alternaria alternata	/	/	30.25 ± 5.1~22.42 ± 2.3
Botryosphaeria dothidea	/	/	45.14 ± 2.4~25.28 ± 2.6
Botrytis cinerea	37.8 ± 0.5	/	50.42 ± 1.8~32.41 ± 3.8
Ceratocystis fimbriata	/	0.0	/
Ceratocystis ulmi	11.1 ± 1.5	/	/
Cercospora asparagi	/	/	33.81 ± 2.97~16.34 ± 1.5
Cercospora beticola	19.5 ± 0.5	/	/
Colletotrichum sp.	/	/	20.94 ± 3.3~8.91 ± 2.4
Colletotrichum lagenarium	0.0	/	/
Fusarium oxysporum	/	34.6	31.12 ± 3.6~12.9 ± 4.5
Fusarium solani	43.2 ± 0.00	16.6	/
Geotrichum candidum,	45.3 ± 0.5	57.0	/
Trichoderma viride	0.0	/	/
Rhizoctonia solani	53.0 ± 1.0	100	/
Sphaerospsis sapinea	/	/	23.85 ± 1.7~9.53 ± 0.6
Sclerotinia sclerotiorum	70.7 ± 1.1	/	/
Valsa sordida	/	/	9.73 ± 2.8~3.15 ± 1.00
Verticillium dahliae	19.4 ± 0.0	0.0	/
Pythium ultimatum	59.1 ± 0.9	13.3	/
Phytophthora cinnamomi	42.0 ± 0.5	/	−19.70 ± 4.19~−5.01 ± 1.14, 25.21 ± 4.3~11.32 ± 4.2
Phytophthora palmivora	5.6 ± 0.5	/	/

* Data reference, ** the values listed as range for the four strains during 72 h, *** no data.

Author Contributions: D.-H.Y. conceived the grant and designed the experiments; D.-H.Y., and T.L. collected the plant samples and identified the plant; X.S. and H.L. performed the experiment; D.-H.Y. organized the manuscript; G.S. advised research and revised manuscript. And G.D. managed laboratory jobs for this work and processed the submission.

Acknowledgments: This work was supported by Fundamental Research Funds of CAF (CAFYBB2017MA010, to D.-H.Y.). We thank Han Xu (Research Institute of Tropical Forestry, Chinese Academy of Forestry) for kindly helping with host plant identification, and thank Kaiying Wang (Research Institute of Forest Ecology, Environment and Protection) for taking part in the work on sample collection and fungal isolation. The authors sincerely appreciate the three peer reviewers' valued comments on the paper.

References

1. Kumar, G.; Chandra, P.; Choudhary, M. Endophytic fungi: A potential source of bioactive compounds. *Chem. Sci. Rev. Lett.* **2017**, *6*, 2373–2381.
2. Goyal, P.; Khanna, A.; Chauhan, A.; Chauhan, G.; Kaushik, P. In vitro evaluation of crude extracts of *Catharanthus roseus* for potential antibacterial activity. *Int. J. Green Pharm.* **2008**, *2*, 178–181. [CrossRef]
3. Ramya, S.; Govindaraji, V.; Kannan, K.N.; Jayakumararaj, R. In vitro evaluation of antibacterial activity using crude extracts of *Catharanthus roseus* L. (G.) Don. *Ethnobot. Leafl.* **2008**, *12*, 1067–1072.

4. Kabesh, K.; Senthilkumar, P.; Ragunathan, R.; Kumar, R.R. Phytochemical analysis of *Catharanthus roseus* plant extract and its antimicrobial activity. *Int. J. Pure Appl. Biosci.* **2015**, *3*, 162–172.
5. Patil, P.J.; Ghosh, J.S. Antimicrobial activity of *Catharanthus roseus*—A detailed study. *Br. J. Pharmacol. Toxicol.* **2010**, *1*, 40–44.
6. Hanafy, M.; Matter, M.; Asker, M.; Rady, M. Production of indole alkaloids in hairy root cultures of *Catharanthus roseus* l. and their antimicrobial activity. *S. Afr. J. Bot.* **2016**, *105*, 9–18. [CrossRef]
7. Lakra, N.S.; Koul, M.; Chandra, R.; Chandra, S. Histological investigations of healthy tissues of *Catharanthus roseus* to localize fungal endophytes. *Int. J. Pharm. Sci. Rev. Res.* **2013**, *20*, 205–209.
8. Pandey, S.S.; Singh, S.; Babu, C.S.V.; Shanker, K.; Srivastava, N.K.; Shukla, A.K.; Kalra, A. Fungal endophytes of *Catharanthus roseus* enhance vindoline content by modulating structural and regulatory genes related to terpenoid indole alkaloid biosynthesis. *Sci. Rep.* **2016**, *6*, 26583. [CrossRef] [PubMed]
9. Shi, Y.; Zhou, M.; Tang, Z.; Rao, L. Isolation and identification on endophytic fungus from *Catharanthus roseus*. *J. Anhui Agric. Sci.* **2008**, *36*, 12712–12713. [CrossRef]
10. Akpotu, M.O.; Eze, P.M.; Abba, C.C.; Umeokoli, B.O.; Nwachukwu, C.U.; Okoye, F.B.C.; Esimone, C.O. Antimicrobial activities of secondary metabolites of endophytic fungi isolated from *Catharanthus roseus*. *J. Health Sci.* **2017**, *7*, 15–22. [CrossRef]
11. Nair, D.N.; Padmavathy, S. Impact of endophytic microorganisms on plants, environment and humans. *Sci. World J.* **2014**, *2014*. [CrossRef] [PubMed]
12. Santos, P.J.C.D.; Savi, D.C.; Gomes, R.R.; Goulin, E.H.; Senkiv, C.D.C.; Tanaka, F.A.O.; Almeida, Á.M.R.; Galli-Terasawa, L.; Kava, V.; Glienke, C. Diaporthe endophytica and D. terebinthifolii from medicinal plants forbiological control of *Phyllosticta citricarpa*. *Microbiol. Res.* **2016**, *186*, 153–160. [CrossRef] [PubMed]
13. Tonial, F.; Maia, B.H.L.N.S.; Sobottka, A.M.; Savi, D.C.; Vicente, V.A.; Gomes, R.R.; Glienke, C. Biological activity of *Diaporthe terebinthifolii* extracts against *Phyllosticta citricarpa*. *FEMS Microbiol. Lett.* **2017**, *364*. [CrossRef] [PubMed]
14. Tanapichatsakul, C.; Monggoot, S.; Gentekaki, E.; Pripdeevech, P. Antibacterial and antioxidant metabolites of *Diaporthe* spp. Isolated from flowers of *Melodorum fruticosum*. *Curr. Microbiol.* **2017**, *75*, 476–483. [CrossRef] [PubMed]
15. Rakshith, D.; Santosh, P.; Satish, S. Isolation and characterization of antimicrobial metabolite producing endophytic *Phomopsis* sp. from *Ficus pumila* Linn. (Moraceae). *Int. J. Chem. Anal. Sci.* **2013**, *4*, 156–160. [CrossRef]
16. Chepkirui, C.; Stadler, M. The genus *Diaporthe*: A rich source of diverse and bioactive metabolites. *Mycol. Prog.* **2017**, *16*, 477–494. [CrossRef]
17. Strobel, G.A.; Dirkse, E.; Sears, J.; Markworth, C. Volatile antimicrobials from *Muscodor albus*, a novel endophytic fungus. *Microbiology* **2001**, *147*, 2943–2950. [CrossRef] [PubMed]
18. Singh, S.K.; Strobel, G.A.; Knighton, B.; Geary, B.; Sears, J.; Ezra, D. An endophytic *Phomopsis* sp. possessing bioactivity and fuel potential with its volatile organic compounds. *Microb. Ecol.* **2011**, *61*, 729–739. [CrossRef] [PubMed]
19. Qadri, M.; Deshidi, R.; Shah, B.A.; Bindu, K.; Vishwakarma, R.A.; Riyaz-Ul-Hassan, S. An endophyte of *Picrorhiza kurroa* Royle ex. Benth, producing menthol, phenylethyl alcohol and 3-hydroxypropionic acid, and other volatile organic compounds. *World J. Microbiol. Biotechnol.* **2015**, *31*, 1647–1654. [CrossRef] [PubMed]
20. Wang, K.; Strobel, G.; Yan, D.-H. The production of 1,8-cineole, a potential biofuel, from an endophytic strain of *Annulohypoxylon* sp. FPYF3050 when grown on agricultural residues. *J. Sustain. Bioenergy Syst.* **2017**, *7*, 65–84. [CrossRef]
21. White, T.J.; Bruns, T.D.; Lee, S.B.; Taylor, J.W.; Innis, M.A.; Gelfand, D.H.; Sninsky, J.J. Amplification and direct sequencing of fungal ribosomal rnagenes for phylogenetics. *PCR Protoc. A Guide Methods Appl.* **1990**, *18*, 315–322. [CrossRef]
22. Alves, A.; Crous, P.; Correia, A.; Phillips, A. Morphological and molecular data reveal cryptic species in *Lasiodiplodia theobromae*. *Fungal Divers.* **2008**, *28*, 1–13.
23. Santos, L.; Alves, A.; Alves, R. Evaluating multi-locus phylogenies for species boundaries determination in the genus *Diaporthe*. *PeerJ* **2017**, *5*, e3120. [CrossRef] [PubMed]
24. Glass, N.; Donaldson, G. Development of primer sets designed for use with the PCR to amplify conserved genes from filamentous ascomycetes. *Appl. Environ. Microbiol.* **1995**, *61*, 1323–1330. [PubMed]

25. O'Donnell, K.; Cigelnik, E. Two divergent intragenomic rDNA ITS2 types within a monophyletic lineage of the fungus *Fusarium arenonorthologous*. *Mol. Phylogenet. Evol.* **1997**, *7*, 103–116. [CrossRef] [PubMed]
26. Udayanga, D.; Castlebury, L.A.; Rossman, A.Y.; Chukeatirote, E.; Hyde, K.D. Insights into the genus *Diaporthe*: Phylogenetic species delimitation in the *D. eres* species complex. *Fungal Divers.* **2014**, *67*, 203–229. [CrossRef]
27. Katoh, K.; Standley, D.M. Mafft multiple sequence alignment software version 7: Improvements in performance and usability. *Mol. Biol. Evol.* **2013**, *30*, 772–780. [CrossRef] [PubMed]
28. Lanfear, R.; Frandsen, P.B.; Wright, A.M.; Senfeld, T.; Calcott, B. Partitionfinder 2: New methods for selecting partitioned models of evolution for molecular and morphological phylogenetic analyses. *Mol. Biol. Evol.* **2017**, *34*, 772–773. [CrossRef] [PubMed]
29. Vaidya, G.; Lohman, D.; Meier, R. Sequencematrix: Concatenation software for the fast assembly of multi-gene datasets with character set and codon information. *Cladistics* **2011**, *27*, 171–180. [CrossRef]
30. Tamura, K.; Stecher, G.; Peterson, D.; Filipski, A.; Kumar, S. Mega6: Molecular evolutionary genetics analysis version 6.0. *Mol. Biol. Evol.* **2013**, *30*, 2725–2729. [CrossRef] [PubMed]
31. Stöver, B.C.; Müller, K.F. Treegraph 2: Combining and visualizing evidence from different phylogenetic analyses. *BMC Bioinform.* **2010**, *11*, 7. [CrossRef] [PubMed]
32. Ayob, F.W.; Simarani, K. Endophytic filamentous fungi from a *Catharanthus roseus*: Identification and its hydrolytic enzymes. *Saudi Pharm. J.* **2016**, *24*, 273–278. [CrossRef] [PubMed]
33. Momsia, P.; Momsia, T. Isolation, frequency distribution and diversity of novel fungal endophytes inhabiting leaves of *Catharanthus roseus*. *Int. J. Life Sci. Biotechnol. Pharm. Res.* **2013**, *2*, 82–87.
34. Kharwar, R.N.; Verma, V.C.; Strobel, G.; Ezra, D. The endophytic fungal complex of *Catharanthus roseus* (L.) g. Don. *Curr. Sci.* **2008**, *95*, 228–233. [CrossRef]
35. Manogaran, S.; Kannan, K.P.; Mathiyalagan, Y. Fungal endophytes from *Phyllanthus acidus* (L.) and *Catharanthus roseus* (L.). *Int. Res. J. Pharm.* **2017**, *8*, 86–89. [CrossRef]
36. Srinivasan, K.; Muthumary, J. Taxol production from *Pestalotiopsis* sp an endophytic fungus isolated from *Catharanthus roseus*. *J. Ecobiotechnol.* **2009**, *1*, 28–31.
37. Palem, P.P.C.; Kuriakose, G.C.; Jayabaskaran, C. An endophytic fungus, *Talaromyces radicus*, isolated from catharanthus roseus, produces vincristine and vinblastine, which induce apoptotic cell death. *PLoS ONE* **2015**, *10*, e0144476. [CrossRef] [PubMed]
38. Ayob, F.W.; Simarani, K.; Abidin, N.Z.; Mohamad, J. First report on a novel nigrospora sphaerica isolated from *Catharanthus roseus* plant with anticarcinogenic properties. *Microb. Biotechnol.* **2017**, *10*, 926–932. [CrossRef] [PubMed]
39. Rakotoniriana, E.F.; Chataigné, G.; Raoelison, G.; Rabemanantsoa, C.; Munaut, F.; El Jaziri, M.; Urveg-Ratsimamanga, S.; Marchand-Brynaert, J.; Corbisier, A.-M.; Declerck, S.; et al. Characterization of an endophytic whorl-forming *Streptomyces* from *Catharanthus roseus* stems producing polyene macrolide antibiotic. *Can. J. Microbiol.* **2012**, *58*, 617–627. [CrossRef] [PubMed]
40. Singh, D.; Rathod, V.; Singh1, A.K.; Haq, M.U.; Mathew, J.; Kulkarni, P. A study on extracellular synthesis of silver nanoparticles from endophytic fungi, isolated from ethanomedicinal plants *Curcuma longa* and *Catharanthus roseus*. *Int. Lett. Nat. Sci.* **2016**, *57*, 58–66. [CrossRef]
41. Tickoo, M.; Farooq, U.; Bhatt, N.; Dhiman, M.; Alam, A.; Khan, M.A.; Jaglan, S. Alterneriol: Secondary metabolites derived from endophytic fungi *Alternaria* spp. isolated from *Catharanthus roseus*. *UJPAH* **2015**, *1*, 15–17.
42. Tang, Z.; Rao, L.; Peng, G.; Zhou, M.; Shi, G.; Liang, Y. Effects of endophytic fungus and its elicitors on cell status and alkaloid synthesis in cell suspension cultures of *Catharanthus roseus*. *J. Med. Plants Res.* **2011**, *5*, 2192–2200.
43. Sunitha, V.H.; Devi, D.N.; Srinivas, C. Extracellular enzymatic activity of endophytic fungal strains isolated from medicinal plants. *World J. Agric. Sci.* **2013**, *9*, 01–09. [CrossRef]
44. Roy, S.; Banerjee, D. Broad spectrum antibacterial activity of granaticinic acid, isolated from *Streptomyces thermoviolaceus* NT1; an endophyte in *Catharanthus roseus* (L.) G. Don. *J. Appl. Pharm. Sci.* **2015**, *5*, 006–011. [CrossRef]
45. De Pinho, P.G.; Goncalves, R.F.; Valent, P.; Pereira, D.M.; Seabra, R.M.; Andrade, P.B.; Sottomayor, M. Volatile composition of *Catharanthus roseus* (L.) G. Don using solid-phase microextraction and gas chromatography/mass spectrometry. *J. Pharm. Biomed. Anal.* **2009**, *49*, 674–685. [CrossRef] [PubMed]

46. Aziz, S.; Saha, K.; Sultana, N.; Khan, M.; Nada, K.; Afroze, M. Comparative studies of volatile components of the essential oil of leaves and flowers of *Catharanthus roseus* growing in bangladesh by GC-MS analysis. *Indian J. Pharm. Biol. Res.* **2015**, *3*, 6–10. [CrossRef]
47. Bier, M.C.J.; Medeiros, A.B.P.; Soccol, C.R. Biotransformation of limonene by an endophytic fungus using synthetic and orange residue-based media. *Fungal Biol.* **2017**, *121*, 137–144. [CrossRef] [PubMed]
48. Bertrand, S.; Schumpp, O.; Bohni, N.; Monod, M.; Gindro, K.; Wolfender, J.-L. De novo production of metabolites by fungal co-culture of *Trichophyton rubrum* and *Bionectria ochroleuca*. *J. Nat. Prod.* **2013**, *76*, 1157–1165. [CrossRef] [PubMed]
49. Dickschat, J.S. Fungal volatiles—A survey from edible mushrooms to moulds. *Nat. Prod. Rep.* **2017**, *34*, 310–328. [CrossRef] [PubMed]
50. Garzoli, S.; Božović, M.; Baldisserotto, A.; Sabatino, M.; Cesa, S.; Pepi, F.; Vicentini, C.B.; Manfredini, S.; Ragno, R. Essential oil extraction, chemical analysis and anti-candida activity of *Foeniculum vulgare* miller—New approaches. *Nat. Prod. Res.* **2018**, *32*, 1254–1259. [CrossRef] [PubMed]
51. Dai, J.; Zhu, L.; Yang, L.; Qiu, J. Chemical composition, antioxidant and antimicrobial activities of essential oil from *Wedelia prostrata*. *EXCLI J.* **2013**, *12*, 479–490. [CrossRef] [PubMed]
52. Cosge, B.; Turker, A.; Ipek, A.; Gurbuz, B.; Arslan, N. Chemical compositions and antibacterial activities of the essential oils from aerial parts and corollas of *Origanum acutidens* (Hand.-Mazz.) Ietswaart, an endemic species to turkey. *Molecules* **2009**, *14*, 1702–1712. [CrossRef] [PubMed]
53. Espinosa-García, F.J.; Langenheim, J.H. Effects of sabinene and γ-terpinene from coastal redwood leaves acting singly or in mixtures on the growth of some of their fungus endophytes. *Biochem. Syst. Ecol.* **1991**, *19*, 643–650. [CrossRef]
54. Verma, R.S.; Joshi, N.; Padalia, R.C.; Singh, V.R.; Goswami, P.; Verma, S.K.; Iqbal, H.; Chanda, D.; Verma, R.K.; Darokar, M.P.; et al. Chemical composition and antibacterial, antifungal, allelopathic and acetylcholinesterase inhibitory activities of cassumunar-ginger. *J. Sci. Food Agric.* **2018**, *98*, 321–327. [CrossRef] [PubMed]
55. Rachitha, P.; Krupashree, K.; Jayashree, G.; Gopalan, N.; Khanum, F. Growth inhibition and morphological alteration of *Fusarium sporotrichioides* by *Mentha piperita* essential oil. *Pharmacogn. Res.* **2017**, *9*, 74–79. [CrossRef]
56. Moghaddam, M.; Pourbaige, M.; Tabar, H.K.; Farhadi, N.; Hosseini, S.M.A. Composition and antifungal activity of peppermint (*Mentha piperita*) essential oil from Iran. *J. Essent. Oil Bear. Plants* **2013**, *16*, 506–512. [CrossRef]
57. Strobel, G. The story of mycodiesel. *Curr. Opin. Microbiol.* **2014**, *19*, 52–58. [CrossRef] [PubMed]

12

Fungi as Endophytes in *Artemisia thuscula*: Juxtaposed Elements of Diversity and Phylogeny

Andreea Cosoveanu *, Samuel Rodriguez Sabina and Raimundo Cabrera

Department Botanica, Ecologia & Fisiologia Vegetal, Universidad de La Laguna, 38206 La Laguna, Tenerife, Spain; samuelrguezsabina@gmail.com (S.R.S.); rcabrera@ull.edu.es (R.C.)

* Correspondence: andreeacosoveanu@gmail.com

Abstract: *Artemisia* is a plant genus highly studied for its medicinal applications. The studies on the associated fungal endophytes are scarce. Ten plants specimens of *Artemisia thuscula* from Tenerife and La Palma were sampled to isolate the endophytic fungi. Identification of the endophytic fungi was based on morphology, Internal Transcribed Spacer (ITS) and Large Subunit (LSU) regions sequencing and indicates 37 fungal species affiliated to 25 fungal genera. Colonization rate varied among plants (CR = 25% to 92.11%). The most dominant colonizers found were *Alternaria alternata* (CF = 18.71%), *Neofusicoccum* sp. (CF = 8.39%) and *Preussia* sp. (CF = 3.23). Tendency for host specificity of most endophytic fungal species was observed. Sorensen–Dice index revealed that of 45 cases in the matrix, 27 of them were of zero similarity. Further, only one case was found to have 57% similarity (TF2 and TF7) and one case with 50% similarity (TF1 and TF4). The rest of the cases had values ranging between 11% and 40% similarity. Diversity indices like Brillouin, Margalef species richness, Simpson index of diversity and Fisher's alpha, revealed plants from La Palma with higher values than plants from Tenerife. Three nutrient media (i.e., potato dextrose agar—PDA, lignocellulose agar—LCA, and tomato juice agar—V8) were used in a case study and revealed no differences in terms of colonization rate when data was averaged. Colonization frequency showed several species with preference for nutrient medium (63% of the species were isolated from only one nutrient medium). For the phylogenetic reconstruction using the Bayesian method, 54 endophytic fungal ITS sequences and associated GenBank sequences were analyzed. Ten orders (Diaporthales, Dothideales, Botryosphaeriales, Hypocreales, Trichosphaeriales, Amphisphaeriales, Xylariales, Capnodiales, Pleosporales and Eurotiales) were recognized. Several arrangements of genera draw the attention, like *Aureobasidium* (Dothideales) and *Aplosporella* (Botryosphaeriales) which are clustered with a recent ancestor (BS = 0.97).

Keywords: *Artemisia*; fungal endophytes; biodiversity; phylogeny

---

1. Introduction

Vascular plants species [1,2], aquatic plants and algae [3,4], mosses and ferns [5,6] examined to date are found to be hosts for endophytic bacteria and fungi [7]. Endophytic microorganisms have been isolated from different parts of plant-like scale primordia, meristem and resin ducts [8,9], leaf segments with midrib and roots, stem, bark, leaf blade, petiole [10], buds [11], and seeds [12]. Successful endophytic colonization is dependent on many factors including plant tissue type, plant genotype, the microbial taxon and strain type, and biotic and abiotic environmental conditions. Fungal endophytes aid plants to withstand and tolerate unfavorable environmental conditions [13,14] and also promote plant growth [15,16]. These inhabitants can produce the same or similar secondary metabolites [17–20] as their host and play vital roles in vivo such as signaling, defense, and regulation of the symbiosis [21]. Mainly investigations are based on their use as biochemical tools and the end products are to be used in pharmaceutics, industry, and agriculture.

Artemisia is a plant highly evaluated for medicinal and biopesticide traits. A survey of the literature shows this plant genus to be in the hot spot among researchers with over 11,200 publications in Scopus library. Even though *Artemisia* is a large plant genus with species producing a variety of interesting and active compounds, its endophytic communities are under investigated. The identification of the fungal endophytes in *Artemisia* spp. is made mainly based on morphological characterization and molecular analysis using nuclear ribosomal DNA sequences, including both the internal transcribed spacers and the 5.8S gene region. To the best of our knowledge, there have been only four studies which investigate the phylogenetic analysis of the *Artemisia* spp. fungal endophytes [22–25]. In terms of diversity, the studies are also scarce but interesting facts are brought to light in terms of diversity and plant colonization. For instance, Yuan et al., 2011 [26] performed a comparative study related to infection frequency between cultivated plants and wild plants of *Artemisia annua*. The results revealed slightly higher infection frequency of the endophytic fungi in cultivated roots (20.9%) than in native roots (16.7%). Further, authors described that the naturally regenerated roots harbored richer fungal genotypes, which supports the hypothesis that wild plant species are predisposed to host rich and novel mycoflora [27]. It is worth mentioning that Qian et al., 2014 [27] reported the presence of *Rhodotorula* sp. and *Fusarium* sp. in *Artemisia argyi* for the first time. The endophytic fungi associated with *Artemisia nilagirica* were investigated and one strain of *Pythium intermedium* (Oomycota) and one strain of *Rhizopus oryzae* (Mucoromycota) were isolated among the majority clade of Ascomycota [28]. Huang et al., 2009 [24] classified 108 fungal isolates obtained from three medicinal plant species *Artemisia capillaris*, *Artemisia indica* and *Artemisia lactiflora* using morphological identification and among the three plant hosts, the highest endophytic colonization rate occurred in *Artemisia capillaris*, which exhibited highest fungal diversity. Five fungal isolates belonging to *Aureobasidium pullulans*, *Ephelis*, *Pestalotiopsis*, and Pleosporaceae, were only recovered from *Artemisia capillaris*. *Xylaria* species was reported to be dominant endophytic fungi in *Artemisia indica*. Seven *Artemisia* species were sampled in two locations (Qichun and Wuhan in China) and 21 fungal endophytic species belonging to: *Diaporthe*, *Colletotrichum*, *Nigrospora*, *Botryosphaeria*, *Aspergillus*, *Penicillium*, *Neofusicoccum*, *Cercospora*, *Rhizoctonia*, *Alternaria*, and *Curvularia* were found [23]. The highest incidences of colonization frequency per plant host revealed *Nigrospora sphaerica* in *Artemisia* sp., *Nigrospora oryzae* in *Artemisia argyi*, *Alternaria alternata* in *Artemisia subulata* and *Artemisia tangutica* and *Botryosphaeria dothidea* in *Artemisia lavandulifolia*. The authors report for the first time *Nigrospora*, *Neofusicoccum* and *Curvularia* species in *Artemisia* spp.

Artemisia thuscula is an endemic plant of Canary Islands and community of endophytes housed inside its plant tissues remains unexplored. With the idea of exploring endemic medicinal plants for useful and underexplored fungal endophytes, we strategically pinned down to *Artemisia thuscula* that has been harboring in western areas of islands i.e., Tenerife and La Palma, for ages. Elements of phylogeny and diversity were framed for the strains obtained from both islands with a case study of Tenerife where diversity was intended to be enhanced by using different nutrient media and stem ages. Questions on host specificity were explored, having one plant species and various collection locations.

2. Materials and Methods

2.1. Plants Sampling

Plants of *Artemisia thuscula* species were collected from Canary Islands (La Palma and Tenerife). 10 plants specimens were sampled in total. Three plants were sampled from La Palma and seven plants were sampled from Tenerife; GPS coordinates are mentioned in Table 1. In situ, plants were observed for their healthy appearance prior to the sampling, only those individuals that did not show symptoms of attack by pest or disease were selected. From each plant only stems segments were cut, labeled and kept in paper bags inside zip-locked bags at T = 4–5 °C until transported to the laboratory and then processed within 24 h. Identification of the plant species was performed using classical morphological examination. The plants were deposited at the University of La Laguna (ULL) herbarium (TFC).

Table 1. Details of collected *Artemisia* species plants.

Plant Species	Plant Code	Collection Place	Country/Island	Herbarium Type Details	GPS UTM Latitude	GPS UTM Longitude
A. thuscula	LP1	El Granel	La Palma	TFC. No. 52658	28°45′47.43″ N	17°45′7.47″ W
A. thuscula	LP2	San Bartolo	La Palma	TFC. No. 52659	28°46′1.08″ N	17°45′26.07″ W
A. thuscula	LP4	Tigalate	La Palma	TFC. No. 52661	28°32′35.45″ N	17°48′41.29″ W
A. thuscula	TF8	El Palmar	Tenerife	TFC. No. 52669	28°20′35.18″ N	16°51′26.57″ W
A. thuscula	TF7	Granadilla	Tenerife	TFC. No. 52668	28°06′54.19″ N	16°34′51.14″ W
A. thuscula	TF4	Caletillas	Tenerife	TFC. No. 52665	28°23′2.03″ N	16°21′54.71″ W
A. thuscula	TF1	Mesa Mota	Tenerife	TFC. No. 52662	28°30′38.75″ N	16°19′20.55″ W
A. thuscula	TF2	Mesa Mota	Tenerife	TFC. No. 52663	28°30′38.75″ N	16°19′20.55″ W
A. thuscula	TF5	San Andres	Tenerife	TFC. No. 52666	28°30′51.01″ N	16°11′41.94″ W
A. thuscula	TF3	Taborno	Tenerife	TFC. No. 52664	28°33′18.36″ N	16°15′53.10″ W

2.2. Fungal Endophyte Isolation

Surface sterilization method was used to suppress epiphytic microorganisms from the plant [23]. Thus, stem fragments were first washed with sterile water, then immersed in 70% ethanol for 1 min, followed by an immersion in 15% sodium hypochlorite for 1 min, again in 70% ethanol for 1 min and lastly were washed with sterile distilled water. To assure a successful sterilization, fragments were rolled on potato dextrose agar (PDA) medium and drops of last step sterilization water were poured on medium, as a control check for complete sterilization. After this process, plant material was dried on sterile blotting sheet, excised in pieces of 2 cm and cut longitudinally with a sterile scalpel. Segments were placed in PDA (Sigma-Aldrich, St. Louis, MI, USA) Petri plates amended with tetracycline (10 mg L^{-1}). Plates incubated with the plant segments were incubated at 25 °C in the dark for two weeks and observed daily for fungal growth. When fungal outgrowth from the plant tissues occurred observations on emerged fungi were made. Only the fungi with different morphological characteristics were subcultured. Eventually, when an endophyte was acquired in pure culture it was preserved in Czapek medium (Fluka Analytical, Sigma-Aldrich), T = 5 °C and in glycerol (≥99.5, Sigma-Aldrich) 20% in deionized H_2O, T = −32 °C and identified. To analyze the fungal diversity, each replicate of the distinct stem fragments was noted. To enhance bioprospection and diversity, variable nutritive media were utilized to incubate stem fragments (with ages less than one year and more than one year) of eight plants from Tenerife. Therefore, V8 tomato juice agar and lignocellulose agar (LCA) [29] media were additionally used. All the reagents were purchased from Sigma-Aldrich, except Agar Agar—GUINAMA (Valencia, Spain) and Potassium chloride—PanReac AppliChem (Barcelona, Spain).

2.3. Fungal Endophyte Collection and Maintenance

Every isolate and its plant origin were dully recorded for calculation of colonization rate from host, counting the same isolate identification only once if it emerges from the same plant segment. After purification of each isolate, it was subjected to microscopical observations followed by molecular analysis to identify at genus and/or species level. Isolates are presently maintained in three types of media: Czapek, T = 5 °C; mineral oil (Sigma-Aldrich), T = 5 °C and glycerol (Sigma-Aldrich) 20% diH_2O, T = −80 °C. For short term use, fungal isolates were maintained on PDA, 25 °C.

2.4. Morphological Identification

Prior to taxonomic identification, a preliminary classification was made to avoid the selection of identical strains arising from the same plant individual, separating isolates into morphotypes. Observations targeted characteristics related to the colony and medium as: colony shape, texture and colour; exudates, medium colour and growth rate. For the microscopic observations, a strain was inoculated onto a PDA Petri plate and a sterile cover slide was attached at two centimeters. Once the growth of the fungus partially covered the cover slide, the slide was removed, inverted on a slide with cotton blue (for the slightly coloured colonies) and observed under microscope.

2.5. Molecular Identification

Out of several procedures for genomic DNA extraction, the most efficient protocol, although time consuming, was the one described by Shu et al., 2014 [20] to which the following modifications were made. Samples were centrifuged for 15 min at 12,000 rpm; after the chloroform (≥99.5, Sigma-Aldrich) procedure the upper phase was mixed with 10% Sodium acetate (ReagentPlus®, ≥99.0%, Sigma-Aldrich) and 60% Isopropyl alcohol (Aldrich ≥ 97.0%, Sigma-Aldrich), incubated for 10 min at −30 °C and centrifuged (10 min, 12,000 rpm). Finally, the pellet was washed twice with 75% ethanol (before maintained at −20 °C) and centrifuged (10 min, 12,000 rpm). The solvent was removed by evaporation, keeping the sample in the laminar flow cabinet. The purified DNA was suspended in 20 μL TE buffer (10 mM Tris-HCl, pH 8.0, 1 mM EDTA); all reagents were purchased from Sigma-Aldrich. RNase A was added, and the sample was incubated for 1 hour at room temperature (long-term storage at −32 °C).

The second protocol for DNA extraction involves no purification of DNA but acceptable results were garnered (around 50% samples succeeded). 20 μL of TE buffer was pipetted into a microtube and glass beads (diameter = 0.4–0.6 mm) were added to make up 3/4 of the reagent's volume. A small quantity of fungal mycelium was added (2–5 mm/2–3 mg) with a needle. Samples were homogenized using FastPrep 24™ 5 G (MP Bio, Santa Ana, California, USA) at 4 m/s, 20 s. Subsequently samples were centrifuged at 13,000 rpm for 1 min and maintained on ice. One μL of the supernatant was used for the PCRs.

The third and fourth protocol involved two genomic DNA extraction kits. First one used was E.Z.N.A. Fungal DNA Kit according to the manufacturer indications (OMEGA bio-tek, Norcross, Georgia, USA) with overall good results (around 80% of the samples succeeded). The second one tested was Fungi/Yeast Genomic DNA Isolation Kit, according to the manufacturer indications (NORGEN Biotek, Thorold, ON, Canada) with overall good results also (approximately 70% of the samples succeeded).

The fourth protocol approaches nucleic acid extraction by application of silica coupled to magnetic particles, which is efficient and automated. Genomic fungal DNA was extracted using Maxwell 16 Mouse Tail DNA purification kit. The Promega kit is designed for automated DNA extraction from tissue samples using the Maxwell™ 16 platform (Promega BioSciences, San Luis Obispo, CA, USA). This protocol was performed at the University Institute of Tropical Diseases and Public Health of the Canary Islands, University of La Laguna.

Molecular identification of the fungal Dicarya strains was performed using ITS1 (5′-TCCGTAGGTGAACCTGCGG-3′) and ITS4 (5′-TCCTCCGCTTATTGATATGC-3′) primer pair to amplify the 5.8S rDNA and the two internal transcribed spacers ITS1 and ITS2 [30] for the majority of the samples and NL-1 (5′-GCA TAT CAA TAA GCG GAG GAA AAG-3′) and NL-4 (5′-GGT CCG TGT TTC AAG ACG G-3′) primer pair to amplify the 5′ end of 28S rDNA spanning domains D1 and D2) [31]. PCRs were performed in a total volume of 25 μL containing 10 ng genomic DNA, 0.5 μM primer, 200 μM dNTPs, 1X Buffer Taq, 0.0125U of Taq DNA Polymerase. For ITS sequences, PCR cycling parameters were carried out according to Shu et al. 2014 [20] with slight modifications: 94 °C for 2.5 min; 40 cycles of 94 °C for 30 s, 58 °C for 30 s, and 72 °C for 1 min; and a final extension at 72 °C for 10 min. For 28S rDNA domain, the PCR conditions were denaturation for 4 min at 95 °C followed by 45 s at 95 °C and then annealing for 45 s at 58 °C, 1 min at 72 °C, followed by an extension at 72 °C for 5 min. The final step was at 16 °C for 5 min. A total of 40 cycles were performed. All PCR products were detected by agarose gel electrophoresis (110V, 35 min, on 2% agarose gels, 1X TAE Buffer) loading 5 μL PCR product, 1 μL Loading Buffer (6X) and 2 μL SYBR Green I (Sigma-Aldrich; dilution 1:10,000). PCR and electrophoresis reagents were purchased from Sigma-Aldrich. PCR products were purified using GenElute™ PCR Clean-Up Kit (Sigma-Aldrich) and sequenced by Sequencing

Services SEGAI (La Laguna, Spain). The sequences were run through the BLASTN search page using Megablast program (National Center for Biotechnology Information; Bethesda MD, USA) where the most identical hits and their accession numbers were obtained. Further, only ITS sequences were used for the phylogenetic analysis, therefore details on 28S sequenced strains are listed in Table 2.

Table 2. *Artemisia* fungal endophytic strains: codes, identities as per morphology, 28S rDNA LSU sequences and their most similar hits from Genbank with accession numbers and values.

EF Code	Assigned Species/Species Complex	GenBank Identified Seq.	Max Score	Total Score	E Value	Max Identity (%)	Accession No.
HLP1	*A. alternata*	*A. alternata*	1107	1107	0	99	KX609781.1
HLP10	*A. alternata*	*A. alternata*	1093	1093	0	99	KX609781.1
HLP31	*Fungus* sp. 1	Uncultured fungus clone	135	724	2.00×10^{-27}	100	KP843503.1
HLP5	*A. alternata*	*A. alternata*	1052	1052	0	99	KF751621.1
HLP6	*Curvularia lunata*	*Cochliobolus lunatus*	1026	1026	0	99	KC616350.1
HLP8	*Neofusicoccum* sp. 1	*N. cryptoaustrale*	511	511	5.00×10^{-141}	92	KX464415.1
HLP9	*Preussia* sp. 1	*P. mimoides*	1000	1000	0	97	KF557659.1
HTF25	*Alternaria* sp. 6	*A. brassicicola*	289	289	5.00×10^{-74}	77	AF397222.1
HTF37	*A. alternata*	*A. alternata*	982	982	0	96	KX609781.1
HTF42	*Neofusicoccum australe*	*N. australe*	1036	1036	0	97	KF766367.1
HTF49	*Neofusicoccum australe*	*N. australe*	1058	1058	0	98	HM176550.1
HTF50	*A. alternata*	*A. alternata*	971	971	0	98	KF543048.1
HTF67	*Chaetomium* sp. 1	*C. coarctatum*	846	846	0	99	KX976729.1
HTF75	*Neofusicoccum australe*	*N. australe*	934	934	0	98	HM176550.1
HTF78	*A. alternata*	*A. alternata*	1051	1098	0	99	FJ839651.1
HTF80	*Camarosporium* sp. 1	*Camarosporium* sp.	1024	1024	0	97	KF733369.1

2.6. Phylogenetic Analysis

ITS sequences [i.e., endophytic fungi—Table 3, their most similar hits from GenBank (NCBI, Bethesda MD, USA) and type sequences of the selected taxa] were aligned with the multiple alignment program ClustalW [32] as implemented in Mega 6.0 (Molecular Evolutionary Genetics Analysis) [33] and indels corrected manually to minimize alignment gaps [34]. Designated outgroup was *Caloscypha fulgens* (GenBank Accession No. DQ491483). After the exclusion of non-overlapping leading/trailing gaps the length of the alignment was 603 bps. Because of the high number of indels, these were recoded as a binary matrix by means of the simple indel coding algorithm [35], appending the fragments to the nucleotide data as additional characters, as implemented in FastGap 1.21 (Department of Biosciences, Aarhus University, Denmark) [36]. This "indel matrix" was used in all Bayesian and maximum likelihood analyses. Formerly, Gblocks program (hosted at www.phylogeny.fr) was used to eliminate poorly aligned positions and divergent regions [37]. Best-fit models were compared in jModelTest 2 according to Bayesian Information Criterion (BIC) [38]. Best fit according to the BIC criterion model (K80 + G) was selected to reconstruct the Bayesian tree. Bayesian Inference analysis was conducted with MrBayes 3.2.3 (hosted by Mobyle SNAP Workbench, North Carolina State University) [39] and run for 1×10^7 generations with a sampling frequency of 100 generations. Of the resulting trees, the first 25,000 trees were discarded as burn-in and the following 75,001 were used to estimate topology and tree parameters. The percentage number of times a node occurred within these 75,001 was interpreted as the posterior probability of the node [40]. Convergence of the runs was indicated by an average standard deviation of split frequencies between duplicate runs of less than 0.01. The consensus trees were drawn using Treegraph 2 software (Institute for Evolution and Biodiversity, University of Munster, Germany) [41] and edited with Adobe Illustrator CS3 (Adobe Systems Incorporated, San Jose, CA, USA).

Table 3. Endophytic fungi isolated from *A. thuscula* and used for the phylogenetic analysis: codes, identity and accession numbers of the ITS sequences.

Strain Code	Identity	Accession No.
HLP12	*Phoma* sp.	MG025848
HLP14	*Preussia* sp.	MG025849
HLP15	*Diaporthe phaseolorum*	MG025850
HLP19	*Alternaria alternata*	MG025851
HLP23	*Diaporthe novem*	MG025852
HLP24	*Nigrospora oryzae*	MG025853
HLP25	*Camarosporium brabeji*	MG025854
HLP27	*Coniothyrium* sp.	MG025855
HLP3	*Aspergillus flavus*	MG025856
HLP32	*Alternaria alternata*	MG025857
HLP37	*Diaporthe* sp.	MG025858
HLP40	*Alternaria alternata*	MG025859
HLP43	*Tremateia* sp.	MG025860
HLP44	*Neoplatysporoides aloicola*	MG025861
HLP45	*Neofusicoccum parvum*	MG025862
HLP46	*Neofusicoccum parvum*	MG025863
HLP7	*Paraphoma chrysanthemicola*	MG025864
HTF23	*Nectria mauritiicola*	MG025865
HTF26	*Stachybotrys longispora*	MG025866
HTF27	*Stemphylium solani*	MG025867
HTF30	*Aplosporella prunicola*	MG025868
HTF31	*Stemphylium solani*	MG025869
HTF40	*Biscogniauxia mediterranea*	MG025870
HTF41	*Alternaria alternata*	MG025871
HTF44	*Alternaria alternata*	MG025872
HTF46	*Alternaria alternata*	MG025873
HTF48	*Phoma* sp.	MG025874
HTF52	*Alternaria alternata*	MG025875
HTF53	*Alternaria alternata*	MG025876
HTF62	*Neofusicoccum parvum*	MG025877
HTF64	*Pestalotiopsis* sp.	MG025878
HTF66	*Aureobasidium pullulans*	MG025879
HTF68	*Stemphylium solani*	MG025880
HTF70	*Cladosporium* sp.	MG025881
HTF74	*Preussia australis*	MG025882
HTF76	*Alternaria alternata*	MG025883
HTF79	*Phoma* sp.	MG025884
HTF81	*Preussia* sp.	MG025885
HTF82	*Biscogniauxia mediterranea*	MG025886
HTF83	*Stemphylium solani*	MG025887
HTF84	*Preussia* sp.	MG025888
HTF85	*Preussia* sp.	MG025889

2.7. Diversity Analysis

The colonization rate (CR%) was calculated as the total number of stem fragments in a sample (plant/nutritive medium) yielding at least one isolate divided by the total number of stem fragments in that sample. Colonization frequency (CF%) was calculated as the total number of fragments in a sample (plant/location) colonized by a species divided by the total number of fragments plated. For the diversity of endophytic fungi, the Margalef index, Brillouin index, Fisher's alpha index and Simpson's dominance index were used. Margalef index [42] measures species richness while Brillouin index combines richness and evenness. The Margalef index was calculated using formula $d = (S - 1)/\ln N$, where S is the number of species and N is the number of individuals in the sample. The Brillouin index [43,44] was calculated using formula: $\mathrm{HB} = (\ln N! - S \ln n_i!)/N$, where N is the total number of individuals, S is the number of taxa and n_i is the number of individuals belonging to i species. Fisher's logarithmic series model [45] is a species-abundant model and describes the relationship between the number of species and the number of individuals of those species. It was calculated using formula $S = a \times \ln(1 + n/a)$, where S is number of taxa, n is the number of individuals and a is the Fisher's alpha. The dominance of Simpson [46] was calculated according to the formula $D = 1 - \sum\left[n_{i(n_i-1)}/N(N-1)\right]$, where n_i is the number of individuals belonging to i species and N is the total number of individuals. The Sorensen–Dice coefficient of similarity [47,48] which expresses the beta diversity was employed to compare the similarity of endophytic fungi communities regarding species composition between two host plants, nutrient media and stem ages. The Sorensen–Dice coefficient is calculated with the formula $QS = 2C/(A + B)$ where A and B are the species numbers in samples A and B, respectively, and C is the number of species shared by the two samples. The Sorensen–Dice coefficient weighs more the joint occurrences than the mismatches and is expressed with values between 0 (no similarity) and 1 (absolute similarity). This index was used to assess host preference and spatial heterogeneity by describing the similarity of endophytic communities within ten host plants at distinct sampling sites. Nevertheless, as the coefficient analyses the presence/absence data, no judgments on abundance or rare taxa can be pursued. A binary matrix was produced and used to calculate the similarity matrix and to plot a dendrogram based on an unweighted paired group method of arithmetic average (UPGMA) cluster analysis. For the diversity indices, PAST software version 3.15 (copyright Hammer & Harper, Natural History Museum, University of Oslo, Norway) was used.

3. Results and Discussion

3.1. Fungal Endophytic Diversity in Artemisia Species

3.1.1. Colonization Rate and Colonization Frequency of Endophytic Fungi in *Artemisia thuscula*

In this study, the employed analyses indicate that 37 fungal species and 25 fungal genera were isolated from 10 plants of *Artemisia thuscula*. Colonization rate (further CR) shows how much a plant can be colonized within predetermined conditions. It is valuable information as different plants showed distinct values of this index; therefore, low values could express plants poor in endophytic fungi culturable in the given conditions.

To calculate the colonization frequency (CF) of fungal endophytes in *Artemisia* species plants, we have considered same fungal endophytic species isolated from two or more plant fragments as being a distinct isolate belonging to the same species. Therefore, if the same species was isolated twice from the same plant fragment, it was considered only one time. This issue is to be expected at isolation moment, when no precise differentiation between the isolates can be defined, and only once purified and further analyzed then only the strain received a final identification. The CF% gives a hint over the distribution and abundance of a certain fungal species in a sample (i.e., plant/location/region). To know the "area" of the distribution and abundance of a certain endophytic fungal species, we have analyzed the data per plant individual or plant location, plant species, and plant region. Regions were grouped here as: La Palma Island and Tenerife Island. This way we can have an overview on where certain fungal species are more abundant or rare, as well as if there is a relation between their distribution and plant-specific parameters.

In *Artemisia thuscula*, only one plant out of 10 had a colonization rate value over 90% (LP2). The lowest values (CR% = 25) were recorded for three plants (TF8, TF7 and TF3). Interestingly, as per variable geographical location there is a considerable variation between La Palma Island and Tenerife Island, with the former having the most colonized plant individuals (Table 4).

Table 4. Colonization rate (CR) of fungal endophytes in *Artemisia thuscula* collected in Canary Islands.

Collection Place	Region	Plant Number	Locality Code	CR%
El Granel	La Palma	*LP1	EG	48.28
El Palmar—Teno	Tenerife	**TF8	EP	25.00
Granadilla	Tenerife	TF7	GR	25.00
Igueste Caletillas	Tenerife	TF4	IC	50.00
Mesa Mota	Tenerife	TF1	MM	62.50
Mesa Mota	Tenerife	TF2	MM	62.50
San Andrés	Tenerife	TF5	SA	62.50
San Bartolo	La Palma	LP2	SB	92.11
Taborno	Tenerife	TF3	TA	25.00
Tigalate	La Palma	LP4	TIG	55.00
AVG				*50.78*
SD				*16.13*

*LP = La Palma; **TF = Tenerife; AVG = average; SD = standard deviation.

Artemisia thuscula cannot escape of the "omnipresence" of *Alternaria alternata*, this species was isolated from eight plants but with relevant differences in the frequency, CF% = 15–50%. A notable presence is remarked here, *Neofusicoccum australe*, isolated from three plants at relatively high values (CF% = 25) when considering that the maximum value is 50. Moreover, the *Neofusicoccum* genus, consisting here of three species was isolated from eight plants, one of which revealed a CF% of 34.21. Interestingly, around 70% of the fungal species in *Artemisia thuscula* were isolated from only one plant each (Table 5). This suggests a host specificity which was also exhibited by the low and moderate values of Sorensen–Dice coefficient when the similarity of the endophytic assemblages was analyzed (see further Diversity indices for endophytic fungi in *Artemisia thuscula*).

Table 5. Colonization frequency on potato dextrose agar (PDA) medium of fungal endophytic species in *Artemisia thuscula* plants.

Plant Code	Locality Code	EF Species	CF%	Plant Code	Locality Code	EF Species	CF%
LP1	EG	*Alternaria alternata*	24.14	LP2	SB	*Preussia* sp. 3	2.63
LP1	EG	*Alternaria* sp. 5	13.79	LP2	SB	*Tremateia* sp. 1	2.63
LP1	EG	*Aspergillus flavus*	3.45	LP4	TIG	*Alternaria alternata*	15.00
LP1	EG	*Aspergillus flavus*	6.90	LP4	TIG	*Curvularia lunata*	5.00
LP1	EG	*Diaporthe novem*	3.45	LP4	TIG	*Neofusicoccum* sp. 1	5.00
LP1	EG	*Fungus* sp. 1	3.45	LP4	TIG	*Paraphoma* cf. *chrysantemicola*	5.00
LP1	EG	*Neofusicoccum parvum*	3.45	LP4	TIG	*Preussia* sp. 1	25.00
LP1	EG	*Nigrospora oryzae*	3.45	TF1	MM	*Alternaria alternata*	25.00
LP1	EG	*Penicillium viridicatum*	3.45	TF1	MM	*Thielavia* sp. 1	8.33
LP1	EG	*Phoma* sp. 3	3.45	TF2	MM	*Alternaria alternata*	50.00
LP1	EG	*Pleosporales* sp. 2	6.90	TF2	MM	*Biscogniauxia mediterrranea*	12.50
LP1	EG	*Preussia* sp. 3	3.45	TF2	MM	*Neofusicoccum australe*	12.50
LP2	SB	*Alternaria alternata*	10.53	TF2	MM	*Phoma* sp. 1	12.50
LP2	SB	*Alternaria* sp. 5	2.63	TF3	TA	*Neofusicoccum australe*	25.00
LP2	SB	*Camarosporium bradgi*	2.63	TF4	IC	*Alternaria alternata*	37.50
LP2	SB	*Coniothyrium* sp. 1	2.63	TF4	IC	*Aureobasidium pullulans*	12.50
LP2	SB	*Diaporthe phaseolorum*	7.89	TF5	SA	*Alternaria alternata*	12.50
LP2	SB	*Diaporthe* sp. 1	5.26	TF5	SA	*Alternaria* sp. 6	12.50
LP2	SB	*Dothideomycetes* sp. 1	2.63	TF5	SA	*Macrophomina phaseolina*	25.00
LP2	SB	*Fungus* sp. 1	2.63	TF5	SA	*Stachybotrys longispora*	12.50
LP2	SB	*Neofusicoccum parvum*	13.16	TF7	GR	*Alternaria alternata*	25.00
LP2	SB	*Neofusicoccum* sp. 3	34.21	TF7	GR	*Neofusicoccum australe*	25.00
LP2	SB	*Neoplatysporoides aloicola*	5.26	TF7	GR	*Stemphylium solani*	25.00
LP2	SB	*Nigrospora* sp. 2	2.63	TF8	EP	*Camarosporium* sp. 1	12.50
LP2	SB	*Pleosporales* sp. 3	2.63	TF8	EP	*Phoma* sp. 1	12.50

EG = El Granel; SB = San Bartolo; TIG = Tigalate; MM = Mesa Mota; TA = Taborno; IC = Igueste Caletillas; SA = San Andres; GR = Granadilla; EP = El Palmar.

34 endophytic fungal species were isolated from *Artemisia thuscula* (Table 6) and their frequency varied within a low range with two exceptions: *Alternaria alternata* (CF% = 18.71) and *Neofusicoccum* sp. 3 (CF% = 8.39).

Table 6. Colonization frequency of fungal endophytic species in *Artemisia thuscula* (overall CF%/plant species).

EF Species	CF%	EF Species	CF%
Alternaria alternata	18.71	*Neofusicoccum parvum*	3.87
Alternaria sp. 5	3.23	*Neofusicoccum* sp. 1	0.65
Alternaria sp. 6	0.65	*Neofusicoccum* sp. 3	8.39
Aspergillus flavus	1.94	*Neoplatysporoides aloicola*	1.29
Aureobasidium pullulans	0.65	*Nigrospora oryzae*	0.65
Biscogniauxia mediterrranea	0.65	*Nigrospora* sp. 2	0.65
Camarosporium bradgi	0.65	*Paraphoma chrysantemicola*	0.65
Camarosporium sp. 1	0.65	*Penicillium viridicatum*	0.65
Coniothyrium sp. 1	0.65	*Phoma* sp. 1	1.29
Curvularia lunata	0.65	*Phoma* sp. 3	0.65
Diaporthe novem	0.65	*Pleosporales* sp. 2	1.29
Diaporthe phaseolorum	1.94	*Pleosporales* sp. 3	0.65
Diaporthe sp. 1	1.29	*Preussia* sp. 1	3.23
Dothideomycetes sp. 1	0.65	*Preussia* sp. 3	1.29
Fungus sp. 1	1.29	*Stachybotrys longispora*	0.65
Macrophomina phaseolina	1.29	*Stemphylium solani*	0.65
Neofusicoccum australe	2.58	*Tremateia* sp. 1	1.94

Studies that are independent of fungal isolation and identification methods often revealed higher numbers of fungal species [49]. We purposely chose the culture method to further select endophytic fungi of high interest according to their biological activities. Our goal was to yield a large number of endophytes, and not to produce a complete species list of fungal endophytes in these *Artemisia* species. Nevertheless, the data obtained gave us an interesting fragment of knowledge about the communities of these microorganisms in their plant hosts.

In terms of endophytic fungal species CF%, the most isolated species was *Alternaria alternata* (CF = 18.71; eight of ten plants), as expected. It is a common saprobe found on various plants and other substrata worldwide [50,51] and has often been isolated as endophyte in previous studies [52–55]. Qian et al., 2014 [27] isolated endophytic fungi from *Artemisia argy* and found Pleosporales to be the most represented group, with three species of *Alternaria* present. It was found as the most predominant species in grasses [56] and various plants families, also [57]. Among dominant endophytic fungal species, we observed taxa like *Neofusicoccum* and *Preussia*. These genera of endophytic fungi were previously isolated from a wide range of host plants including *Artemisia* spp. [24,57–60].

Interestingly, it was observed a tendency on host specificity of most endophytic fungal species. In *Cirsium arvense* similarity in endophytic communities decreased with increasing intersite distance [61] while in *Holcus lanatus* the similarity between leaf and root myco-assemblages at the same location was lower than that observed in leaves at different locations [58]. Further, in leaf fungal communities the average number of species shared by any pair of location was 3.13 and in root assemblages was 1.73 out of an average of 12.2 species identified at each location [58].

Despite the dominant species, the rest of the endophytic fungal species reflect an unequal distribution of a certain endophytic species among plant individuals. This same issue was previously observed [58] but no definitive answer has been found. Some hypotheses were proposed like ubiquitous taxa with spatial dominance or selection of certain dependent on culture conditions [58]. In the case study on *Artemisia thuscula* (see Section 3.1.3) taxa such as *Preussia, Pestalotiopsis, Aplosporella, Chaetomium* and *Cladosporium* were isolated from only one nutrient medium out of the three media tested. Nevertheless, this is not a unique parameter, which should account for the determination of an endophytic taxa preference for a nutrient medium. One of the major variables which we consider is the rest of the community involved and their role in the interaction when the isolation performed. That is, which are the other taxa living in the same "space" (i.e., plated plant fragment) and we must consider if there are (i) fast-growing taxa versus slow growing taxa; (ii) nutrient deficiency or promoting medium for certain taxa, as well as (iii) the interaction between the taxa (i.e. antagonism).

3.1.2. Diversity Indices for Endophytic Fungi in *Artemisia thuscula*

In the La Palma results of diversity, Margalef index revealed the highest value for species richness in San Bartolo (Margalef = 4.24) followed by El Granel (Margalef = 3.69). The Brillouin index agrees that the highest diversity is found in San Bartolo (Brillouin = 1.8) but Fisher's alpha index shows a higher abundance of rare species in El Granel (Fisher's alpha = 18.6) than in San Bartolo (Fisher's alpha = 13.9). Diversity regarded as evenness was found to be similar in both localities (Simpson's index: El Granel = 0.88 and San Bartolo = 0.87). In La Palma Island, San Bartolo locality was revealed as having the highest value for species richness and diversity. Yet, El Granel was shown as having a higher abundance of rare species (Fisher's alpha: El Granel = 18.6 and San Bartolo = 13.9) and a higher value of evenness than San Bartolo (Simpson's index: El Granel = 0.88 and San Bartolo = 0.87). In Tenerife, the locality San Andres showed by far the highest diversity in all previously mentioned terms and all the indices confirm it (Table 7).

Table 7. Diversity indices of fungal endophytic species per plant individual/locality.

Locality Code	Taxa No.	Strains No.	Simpson 1-D	Brillouin	Margalef	Fisher Alpha
LP1/EG	11	15	0.89	1.65	3.69	18.60
LP2/SB	15	27	0.87	1.85	4.25	13.90
LP4/TIG	5	5	0.80	0.96	2.49	0.00
TF1/MM	5	10	0.60	0.85	1.74	3.98
TF3/TA	1	2	0.00	0.00	0.00	0.80
TF4/IC	2	3	0.44	0.37	0.91	2.62
TF5/SA	4	5	0.72	0.82	1.86	9.28
TF7/GR	3	3	0.67	0.60	1.82	0.00
TF8/EP	2	2	0.50	0.35	1.44	0.00

LP1/EG = El Granel; LP2/SB = San Bartolo; LP4/TIG = Tigalate; TF1/MM = Mesa Mota; TF3/TA = Taborno; TF4/IC = Igueste Caletillas; TF5/SA = San Andres; TF7/GR = Granadilla; 14/EP = El Palmar.

Sorensen–Dice index revealed that of 45 cases in the matrix, 27 of them were of zero similarity. Further, only one case was found to have 57% similarity (TF2 versus TF7) and one case with 50% similarity (TF1 versus TF4). The rest of the cases had values ranging between 11% and 40% similarity. These different similarity values may be due to distance among hosts, soil composition and/or climatic conditions. When the distance was plotted (UPGMA), the Sorensen–Dice coefficient clustered plants LP1 and LP2 with maximum bootstrap support (BPP = 100), although these plants had only 38% similarity in between. Nevertheless, this is to be considered a high value of similarity in the given matrix and one of the reasons for obtaining it might be the proximity of the collection places (approx. 5 km) between the host plants and similar altitudes and climate. Further clusters were formed like LP4 and TF1; TF2 and TF3; TF4, TF5 and TF7 (Figure 1). As we expected (from CF and CR values) TF8 is the most different host plant, the backbone of the dendrogram divides into this branch and the other branches which form various clusters of similarity. Also, cluster LP1 and LP2 is a sister cluster of the other clusters which were exhibited as more related in terms of similarity.

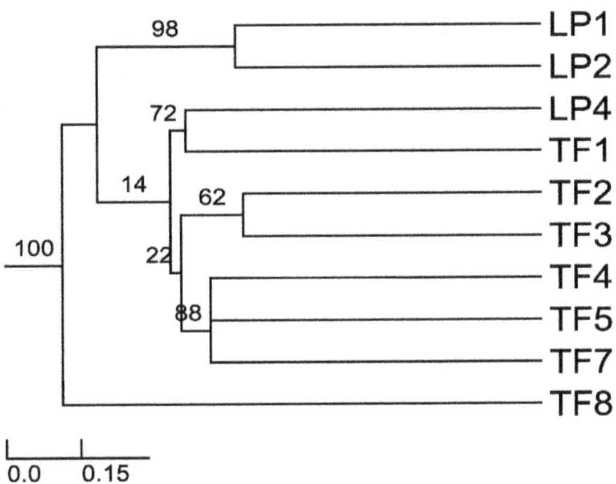

Figure 1. Sorensen's similarity coefficient for the endophytic fungi isolated from *A. thuscula*: unweighted paired group method of arithmetic average (UPGMA) dendrogram plot. The results were obtained with 95% of confidence and bootstrap values calculated from 1000 iterations.

3.1.3. Case Study: *Artemisia thuscula* of Tenerife, Endophytic Fungi Isolated from Two Types of Stems on Three Media: Colonization Frequency and Colonization Rate

In this study, we can observe throughout various individual plants from the same species (i.e., *Artemisia thuscula*) the relevance of nutrient media and the age of the stem as the selected organ to yield endophytic fungi. When averaged the colonization rates of the three nutrient media selected (PDA, LCA, and V8) do not show relevant differences (CR% = 33.93. 33.93 and 37.50, respectively). Neither do the differences of age in stems; stems with the age < 1 year have CR% = 30.95 and stems with age > 1 year have CR% = 36.90.

Differences may be observed (Table 8) when comparing different plants, as for instance TF3 and TF4 had the lowest colonization rates (CR% = 16.67) and no endophytic fungi was isolated from V8 or stems with age of more than 1 year for TF4 and TF3, respectively. In addition, there is no higher value than 58.33 of colonization rate, as observed in other individuals of *Artemisia*.

Table 8. *Artemisia thuscula* colonization rate (CR%) of endophytic fungi per plant. per medium and per stem age.

Plant Code	CR%/Plant	CR%/PDA	CR%/LCA	CR%/V8	CR%/Stem	
					Age < 1 Year	Age > 1 Year
TF3	16.67	25	12.5	12.5	33.33	0
TF4	16.67	37.5	12.5	0	8.33	25
TF5	58.33	62.5	75	37.5	58.33	41.67
TF7	45.83	25	50	62.5	50	41.67
TF8	37.5	25	37.5	50	8.33	66.67
AVG	35.12	33.93	33.93	37.50	30.95	36.90

If colonization frequency data is segregated into plants sampled (Table 9), we observe that *Alternaria alternata* is the major colonizer in three out of seven plants, namely TF2, TF4, and TF5. Plants had different yields considering number of endophytic fungal species, ranging between three (TF3) and eight (TF7).

Among the major colonizers we observed *Neofusicoccum austral* and *Neofusicoccum parvum* in TF3 (CF% = 8.33), *Chaetomium* sp. 1 and *Phoma* sp. 1 in TF7 and *Phoma* with two different species in TF8 (CF% = 16.67; Table 8). Myrchiang et al., 2014 [28] investigated the endophytic fungi associated with *Artemisia nilagirica* and comparing the colonization of three organs (i.e., root, stem and leaf), the authors obtained the highest diversity in the roots (i.e., 14 species), less in stem (i.e., 10 species) and the smallest number in the leaves (i.e., 6 species). Similarly, in *Artemisia thuscula* Cosoveanu et al., 2012 [62] isolated 29 distinct morphotypes: 20 from roots, 7 from stem and 2 from leaves. In addition, Myrchiang et al., 2014 [28] observed that from all fungal endophytic species, only *Phoma eupyrena* was found to be a common occurrence in all plants sample, the other species having a certain preference for one or maximum two organs.

Table 9. *Artemisia thuscula* colonization frequency (CF%) of endophytic fungi species per plant, per medium and per stem age.

EF Species	Plant	CF%/Plant	CF%/Medium			CF%/Stem Age	
			PDA	LCA	V8	≤1 year old	>1 year old
Alternaria alternata	TF2	50	50	37.5	62.5	41.67	58.33
Alternaria sp. 4		4.17		12.5			8.33
Biscogniauxia mediterrranea		4.17	12.5			8.33	
Neofusicoccum australe		16.67	12.5	12.5	12.5	16.67	8.33
Pestalotiopsis sp.		4.17		12.5		8.33	
Phoma sp. 1		4.17	12.5				8.33
Neofusicoccum australe	TF3	8.33	25			16.67	
Neofusicoccum parvum		8.33		12.5	12.5	16.67	
Pestalotiopsis sp.		4.17		12.5		8.33	
Alternaria alternata	TF4	12.5	37.5			16.67	8.33
Aureobasidium pullulans		4.17	12.5			8.33	
Nectria mauritiicola		8.33		25			16.67
Alternaria alternata	TF5	37.5	12.5	50	50	8.33	66.67
Alternaria sp. 6		4.17	12.5			8.33	
Aplosporella prunicola		4.17		12.5		8.33	
Aureobasidium pullulans		4.17		12.5			8.33
Macrophomina phaseolina		8.33	25				16.67
Neofusicoccum australe		4.17		12.5		8.33	
Stachybotrys longispora		4.17	12.5			8.33	
Stemphylium solani		8.33		12.5	12.5	16.67	
Alternaria alternata	TF7	8.33	8.33	12.5	12.5		16.67
Aureobasidium pullulans		12.5		12.5		8.33	
Chaetomium sp. 1		16.67			25		16.67
Cladosporium sp. 1		12.5			25	16.67	
Neofusicoccum australe		8.33			12.5		8.33
Phoma sp. 1		16.67		50		33.33	
Preussia australis		8.33	12.5				8.33
Stemphylium solani		4.17	12.5			8.33	
Biscogniauxia mediterrranea	TF8	8.33			12.5		8.33
Camarosporium sp. 1		8.33	12.5				8.33
Phoma sp. 1		16.67	12.5			8.33	
Preussia sp. 2		16.67			37.5		25
Preussia sp. 3		12.5		12.5			8.33
Preussia sp. 5		12.5		25			16.67
Stemphylium solani		4.17		12.5			8.33

Comparing different plant individuals of the same species and observing the distribution of fungal endophytes provides insights to determine the occurrence of a certain species. For instance, in TF2 four fungal species were isolated only from one nutrient medium, namely *Biscogniauxia mediterranea* in PDA, *Alternaria* sp. on LCA, *Phoma* sp. on PDA and *Pestalotiopsis* sp. on LCA (Table 9). Furthermore, we may observe that the same species of *Phoma* sp. 1 was also isolated from TF8 on PDA, similar to *Pestalotiopsis* isolated from TF3 on LCA while *Biscogniauxia mediterranea* was isolated on V8 from TF8.

When the distribution of endophytic fungi species is observed in terms of colonization frequency per total number of the studied plants (Figure 2), data showed several species like *Aplosporella prunicola*, *Camarosporium* sp., *Chaetomium* sp., *Cladosporium* sp, *Nectria mauritiicola* and others with certain "preference" for nutrient medium. It is well known that fungi have specific carbon and nitrogen requirements for sporulation [63–65]. However, the requirements for fungal growth are less stringent but not less important when isolation is pursued. Nutrient - rich media result in selective isolation for fast-growing fungi, overlooking slow growing species if present [66]. Osono and Takeda [29] stated that LCA due to its low glucose content suppresses the overgrowth of fast-growing species. 22 species of fungal endophytes were isolated from all *Artemisia thuscula* plants in this case study and 14 species (63%) were isolated only from one nutrient medium. Additionally, 12 fungal species were isolated from stems older than 1 year and seven were isolated from stems younger than one year. Seven fungal species are to be considered rare, as their colonization frequency value is the lowest one, throughout the data set (CF% = 0.60; Figure 2).

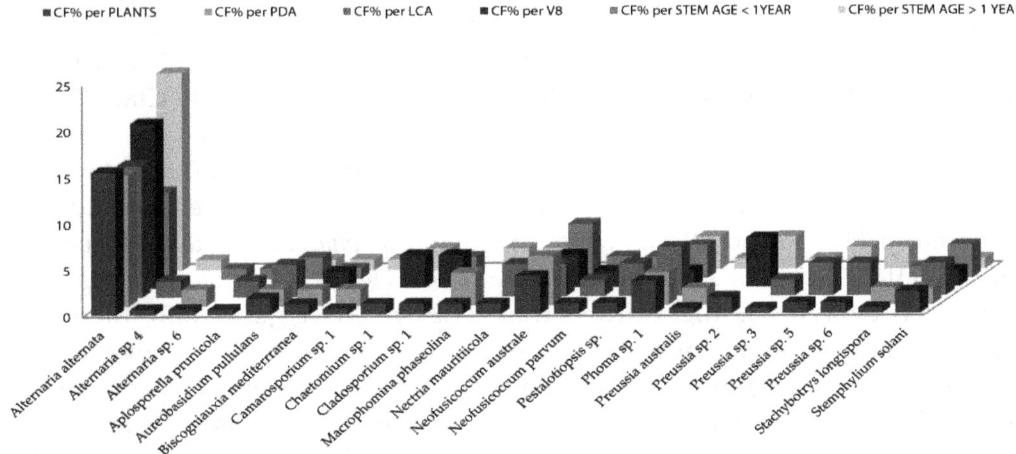

Figure 2. *Artemisia thuscula* colonization frequencies of endophytic fungi species per total number of plant individuals, per media (PDA, lignocellulose agar (LCA) and V8) and per stem age (>1 year and <1year). The *y*-axis data correspond to colonization frequency percentage.

Further, Sorensen–Dice similarity coefficient reveals proximate values among the similarities of the endophytic communities isolated on the three tested media (Table 10). Yet, none of them overpassed 52% similarity (i.e., LCA versus V8). As for the stem ages, the index showed a value of 43% similarity. Evidence for tissue specificity was previously demonstrated for phloem and xylem tissue, where the value of endophytic similarity reached 36% in roots of *Sophora tonkinensis* [67]. This suggests the necessity to broad both culture media and diversity of tissues to obtain a higher richness of endophytic fungal taxa.

Table 10. *A. thuscula* fungal endophytes isolated on different nutrient media and stem age: Sorensen–Dice coefficient of similarity.

	LCA	V8	≤1 Year	>1 Year
PDA	0.42	0.48	0.64	0.71
LCA		0.52	0.67	0.67
V8			0.67	0.59
≤1 year				0.43

Among the singleton species that occur only in the *Artemisia thuscula* plant individuals selected for this case study (i.e., limited to Tenerife) we have: *Aplosporella prunicola*, *Camarosporium* sp. 1, *Macrophomina phaseolina*, *Chaetomium* sp. 1, *Nectria mauritiicola*, *Neofusicoccum australe*, *Pestalotiopsis* sp., *Phoma* sp. 1 and *Stachybotrys longispora*. Except *Phoma* sp., all endophytic fungal species previously mentioned were isolated from only one nutrient medium. In addition, except *Camarosporium* sp. (isolated also from *Artemisia thuscula* in Palma Island), all endophytic fungal species previously named were isolated only from *Artemisia thuscula* in Tenerife Island.

3.2. Phylogenetic Relations

54 endophytic fungal ITS sequences and the associated GenBank sequences were used for the phylogenetic analysis (Table 3; sequences of strains HLP16, HLP22, HLP28, HLP33, HLP48A, HLP4,

HTF29, HTF33, HTF43, and HTF61 are not listed and are only available at request). The dataset consists of 603 characters after alignment, 43 characters are conserved, and 447 characters are parsimony informative, while 557 are variable characters. Bayesian Posterior Probabilities (BPPs) given below each node are shown on the upper branches.

Ten orders (Diaporthales, Dothideales, Botryosphaeriales, Hypocreales, Trichosphaeriales, Amphisphaeriales, Xylariales, Capnodiales, Pleosporales and Eurotiales) are recognized (Figure 3). The phylogenetic tree divides the taxa in five main clades, leaving *Diaporthe* sequences unclustered. Clade 1 consists of Dothideales and Botryosphaeriales (BPP = 0.98), Clade 2 groups Hypocreales, Trichosphaeriales, Amphisphaeriales and Xylariales (BPP =0.88), Clade 3 and Clade 4 contain Capnodiales (BPP = 0.79) and Pleosporales (BPP = 0.63), respectively while Clade 5 accommodates Eurotiales (BPP = 0.62).

Interestingly, *Diaporthe* sequences are not clustered but several show different branch lengths. Yet, taxa *D. novem* and *D. phaseolorum* do not differentiate. Endophytic fungi were basically identified using morphology; therefore, HLP15 and HLP23 were considered *D. phaseolorum* and *D. novem*, respectively while structures of HLP37 did not allow an accurate species level identification. Apparently, the ITS region in *Diaporthe* is evolving at higher rates than TEF1 or MAT genes [68], therefore presenting a wider variation than advisable for species boundaries. Thus, a slowly evolving gene region should be used in order to establish species limits [69]. Nevertheless, ITS sequence data can be used for reliable identification of phylogenetic relationships as long as they are interpreted with care [69]. Several arrangements of genera draw the attention, like *Aureobasidium* (Dothideales) and *Aplosporella* (Botryosphaeriales) which are shown with an immediate common ancestor (BPP = 0.97). *Aplosporella* has over 300 species and appears to be heterogenous; therefore not all species are likely to belong in Botryosphaeriaceae [70]. The ascomycete genus *Aureobasidium* is a member of the family Aureobasidiaceae within the class of the Dothideomycetes [71]. Dothideomycetidae subclass was emended by Schoch et al., 2006 [72] and a new subclass was proposed, Pleosporomycetidae, with an additional order, the Botryosphariales.

Penicillium and *Aspergillus* sequences form two sister clades as expected (BPP = 0.60). Three species of *Neofusicoccum* are clustered with relevant support (BPP = 0.89) while *N. parvum* is drawn outside. Hypocreales taxa are split in two sister clusters along with *Stachybotrys*, *Grandibotrys*, *Melanopsamma* and *Sirastachys* in one sister clade although with no relevant support (BPP = 0.55) and *Nectria*, *Sarocladium* and *Corallomycetela* as another sister clade (BPP = 0.87). Also, internal clustering is revealed between several taxa of the mentioned genera. Trichosphaeriales and Amphisphaeriales are shown having a common recent ancestor (BPP = 0.98). Hypocreales is recognized as monophyletic [73]. The order Hypocreales incorporates Nectriaceae and Stachybotriaceae beyond other six families [74]. Maharachchikumbura et al., 2014 [75] found using a combined LSU, SSU, TEF and RPB2 sequences data that *Stachybotrys* and related taxa (Stachybotriaceae) form a sister cluster of *Nectria* and related taxa (Nectriaceae). The results obtained with the ITS region are in accordance with the combined inference obtained by Maharachchikumbura et al., [75]. The Nectriaceae group (BPP = 0.87) comprises *Nectria* (Nectriaceae), *Sarocladium* (Hypocreomycetidae) and *Corallomycetella*—shown to comprise two distinct clades in Nectriaceae [76]. The second cluster joints *Stachybotrys*, *Grandibotrys*, *Sirastachys*, *Stachybotrys* (Stachybotriaceae, Hypocreomycetidae) and *Melanopsamma* (Chaetosphaeriaceae, Sordariomycetidae). *Melanopsamma pomiformis* was recently excluded from the genus [77] and it was linked to the asexual morph *Stachybotrys albipes* [78]. Strains of Sordariomycetes clustered into six subclasses among which Diaporthomycetidae, Xylariomycetidae and Hypocreomycetidae [75]. Our Bayesian analysis resulted in a monophyletic clade (Clade 2) which accommodates Hypocreales (Hypocreomycetidae), Trichosphaeriales (Diaporthomycetidae), Amphisphaeriales (Xylariomycetidae) and Xylariales (Xylariomycetidae). Yet, Diaporthales taxa (*Diaporthe* spp.) were left outside this clade. A resulting parsimonious tree of multi-locus based (LSU, ITS, and *TEF1*) sequences shows that the genus *Diaporthe* has paraphyletic origins [79]. Xylariales and Amphisphaeriales were found as sister clusters in Xylariomycetidae sharing a common ancestor [80].

Yet, the clade which accommodates Xylariomycetidae is a sister clade of Diaporthomycetidae (Diaporthales) and Hypocreomycetidae (Hypocreales).

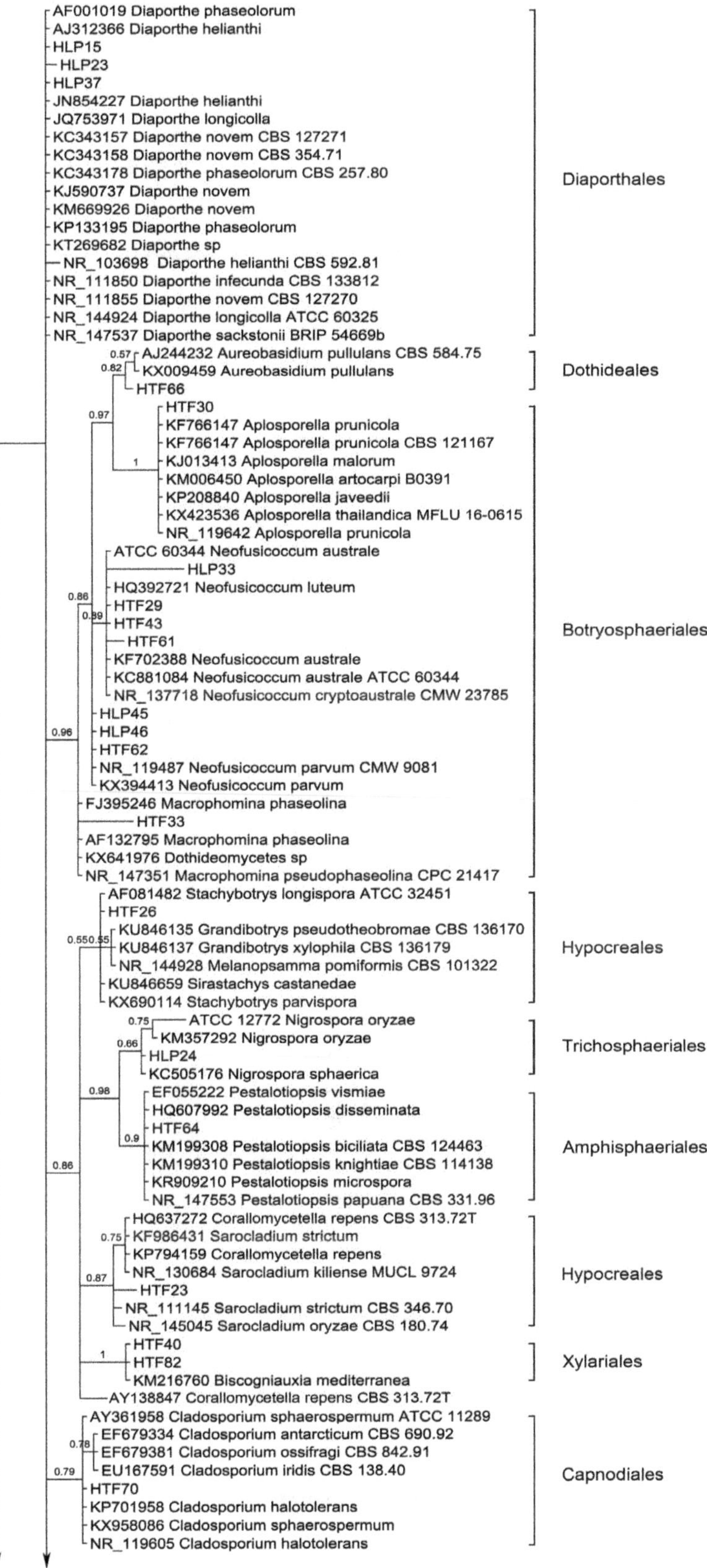

Figure 3. *Cont.*

AF229476 Alternaria tenuissima ATCC 16423
AF261661 A alternata
AF314580 A alternata
EF076750 Stemphylium solani
GU584954 Pleospora herbarum
HTF27
HTF31
HTF68
HTF83
KF479193 Stemphylium globuliferum
0.53
KU850555 Stemphylium vesicarium CBS 155.24
KU850556 Stemphylium vesicarium CBS 157.24
KU850575 Stemphylium mali CBS 122640
KU850618 Stemphylium majusculum CBS 717.68
0.75
KU850627 Stemphylium solani CBS 116586
KX426693 Pleospora herbarum
LN896693 Stemphylium vesicarium
MG065802 Stemphylium vesicarium
0.77
KR912323 A infectoria
NR_131263 Alternaria infectoria CBS 210.86
FJ025207 A alternata
FJ755198 A tenuissima
GU237713 Didymella pinodella CBS 100580
HLP12
HTF48
HTF79
KF293763 Phoma fungicola
KJ173541 Phoma sp
KM030324 Peyronellaea pinodella
KM507779 Phoma pedeiae
KU554585 Didymella sp
KU728479 Briansuttonomyces eucalypti CBS 114879
0.93
KX065025 Phoma moricola
KX079485 Dothiorella gregaria
KC357255 Phoma fungicola
NR_135942 Ascochyta phacae CBS 184.55
NR_135963 Didymella heteroderae CBS 109.92
NR_135968 Nothophoma infossa CBS 123395
0.7
0.89
NR_135984 Didymella pedeiae CBS 124517
NR_135991 Didymella dimorpha CBS 346.82
NR_135992 Nothophoma anigozanthi CBS 381.91
NR_136125 Didymella rosea BRIP 50788
NR_137836 Peyronellaea prosopidis CBS 136414
0.88
HLP7
KC180716 Paraphoma cf chrysanthemicola
KF251165 Paraphoma chrysanthemicola CBS 172.70
1
HLP16
0.91
HLP22
HLP48
HLP19
HLP32
HLP40
HQ647312 1 A tenuissima
HTF22
HTF41
HTF44
HTF46
HTF52
HTF53
0.89
HTF58
HTF76
JF802119 A arborescens
JX867224 A alternata
KC415810 Alternaria arborescens
Pleosporales
KF293964 A alternata
KF887098 A tenuissima
KU180457 A longipes
KU324783 A tenuissima
KU982603 A ochroleuca
KU997022 A alternata
KX685356 A arborescens
KY039170 A longipes
KY197939 A tenuissima
LT604489 A tenuissima
MG562516 Alternaria alternata
NR_111832 Alternaria eichhorniae ATCC 22255
NR_135927 Alternaria arborescens CBS 102605
NR_137143 Alternaria destruens ATCC 204363
HLP44
KR476719 Neoplatysporoides aloicola CBS 139901
NR_145200 Libertasomyces myopori CBS 141302
AY510422 Preussia minimoides
AY943053 Sporormiella isomera
DQ885897 Paracamarosporium hawaiiensis CBS 12025
EU295650 Paraconiothyrium africanum
EU552105 Camarosporium brabeji
HLP25
HLP27
HLP28
HQ115651 Coniothyrium sp
0.51
KF010841 Microdiplodia sp
KF367518 Paraconiothyrium sp
JQ346215 Camarosporium brabeji
0.63
KX869961 Didymosphaeria variabile
KY979754 Paraconiothyrium hakeae CBS 142521
MG065737 Pseudocamarosporium sp
MG065739 Camarosporium sp
NR_137669 Paraconiothyrium estuarinum CBS 109850
0.78
NR_145167 Paraphaeosphaeria sardoa CBS 501.71
DQ885900 Camarosporium mamanes CBS 120031
EU552106 Camarosporium leucadendri
GQ254687 Leptosphaerulina chartarum
HLP43
0.55
KM979793 Coniothyrium sp
KX274241 Tremateia sp
KX274240 Tremateia sp
KY984340 Helminthosporium quercinum
HLP14
HTF81
0.9
HTF85
KP401890 Preussia sp
KX908900 Dothideomycetes sp
HQ130664 Sporormiella sp
HTF21
HTF74
HTF84
KF557658 Preussia minimoides
KP991588 Dothideomycetes sp
KU295583 Preussia australis
KU869522 Preussia sp
HQ608018 Preussia australis CY188
GU183123 Preussia minimoides
AY943052 Preussia australis ATCC 22797
KX650546 Hobus wogradensis
0.56
KX650550 Nigrograna fuscidula
KX650569 Thyridaria broussonetiae
NR_077168 Preussia flanaganii CBS 112.73
NR_103588 Preussia isomera CBS 318.65
AF033475 Penicillium polonicum NRRL 447
0.8
FJ613113 Penicillium viridicatum
JF922035 Penicillium chrysogenum CBS 306.48
KX958079 Penicillium polonicum
HLP3
0.62
JX427022 Aspergillus flavus
Eurotiales
0.7
KX650844 Aspergillus oryzae
NR_111041 Aspergillus flavus ATCC 16883
NR_135395 Aspergillus oryzae NRRL 447
HLP4
DQ491483 Caloscypha fulgens
0.0 0.2

Figure 3. Bayesian phylogenetic tree based on ITS rDNA sequence variants of the endophytic fungi isolated from *A. thuscula* and their associated external GenBank hits. The tree was rooted with *Caloscypha fulgens* sequence as outgroup. The Bayesian clade-credibility values (posterior probabilities) are indicated at internodes (BPP). The scale bar represents the expected changes per site. Sequences coded with HLP/HTF were obtained from endophytic fungi, sequences coded with taxa names are associated external sequences and the ones coded with CBS/L/ATCC/NRRL/B/MFLU/CMW/MUCL/BRIP were obtained from type strains.

Cladosporium sequences are clustered (BPP = 0.79) and different branch lengths between species are revealed, grouping *C. ossifragi*, *C. antarcticum* and *C. iridis* (BPP = 0.78). Conversely, *Aplosporella* sequences do not differentiate in between, showing all species with same branch lengths.

Stemphylium sequences are grouped but support does not avail this grouping (BPP = 0.53). *Phoma*-like sequences are clustered as expected (BPP = 0.93) showing higher differences between *Phoma*, *Didymella*, *Dothiorella* and *Notophoma* on one side (BPP = 0.93) and *Paraphoma chrysantemicola* on the other side (BPP = 0.88). It is curious that several sequences of endophytes are grouped in a sister clade of *Alternaria* clade, *Phoma*-like clade and *Stemphylium* clade with high probability (BPP = 1), indicating different branch lengths. *Alternaria* sequences are not grouped in a single cluster but different branch lengths are drawn among the species. Similarly, *Preussia* and *Sporormiella* taxa are spread. *Coniothyrium*-like sequences are clustered, but support has an average value, BPP = 0.78. Coniothyriaceae and Camarosporiaceae grouping as well as the *Coniothyrium*-like sequences cluster and its sister cluster of Pleosporaceae is supported by the findings of Wijayawardene et al., 2014 [81]. Mainly the sequences obtained from the endophytic strains are grouped with the external sequences as expected (i.e., morphological identification) but several are left unclustered. For instance, inside the group of Pleosporales three endophytic sequences (HLP16, HLP22 and HLP48A) appear as more related, forming a strong-supported cluster (BPP = 1). This apparently new lineage should be confirmed with another phylogenetic study based on large subunit and small subunit nuclear rDNA regions, where only Pleosporales taxa would be included. In the present study none of the methods used like the morphology (absence of the sporulating structures), BLAST alignment (values of similarity with GenBank provided sequences did not exceed 86%, 88% and 84% for HLP16, HLP22, and HLP48A, respectively) and the ITS inference, could provide their proper identification or genetic stronger alliances inside Pleosporales.

4. Conclusions

The present study suggests culturable endophytic species have specificity for a plant host and "preference" for nutrient medium. Therefore, this study indicates the apparent necessity of using different culture media so as to obtain a higher diversity of species.

Acknowledgments: We thank Kanika Chowdhary (Indian Institute of Technology) and Mariano Hernandez (Dept. Bioquimica, Microbiologia, Biologia Celular y Genetica, Universidad of La Laguna) for their valuable contributions. Two anonymous reviewers are thanked for their comments and suggestions on the draft version of this script. The authors are grateful to Cristian Predescu and Alin Gayraud for the graphical elements - phylogenetic tree and abstract, respectively. Andreea Cosoveanu was supported by the pre-doc contract as research personnel in formation (sp. "Contratos predoctorales para la formación de doctores"), Call 2015 by Universidad de La Laguna and La Caixa.

Author Contributions: Andreea Cosoveanu and Raimundo Cabrera conceived and designed the experiments; Andreea Cosoveanu and Samuel Rodriguez Sabina performed the experiments; Andreea Cosoveanu and Raimundo Cabrera analyzed the data; Raimundo Cabrera contributed reagents/materials/analysis tools; Andreea Cosoveanu and Raimundo Cabrera wrote the paper.

References

1. Arnold, A.E.; Maynard, Z.; Gilbert, G.; Coley, P.; Kursar, T. Are Tropical Fungal Endoyphytes Hyperdiverse? *Ecol. Lett.* **2000**, *3*, 267–274. [CrossRef]
2. Sturz, A.V.; Christie, B.R.; Nowak, J. Critical Reviews in Plant Sciences Bacterial Endophytes: Potential Role in Developing Sustainable Systems of Crop Production. *CRC Crit. Rev. Plant Sci.* **2000**, *19*, 1–30. [CrossRef]
3. Smith, C.S.; Chand, T.; Harris, R.F.; Andrews, J.H. Colonization of a Submersed Aquatic Plant, Eurasian Water Milfoil (*Myriophyllum spicatum*), by Fungi under Controlled Conditions. *Appl. Environ. Microbiol.* **1989**, *55*, 2326–2332. [PubMed]
4. Stanley, S.J. Observations on the Seasonal Occurrence of Marine Endophytic and Parasitic Fungi. *Can. J. Bot.* **1992**, *70*, 2089–2096. [CrossRef]
5. Petrini, O.; Fisher, P.J.; Petrini, L.E. Fungal Endophyte of Bracken (*Pteridium aquilinum*) with Some Reflections on Their Use in Biological Control. *Sydowia* **1992**, *44*, 282–293.
6. Raviraja, N.S.; Sridhar, K.R.; Barlocher, F. Endophytic Aquatic Hyphomycetes of Roots of Plantation Crops and Ferns from India. *Sydowia* **1996**, *48*, 152–160.
7. Nisa, H.; Kamili, A.N.; Nawchoo, I.A.; Shafi, S.; Shameem, N.; Bandh, S.A. Fungal Endophytes as Prolific Source of Phytochemicals and Other Bioactive Natural Products: A Review. *Microb. Pathog.* **2015**, *82*, 50–59. [CrossRef] [PubMed]
8. Pirttilä, A.M.; Laukkanen, H.; Pospiech, H. Detection of Intracellular Bacteria in the Buds of Scotch Pine (*Pinus sylvestris* L.) by In Situ Hybridization. *Appl. Environ. Microbiol.* **2000**, *66*, 3073–3077. [CrossRef] [PubMed]
9. Pirttilä, A.M.; Pospiech, H.; Laukkanen, H.; Myllylä, R.; Hohtola, A. Two Endophytic Fungi in Different Tissues of Scots Pine Buds (*Pinus sylvestris* L.). *Microb. Ecol.* **2003**, *45*, 53–62. [CrossRef] [PubMed]
10. Hata, K.; Sone, K. Isolation of Endophytes from Leaves of *Neolitsea sericea* in Broadleaf and Conifer Stands. *Mycoscience* **2008**, *49*, 229–232. [CrossRef]
11. Pirttilä, A.M.; Podolich, O.; Koskimäki, J.J.; Hohtola, E.; Hohtola, A. Role of Origin and Endophyte Infection in Browning of Bud-Derived Tissue Cultures of Scots Pine (*Pinus sylvestris* L.). *Plant Cell Tissue Organ Cult.* **2008**, *95*, 47–55. [CrossRef]
12. Panaccione, D.G.; Beaulieu, W.T.; Cook, D. Bioactive Alkaloids in Vertically Transmitted Fungal Endophytes. *Funct. Ecol.* **2014**, *28*, 299–314. [CrossRef]
13. Porras-Alfaro, A.; Herrera, J.; Sinsabaugh, R.L.; Odenbach, K.J.; Lowrey, T.; Natvig, D.O. Novel Root Fungal Consortium Associated with a Dominant Desert Grass. *Appl. Environ. Microbiol.* **2008**, *74*, 2805–2813. [CrossRef] [PubMed]
14. Maciá-Vicente, J.G.; Ferraro, V.; Burruano, S.; Lopez-Llorca, L.V. Fungal Assemblages Associated with Roots of Halophytic and Non-Halophytic Plant Species Vary Differentially Along a Salinity Gradient. *Microb. Ecol.* **2012**, *64*, 668–679. [CrossRef] [PubMed]
15. Ernst, M.; Mendgen, K.W.; Wirsel, S.G.R. Endophytic Fungal Mutualists: Seed-Borne *Stagonospora* spp. Enhance Reed Biomass Production in Axenic Microcosms. *Mol. Plant-Microbe Interact.* **2003**, *16*, 580–587. [CrossRef] [PubMed]
16. Rodriguez, R.; Freeman, D. Symbiotic Regulation of Plant Growth, Development and Reproduction. *Commun. Integr. Biol.* **2009**, *2*, 141–143. [CrossRef] [PubMed]
17. Kusari, S.; Lamshoft, M.; Zühlke, S.; Spiteller, M. An Endophytic Fungus from *Hypericum perforatum* That Produces Hypericin. *J. Nat. Prod.* **2008**, *71*, 159–162. [CrossRef] [PubMed]
18. Stierle, A.; Strobel, G.; Stierle, D. Taxol and Taxane Production by *Taxomyces andreanae*, an Endophytic Fungus of Pacific Yew. *Science* **1993**, *260*, 214–216. [CrossRef] [PubMed]
19. Stierle, D.B.; Stierle, A.A.; Bugni, T. Sequoiamonascins A-D: Novel Anticancer Metabolites Isolated from a Redwood Endophyte. *J. Org. Chem.* **2003**, *68*, 4966–4969. [CrossRef] [PubMed]
20. Shu, S.; Zhao, X.; Wang, W.; Zhang, G.; Cosoveanu, A.; Ahn, Y.; Wang, M. Identification of a Novel Endophytic Fungus from *Huperzia serrata* Which Produces Huperzine A. *World J. Microbiol. Biotechnol.* **2014**, *30*, 3101–3109. [CrossRef] [PubMed]
21. Schulz, B.; Boyle, C. The Endophytic Continuum. *Mycol. Res.* **1998**, *109*, 661–686. [CrossRef]

22. Gashgari, R.; Gherbawy, Y.; Ameen, F.; Alsharari, S. Molecular Characterization and Analysis of Antimicrobial Activity of Endophytic Fungi from Medicinal Plants in Saudi Arabia. *Jundishapur J. Microbiol.* **2016**, *9*, 1–8. [CrossRef] [PubMed]
23. Cosoveanu, A.; Hernandez, M.; Iacomi-Vasilescu, B.; Zhang, X.; Shu, S.; Wang, M.; Cabrera, R. Fungi as Endophytes in Chinese *Artemisia* spp.: Juxtaposed Elements of Phylogeny, Diversity and Bioactivity. *Mycosphere* **2016**, *7*, 102–117. [CrossRef]
24. Huang, W.Y.; Cai, Y.Z.; Surveswaran, S.; Hyde, K.D.; Corke, H.; Sun, M. Molecular Phylogenetic Identification of Endophytic Fungi Isolated from Three *Artemisia* Species. *Fungal Divers.* **2009**, *36*, 69–88.
25. Qadri, M.; Johri, S.; Shah, B.A.; Khajuria, A.; Sidiq, T.; Lattoo, S.K.; Abdin, M.Z.; Riyaz-Ul-Hassan, S. Identification and Bioactive Potential of Endophytic Fungi Isolated from Selected Plants of the Western Himalayas. *SpringerPlus* **2013**, *2*, 8. [CrossRef] [PubMed]
26. Yuan, Z.L.; Chen, Y.C.; Ma, X.J. Symbiotic Fungi in Roots of *Artemisia annua* with Special Reference to Endophytic Colonizers. *Plant Biosyst.* **2011**, *145*, 495–502. [CrossRef]
27. Qian, Y.; Kang, J.; Geng, K.; Wang, L.; Lei, B. Endophytic Fungi from *Artemisia argyi* Levl. et Vant. and Their Bioactivity. *Chiang Mai J. Sci.* **2014**, *41*, 910–921.
28. Myrchiang, P.; Dkhar, M.S.; Devi, H.R. Studies on Endophytic Fungi Associated with Medicinally Important Aromatic Plant *Artemisia nilagirica* (C.B. Clarke) Pamp. and Their Antagonistic Activity against *Phytophthora infestans*. *J. Adv. Lab. Res. Biol.* **2014**, *5*, 112–119.
29. Osono, T.; Ishii, Y.; Takeda, H.; Seramethakun, T.; Khamyong, S.; To-Anun, C.; Hirose, D.; Tokumasu, S.; Kakishima, M. Fungal Succession and Lignin Decomposition on *Shorea obtusa* Leaves in a Tropical Seasonal Forest in Northern Thailand. *Fungal Divers.* **2009**, *36*, 101–119.
30. White, T.J.; Bruns, T.; Lee, S.; Taylor, J. Amplification and Direct Sequencing of Fungal Ribosomal RNA Genes for Phylogenetics. In *PCR Protocols: A Guide to Methods and Applications*; Innis, M.A., Gelfand, D.H., Sninsky, J.J., White, T.J., Eds.; Academic Press, Inc.: New York, NY, USA, 1990; pp. 315–322.
31. O'Donnell, K. *Fusarium* and Its near Relatives. In *The Fungal Holomorph: Mitotic, Meiotic and Pleomorphic Speciation in Fungal Systematics*; Reynolds, R., Taylor, J.W., Eds.; CBA International: Wallingford, UK, 1993; pp. 225–233.
32. Thompson, J.D.; Higgins, D.G.; Gibson, T.J. CLUSTAL W: Improving the Sensitivity of Progressive Multiple Sequence Alignment through Sequence Weighting, Position-Specific Gap Penalties and Weight Matrix Choice. *Nucleic Acids Res.* **1994**, *22*, 4673–4680. [CrossRef] [PubMed]
33. Tamura, K.; Stecher, G.; Peterson, D.; Filipski, A.; Kumar, S. MEGA6: Molecular Evolutionary Genetics Analysis Version 6.0. *Mol. Biol. Evol.* **2013**, *30*, 2725–2729. [CrossRef] [PubMed]
34. Foronda, P.; López-González, M.; Hernández, M.; Haukisalmi, V.; Feliu, C. Distribution and Genetic Variation of *Hymenolepidid cestodes* in Murid Rodents on the Canary Islands (Spain). *Parasit. Vectors* **2011**, *4*, 185. [CrossRef] [PubMed]
35. Simmons, M.P.; Ochoterena, H. Gaps as Characters in Sequence-Based Phylogenetic Analyses. *Syst. Biol.* **2000**, *49*, 369–381. [CrossRef] [PubMed]
36. Borchsenius, F. *FastGap*, version 1.2; Department of Biosciences, Aarhus University: Aarhus, Denmark, 2009.
37. Dereeper, A.; Guignon, V.; Blanc, G.; Audic, S.; Buffet, S.; Chevenet, F.; Dufayard, J.-F.; Guindon, S.; Lefort, V.; Lescot, M.; et al. Phylogeny.fr: Robust Phylogenetic Analysis for the Non-Specialist. *Nucleic Acids Res.* **2008**, *36*, W465–W469. [CrossRef] [PubMed]
38. Darriba, D.; Taboada, G.L.; Doallo, R.; Posada, D. jModelTest 2: More Models, New Heuristics and Parallel Computing. *Nat. Methods* **2012**, *9*, 772. [CrossRef] [PubMed]
39. Huelsenbeck, J.P.; Ronquist, F. MRBAYES: Bayesian Inference of Phylogenetic Trees. *Bioinformatics* **2001**, *17*, 754–755. [CrossRef] [PubMed]
40. Ramírez-Bahena, M.H.; Hernández, M.; Peix, A.; Velázquez, E.; León-Barrios, M. Mesorhizobial Strains Nodulating *Anagyris latifolia* and *Lotus berthelotii* in Tamadaya Ravine (Tenerife, Canary Islands) Are Two Symbiovars of the Same Species, *Mesorhizobium tamadayense* sp. nov. *Syst. Appl. Microbiol.* **2012**, *35*, 334–341. [CrossRef] [PubMed]
41. Stöver, B.C.; Müller, K.F. TreeGraph 2: Combining and Visualizing Evidence from Different Phylogenetic Analyses. *BMC Bioinform.* **2010**, *11*, 7. [CrossRef] [PubMed]
42. Margalef, R.D. Information Theory In Ecology. *Gen. Syst.* **1958**, *3*, 36–71.
43. Pielou, E.C. *An Introduction to Mathematical Ecology*; Wiley-Interscience: New York, NY, USA, 1969.

44. Pielou, E.C. *Ecological Diversity*; Wiley: New York, NY, USA, 1975.
45. Fisher, R.A.; Corbet, S.A.; Williams, C.B. The Relation between the Number of Species and the Number of Individuals in a Random Sample of an Animal Population. *J. Anim. Ecol.* **1943**, *12*, 42–58. [CrossRef]
46. Simpson, E.H. Measurement of Diversity. *Nature* **1949**, *163*, 688. [CrossRef]
47. Sorensen, T. *A Method of Establishing Groups of Equal Amplitude in Plant Sociology Based on Similarity of Species Content and Its Application to Analyses of the Vegetation on Danish Commons*; I kommission hos E. Munksgaard: København, Danmark, 1948.
48. Dice, L. Measures of the Amount of Ecologic Association Between Species. *Ecology* **1945**, *26*, 297–302. [CrossRef]
49. Arnold, A.E. Understanding the Diversity of Foliar Endophytic Fungi: Progress, Challenges, and Frontiers. *Fungal Biol. Rev.* **2007**, *21*, 51–66. [CrossRef]
50. Grunden, E.; Chen, W.D.; Crane, J.L. Fungi Colonizing Microsclerotia of *Verticillium dahliae* in Urban Environments. *Fungal Divers.* **2001**, *8*, 129–141.
51. Tokumasu, S.; Aoiki, T. A New Approach to Studying Microfungal Succession on Decaying Pine Needles in an Oceanic Subtropical Region in Japan. *Fungal Divers.* **2002**, *10*, 167–183.
52. Guo, L.; Xu, L.; Zheng, W.; Hyde, K.D. Genetic Variation of *Alternaria alternata*, an Endophytic Fungus Isolated from *Pinus tabulaeformis* as Determined by Random Amplified Microsatellites (RAMS). *Fungal Divers.* **2004**, *16*, 53–67.
53. Kumaresan, V.; Suryanarayanan, T.S. Endophyte Assemblages in Young, Mature and Senescent Leaves of *Rhizophora apiculata*: Evidence for the Role of Endophytes in Mangrove Litter Degradation. *Fungal Divers.* **2002**, *9*, 81–91.
54. Taylor, J.E.; Hyde, K.D.; Jones, E.B.G. Endophytic fungi associated with the temperate palm, *Trachycarpus fortunei*, within and outside its natural geographic range. *New Phytol.* **1999**, *142*, 335–346. [CrossRef]
55. Romero, A.; Carrion, G.; Rico-Gray, V. Fungal Latent Pathogens and Endophytes. *Fungal Divers.* **2001**, *7*, 81–87.
56. Wearn, J.A.; Sutton, B.C.; Morley, N.J.; Gange, A.C. Species and Organ Specificity of Fungal Endophytes in Herbaceous Grassland Plants. *J. Ecol.* **2012**, *100*, 1085–1092. [CrossRef]
57. Sun, Y.; Wang, Q.; Lu, X.; Okane, I.; Kakishima, M. Endophytic Fungal Community in Stems and Leaves of Plants from Desert Areas in China. *Mycol. Prog.* **2012**, *11*, 781–790. [CrossRef]
58. Sánchez Márquez, S.; Bills, G.F.; Acuña, L.D.; Zabalgogeazcoa, I. Endophytic Mycobiota of Leaves and Roots of the Grass *Holcus lanatus*. *Fungal Divers.* **2010**, *41*, 115–123. [CrossRef]
59. Cannon, P.F.; Simmons, C.M. Diversity and Host Preference of Leaf Endophytic Fungi in the Iwokrama Forest Reserve, Guyana. *Mycologia* **2002**, *94*, 210–220. [CrossRef] [PubMed]
60. Rodriguez, R.J.; White, J.F.; Arnold, A.E.; Redman, R.S. Fungal Endophytes: Diversity and Functional Roles. *New Phytol.* **2009**, *182*, 314–330. [CrossRef] [PubMed]
61. Gange, A.C.; Dey, S.; Currie, A.F.; Sutton, B.C. Site- and species-specific differences in endophyte occurrence in two herbaceous plants. *J. Ecol.* **2007**, *95*, 614–622. [CrossRef]
62. Cosoveanu, A.; Da Silva, E.; Gimenez Marino, C.; Nunez Trujillo, G.; Gonzales Coloma, A.; Frias Viera, I.; Cabrera, R. *Artemisia thuscula* Cav.: Antibacterial, Antifungal Activity of the Plant Extracts and Associated Endophytes. *J. Hortic. For. Biotechnol.* **2012**, *16*, 87–90.
63. Su, Y.-Y.; Qi, Y.-L.; Cai, L. Induction of Sporulation in Plant Pathogenic Fungi Induction of Sporulation in Plant Pathogenic Fungi. *Mycology* **2012**, *33*, 195–200. [CrossRef]
64. Sharma, G.; Pandey, R.R. Influence of Culture Media on Growth, Colony Character and Sporulation of Fungi Isolated from Decaying Vegetable Wastes. *J. Yeast Fungal Res.* **2010**, *1*, 157–164.
65. Kumara, K.W.; Rawal, R.D. Influence of Carbon, Nitrogen, Temperature and pH on the Growth and Sporulation of Some Indian Isolates of *Colletotrichum gloeosporioides* Causing Anthracnose Disease of Papaya (*Carrica papaya* L.). *Trop. Agric. Res. Ext.* **2008**, *11*, 7–12. [CrossRef]
66. Osono, T. *Fungal Decomposition of Leaf Litter in a Cool Temperate Forest*; Kyoto University: Kyoto, Japan, 2002.
67. Yao, Y.Q.; Lan, F.; Qiao, Y.M.; Wei, J.G.; Huang, R.S.; Li, L.B. Endophytic Fungi Harbored in the Root of *Sophora tonkinensis* Gapnep: Diversity and Biocontrol Potential against Phytopathogens. *MicrobiologyOpen* **2017**, *6*. [CrossRef] [PubMed]

68. Santos, J.M.; Correia, V.G.; Phillips, A.J.L. Primers for Mating-Type Diagnosis in *Diaporthe* and *Phomopsis*: Their Use in Teleomorph Induction in Vitro and Biological Species Definition. *Fungal Biol.* **2010**, *114*, 255–270. [CrossRef] [PubMed]
69. Udayanga, D.; Liu, X.; McKenzie, E.H.C.; Chukeatirote, E.; Bahkali, A.H.A.; Hyde, K.D. The Genus *Phomopsis*: Biology, Applications, Species Concepts and Names of Common Phytopathogens. *Fungal Divers.* **2011**, *50*, 189–225. [CrossRef]
70. Hyde, K.D.; Mckenzie, E.H.C.; KoKo, T.W. Towards Incorporating Anamorphic Fungi in a Natural Classification–checklist and Notes for 2011. *Mycosphere* **2012**, *3*, 157–228. [CrossRef]
71. Thambugala, K.M.; Ariyawansa, H.A.; Li, Y.-M.; Boonmee, S.; Hongsanan, S.; Tian, Q.; Singtripop, C.; Bhat, D.J.; Camporesi, E.; Jayawardena, R.; et al. Dothideales. *Fungal Divers.* **2014**, *68*, 105–158. [CrossRef]
72. Schoch, C.L.; Shoemaker, R.A.; Seifert, K.A.; Hambleton, S.; Spatafora, J.W.; Crous, P.W. A Multigene Phylogeny of the Dothideomycetes Using Four Nuclear Loci. *Mycologia* **2006**, *98*, 1041–1052. [CrossRef] [PubMed]
73. Eriksson, O.E. Outline of Ascomycota—2006. *Myconet* **2006**, *12*, 1–82.
74. Eriksson, O.E.; Winka, K. Supraordinal Taxa of Ascomycota. *Myconet* **1997**, *1*, 1–16.
75. Maharachchikumbura, S.S.N.; Hyde, K.D.; Jones, E.B.G.; McKenzie, E.H.C.; Huang, S.K.; Abdel-Wahab, M.A.; Daranagama, D.A.; Dayarathne, M.; D'souza, M.J.; Goonasekara, I.D.; et al. Towards a Natural Classification and Backbone Tree for Sordariomycetes. *Fungal Divers.* **2015**, *72*, 199–301. [CrossRef]
76. Herrera, C.S.; Rossman, A.Y.; Samuels, G.J.; Lechat, C.; Chaverri, P. Revision of the Genus *Corallomycetella* with *Corallonectria* Gen. Nov. for *C. jatrophae* (Nectriaceae, Hypocreales). *Mycosystema* **2013**, *32*, 518–544.
77. Wang, Y.-Z. Three Species Excluded from *Melanopsamma* (*Ascomycetes*). *Mycotaxon* **2011**, *115*, 365–368. [CrossRef]
78. Castlebury, L.A.; Rossman, A.Y.; Sung, G.H.; Hyten, A.S.; Spatafora, J.W. Stachybotrys, Multigene Phylogeny Reveals New Lineage for *Chartarum*, the Indoor Air Fungus. *Mycol. Res.* **2014**, *108*, 864–872. [CrossRef]
79. Gao, Y.; Liu, F.; Duan, W.; Crous, P.W.; Cai, L. *Diaporthe* Is Paraphyletic. *IMA Fungus* **2017**, *8*, 153–187. [CrossRef] [PubMed]
80. Hongsanan, S.; Maharachchikumbura, S.S.N.; Hyde, K.D.; Samarakoon, M.C.; Jeewon, R.; Zhao, Q.; Al-Sadi, A.M.; Bahkali, A.H. An Updated Phylogeny of Sordariomycetes Based on Phylogenetic and Molecular Clock Evidence. *Fungal Divers.* **2017**, *84*, 25–41. [CrossRef]
81. Wijayawardene, N.N.; Jayarama Bhat, D.; Hyde, K.D.; Camporesi, E.; Chethana, K.W.T.; Tangthirasunun, N.; Wang, Y. *Camarosporium* Sensu Stricto in Pleosporinae, Pleosporales with Two New Species. *Phytotaxa* **2014**, *183*, 16–26. [CrossRef]

PERMISSIONS

All chapters in this book were first published in MDPI; hereby published with permission under the Creative Commons Attribution License or equivalent. Every chapter published in this book has been scrutinized by our experts. Their significance has been extensively debated. The topics covered herein carry significant findings which will fuel the growth of the discipline. They may even be implemented as practical applications or may be referred to as a beginning point for another development.

The contributors of this book come from diverse backgrounds, making this book a truly international effort. This book will bring forth new frontiers with its revolutionizing research information and detailed analysis of the nascent developments around the world.

We would like to thank all the contributing authors for lending their expertise to make the book truly unique. They have played a crucial role in the development of this book. Without their invaluable contributions this book wouldn't have been possible. They have made vital efforts to compile up to date information on the varied aspects of this subject to make this book a valuable addition to the collection of many professionals and students.

This book was conceptualized with the vision of imparting up-to-date information and advanced data in this field. To ensure the same, a matchless editorial board was set up. Every individual on the board went through rigorous rounds of assessment to prove their worth. After which they invested a large part of their time researching and compiling the most relevant data for our readers.

The editorial board has been involved in producing this book since its inception. They have spent rigorous hours researching and exploring the diverse topics which have resulted in the successful publishing of this book. They have passed on their knowledge of decades through this book. To expedite this challenging task, the publisher supported the team at every step. A small team of assistant editors was also appointed to further simplify the editing procedure and attain best results for the readers.

Apart from the editorial board, the designing team has also invested a significant amount of their time in understanding the subject and creating the most relevant covers. They scrutinized every image to scout for the most suitable representation of the subject and create an appropriate cover for the book.

The publishing team has been an ardent support to the editorial, designing and production team. Their endless efforts to recruit the best for this project, has resulted in the accomplishment of this book. They are a veteran in the field of academics and their pool of knowledge is as vast as their experience in printing. Their expertise and guidance has proved useful at every step. Their uncompromising quality standards have made this book an exceptional effort. Their encouragement from time to time has been an inspiration for everyone.

The publisher and the editorial board hope that this book will prove to be a valuable piece of knowledge for researchers, students, practitioners and scholars across the globe.

LIST OF CONTRIBUTORS

Rosario Nicoletti
Council for Agricultural Research and Economics, Research Centre for Olive, Fruit and Citrus Crops, 81100 Caserta, Italy
Department of Agricultural Sciences, University of Naples Federico II, 80055 Portici, Italy

Claudio Di Vaio and Chiara Cirillo
Department of Agricultural Sciences, University of Naples Federico II, 80055 Portici, Italy

Daniela Costa, Rui M. Tavares and Teresa Lino-Neto
BioSystems & Integrative Sciences Institute (BioISI), Plant Functional Biology Centre, University of Minho, Campus de Gualtar, 4710-057 Braga, Portugal

Paula Baptista
Centro de Investigação de Montanha (CIMO), Instituto Politécnico de Bragança, Campus de Santa Apolónia, 5300-253 Bragança, Portugal

Sunil K. Deshmukh and Manish K. Gupta
TERI-Deakin Nano Biotechnology Centre, The Energy and Resources Institute (TERI), Darbari Seth Block, IHC Complex, Lodhi Road, New Delhi 110003, India

Ved Prakash
Department of Biotechnology, Motilal Nehru National Institute of Technology, Allahabad 211004, India

M. Sudhakara Reddy
Department of Biotechnology, Thapar Institute of Engineering & Technology, Patiala, Punjab 147004, India

Brian R. Murphy and Trevor R. Hodkinson
School of Natural Sciences & Trinity Centre for Biodiversity Research, Trinity College Dublin, The University of Dublin, College Green, Dublin 2, Ireland

Fiona M. Doohan
UCD School of Biology & Environmental Science and UCD Earth Institute, University College Dublin, Dublin 4, Ireland

Tyler Maxwell, Yuemin Wang, Matthew Rex and James K. Harper
Department of Chemistry, University of Central Florida, 4111 Libra Drive, Orlando, FL 32816, USA

Richard G. Blair
Florida Space Institute, University of Central Florida, 12354 Research Parkway, Suite 214, Orlando, FL 32826, USA

Andrew H. Kettring and Sean D. Moore
Burnett School of Biomedical Sciences, University of Central Florida, 4110 Libra Dr., Orlando, FL 32816, USA

Sanjai Saxena
Department of Biotechnology, Thapar Institute of Engineering & Technology, Deemed to be a University, Patiala, Punjab 147004, India

Gary Strobel
Department of Plant Sciences, Montana State University, Bozeman, MT 59717, USA

Joël E. T. Ateba, Brice M. Mba'ning and Etienne Tsamo
Department of Organic Chemistry, Faculty of Science, University of Yaoundé 1, 237 Yaoundé, Cameroon

Rufin M. K. Toghueo and Fabrice F. Boyom
Department of Biochemistry, Faculty of Science, University of Yaoundé I, Cameroon; 237 Yaoundé, Cameroon

Angelbert F. Awantu
Department of Chemistry, Faculty of Science, University of Bamenda, 237 Bambili, Cameroon

Sebastian Gohlke and Norbert Sewald
Department of Chemistry, Organic and Bioorganic Chemistry, Bielefeld University, D-33501 Bielefeld, Germany

Dinkar Sahal
International Centre for Genetic Engineering and Biotechnology, Aruna Asaf Ali Marg, New Delhi 110067, India

Edson Rodrigues-Filho
Departamento de Química-Universidade Federal de São Carlos-CP 676, São Carlos-SP 13565-905, Brazil

Bruno N. Lenta
Department of Chemistry, Higher Teacher Training College, University of Yaoundé 1, 237 Yaoundé, Cameroon

Eyob Chukalo Chutulo
Department of Biology, Wolaita Sodo University, Wolaita Sodo 138, Ethiopia
Department of Biosciences, Mangalore University, Mangalagangothri 574199, India

Raju Krishna Chalannavar
Department of Applied Botany, Mangalore University, Mangalagangothri 574199, India

Dong-Hui Yan, Xiaoyu Song, Hongchang Li and Guiming Dou
Research Institute of Forest Ecology, Environment and Protection, Chinese Academy of Forestry, Hai Dian District, Beijing 100091, China

Tushou Luo
Research Institute of Tropical Forestry, Chinese Academy of Forestry, Tianhe District, Guangzhou 510520, China

Andreea Cosoveanu, Samuel Rodriguez Sabina and Raimundo Cabrera
Department Botanica, Ecologia & Fisiologia Vegetal, Universidad de La Laguna, 38206 La Laguna, Tenerife, Spain

Index

A

Acetic Acid, 143-144, 172

Acetonitrile, 106-107

Atmosphere, 184

B

Beta Diversity, 94, 223

Binomial, 107, 110

Biocontrol, 12, 15-16, 18-22, 26, 34-35, 38-41, 95, 102, 159, 202, 236

Bioinformatics, 110, 117, 235

Bioinformatics Analysis, 110

Biosphere, 198

C

Candida Albicans, 120-121, 126, 129, 135, 195, 212

Carbohydrates, 90, 105, 108, 113, 161

Carbon Monoxide, 82, 173

Carrier Gas, 85, 107, 207

Caspase, 50, 58

Cell Cycle, 44, 46, 75, 183

Cell Death, 31, 40, 48, 50, 74, 170, 215

Cell Lines, 44-46, 48, 50, 52-61, 74, 158, 169

Cell Proliferation, 183

Cellulose, 81, 89-90, 104-115, 171, 178

Chemotherapy, 43, 194

Chirality, 155

Chlorine, 23, 182

D

Diabetes, 181

Diagnostics, 199

Disease Development, 97

E

Erythrocytes, 183

Escherichia Coli, 195, 212

Essential Oil, 212, 216

F

Fluorescein Isothiocyanate, 50

G

Gas Chromatography, 81, 84, 91, 107-108, 215

Gel Electrophoresis, 107, 220

Gonorrhea, 181

Gram-positive Bacteria, 169, 178

H

Helium, 85, 107, 207

Herbarium, 181, 218-219

Heterogeneity, 223

I

Infectious Diseases, 196

Ingredient, 98

Intellectual Property, 95, 167

Inventory, 11, 17

Isothiocyanate, 50

K

Kinase, 38, 47, 50, 74, 169

L

Latency, 12, 14

Leaf Extract, 14

Lesions, 11, 18

Lichens, 13, 175

Liquid Chromatography, 174

Lung Cancer, 75, 78

Lysate, 107

Lysis, 183

M

Magnesium, 106-107

Morphology, 18, 20, 92, 116, 158, 182, 217, 221, 230, 233

Mortality, 22, 181, 196

N

Nanoparticles, 215

O

Operational Taxonomic Units, 24

Organic Analysis, 159

P

Pathogens, 2, 11-14, 16, 18-22, 24, 26, 31-38, 91, 120, 131, 139, 153-154, 157-158, 163-165, 170, 175, 190-191, 194-196, 201-202, 209, 211-213, 236

Pathology, 161, 176

Penetration, 32, 38

Peptides, 44, 52, 77, 120, 161

Phosphorylation, 43, 47
Phylogenetic Analysis, 102, 115, 193, 196, 203-204, 211, 218, 221-222, 229, 235
Phytochemical, 122, 197, 214
Phytochemicals, 234
Plant Physiology, 11
Plasmodium Falciparum, 180-181, 183, 185, 187-188
Polyketides, 52, 75, 77-78, 131, 152, 157
Python, 117

Q
Quantitative Analysis, 173
Quorum Sensing, 169

R
Reactive Oxygen Species, 2, 40, 196
Red Blood Cells, 183
Resolution, 36, 54, 129
Retention Time, 210
Rhizopus, 10, 125, 129, 218
Rhodotorula, 10, 13, 218
Ribosomal Rna, 36, 107, 235

S
Saccharomyces, 122, 128-129, 133, 135, 142, 158
Salinity, 2, 95-96, 234
Signal Transduction, 74
Soil Erosion, 2
Solid-phase Microextraction, 84, 90, 201-202, 210, 215

T
Tumor Cell Lines, 56
Tumor Cells, 51, 76
Tyrosinase, 194, 199

U
Ultraviolet Light, 203

V
Variables, 95-96, 226
Volatile Organic Compound, 80, 92, 116, 179, 187

W
Western Blotting, 50
Wound Healing, 194

X
X-ray Crystallography, 170, 195
X-ray Diffraction, 168

Printed in the USA
CPSIA information can be obtained
at www.ICGtesting.com
JSHW051408091023
49903JS00006B/339